The Physics and Chemistry of SiO_2 and the Si-SiO_2 Interface 2

The Physics and Chemistry of SiO_2 and the Si-SiO_2 Interface 2

Edited by

C. Robert Helms and Bruce E. Deal

Stanford University
Stanford, California

Springer Science+Business Media, LLC

CONTENTS

I. THERMAL OXIDATION MECHANISMS AND MODELING

II. NOVEL OXIDATION METHODS AND CHARACTERIZATION

VI. NOVEL STRUCTURES, PROCESSES, AND PHENOMENA

VII. DEFECTS AND HOT-CARRIER INDUCED DAMAGE IN Si-SiO_2 SYSTEMS

VIII. RADIATION AND HYDROGEN INDUCED EFFECTS IN SILICON-SILICON DIOXIDE SYSTEMS

I. THERMAL OXIDATION MECHANISMS AND MODELING

INTRODUCTION

E. A. Irene
Dept. of Chemistry
University of North Carolina, Chapel Hill, NC 27599

M. Hirose
Dept. of Electrical Engineering
Hiroshima University, Higashi-Hiroshima 724, Japan

Five papers, one invited and four contributed, are included in this chapter. The subject of modeling of Si oxidation has been open and active since about 1965 when the Deal and Grove linear parabolic, L-P, oxidation model for Si thermal oxidation was published. While this model is still widely accepted, the basic ideas that underlie the model have been expanded and over the intervening years some kinetic differences have been noticed leading to attempts at modifying the L-P model. This first chapter is intended to collect and update studies on mechanisms and modeling since the last symposium.

The invited review by Stoneham covers many experimental and theoretical results including stress effects, roughness, interface, electronics, and more. The relation of these effects to mechanisms and modeling is not made, and this likely points to the fact that such relations remain tenuous. It seems clear that while new facts about Si oxidation and the Si-SiO_2 interface are being uncovered, there are either not yet enough facts for the theorists or not enough theorists. Most of my fellow experimentalists believe the former. After the Stoneham review the other four papers deal with experimental facts about Si oxidation.

The paper by Trimaille et al employing 0^{18} tracer studies of Si oxidation concludes that the 0^{18} found in earlier experiments at the outer SiO_2 surface after oxidation is not related to the oxidation kinetics at the Si-SiO_2 interface, as was previously thought by some. The presentation by Kao and Doremus relates the ellipsometric SiO_2 thickness to TEM derived thicknesses and finds that the TEM thicknesses are systematically smaller for thin oxides, in substantial agreement with publications several years before. The results were

(e.g. under excitation by low-energy electron beams it seems likely that charged oxygen ions are formed and, for them, the image interactions at the interface between the Si and its oxide are important. The scanning tunnelling microscope is proving a powerful tool for the early stages (perhaps to one or two monolayers coverage), but is of less use at the intermediate thicknesses which determine operating behaviour.

Many observations imply there is a difference between the oxide close to the silicon and the oxide further towards the outer surface. In certain cases there is unambiguous evidence for a lower oxide; in special circumstances, notably when the silicon surface is flat on an atomic scale, it appears possible to grow a crystalline layer, rather than an amorphous layer, though the precise crystalline form is in doubt (indeed, since it has finite thickness and is under stress, it may not correspond to a bulk phase at all). In other cases still the differences are variously attributed to stress (e.g. some altered form of silica, perhaps with different ring structures), to roughness, or to defects (different local coordination). Oxide layers respond to stress rather like glasses: they can be viscous, and indeed several workers have modelled the kinetics by including known data for bulk oxide viscoelastic response. There are, of course many forms of silicon dioxide, and the enthalpy and molecular volume are related in systematic ways. For many purposes the oxide can be thought of as close-packed oxygens with interstitial cations, and indeed silica forms an interesting limiting case of many silicate glasses. For these glasses, the systematics of densification behaviour (both in rate and in final volume per oxygen) are common to simple glasses and to glasses as complex as those used for vitrification of radioactive waste.

However, it is more the oxide quality than the oxide structure which matters, and the issues of breakdown and noise are important. This, it is emerging, raises issues of chemical impurities, especially metal contamination and also the role of carbon. Breakdown is determined at least partly by the field enhancements resulting from interface roughness. This roughness is itself affected by stress, as can be seen by looking at the inhibition of oxidation of Si particles as the particles become smaller. Other effects concern defect generation, perhaps as a precursor to breakdown, and similar to the "forming" processes reported in many oxide thin films. Such effects are striking in that they are frequently asymmetric, with different behaviour depending on whether the Si substrate is the cathode or the anode.

Chemical impurities can play a role, even at the level of one per 1000 surface atoms. Among the results to emerge recently [6, 7, 8, 9] are the following. First, the cleaning procedure can affect the roughness substantially. Thus Ca metal contamination roughens Si on heating in an inert atmosphere, and leads to a rough oxide which breaks down more readily. Fe at similar levels leads to defect formation; the Fe is subsequently incorporated into the oxide. Metals like Cu and Zn readily enter the Si, and have negligible surface effects. Secondly, the cleaning procedure is especially important for UHV-cleaned Si, in which the initial cleaning stage is followed by driving off the "native" oxide in UHV at an elevated temperature. HF cleaning leads to SiC microcrystals, which may affect breakdown; the Shiraki clean leads to fewer SiC microcrystals, but poor capacitors. UV ozone cleaning, followed by driving off the native oxide, avoids the production of SiC. Anneal at higher temperatures gives smoother oxide and better capacitance behaviour. This can be seen by comparing breakdown statistics with roughness measured in two ways. Atomic force microscopy measures areas of a few microns square with a characteristic length scale of the order of nanometres; AFM monitors the oxide / vacuum interface. Optical haze data have a characteristic length scale of not much better than a micron.

However, because of the high Si dielectric constant, haze data monitor the Si / oxide interface, and it is this interface which determines breakdown.

Defect generation is of significance both in wear-out phenomena and in optical and electron beam processing. One component of what happens is self-trapping, which leads to energy localisation. If relatively low-energy excitation is to cause defect production or desorption, the energy has to be localised on a very small number of sites (one or two) so as to liberate an ion. In many systems (such as the alkali halides) the self-trapped exciton is the key to defect processes based on optical or low-energy electron excitation [10]. The characteristic blue luminescence of quartz and of other silicas appears to be intrinsic. Recent theory indicates it arises from recombination of the self-trapped exciton, and I shall discuss the way theory explains a range of observed properties [11]. This theory is based on methods which allow atomic positions to relax at the same time as electronic self-consistency is achieved, and I shall remark on other work by parallel methods. There is a problem here, in that self-trapping in quartz is fast and efficient; there are signs of similar rapid localisation in amorphous bulk oxides. Yet other data in thin oxide films are more consistent with the free exciton state surviving so that the exciton can traverse the film before giving up its excitation.

What emerges from these studies is the several ways in which theory aids the understanding of defect processes in silicon, its oxidation, and its oxides. At one level there is the Deal-Grove model, providing a framework; at a second level, there are calculations of transport and of defect properties using semi-empirical schemes to give a useful quantitative guide; at a third level, theory approaches the a priori, and examines excited states which are currently almost inaccessible by experiment. The wealth of behaviour, and the complexity of the phenomena which affect oxide quality most, however, ensure that it is good experiments combined with theory that will give enhanced performance consistent with technological demand.

References

1. N.F. Mott, S. Rigo, F. Rochet and A.M. Stoneham. 1989 Phil. Mag. B60 189.

2. A.M. Stoneham. 1991 p 19 of Insulating Films on Semiconductors 1991 (edited W. Eccleston and M. Uren); Bristol: Adam Hilger.

3. A.M. Stoneham. 1991 in "Ultra Clean Silicon Processing", in press: Springer Verlag.

4. See also the two special issues of Phil. Mag. B55 (1987).

5. M.I. Heggie, R. Jones, C.D. Latham, C. Maynard and P. Tole. 1992, Phil. Mag. 65, 463

6. S. Verhavebeke, M. Meuris, A. Kelleher, M. M. Heyns, R. F. de Keersmaecker, M.P. Murrell and C.J. Sofield, 1991, Proc. Electrochem. Soc. Fall Meeting, Phoenix, Arizona, in press.

7. C.J. Sofield, M.P. Murrell, S. Verhavebeke and M.E. Welland. 1992, Proc. MRS Spring Meeting, in press.

8. S. Verhavebeke, M. Meuris, P.W. Martens, M.M. Heyns, A. Philipassian, D. Graf and A. Schnegg. p71 of Int. Electron. Reviews Meeting, Washington, 1991.

9. S. Verhavebeke, P.W. Mertens, M. Meuris and M.M. Heyns. 1992, Proc. MRS Spring Meeting, in press.

10. W. Hayes and A.M. Stoneham 1985 Defects and Defect Processes in Non-Metallic Solids; New York: John Wiley.

11. A.J. Fisher, W. Hayes and A.M. Stoneham. 1990 Phys. Rev. Lett. 64 2667 and A.J. Fisher, W. Hayes and A.M. Stoneham. 1990 J. Phys. Cond. Mat. 2 6707.

12. D.C. Allan and M.P. Teter. 1990 J. Amer. Ceram. Soc. 73 3247.

USE OF ^{18}O LABELLING TO STUDY GROWTH MECHANISMS IN DRY OXIDATION OF SILICON

Isabelle Trimaille,[1] Stan I. Raider,[2] Jean-Jacques Ganem,[1] Serge Rigo,[1] and Nicholas A. Penebre[2]

[1]Groupe de Physique des Solides
Universités Paris 7 et Paris 6
Tour 23, 2 Place Jussieu
75251 Paris Cedex 05, France
[2]IBM Research Division
T.J. Watson Research Center
P.O. Box 218
Yorktown Heights, NY USA 10598

ABSTRACT

Oxygen fixation during dry oxidation of silicon was investigated using ^{18}O labelling. We found that the amount of oxygen fixed near the outer surface of the oxide was not notably influenced by the silicon oxidation rate at the Si-SiO_2 interface. To show this, we oxidized in $^{18}O_2$ nitrogen-implanted $Si^{16}O_2$-Si structures (for which interfacial oxidation is inhibited) and unimplanted $Si^{16}O_2$-Si structures. We compared ^{18}O fixation near the outer surface in both cases.

INTRODUCTION

^{18}O labelling has been used to establish atomic transport mechanisms in silicon oxidation (Rochet et al., 1984; Trimaille and Rigo, 1989). The method consists of oxidizing silicon with ^{16}O and then with ^{18}O. The oxygen isotopic profiles in the oxide film are determined and compared to theoretical profiles which correspond to possible growth mechanisms. These mechanisms have been reviewed by Rigo (1988). The main result of these studies is that ^{18}O is found at the oxide-silicon interface and at the external surface (see figure 1). The fixation of ^{18}O at the interface is due to growth by the Deal and Grove mechanism; in this case there is an abrupt $^{18}O/^{16}O$ boundary. The fixation of ^{18}O at the external surface exhibits a profile very close to a complementary error function. This is due to a parallel atomic transport which is related to network defects. This transport leads mostly to exchange near the surface between the silica network and the oxidizing species, but also (to a lesser extent) to growth of oxide (Trimaille and Rigo, 1989).

The parallel growth mechanism near an external surface has been analyzed (Trimaille and Rigo, 1989). N_{sg}, the number of oxygen atoms per cm^2 fixed during the second oxidation

The Physics and Chemistry of SiO_2 and the Si-SiO_2 Interface 2
Edited by C.R. Helms and B.E. Deal, Plenum Press, New York, 1993

Two processing conditions were used.

For the first set, the $^{18}O_2$ oxidations were made at 1030°C, under a 0.2 bar gas pressure, for 21.5 h.

For the second set, the temperature of the gas was 930°C, under pressures ranging from 10 to 100 mbar. The oxidation times ranged from 2 to 18 h.

Nuclear Analysis

The principles of nuclear microanalysis have been detailed by Amsel et al. (1971). To measure the amounts of nitrogen and oxygen, we used the following nuclear reactions with a detection angle of 150°:

- $^{14}N(d,\alpha_0)^{12}C$ induced by a deuteron beam of 1.45 MeV
- $^{16}O(d,p_1)^{17}O^*$ induced by a deuteron beam of 0.85 MeV
- $^{18}O(p,\alpha)^{15}N$ induced by a proton beam of 0.73 MeV.

Absolute values of the amounts of atomic quantities were determined by comparison with standard references.

Nitrogen measurements. The nitrogen quantities are small (a few times 10^{15} atoms per cm^2), and the differential cross-section of the $^{14}N(d,\alpha_0)^{12}C$ nuclear reaction is less than 1 mb/st (Amsel et al., 1971; Ganem, 1992), under our experimental conditions. The estimation of the absolute quantities of nitrogen is consequently difficult due to the overlapping of the nuclear reactions induced by the deuteron beam, between the silicon atoms of the substrate and the nitrogen atoms. The relative precision of the measurements is within 20%, and the absolute precision within 50%. These measurements were performed in order to evaluate roughly the amount of nitrogen atoms lost during $^{18}O_2$ oxidations.

$^{18}O_2$ distribution. In order to determine how the total amount of ^{18}O atoms is distributed among both interfaces of the oxide, we chemically etched half of each sample to remove 5 to 10 nm of oxide. The external region rich in ^{18}O is etched but not the interfacial one. The amounts of ^{18}O atoms and ^{16}O atoms were measured on both etched and unetched samples. From the difference in the amounts of ^{18}O in the unetched and etched sample, we obtained the amount of ^{18}O atoms fixed at the external surface.

One implanted sample (SR 7) was submitted to three different dissolution times to make sure that: (i) the fixation of ^{18}O in the bulk of the oxide was that of natural oxygen only; (ii) the external ^{18}O layer was actually etched.

RESULTS AND DISCUSSION

The results are presented in table 1 for reference samples (SR 1 and SR 16) and for various oxidations in $^{18}O_2$ gas.

$^{18}O_2$ Oxidations at 1030°C

The first set of wafers (SR 3 (implanted), SR 18 (non-implanted)) underwent an $^{18}O_2$ oxidation at 1030°C, for 21.5 h under a pressure of 0.2 bar. Under these conditions, the implanted wafer is no longer oxidation resistant (see Table 1). These wafers were used to estimate the homogeneity of ^{18}O integration on the wafers, which is an important parameter for our subsequent calculations of ^{18}O atoms distribution. By measuring the ^{18}O quantities on four different spots on SR 3, we found an inhomogeneity up to 10% in comparison with the value at the center of the wafer. On SR 18, the ^{18}O quantities were identical on several spots of the wafer. No inhomogeneity was observed in the ^{16}O and ^{14}N quantities on SR 3 and SR 18.

$^{18}O_2$ Oxidations at 930°C

Figure 3 shows the amounts of oxygen atoms measured on sample SR 7 (implanted sample), and on the same sample etched for 3 different dissolution times. The three etch times were sufficiently long to remove the external ^{18}O layer. The ^{18}O labelling of natural oxygen is 0.2% and the data corresponding to residual ^{18}O on etched samples can be fitted with a straight line of slope 0.002. This shows that there is no ^{18}O fixation in the silica bulk during $^{18}O_2$ oxidation for an implanted sample. From these measurements, we deduced the part of ^{18}O fixed near the interface and the part of ^{18}O fixed near the sample surface (see Table 1). For the other samples only 1 or 2 etch times were performed. Especially in the case of non-implanted wafers, this differential method leads to a lack of precision in the determination of the amount of ^{18}O atoms fixed near the external surface since the amounts of ^{18}O atoms in the unetched and etched samples are of the same order of magnitude.

Table 1. ^{14}N, ^{16}O and ^{18}O atomic quantities given as 10^{15} at.cm^{-2} measured on the implanted (SR 1 to SR 7) and non-implanted (SR 16 to SR 19) wafers, for reference samples (SR 1 and SR 16) and for samples oxidized in $^{18}O_2$ gas for various durations, pressures of oxygen and temperatures.

sample	$^{18}O_2$ treatment	^{14}N	^{16}O	^{18}O	^{18}O fixed near surface (N_s)	^{18}O fixed near interface	total oxide thickness (nm)[1]
SR 1		8	91				20
SR 16		-	90				20
SR 3	200 mbar / 21.5 h	5	72	35			24
SR 18	1030°C	-	82	568			147
SR 5	10 mbar / 18 h	6	86	12	9	3	22
SR 17b	930°C	-	83	42	9	33	28
SR 7	100 mbar / 2 h	6	92	8.1	6.2	1.7	23
SR 19	930°C	-	92	36	6	30	29
SR 6	100 mbar / 5 h	6	88	13	10	3	23
SR 17a	930°C	-	89	84	10	74	39

[1] The oxide thicknesses are obtained assuming stoichiometric SiO_2 with a density of 2.21 g.cm^{-3}.

Under the conditions of the experiment, the interfacial oxidation of the implanted wafers is widely inhibited, although the isotopic tracer method evidences a very slight oxidation occurring at the interface of these wafers. On the contrary the non-implanted annealed samples are mostly oxidized at the interface. We cannot exclude the possibility that an oxidation reaction occurs in the nitrogen implanted samples between $^{18}O_2$ and impurities located at the interface (Raider et al, 1991; Raider, 1992).

The amounts of ^{18}O atoms fixed near the external surface (^{18}O FNES) are the same for implanted and non implanted wafers, within the precision of the experimental differential method we used. In the case of implanted wafers, the amounts of ^{18}O FNES are in agreement with a $t^{1/2}$ law, which means that fixation of oxygen is dominated by exchange (Rochet, 1981) as it is for non-implanted samples.

Moreover, the amounts of ^{18}O atoms fixed at the Si-SiO_2 interface (that is to say the oxidation rate at the interface), varies up to more than twenty times between implanted and non-implanted samples, whereas the amounts of ^{18}O atoms fixed near the external surface (^{18}O FNES) are the same. Previous results (Rochet et al., 1984), regarding the variation of ^{18}O FNES versus interfacial oxidation rate (due to different oxide thicknesses), indicate that

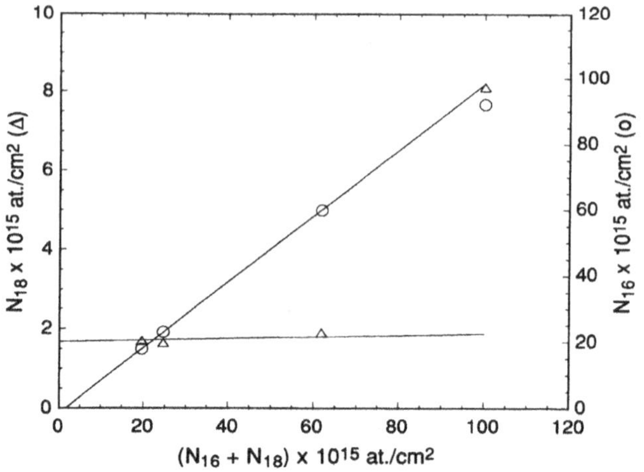

Figure 3. Amounts of ^{18}O (Δ) and ^{16}O (O) atoms on sample SR 7 and on the same sample etched for three different dissolution times as a function of the total oxygen content. This implanted sample underwent oxidation at 930°C for 2 hours under a 200 mbar pressure of $^{18}O_2$. The ^{18}O and ^{16}O data corresponding to etched samples are fitted by straight lines of slopes 0.002 and 0.998, indicating that the ^{18}O labelling in the bulk of the silica network is that of natural oxygen.

there would be a significant difference in ^{18}O FNES if there was a dependence upon interfacial oxidation rate (of the order of four times).

Since the proportion of the ^{18}O FNES which contributes to growth is itself proportional to ^{18}O FNES (see equation (1)), these results show that the dependence upon oxide thickness of the parallel growth mechanism is not correlated to the oxidation rate at the interface.

However, inhibition of interfacial oxidation may slightly affect the amount and the depth profile of ^{18}O atoms fixed near the surface. The differential method we used is inadequate to address this question. A finer analysis giving the ^{18}O depth profiling, could be done using the narrow resonance in the $^{18}O(p,\alpha)^{15}N$ nuclear reaction, at 152 keV, discovered by Lorenz-Wirzba et al. (1979). Recent and precise measurements demonstrated that the most probable value for the width of the resonance is 50 eV, allowing nanometric or even subnanometric near surface depth sensitivity (Battistig et al., 1992). But the ultimate sensitivity requires operation in ultra high vacuum to avoid energy straggling induced by hydrocarbon contamination on the sample surface (Battistig et al., 1992). Theoretical calculations under the ideal conditions have been made showing that in the case of samples similar to ours we could distinguish error function type profiles from rectangular ones (Battistig et al., 1992). Diffusion coefficients can be measured directly with this technique which would improve our understanding of the mechanism involved in fixing ^{18}O at an external SiO_2 surface.

CONCLUSIONS

We have formed a silicon oxidation barrier by implanting nitrogen into Si-SiO_2 structure and have studied the distribution of ^{18}O in these structures after oxidation with $^{18}O_2$ at 930°C. The ^{18}O distributed at an outer SiO_2 surface in implanted and unimplanted substrates are similar whereas the ^{18}O at a Si-SiO_2 interface is distributed primarily in the non-implanted control sample. These results indicate that the oxygen isotopic tracer distribution at the outer SiO_2 surface is not notably influenced by the silicon oxidation rate at a Si-SiO_2 interface.

Use of the narrow resonance in the $^{18}O(p,\alpha)^{15}N$ nuclear reaction at 152 keV would increase the sensitivity to oxygen fixation near surfaces or interfaces and further improve modeling thin (< 10 nm) film oxidation mechanisms.

Acknowledgments

One of the authors (I.T.) thanks A. L'Hoir for helpful discussions. Valuable discussions with G. Battistig are gratefully acknowledged.

This work was supported by the Centre National de la Recherche Scientifique, GDR N° 86.

References

Amsel, G., Nadai, J.P., d'Artemare, E., David, D., Girard, E., and Moulin, J., 1971, Microanalysis by the direct observation of nuclear reactions using a 2 MeV Van de Graaff, *Nucl. Instrum. Methods* 92:481.

Battistig, G., Amsel, G., d'Artemare, E., and Vickridge,Y., 1992, A very narrow resonance in $^{18}O(p,\alpha)^{15}N$ near 150 keV. Application to isotopic tracing. II. High resolution depth profiling of ^{18}O, *Nucl. Instrum. Methods* B66:1.

Ganem, J.J., 1992, Mécanismes de croissance de diélectriques par nitruration et par oxydation thermiques rapides du silicium, *Thèse de l'Université Paris 7.*

Habraken, F.H.P.M., Kuiper, A. E. T., and Tamminga, Y., 1982, Thermal nitridation of silicon dioxide films, *J. Appl. Phys.* 53:10.

Josquin, W.J.M.J., and Tamminga, Y., 1982, The oxidation inhibition in nitrogen-implanted silicon, *J. Electrochem. Soc.* 129:1803.

Kim, M.J., and Ghezzo, M., 1984, Characterization of implanted nitride for VLSI applications, *J. Electrochem. Soc.* 131:1934.

Lorenz-Wirzba, H., Schmalbrock, P., Trautvetter, H.P., Wiescher, M., Rolfs, C., and Rodney, W.S., 1979, The $^{18}O(p,\alpha)^{15}N$ reaction at stellar energies, *Nucl. Phys.* A313:346.

Murarka, S.P., Chang, C.C., and Adams, A.C., 1979, Thermal nitridation of silicon in ammonia gas: composition and oxidation resistance of the resulting films, *J. Electrochem. Soc.* 126:996.

Raider, S.I., Gdula, R.A., and Petrak, J.R., 1975, Nitrogen reaction at a silicon-silicon dioxide interface, *Appl. Phys. Lett.* 27:150.

Raider, S.I., Herd, S.R., and Walkup, R.E., 1991, SiO_2 film decomposition reaction initiated by carbon impurities located at a Si-SiO_2 interface, *Appl. Phys. Lett.* 59:2424.

Raider, S.I., 1992, Carbon impurities at Si-SiO_2 interface, *This meeting (Second Symposium on the Physics and Chemistry of SiO_2 and the Si-SiO_2 Interface).*

Rigo S., 1988, Si oxidation mechanisms as studied by oxygen tracer methods, *in* "The Physics and Chemistry of SiO_2 and the Si-SiO_2 Interface," C.R. Helms and B.E. Deal, ed., Plenum Press, New York.

Rochet, F., 1981, Etude par traçage isotopique à l'oxygène 18 des mécanismes de transport atomique dans les couches minces d'oxyde de silicium et d'oxyde de tantale sur silicium, *Thèse de Doctorat de 3è cycle de l'Université Paris 6.*

Rochet, F., Agius, B., and Rigo, S., 1984, An ^{18}O study of the oxidation mechanism of silicon in dry oxygen, *J. Electrochem. Soc.* 131:914.

Schott, K., Hofmann, K.C., and Schultz M., 1988, Blocking of silicon oxidation by low-dose nitrogen implantation, *Appl. Phys.* A45:73.

Trimaille, I., and Rigo S., 1989, Use of ^{18}O isotopic labelling to study thermal dry oxidation of silicon as a function of temperature and pressure, *Appl. Surf. Science* 39:65.

Ziegler, J.F., Biersack, J.P., Cuomo, G., 1984-1989, The stopping and range of ions in matter, TRIM-89 code.

refraction and increasing compressive stress, Ref. 15 suggested that diffusion of the oxidizing species is inhibited by compressive stress. However, in this study we report on the strain profiles present after an intermediate high temperature anneal followed by the second oxidation step and show the extent and position of a relaxed region giving rise to the observed increase in oxidation rate.

EXPERIMENTAL PROCEDURES AND RESULTS

P-type (100) silicon wafers with resistivities in the range of 10-30 Ω-cm were thermally oxidized in two steps at 850°C in a dry O_2 ambient with an intermediate annealing step. First, 300 Å of SiO_2 was grown followed by a furnace anneal in Ar at either 850°C (385 min) or 1050°C (19 min). The second oxidation step at 850°C added on average 270 Å of SiO_2. The samples were then incrementally etched back and infrared spectroscopy was used to determine the frequency of the dominant SiO_2 absorption band as a function of position in the oxide film.

During the second oxidation step, the films annealed in at 850°C and 1050°C, grew 240 Å and 309 Å, respectively, as is illustrated in Fig. 1. These thicknesses are very similar to those obtained in a previous experiment using N_2 as the annealing ambient,[20] from which we conclude that any possible Si-N bonding, as for example Si_3N_4 formation, at the Si-SiO_2 interface did not cause the thickness differences observed for the N_2 anneals. Up to 10 Å of the thickness difference between the 850°C and 1050°C anneals is explained by the volume expansion that occurs during high temperature annealing.[4,10,21] The remaining difference between the films annealed at 850°C and 1050°C then corresponds to an increase in the average oxidation rate of about 20%. Figure 2 shows the bond-stretching frequency as a function of SiO_2 thickness for films grown at 850°C and 1050°C. The bond-stretching frequency decreases as one approaches the Si-SiO_2 interface, indicating a highly strained SiO_2 region at that interface.[1-2] This region is a result of the thermal oxidation process, since the SiO_2 occupies a 120% larger molar volume than the Si consumed to grow it,[4] and since complete strain relaxation is prevented by the oxide on top of it.[12]

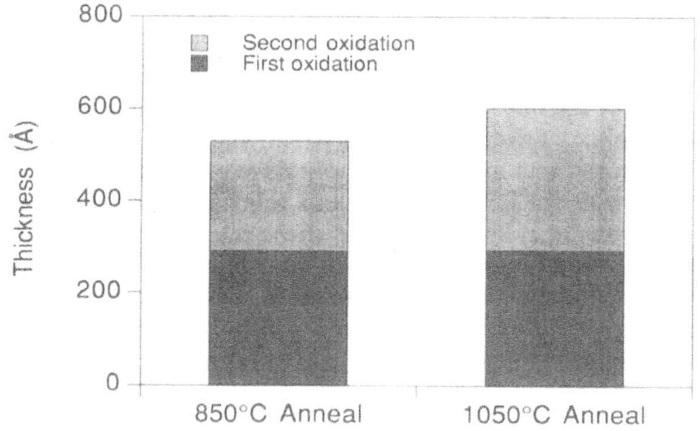

Figure 1. Oxide thickness before and after the second oxidation step. The films were annealed at 850°C, 950°C, and 1050°C, respectively.

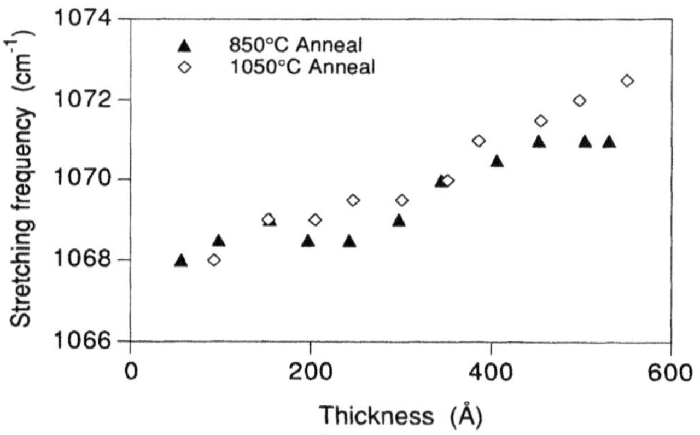

Figure 2. Bond stretching frequency as a function of oxide thickness. The films were annealed at 850°C and 1050°C, respectively.

DISCUSSION

Small differences in the bond-stretching frequency were observed in the part of the oxides subjected to furnace anneals (300 - 600 Å), as Fig. 2 illustrates. A self-consistent method was used to deconvolve the thickness-averaged results for 50 Å layers of the oxide.[12] This method involves fitting a curve to the thickness averaged data set and requiring that the average stretching frequency of the first n layers is equal to the thickness-averaged value at n∗50 Å. Applying this deconvolution procedure to the thickness-averaged data points for films annealed at 850°C and 1050°C, respectively, yields the result shown in Fig. 3. Since different amounts of oxide were added during the second oxidation step, the Si-SiO_2 interfaces were at different depths from the SiO_2 surface depending on the annealing temperature. We have therefore chosen to plot the data as a function of depth instead of as a function of distance from the interface, as is shown in Fig. 3.

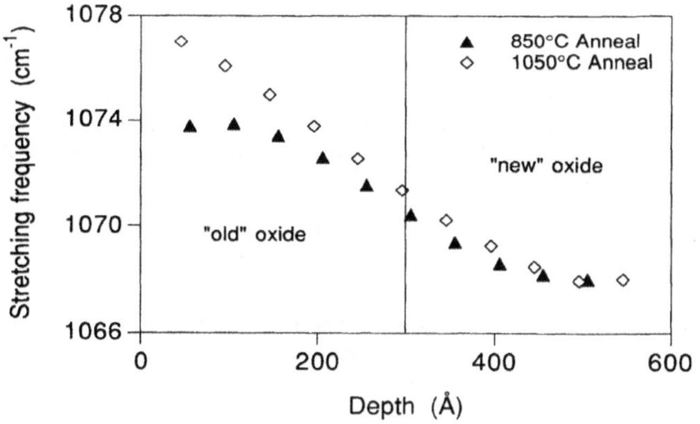

Figure 3. Deconvolved bond-stretching frequency as a function of distance from the oxide surface.

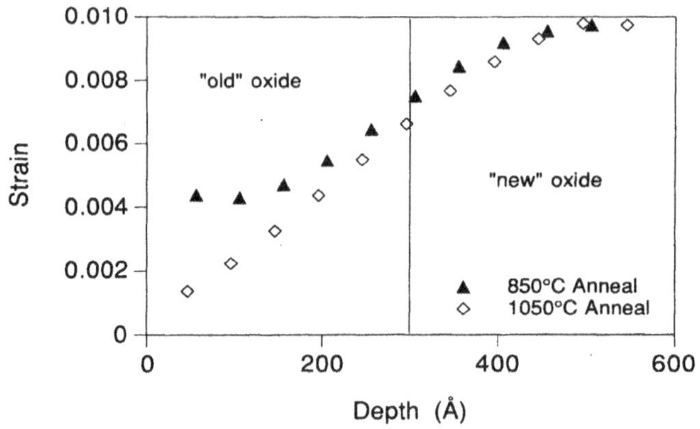

Figure 4. Deconvolved strain as a function of distance from the oxide surface. The strain was calculated by using data from Fig. 3 in Eqn. (1).

Using a previously developed relationship between the bond-stretching frequency and the local atomic compressive strain, ε:[22]

$$\varepsilon = 1 - \frac{\nu}{\nu_1}\,, \tag{1}$$

and the data from Fig. 3, we have also plotted the deconvolved compressive strain as a function of the distance from the SiO_2 surface (depth), which is displayed in Fig. 4. Thus, Fig. 4 illustrates that the "new" oxide layers grown after annealing, have essentially the same level of strain, independent of annealing temperature. A strong dependence on annealing temperature was observed in the top 300 Å of SiO_2 ("old" oxide), with films annealed at higher temperatures showing the lowest levels of strain.

The diffusion of oxidizing species through the oxide to the interface is described by Fick's law:

$$F = -D\frac{dC}{dx}\,, \tag{2}$$

where F is the flux of oxidizing species, D the diffusion coefficient for these species in the oxide, and C(x) the concentration of oxidizing species as a function of distance from the oxide surface, x. If the diffusion coefficient varies across the oxide thickness, we can divide the oxide into layers Δx thick, each with a diffusion coefficient D_n. The flux through the n:th layer can therefore be expressed as:

$$F_n = -D_n\frac{\Delta C_n}{\Delta x}\,, \tag{3}$$

where ΔC_n is difference in concentration between $(n-1)*\Delta x$ and $n*\Delta x$. Under steady state conditions, when there is no build-up or depletion of oxidizing species, the flux must be the same through all layers. Comparing the flux in the layers closest to the surface and interface, we have

$$F = -D_s\frac{\Delta C_s}{\Delta x} = -D_i\frac{\Delta C_i}{\Delta x}\,. \tag{4}$$

When the oxidation is diffusion controlled, the concentration of oxidizing species is zero at the growth interface. As Eqn. (4) illustrates, an increased flux through the surface layer, while the diffusion coefficient at the interface remains constant, therefore translates into an increased concentration at a distance Δx from the interface. This was observed by Imai et al.[19], although it was interpreted as an increased solubility for the oxidizing species.

It has been shown that a compressive strain reduces the diffusion coefficient of water in vitreous silica, whereas a tensile strain increases it.[23] A model for the diffusion coefficient of the oxidizing species in amorphous silica has been proposed by Doremus.[24] According to this model the diffusion coefficient depends exponentially on the compressive strain, ε:

$$D = D_o \exp(-k\varepsilon) \quad , \quad \text{where} \tag{5}$$

$$k = \frac{EV^*}{RT} \quad . \tag{6}$$

As Eqn. (5) illustrates, D_o is the diffusion coefficient corresponding to zero strain. In the expression for the dimensionless coefficient, k, E is Young's modulus and V* an activation volume which may vary as a function of temperature. Since we measured a reduction of the compressive strain in the film annealed at 1050°C relative to the film annealed at 850°C, Eqn. (5) shows that this corresponds to an increase in the diffusion coefficient. It therefore appears that the observed increase in flux in these experiments was achieved by increasing the diffusion coefficient in the top 300 Å of oxide, but it should be pointed out that an increased concentration of oxidizing species close to the interface can be obtained by increasing the flux through any oxide layer.

According to Nicollian et al.,[25] a possible explanation for the increased oxidation rate for films annealed at 950°C and 1050°C, could be a lower activation energy for creating the free volume (lower strain) at the beginning of the second oxidation. As a result, the first few tens of Ångströms of "new" oxide would be more relaxed for films annealed at higher temperatures. This would show up in Fig. 4 at or below a depth of 300 Å. However, no differences in strain were observed in this region or anywhere else in the "new" oxide. In addition, previous studies have shown that the extrapolated oxide stress at the Si-SiO_2 interface is approximately the same for films grown at 850°C and 1050°C, and that it remains the same even if the films are annealed.[6,26] These two observations indicate that the increased oxidation rate is not due to an enhanced reaction rate at the growth interface, but is related to properties of the "old" 300 Å oxide, which is in agreement with work by Landsberger et al.[18] Therefore, our results support an oxidation model based on bulk properties, such as the strain dependent diffusion controlled model.

In this interpretation, an intermediate annealing step enhances the diffusion through the top 300 Å of oxide enough to increase the thickness-averaged diffusion, and as a result increases the oxidation rate. As we have reported,[12] an oxide layer at the Si-SiO_2 interface is prevented from full relaxation by any previously grown oxide. Also, the first oxide layer grown, is not subject to the 120% molar volume mismatch between Si and SiO_2, since no oxide exists at this stage. Assuming that the oxidation is diffusion controlled even for thin films (≤300 Å), the influence of strain could explain the initial rapid growth since the interfacial strain due to the molar volume mismatch has to build up to an equilibrium value as the oxide thickness increases. Therefore, the initial thickness-averaged diffusion coefficient would be greater than for thicker films.

Incorporation of strain dependent diffusion into the oxidation model presents a great challenge, since diffusion through a given layer is determined by its position relative to both the growth interface and the oxide surface in addition to the oxidation temperature. Oxidation models incorporating strain dependent diffusion and strain relaxation have been proposed by Sarti et al.[27] and Fargeix et al.,[28] and one such model has been used to analyze the two-step

OXIDATION OF SILICON IN OXYGEN: MEASUREMENT OF FILM THICKNESS AND KINETICS

S.C. Kao and R.H. Doremus

Materials Engineering Department
Rensselaer Polytechnic Institute
Troy, New York 12180-3590

ABSTRACT

Ellipsometry, transmission electron microscopy(TEM) and step-profile measurement were used to study the dry oxidation kinetics of silicon at temperatures from 800 to 1100°C. For oxide films thicker than 800 Å, all three thickness measurements agreed within experimental error. For oxide films thinner than 350 Å, the ellipsometry gave higher thickness values than the TEM measurements. Thickness measurements by TEM below 500 Å were combined with measurements on thicker films by all three measurements and fitted a linear-parabolic relationship throughout the measured thickness range. Previous deviations from linear-parabolic behavior resulted from inaccurate thickness measurements by ellipsometry for film thinner than 300 Å. The oxidation kinetics are modeled as resulting from the diffusion of molecular oxygen through two different films, the main film of silicon dioxide and a thin interfacial film in which oxygen diffuses more slowly. If this film has constant thickness, a linear term reults; the parabolic term for the growth of the main film is reduced because of compressive film stress.

INTRODUCTION

The square of the thickness of a reaction layer growing on the surface of a solid often is proportional to time t:

$$L^2 = Bt \tag{1}$$

where B is a parabolic rate coefficient. Tammann[1] and Pilling and Bedworth[2] showed that this relation results when the rate of growth of the layer is controlled by diffusion in the layer. Some layers grow by a combination of a linear and a parabolic process:

The Physics and Chemistry of SiO_2 and the Si-SiO_2 Interface 2
Edited by C.R. Helms and B.E. Deal, Plenum Press, New York, 1993

$$L^2 + AL = Bt \quad (2)$$

in which A is a coefficient independent of L and t at constant temperature. Evans[3] showed that this linear parabolic equation can result from a combination of diffusion in the layer and a chemical reaction at the layer-substrate interface. These and other kinetic equations for formation of oxide layers are summarized by Kubaschewski and Hopkins.[4] Strain in the layer[5] or an additional surface layer of constant thickness can also give rise to the linear-parabolic eq. 2, as shown in this paper.

If there is a thin initial reaction layer of thickness L_0, eq. 2 becomes:

$$L^2 - L_0^2 + A(L - L_0) = Bt \quad (3)$$

This is the same equation as that of Deal and Grove, with $\tau = -\left(L_0^2 + AL_0\right)/B$.

Experimental measurements of the growth of an oxide layer on silicon in oxygen fit eq. 1 in the temperature range from 1200°C to 1400°C,[6,7] and the permeation rate of oxygen in the oxide deduced from the oxidation experiments is close in magnitude and temperature dependence to the measurements of Norton[8] on bulk vitreous silica.[14] At temperatures below 1200°C previous thickness-time data for silicon in oxygen have not fit any of eqs. 1-3; various three parameter equations have been proposed to fit the experimental thickness-time relations.[6,10-13] The parabolic growth parameter B calculated for measurements below 1200°C is less than expected from the data of Norton, and this deviation increases the lower the growth temperature.[14] The oxide-thickness-time data for wet oxidation of silicon fit[6] the equation 2 at growth temperatures below 1200°C; again the parabolic parameter is progressively lower than expected from the diffusion of water in silica.[5,9]

The thickness of oxide layers on silicon in these kinetic studies have been measured from optical polarization with an ellipsometer. Carim and Sinclair measured the thicknesses of oxide films on silicon after oxidation in oxygen in the transmission electron microscope, and found that for films thinner than 0.02 μm, the ellipsometer measurements gave higher film thicknesses in a range of 15 to 25%, than those measured by electron microscopy.[15,16] Ravindra et al. also found these differences[17]. In the present work we have confirmed these results; for film thicknesses above about 0.1 μm the two measurements agree, but for thinner films they deviate. Thicknesses of films greater than about 0.1 μm were also measured by a step-profile method. We have found that the electron microscopy measurements agree with the two-parameter linear-parabolic equation (eq. 2) for oxidation of silicon in oxygen below 1200°C. The parabolic coefficients B still deviate increasingly from those calculated from Norton's data as the growth temperature is lower.

To explain these results we suggest two effects. The decrease of B from the value expected from oxygen (or water) diffusion in the film is caused by stress in the film. The linear term in eq. 2 results from a thin, second oxide film of constant thickness at the oxide-silicon-interface that has a lower oxide content than the main SiO_2 film.

EXPERIMENTAL METHODS

The single-crystal silicon wafers used in this study were Czochralski grown, (100) oriented, 7.6 cm in diameter, p (boron) doped and had a resistivity range of 5-10 ohm-cm. Prior to oxidation, the wafers were cleaned thoroughly with the standard RCA procedure followed by immersion in HF solution to remove the native oxide.

Four groups of samples were oxidized at 800, 900, 1000 and 1100 °C. The first group, grown at 800°C, was oxidized in one atmosphere pure oxygen. The other groups were oxidized in ~10% oxygen in nitrogen ambient. Silicon wafers were inserted into the furnace held at 800°C and heated at a rate of 10°C/min. to the desired temperature. One wafer from each group after each oxidation run was examined in the TEM.

Film thicknesses were measured with a Rudolph Research Auto EL-2 ellipsometer equipped with a He-Ne laser with a wavelength of 6328 Å and an angle of incidence of 70°. Delta and psi values were measured on at least five different sites on each wafer. The oxide thickness and refractive index on three wafers in the thin (< 0.06 μm) range and up to six different wafers on thicker films were calculated and averaged. The variation in oxide thickness was less than 1% for each wafer and between different wafers. This uniformity ensures the validity of the TEM measurements which were made on small spot on a wafer.

For the TEM measurements a film of polysilicon about 0.15 μm thick was deposited on the oxide surface as protection and for contrast in the TEM measurements. Cross-section specimens were prepared using the technique described by Bravman and Sinclair.[18] The images were viewed edge-on along the <100> zone axis.

An α-200 step profiler with a resolution of 5 Å was used to measure the step height of oxide thicker than 90 nm. To produce a sharp step on the samples for step profile measurement, a mask with suitable line pattern was used. With standard photolithography procedure followed by etching in HF solution, the patterned oxide appeared on the wafer.

EXPERIMENTAL RESULTS

Previous investigators have reported initial oxide thicknesses from about 1 to 3 nm. In the present work the thickness of oxide measured right after cleaning with HF was about 0.8 to 1.0 nm. However, after the wafer was inserted into the furnace and the furnace heated to oxidizing temperature, the oxide thickness grew to about 2 or 3 nm. In analyzing the thickness-time data, the thickness at time zero was considered as an additional parameter (L_0 in eq. 3, see Table 3).

Oxide thicknesses measured by electron microscopy and calculated from ellipsometric measurements for growth temperatures from 800°C to 1100°C are compared in Tables 1 and 2. At all temperatures the three measurements agree within experimental error for films thicker than about 50 nm. For thinner films the agreement between TEM and ellipsometry thicknesses is poorer, especially at thickness below about 30 nm, as shown in Figures 1 and 2. Therefore for data analysis only the electron microscopy measurements were used below 50 nm, whereas for thicker films, measurements by the ellipsometry were used.

To analyze these selected thickness-time data, eq. 3 was used with three parameters - B, A, and L_0. This equation can be rearranged to the form

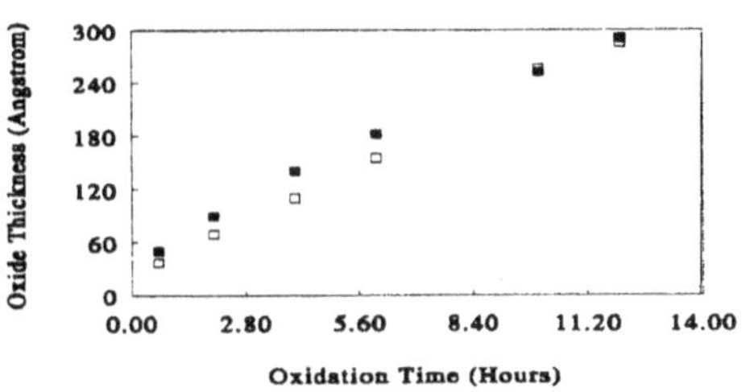

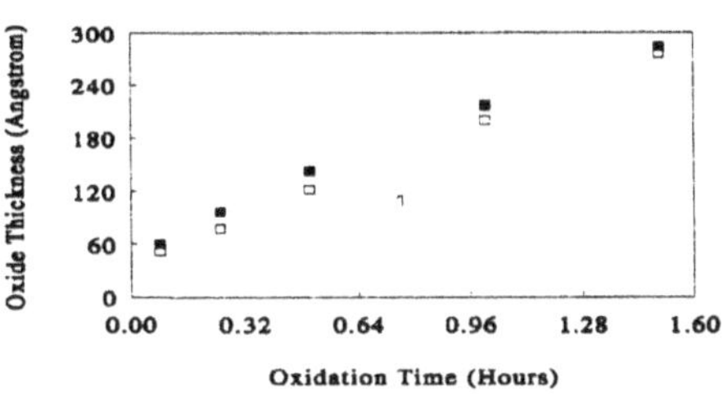

Figure 1. (Left, 800°C group) and **Figure 2.** (Right, 1000°C group) compare the TEM and ellipsometric thickness measurements.

parabolic behavior result from inaccurate thickness measurements by ellipsometry for films thinner than about 50 nm.

Some workers[12,13] have found that thickness-time data for the oxidation of silicon in oxygen over certain thickness ranges can be approximated by a power law:

$$L^m = kt \tag{5}$$

in which m and k are constants. The reason that this equation is followed can be seen by taking the time derivative of eq. 2 and multiplying both sides by t/L:

$$\frac{t}{L}\left(\frac{dL}{dt}\right) = \frac{d(\log L)}{d(\log t)} = \frac{Bt}{(2L+A)L} = \frac{L+A}{2L+A} \tag{6}$$

Thus the slope of a log L - log t plot is nearly constant over a range of thickness values, and this slope is equal to 1/m in eq. 4. When L >> A, m = 1.5; when 2L >> A, m = 1.33; etc. The (erroneous) larger thickness values found by ellipsometry for thin layers make this fit look better than it actually is.

There is good evidence that the rate of growth of the oxide layer on silicon is controlled entirely by oxygen transport through the film, because the reaction of water or oxygen at the oxide-silicon interface to form oxide is rapid.[19] Thus the linear term in eqs. 2 and 3 is not related to the rate of an interfacial reaction.

If the oxidation rate is controlled by diffusion of oxidant in a uniform film, the coefficient B is given by[6,9]:

$$B = \frac{2D_1C_1}{\rho} \tag{7}$$

in which D_1 is the diffusion coefficient of oxidant, C_1 is its concentration in the oxide at the gas-oxide interface, and ρ is the concentration of oxygen in the oxide. At temperatures of about 1200°C and above, the values of B are within experimental error of those calculated from measurements of diffusion of oxygen and water in vitreous silica.[9] The mechanism of diffusion is by solubility and transport of oxygen and water molecules in the silica.[9] Oxidation rates of some silicides in oxygen at temperatures below 1200°C also fit the simple parabolic eq. 1, and the B values calculated from the data of Norton.[8,9] Therefore there are two questions about the oxidation kinetics of silicon: 1) Why do the parabolic coefficients B becomes smaller than expected from diffusion data at lower temperatures? and 2) What is the origin of the linear term A?

We propose that the reduction in B values is caused by stress (strain) in the oxide film. Strain in silica films on silicon reduces the rate of molecular transport in them.[20] The refractive indices of films grown above about 1100°C are close to that of amorphous silica, whereas at lower temperatures the refractive indices (and densities) of films are greater than for bulk amorphous silica.[21,22] The changes in reactive index become greater at lower growth temperatures, corresponding to greater decreases in B values compared to those expected from Norton's data. All these results strongly suggest that strain plays an important role in the reductions in B values.

A prediction from these considerations is that the refractive indices of SiO_2 films with permeations close to those measured for bulk silica should be close to that of bulk silica. Increasing deviations from expected B values should be accompanied by increasing deviations in the refractive indices of the films involved. This prediction is perhaps the most direct way to test the importance of strain in the oxidation of silicon and silicides.

In addition we propose that the linear term results from a thin nonstoichiometric oxide layer at the oxide silicon interface. Such a "blocking layer" has been proposed by others.[23-25]. Permeation of oxygen or water through a more dense silicon-rich oxide, such as SiO_x, where $1 < x < 2$, should be slower than through amorphous SiO_2. If this SiO_x layer has a constant thickness L_2, then the flux J through the combined oxide films is:

$$J = \frac{D_1(C_1 - C_2)}{L} = \frac{D_2C_2}{L_2} \qquad (8)$$

in which L is the thickness of SiO_2, D_1 the diffusion coefficient of oxygen or water molecules in it, C_1, the concentration of water or oxygen molecules dissolved in the SiO_2 at the gas-SiO_2 interface, D_2 the diffusion coefficient of oxygen or water molecules in the SiO_x layer, and C_2 the concentration of water or oxygen at the SiO_2-SiO_x interface. The two-layer system and gas profiles are shown in Fig. 5. From eq. 8 the concentration C_2 is

$$C_2 = \frac{D_1C_1L_2}{LD_2 + L_2D_1} \qquad (9)$$

Thus if permeation through the SiO_x layer is fast and $D_2/L_2 >> D_1/L$, or if this layer is absent, permeation through the SiO_2 controls the rate of oxidation; whereas if permeation through the SiO_x is slow, $D_2/L_2 << D_1/L$, and the rate of oxidation is constant (linear kinetics). Substituting the value of C_2 from eq. 9 into eq. 8, the flux is:

$$J = \frac{\rho dL}{dt} = \frac{D_1C_1}{L + \frac{D_1L_2}{D_2}} \qquad (10)$$

From dL/dt from eq. 2, B is still given by eq. 6, and

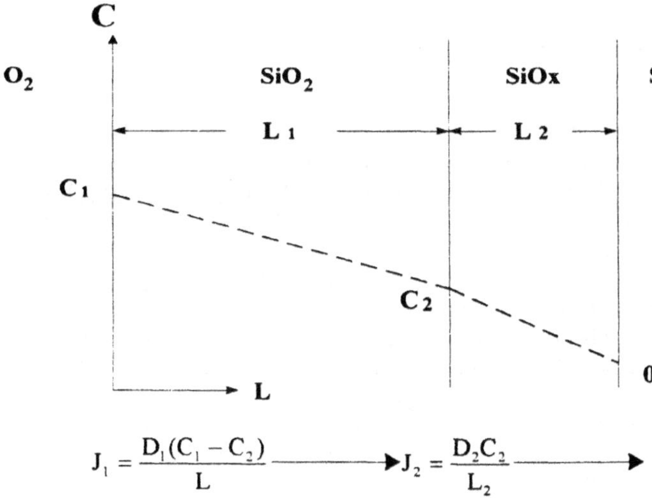

Figure 5. The two layer model and the gas profiles for the oxidation of silicon.

$$A = \frac{2L_2D_1}{D_2} \qquad (11)$$

Again if the SiO_x film is absent ($L_2 = 0$) or permeation through it rapid (large D_2/L_2), A is small and the kinetics are parabolic (eq. 1). There is some evidence for a thin different layer at the SiO_2-Si interface[26-28]. In any event a thin non-stoichiometric SiO_x layer of constant thickness can explain the linear contribution in eq. 2.

ACKNOWLEDGMENT

This work was supported by NSF-MRG contract no. DMR-8510617.

REFERENCES

1. G. Tammann, Z. Anorg. Chem. **111**, 78 (1920)
2. N. B. Pilling and R. E. Bedworth, J. Inst. Met. **29**, 529 (1923)
3. U. R. Evans, Trans. Electrochem. Soc. **46**, 247 (1924)
4. O. Kubaschewski and B. E. Hopkins, **Oxidation of Metals and Alloys**, Butterworths, London, 1953. p. 37ff
5. R. H. Doremus, Thin Solid Films **122**, 191 (1984)
6. B. E. Deal and A. S. Grove, J. Appl. Phys. **36**, 3770 (1965)
7. G. H. Schiroky, J. Matls. Sci. **22**, 3595 (1987)
8. F. J. Norton, Nature **191**, 701 (1961)
9. R. H. Doremus, J. Phys. Chem. **80**, 1773 (1976)
10. C. J. Han and C. R. Helms, J. Appl. Phys. **59**, 1767 (1986)
11. J. Blanc, Phil Mag. **B55**, 685 (1987)
12. R. H. Doremus and A. Szewczyk, J. Matls. Sci. **22**, 2887 (1987)
13. A.Reisman, E.H. Nicollian, C.K. Williams and C.J. Merz, J. Electronic Matls., **16**, 45 (1987)
14. R.H. Doremus in **Electronic Materials and Processing**, ASM, Metals Park, OH, 1989, p37
15. A. H. Carim and R. Sinclair, J. Electrochem. Soc. **134**, 741 (1987)
16. A. H. Carim and R. Sinclair, ibid, **137**, 279 (1990)
17. N.M. Ravindra, J.Narayan, D. Fathy, J.K. Srivastava and E.A.Irene, J. Mater. Res., **2**, 216(1987)
18. J. C. Bravman and R. Sinclair, J. Electromicr. Technique **1**, 53 (1984)
19. R. H. Doremus, J. Appl. Phys. **66**, 4441 (1989)
20. C. K. Huang, R. J. Jacodine and S. Butler, in **Silicon Nitride and Silicon Dioxide Thin Insulating Films**, V. J. Kapoor and K. T. Hankins, eds., Electrochemical Soc., Pennington, N.J.,1987, p. 343
21. E. A. Taft, J. Electr. Soc. **125**, 968 (1978); **127**, 993 (1980); **132**, 2486 (1985)
22. E. A. Irene, Critical Revs. in Solid Stat and Matls. Sci. **14**, 175 (1988)
23. A.Fargeix, G. Ghibaudo and G. Kamarinos, J. Appl. Phys., **54**, 2878(1983)
24. N.F. Mott, S. Rigo, F. Rochet and A.M. Stoneham, Phil. Mag., **B60**, 189(1989)
25. W.A. Tiller, J. Electrochem. Soc., **130**, 501(1983)
26. P.J.Grunthaner,M.H.Hecht,F.J.Grunthanerand N.M.Johnson,J.Appl.Phys.,**61**,629(1987)
27. G. Hollinger and F.J. Himpsel, Appl. Phys. Lett., **44**, 93(1984)
28. T. Hattori and T. Suzuki, Appl. Phys. Lett., **43**, 470(1983)

MODELING PROCESS-DEPENDENT THERMAL SILICON DIOXIDE (SiO_2) FILMS ON SILICON

H.F. Wei[1], A.K. Henning [1], J. Slinkman[2], and J.L. Rogers[2]

[1]Thayer School of Engineering, Dartmouth College, Hanover, NH 03755
[2]IBM General Technology Division, Essex Junction, VT 05452

ABSTRACT

Though extensive work has been done in the Si/SiO_2 system, no process-dependent two-layer SiO_2 film model has ever been established, due largely to the lack of motivation for such a model. This study attempts to model correctly the process dependence of thermal SiO_2 film physical structures and their associated densities, as well as high frequency dielectric constants, so as to provide a foundation for a ULSI process-dependent device reliability simulator. By exploring the characteristic signature of ellipsometric data reduced using a one-layer film model, and comparing it to a two-layer model, we establish a process-dependent, two-layer model for thermal SiO_2 films. Internal consistency in this model is demonstrated using three intrinsic-stress-related phenomena in thermal SiO_2 films on Si. Both the interfacial layer and bulk film are characterized quantitatively for 38 samples, dry-oxidized at four temperatures, leading to three empirical equations describing interlayer thickness, bulk layer density, and bulk layer optical frequency dielectric constant, as functions of oxidation temperature. The interfacial layer refractive index is taken to be independent of oxidation time, and found to be independent of oxidation temperature. The oxidation-temperature-dependent index of refraction of bulk SiO_2 films obtained using the proposed model agrees well with independent one-layer model data on oxides which have thicknesses around the first half-cycle of ellipsometry thickness, for which the interlayer effect is minimal. It is also found that interlayer thickness has a relatively weak dependence on oxidation temperature, which supports the strain energy model for interlayer formation. Application of the thermal SiO_2 film model to Si-device dielectric characterization using fixed index ellipsometry is also discussed, based on recent, new understanding of the ellipsometry equation.

INTRODUCTION

Semiconductor device scaling has followed the set of ideal scaling laws proposed by Dennard and co-workers in 1974 [1], increasing system performance and functionality with higher layout density and lower power-speed product. Projections indicate that by the turn of the century memory and microcomputer chips may have 100 million transistors with 0.2 μm channels and 50 Å gate oxides [2]. On the other hand, chip reliability has improved over the last two decades, driven by customer demand and stiff competition. Projections indicate that by the year 2000 VLSI chips will have failure rates of less than 10 FIT (failure in time) [2]. Both factors have pushed VLSI technologies close to fundamental reliability limits.

One of these limits on device scaling is attributed to hot carrier effects caused by less-than-

The Physics and Chemistry of SiO_2 and the Si-SiO_2 Interface 2
Edited by C.R. Helms and B.E. Deal, Plenum Press, New York, 1993

Establishment and verification of a two-layer thermal SiO_2 film model

Ideally, any assumptions made in establishing the film model should be based on experimental observations. At the same time, a model based on these assumptions should not lead to predictions which contradict other experimental observations. There are three distinct intrinsic-stress-related phenomena in thermal SiO_2 films on Si [18,19] which lead to assumptions used in our two-layer model:

(1) Intrinsic stress at the Si/SiO_2 interface is oxidation temperature independent;

(2) Intrinsic stress decreases quickly with increasing oxide thickness for oxides grown at the same temperature, finally becoming nearly constant in thick (> 300 Å) oxides;

(3) The magnitude of this constant stress in thick oxides is oxidation-temperature-dependent: the higher the oxidation temperature, the lower this constant stress.

Using index of refraction as an estimator of oxide density and intrinsic stress magnitude, we make the following assumptions:

Assumption I: Based on phenomenon (2) above, we assume the bulk oxide film has the same refractive index (n_1) among samples in each group (notice all samples in this study are grouped according to oxidation temperature). That is, n_1 is oxidation-time-independent at a fixed oxidation temperature;

Assumption II: We further assume both the interlayer index of refraction n_2 and thickness d_2 is also oxidation-time-independent at a fixed oxidation temperature;

Assumption III: For simplicity, we assume both the interlayer and bulk films are uniform; that is, there is no refractive index gradient within each layer.

The d_2 (interlayer thickness) standard deviation, or variance estimator, is examined as a function of n_1 (bulk oxide refractive index) and n_2 (interlayer refractive index) for all four groups of samples. The d_2 variance estimation procedure for a given group of samples is described as follows, using the three assumptions made above:

[i] Fix n_2; [ii] For a given (variable, $1.3 < n_1 < 2$) n_1, calculate d_1 and d_2 for every sample in the group using the ellipsometric data measured with the research grade ellipsometer, then plot the whole sample group d_2 variance estimator versus n_1;

[iii] Repeat steps [i] and [ii] for a different n_2 ($1.3 < n_2 < 4$).

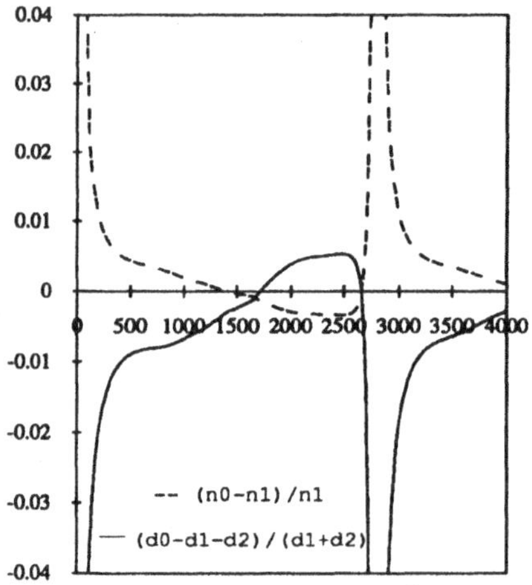

Fig.3 Relative error in refractive index and thickness estimations when a two-layer structure is interpreted by a one-layer model.

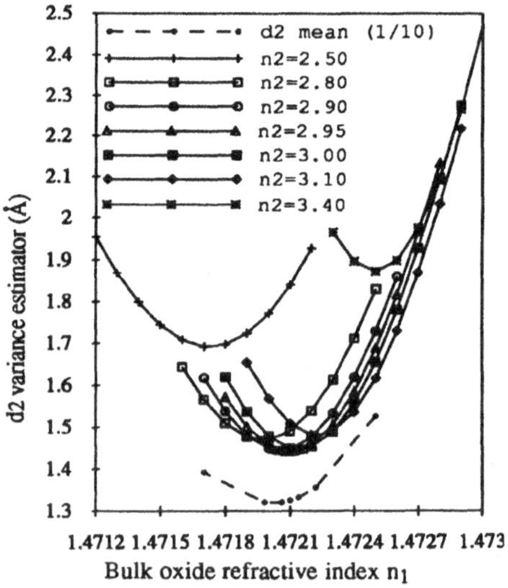

Fig.4 n_2-dependent d_2 variance estimator and d_2 mean (scaled by 10) at the d_2 variance estimation curve minimum for sample group I.

Notice the d_2 variance thus formed for each sample group is a highly non-linear function of n_1 and n_2, using the two-layer model (see Appendix). Based on this procedure, we expect the following outcomes:

A. The value of n_2 obtained should be the same for all 38 samples based on intrinsic-stress phenomenon (1), above;

B. The value of n_1 obtained should be oxidation-temperature-dependent: the higher the oxidation temperature, the lower the n_1, based on intrinsic-stress phenomenon (3);

C. On the plane of n_1 and n_2 which is of physical significance ($1.3<n_1<2$ and $1.3<n_2<4$) the resulting d_2 variance minimum should be unique, and the magnitude of the minimum should not be more than one mono-layer thickness of a-SiO_2 (3.3 Å).

Figure 4 shows the d_2 variance estimation process for group I (800 °C), where only the d_2 variance minimum portion was plotted and the d_2 mean (scaled by 10) was plotted only at the variance estimation curve minimum. Evidently there is a best set of n_1 and n_2 which minimizes the d_2 variance down to 1.45 Å. Thus, the solution of n_1 and n_2 is unique to group I. Incidentally, both d_2 mean and variance are minimized at the same n_1 value. That the d_2 variance minimum (1.45 Å) is less than one mono-layer thickness of a-SiO_2 (3.3 Å) supports Assumption II, that interlayer thickness is essentially the same for sample group I. The d_2 variance minima for the remaining sample groups are found to be between 0.6 Å to 1.4 Å.

The standard deviation of oxide surface roughness of sample #1 was found to be 1.8 Å in a total scanning distance of 8000 Å using the NanoScope III atomic force microscope. It is likely the roughness of the Si-substrate-to-oxide interlayer interface, as well as the interlayer-to-bulk-film interface, are also of the same order (1.8 Å). This may explain why the d_2 variance can not be minimized to zero for any physically significant combination of n_1 and n_2.

Figure 5 shows the d_2 mean for the four groups of samples, using the d_2 standard deviation to mark the upper and lower limit of the error bars. The interlayer refractive index n_2 which minimizes the d_2 standard deviation within each sample group was found to be 2.95 for all four groups of samples. This result is expected based on the growth-temperature-independent interface stress data of [18,19]. Though the d_2 means obtained in this study differ from Taft's reported values at wavelength 5461 Å (d_2=7~8 Å for dry 900°C and 4 Å for dry 1200°C, n_2= 2.8) [15], and Aspnes' reported values at 5461 Å ($d_2=7\pm2$ Å for dry 1000 °C, $n_2=3.2\pm0.5$)

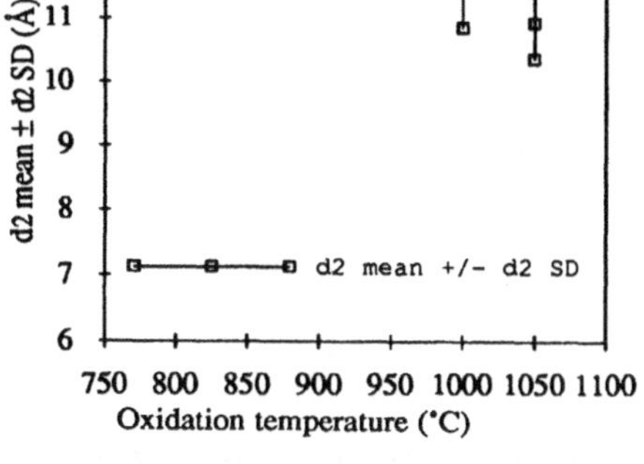

Fig.5 Interlayer thickness d_2 mean ± d_2 standard deviation for all four sample groups.

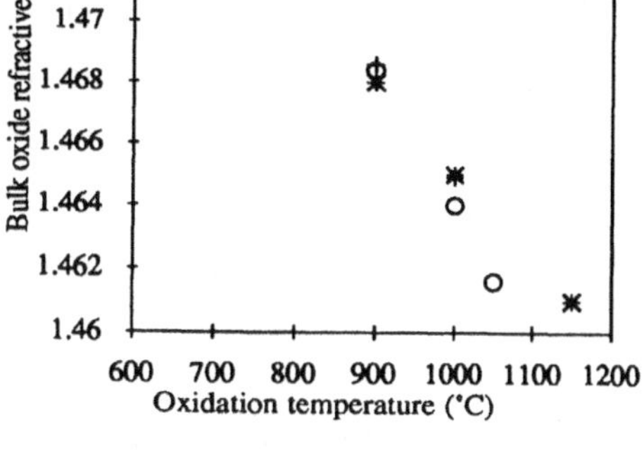

Fig.6 Oxidation-temperature-dependent bulk oxide refractive index n_1 at wavelength 6328 Å.

[16], the d_2 means of this study are close to those found in MOS solar cell open circuit voltage experiments (the oxide non-stoichiometric transition thickness was found to be 13~14 Å) [20].

The relatively weak dependence of interlayer thickness on oxidation temperature in Figure 5 supports the strain energy argument for the interlayer: a large lattice mismatch at the Si/SiO_2 interface can favor an intermediate layer so as to reduce the total free energy, and the interlayer thus formed should be a very slow function of oxidation temperature [21].

Having verified the internal consistency in this two-layer, thermal SiO_2 film model, we turn our attention to the bulk oxide film refractive index (n_1) for the four groups of samples. This was found to be a near-linear function of oxidation temperature, as shown in Figure 6 (n_2 is 2.95). Independent data for thick oxides (1000~1400 Å) based on a one-layer model [22, 23] are also plotted in Figure 6. According to the prediction of Figure 3 (that for oxide thickness around 1400 Å the interlayer effect is a minimum when the one-layer model is used to interpret a two-layer structure), it is expected all these data agree with each other.

Further evaluations of the two-layer thermal SiO_2 film model

A simulation experiment in Figure 7 further verifies the validity of the results in Figures 5 and 6 obtained by the technique of d_2 variance estimation. Three sets of thickness-dependent refractive index curves are depicted in Figure 7. Each set consists of three hypothetical two-layer simulations, assuming interlayers of constant thickness (d_2) and refractive index (n_2 = 2.8), and with bulk films of different refractive index n_1. Interlayer thickness differs between the three sets of simulations. The state of polarization (Δ and ψ) is generated as a function of bulk film thickness d_1, using the two-layer model. The one-layer model is then used to interpret this state of polarization in terms of the thickness-dependent refractive index curves plotted in Figure 7, similar to the procedure used in Figure 3. The simulation curves using different n_2 (varied from 2.6 to 3.0) show little shift from those in Figure 7, compared with the effects caused by d_2 changes. However, the state of polarization (Δ and ψ) generated using a two-layer model in which n_2 is varied from 2.6 to 3.0, does indicate shifts though much smaller than that caused by the d_2 changes, but large enough (more than 0.01°) to be detected. Thus, the range of n_2 (2.8±0.2) should not be interpreted as an uncertainty range.

Four features can be generalized from Figure 7:

(1) The thickness-dependent refractive index (two-layer structure interpreted by a one-layer

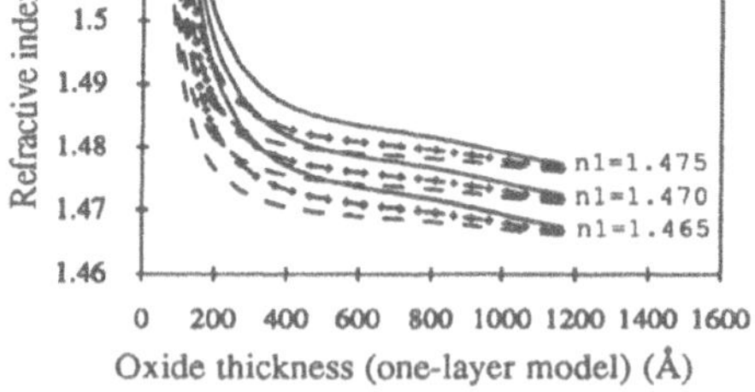

Fig.7 Thickness-dependent refractive index for three sets of hypothetical two-layer structures as interpreted by a one-layer model.

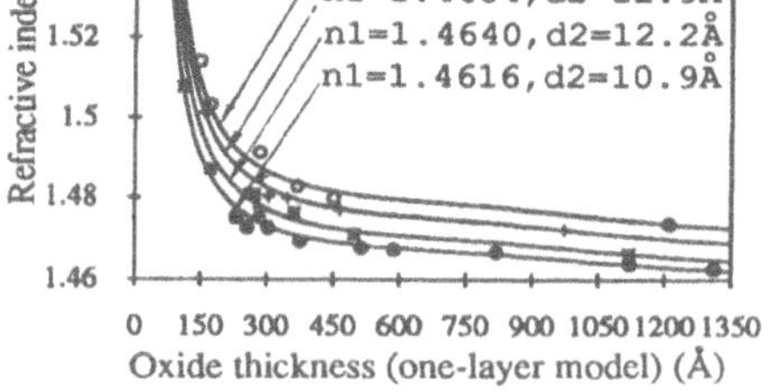

Fig.8 Thickness-dependent refractive index simulations of two-layer structures as interpreted by a one-layer model, compared with Fig. 1.

model) is relatively insensitive to interlayer refractive index n_2 (compared to its sensitivity to d_2 and n_1), for the range $n_2 = 2.6 \sim 3.0$ and the axis scale used in Figure 7;
(2) Differences in bulk film refractive index n_1 cause marked distinctions in the thickness-dependent refractive index curves using the one-layer model interpretation;
(3) All thickness-dependent refractive index curves merge together for oxides thinner than 150 Å, should the two-layer structures these curves represent have the same interfacial layer (n_2, d_2), but different n_1;
(4) For oxide thickness less than 150 Å, the thickness-dependent refractive index curves are distinct only when the two-layer structures these curves represent have interfacial layer (d_2) thicknesses differing by more than around one mono-layer of a-SiO_2.

The relative insensitivity of the d_2 variance estimator and the minimum of n_1 versus n_2 found in Figure 4, agree with feature (1). According to feature (2), distinctions for oxides thicker than 150 Å in Figure 1 imply each sample group should have a distinct bulk film refractive index n_1. Based on features (3) and (4), the convergent behavior for oxide thickness less than 150 Å in Figure 1 implies the interlayer thickness should not differ by more than one a-SiO_2 monolayer for all 38 samples studied. Both conclusions agree with the experimental results in Figures 5 and 6.

Consequently, we compare thickness-dependent refractive index simulations of two-layer structures as interpreted by a one-layer model, with the experimental curves of Figure 1, as shown in Figure 8 (the values of n_2, d_2, and n_1 used in the simulations are from results of Figures 5 and 6). Note that the simulations in Figure 8 are NOT obtained by direct fitting of the experimental curves of Figure 1. Instead, they are obtained using the d_2 variance estimations as shown in Figure 4. The fine match between simulations and experimental curves shown in Figure 8 in turn supports the validity of the d_2 variance estimation procedure using the assumptions inherent in the two-layer thermal SiO_2 film model.

It is worth noting that we have assumed the interlayer is also non-absorbing at 6328 Å in our ellipsometric data reductions. The assumption that $k_2 \approx 0$ is indeed a valid one, since by setting $k_2 = 0.009$ (half of the extinction coefficient of Si at 6328 Å) in the ellipsometric program using the n_2, d_1, d_2, and n_1 ranges in this study, the change in the state of polarization (Δ and ψ) generated is less than 0.01° in most cases, which is beyond the resolution of the research grade ellipsometer used.

Empirical equations in the two-layer thermal SiO_2 film model

Fitting the data of Figure 5 yields the following empirical equation for interlayer thickness d_2 (Å) as a function of oxidation temperature T (°C):

$$d_2 = -11.035 + 6.1146 \times 10^{-2}\,T - 3.8181 \times 10^{-5}\,T^2 \tag{1}$$

Film density can be extracted from the refractive index, using the Lorentz-Lorenz relation given by:

$$\left[\frac{n^2 - 1}{n^2 + 2}\right] \frac{M}{\rho} = \frac{4\pi}{3} N\alpha \tag{2}$$

where n is the index of refraction, M is the molecular weight, ρ is the mass density, and N is Avogadro's number. α is the total polarizability of the molecule, with $\alpha = \alpha_e$ (electron polarizability) $= 2.95 \times 10^{-24}$ cm^3 [23]. Using this expression, bulk film density ρ (g/cm^3) was calculated as shown in Figure 9. In the absence of specific composition data for the interlayer, calculation of its density is not possible.

In a nonferromagnetic ($\mu_r \approx 1$) and insulating ($\sigma \approx 0$) material such as a-SiO_2, the relationship between high frequency relative complex dielectric constant and complex refractive index is given by [24]:

$$\varepsilon_1 - i\varepsilon_2 \approx (n - ik)^2 \tag{3}$$

where $r_{s,i}$ and $r_{p,i}$ are the Fresnel reflection coefficients at the i-th interface for s- and p-polarized light (the ambient-to-film interface corresponds to i = 1; the film-to-substrate interface corresponds to i=2); N_0, N_1, and N_2 are complex refractive indexes for the ambient (air), film, and the substrate (N = n − ik, where n is index of refraction [the real part of N] and k is the extinction coefficient [the imaginary part of N]); θ_0 is the angle of incidence at the first interface (ambient/film), θ_1 and θ_2 are angles of refraction; α is the phase factor in the film for s- or p-polarized light, d is the film thickness, λ is the wave length. Equation (5) is Snell's law.

For the two-layer film model, along with equations (1), (2), and (3), the following equations are also coupled [25]:

$$\alpha_1 = \frac{2\pi d_1}{\lambda}\sqrt{N_1^2 - N_0^2 \sin^2\theta_0} \quad (8), \qquad \alpha_2 = \frac{2\pi d_2}{\lambda}\sqrt{N_2^2 - N_1^2 \sin^2\theta_1} \quad (9)$$

$$N_0 \sin\theta_0 = N_1 \sin\theta_1 = N_2 \sin\theta_2 = N_3 \sin\theta_3 \quad (10)$$

$$R_s = \frac{r_{s,1} + r_{s,2}\, e^{-2i\alpha_1} + r_{s,3}\, e^{-2i(\alpha_1+\alpha_2)} + r_{s,1} r_{s,2} r_{s,3}\, e^{-2i\alpha_2}}{1 + r_{s,1} r_{s,2}\, e^{-2i\alpha_1} + r_{s,1} r_{s,3}\, e^{-2i(\alpha_1+\alpha_2)} + r_{s,2} r_{s,3}\, e^{-2i\alpha_2}} \quad (11)$$

$$R_p = \frac{r_{p,1} + r_{p,2}\, e^{-2i\alpha_1} + r_{p,3}\, e^{-2i(\alpha_1+\alpha_2)} + r_{p,1} r_{p,2} r_{p,3}\, e^{-2i\alpha_2}}{1 + r_{p,1} r_{p,2}\, e^{-2i\alpha_1} + r_{p,1} r_{p,3}\, e^{-2i(\alpha_1+\alpha_2)} + r_{p,2} r_{p,3}\, e^{-2i\alpha_2}} \quad (12)$$

where $r_{s,i}$ or $r_{p,i}$ is the Fresnel reflection coefficient at the ith interface for s- or p-polarized light (the ambient-to-film interface corresponds to i = 1; the bulk-film-to-interlayer interface corresponds to i = 2; the interlayer-to-substrate interface corresponds to i = 3); N_0, N_1, N_2, and N_3 are complex refractive indexes for the ambient (air), bulk film, interlayer, and the substrate; θ_0 is the angle of incidence at the first interface (ambient/bulk-film), θ_1, θ_2, and θ_3 are angles of refraction; α_1 and α_2 are the phase factors in the bulk film and interlayer respectively, for s- or p-polarized light. d_1 and d_2 are thicknesses of the bulk film and interlayer.

ACKNOWLEDGEMENTS

The authors thank D. Baker of IBM GTD Essex Junction for wafer oxidation preparations. This work was supported by an IBM-Shared University Research Grant through the IBM GTD, Essex Junction, VT 05452-4299.

REFERENCES

[1] R.H. Dennard, F.H. Gaensslen, H.N. Yu, V.L. Rideout, E. Bassous, and A.R. LeBlanc, *IEEE J Solid-State Circuits* **SC-9**(5):256, 1974.
[2] D.L. Crook, **28th** IRPS/IEEE, p.2, 1990.
[3] A.K. Henning, "Hot Carrier Effects in CMOS Field Effect Transistors at Cryogenic Temperatures," Ph.D. dissertation, Stanford University, 1987.
[4] D.M. Erb, H.G. Dill, and T.N. Toombs, *IEEE Trans Elec Dev* **ED-18**:105, 1971.
[5] T.H. Ning, *Solid-State Electronics* **21**(1):273, 1978.
[6] C.M. Wang, J.J. Tzou, and C.Y. Yang, **27**th IRPS/IEEE, p.110, 1989.
[7] J. Chung, M.C. Jeng, J.E. Moon, P.K. Ko, and C.Hu, **27**th IRPS/IEEE, p.92, 1989.
[8] J.C. Panner, E.W. Conrad, and J.L. Rogers, *Thin Solid Films* **206**(1-2):381, 1991.
[9] H.F. Wei, A.K. Henning, J. Slinkman, and W.R. Hunter, *J Electrochem Soc* **139**(6):1783, 1992.
[10] J.H. Ho, C.L. Lee, C.W. Jen, and T.F. Lei, *Solid-State Electronics* **330**(9):973, 1987.
[11] A. Kalnitsky, S.P. Tay, J.P. Ellul, S. Chongsawangvirod, J.W. Andrews, and E.A. Irene, *J Electrochem Soc* **137**(1):234, 1990.

[12] S. Chongsawangvirod, E.A. Irene, A. Kalnitsky, S.P. Tay, and J.P. Ellul, *J Electrochem Soc* **137**(11):3536, 1990.
[13] T.S. Chao, C.L. Lee, and T.F. Lei, *J Electrochem Soc* **138**(6):1756, 1991.
[14] L.M. Landsberger and W.A. Tiller, *Appl Phys Lett* **51**(18):1416, 1987.
[15] E. Taft and L. Cordes, *J Electrochem Soc* **126**(1):131, 1979.
[16] D.E. Aspnes and J.B. Theeten, *J Electrochem Soc* **127**(6):1359, 1980.
[17] F.J. Grunthaner and P.J. Grunthaner, *Mater Sci Repts* **1**(2-3):65, 1986.
[18] G. Lucovsky, J.T. Fitch, E. Kobeda, and E.A. Irene, in "The Physics and Chemistry of SiO_2 and the Si-SiO_2 Interface," eds. C. R. Helms and B. E. Deal (Plenum, New York, 1988), p. 139.
[19] E. Kobeda and E.A. Irene, *J Vac Sci Technol* **B6**:574, 1988.
[20] R. Singh, K. Rajkanan, and J. Shewchun, in "Physics of SiO_2 and its Interfaces," ed. S.T. Pantelides (Pergamon, New York, 1978), p. 396.
[21] N.F. Mott, S. Rigo, F. Rochet, and A.M. Stoneham, *Phil Mag B*, **60**(2):189, 1989.
[22] E.A. Irene, E. Tierney, and J. Angilello, *J Electrochem Soc* **129**(11):2594, 1982.
[23] N.M. Ravindra, J. Narayan, D. Fathy, J.K. Srivastava, and E.A. Irene, *J Mater Res* **2**(2):216, 1987.
[24] R.H. Bube, "Electrons in Solids" (Academic Press, New York, 1981).
[25] R.M.A. Azzam and N.M. Bashara, "Ellipsometry and Polarized Light" (North-Holland, Amsterdam, 1977).

II. NOVEL OXIDATION METHODS AND CHARACTERIZATION

INTRODUCTION

S. Rigo
Groupe de Physique des Solides
Universites Paris 7 et Paris 6, Tour 23, 2 Place Jussieu
75251 Paris Cedex 05, France

B. E. Deal
Department of Electrical Engineering
Stanford University, Stanford, CA 94305

The trend to lower processing temperatures in the fabrication of high density ULSI devices has encouraged the development of enhanced thermal oxidation procedures. These have included chemical, electrical, physical, and optical excitation methods. Such excitation can result in modifications of physical as well as electrical properties of the resulting oxides, in addition to effects on the oxidation kinetics. Several examples of enhanced oxidation are described in the first half of this chapter.

R. J. Jaccodine reviews the chemically enhanced thermal oxidation by chloride and fluorine compounds. The most striking results are the following: 1) some ppm of fluorine compounds (NF_3, $C_2H_3C_{12}F$) in the O_2 gas induce a significant increase of the oxide growth rate (a factor of 10 at 1000°C for 500 ppm), 2) a very fast shrinkage of already grown stacking faults occurs during O_2/NF_3 oxidation. This indicates that the basic difference between this process and classical oxidation is the creation of a vacancy-rich interface.

J. Joseph, Y. Z. Hu and E. A. Irene present an in-situ spectroscopic ellipsometry study of the electron cyclotron resonance (ECR) plasma oxidation of silicon and interfacial damage. The ECR source was operating at 300 W, 2.45 GHz with an oxygen pressure of 4 10^{-4} Torr at oxidation temperatures between 80 and 400°C. Accelerated growth under positive substrate bias indicates that a negative atomic species dominates the growth above an oxide thickness of 4 nm. Kinetics results are compatible with the Cabrera-Mott theory, however this assumption needs to be checked. The damage layer appears to be composed of SiO_2 with amorphous Si and due to the oxidation reaction at low temperature (<500°C) rather than the ions from the plasma.

L. M. Landsberger discusses the mechanisms of oxidation rate enhancement in negative-point oxygen discharge at atmospheric O_2 pressure and 600-900°C while creating SiO_2 relaxed films. A simple model is developed and includes the effect of relaxation of oxide by O^- ions, enhanced oxygen diffusion, and O^- flux. Reactions and atomic transport mechanisms in the oxide corresponding to the O^- flux remain debated questions.

Conventional oxidation at high temperature of $Si_{1-x}Ge_x$ produces oxides which, due to the segregation of Ge at the oxide/substrate interface, are useless to device applications. C. Caragianis, Y. Shigesato, and D. C. Paine show that oxidation at very high pressure (500-700 atm) allows the growth of oxide at low temperatures (550°C). This produces high quality oxide which incorporates Ge into the oxide.

The above results all tend to indicate that enhanced thermal oxidation at reduced temperature may be beneficial for future fabrication of advanced device structures. In addition, such techniques can help provide a valuable insight into mechanisms of thermal oxidation.

An important aspect of developing new methods of oxidation is that of characterization of the silicon surface and the oxidized silicon structure. In the second part of this chapter, several methods of such characterization are reported. These include spectrographic ellipsometry of thermally oxidized silicon by E. A. Irene and V. A. Yakovlav; infrared spectroscopy of low temperature plasma deposited oxides over thermal oxides by S. Fujimura and co-workers; infrared spectroscopy and ellipsometry of both thermal oxides and remote plasma deposited oxides by C. E. Shearon and co-workers; and transmission electron microscopy (TEM) of thermally oxidized amorphous silicon by M. Reiche. In investigations involving optical methods of analysis, some differences of opinion exist regarding interpretation of results. Therefore, various conclusions have been reached regarding properties of oxide-silicon interfaces. Undoubtedly, more experiments and data will be forthcoming to provide a better understanding of $Si\text{-}SiO_2$ interface structure. Other analysis tools, such as scanning tunneling microscopy and its offspring, may help in this respect as well.

NEW APPROACH TO CHEMICALLY ENHANCED OXIDATION - A REVIEW

Ralph J. Jaccodine

Sherman Fairchild Laboratory 161
Lehigh University
Bethlehem, PA 18105

ABSTRACT

This paper is a review of the effect of small additions (ppm) of fluorine to the dry oxidation process of silicon. It deals with the various critical areas of the oxidation growth such as A) growth kinetics, B) interfacial strain, C) stacking fault shrinkage, D) effects on oxidation enhanced or retarded diffusion (OED/ORD) and electrical characterization.

INTRODUCTION

The addition of a halogenic specie during dry oxidation results in a significant improvement in the electronic properties of the oxide and the underlying silicon. The addition of percent concentrations of a chlorine bearing compound to the oxidant, therefore, has been widely practiced in silicon integrated circuit processing technology.[1,2]

Interest in a new fluorinated addition process arose in the early 1980's[3] when preliminary work showed that parts per million of a fluorine compound added to the oxidant stream resulted in greatly enhanced oxidation rate as well as improved oxide quality.[4,5] This paper will briefly review the several facets of the fluorine addition oxidation process dealing with the A) growth kinetics, B) interfacial strain, C) stacking fault shrinkage, D) oxidation enhanced or retarded diffusions (OED/ORD), and finally E) electrical characterization measurements.

EXPERIMENTAL PROCEDURES

The aim of the experimental work was to first determine the appropriate growth conditions, times and temperatures and fluorine volume percent (v/o) additions to establish the oxide growth kinetics. Initially two different classes of fluorine compounds were selected, The alphatic liquid hydrocarbon, dichlorofluoroethane ($C_2H_3Cl_2F$), has an appropriate vapor pressure at room temperature for delivery by a gas bubbler system. Nitrogen fluoride (NF_3) is a gas that can be easily diluted and delivery directly into an oxidation gas system. The former source is easy to handle and combines both fluorine and chlorine species whereas the NF_3 has only fluorine and contains none of the additional carbon and hydrogen species as does the liquid source.

The SOLGAS program[6] was used to calculate the partial pressure of the possible active oxidizing species under thermal oxidation conditions to ascertain which of the many possible chemical sources of F and Cl would be suitable. This program was also used for assessing which specific species correlate with the enhanced oxidation rate.

The silicon wafers used were chem-mechanically polished p-type, Czochralski-grown crystals of (100) orientation, with a resistivity of 2-10 Ω/cm. After a standard RCA cleaning, the oxidations were carried out in a double walled, quartz furnace for various times at 900°C and 1000°C. Oxides were grown by inserting wafers into the furnace tube in flowing N_2 and then switching over to the oxidant supply. The oxidizing gas flow for $O_2/C_2H_3Cl_2F$ oxidations were composed of 1 liter/min of oxygen (O_2) along with a much smaller flow of $C_2H_3Cl_2F$. The bubbler filled with the $C_2H_3Cl_2F$ was kept at room temperature, while oxygen was bubbled through at rates from 1 to 10 ml/min. This corresponded to fluorine additions from 0.011 to 0.11 v/o. NF_3 is a gaseous fluorine source, and its gas flow was monitored directly by micro-flowmeter before introduction at into the oxidation furnace. Oxide thickness and refractive index were determined by a Rudolph Research Auto EL-II ellipsometer using a helium-neon laser. Physical integrity of the oxides was observed by optical and scanning electron microscopy (SEM).

A. Oxidation Kinetics

The fluorine oxidations were carried out by varying parameters - oxidation temperature, time, fluorine concentration, and the type of fluorine additive.

In FIGURE 1(a), we plot oxide thickness vs. oxidation time for oxidations at 1000°C with 0.055 v/o $C_2H_3Cl_2F$ and 0.044 v/o NF_3 addition.

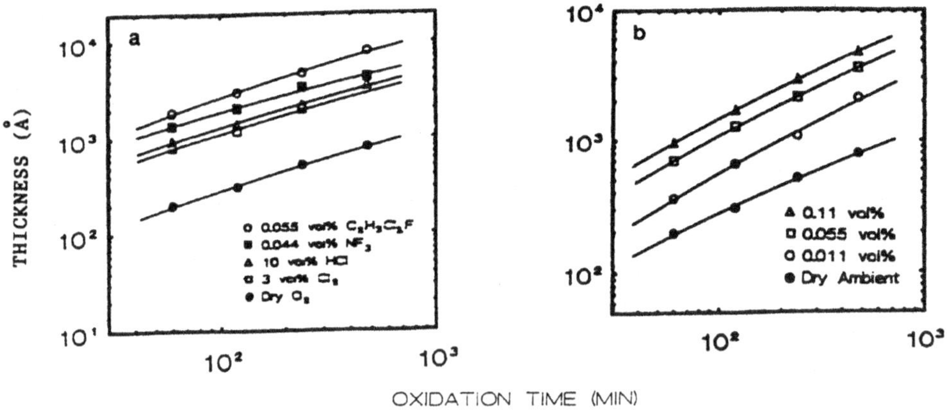

Figure 1. Thickness vs. time (100) at 1000°C (a) and 900°C (b) for $O_2/C_2H_3Cl_2F$. Solid line is least square fit with the power of time.

These data were chosen so that a direct comparison with the published data for films grown in 10 v/o HCl and 3 v/o Cl_2[7,8] could be made. It is observed that the resultant oxide is thicker for films grown using the fluorine compounds despite the two orders of magnitude difference in the volume percent addition.

FIGURE 1(b) represents oxide thickness vs. oxidation data for increasing concentration of $C_2H_3Cl_2F$ at 900°C. It can be seen that increasing the amount of source acts to enhance the oxidation rate. It should be noted that at higher concentrations than 0.055 vol. % and higher temperatures >1000°C poorer quality oxides are observed with this particular source.

Comparable data of oxide thickness vs. oxidation time for increasing volume concentration of NF_3. It is found that the NF_3 species is also effective in enhancing the oxidation rate as can be expected from its chemistry. It should also be noted that optically clear pinhole free oxides result when using NF_3 over the entire concentration, oxidation time-temperature range of these studies.

The experimental data was treated in terms of the linear-parabolic model[9] and the more recently proposed power of time model[10,11]. The following two relationships were used to extract the appropriate parameters:

$$X = a(t_g + t_o)^b \tag{1}$$

$$X = A\{[1 + 4B(t + \tau)/A^2]^{0.5} - 1\}/2 \tag{2}$$

where X is measured oxide thickness. For the power of time model, a and b are coefficient and exponent, respectively, while t_g and t_o are the oxidation times required to grow the final measured oxide and the "pre-existing" oxide. The constants a, b, and t_o were determined by least squares analysis. Since these oxidations were all performed at long oxidation time intervals, a simple least squares fit the data very well. The A and B in the Deal-Grove Equation 2 are the constants in the linear-parabolic model, t is the oxidation time, and τ is a fitting parameter which accounts for initial rapid growth regime. The constants A and B were determined from X vs. $(t + \tau)/X$ plots.

The resultant calculated constants for both these models are given in Table I and II. The solid line in FIGURE 1(b) are the least squares fit with the power of time model. This fit was superior to the linear-parabolic model and therefore was used. As pointed out by Blanc[12] and others, reasonable physical models can lead to equivalent mathematical expression for the thickness as a function of time and that a large class of models exhibit similar mathematical structure.

When the data are compared to previously published results of the oxidation of Si using 1-10 v/o HCl, several interesting differences are evident apart from the nature of the impurity. For all cases of fluorine oxidation both the linear and parabolic constants increase with increasing fluorine concentration, whereas in the comparable HCl data only the parabolic rate constant increases over the full HCl concentration range. In the case of oxidation data at 900°C and 1000°C, an almost order of magnitude increase occurs in these parameters in spite of the very much lower fluorine concentration. As increasing amounts of NF_3 are added to the oxidation, the optical refractive index was found to decrease from 1.46 to 1.451. This leads one to infer that the fluorine incorporation decreases the density of the oxide and also has an influence on the viscoelastic properties of the resultant oxide much as water vapor does. In fact, this correlates with some of the experience of workers in the area of silica optical fibers.[13]

B. Interfacial Strain Measurement

The measurements of thin film strain is usually done by using a straight forward procedure based on beam bending. Strains in a grown or deposited film on a thin substrate will result in bending of the

Table I. Effect of fluorine additive on the constants from the power of time model oxidation at 900° and 1000° C

T(°C)	Ambient	a	b	t_o (min)	Error
900°	Dry O_2	9.30	0.72	6.5	0.0008
	0.011 NF_3	46.77	0.66	1.6	0.0003
	0.022 NF_3	46.77	0.70	1.6	0.0097
	0.044 NF_3	63.09	0.64	5.0	0.0141
	0.011 $C_2H_3Cl_2F$	52.48	0.65	5.0	0.0022
	0.055 $C_2H_3Cl_2F$	107.15	0.56	5.0	0.0015
	0.11 $C_2H_3Cl_2F$	147.91	0.53	5.0	0.0013
1000°	Dry O_2	128.82	0.45	0.3	0.0034
	0.011 NF_3	131.82	0.50	35.0	0.0001
	0.022 NF_3	169.82	0.47	30.0	0.0239
	0.044 NF_3	338.84	0.36	30.0	0.0107
	0.011 $C_2H_3Cl_2F$	208.92	0.45	30.0	0.0029
	0.055 $C_2H_3Cl_2F$	213.79	0.51	24.0	0.0016

Table II. Effect of fluorine additive on the constants from the linear-parabolic models during oxidation at 900° and 1000° C

T(°C)	Ambient	B (μm^2/h)	A/B (μm/h)	τ(h)
900°	Dry O_2	0.0021	0.0122	1.4
	0.011 NF_3	0.0047	0.0127	1.2
	0.022 NF_3	0.0059	0.0152	1.0
	0.044 NF_3	0.0081	0.0194	0.8
	0.011 $C_2H_3Cl_2F$	0.0258	0.0280	1.0
	0.055 $C_2H_3Cl_2F$	0.0438	0.0629	0.5
	0.11 $C_2H_3Cl_2F$	0.0724	0.0946	0.25
1000°	Dry O_2	0.0073	0.0709	0.4
	0.011 NF_3	0.0192	0.1053	0.4
	0.022 NF_3	0.0203	0.2203	0.4
	0.044 NF_3	0.0278	0.2530	0.4
	0.011 $C_2H_3Cl_2F$	0.0471	0.1162	0.2
	0.055 $C_2H_3Cl_2F$	0.1105	0.1162	0.1

substrate; the radius of curvature can be measured using a commercial laser wafer-flatness measuring equipment. This equipment allows the direct measurement of wafer deflection or bowing and thus the radius of curvature of the bent beam. The film stress can be approximately determined by using linear elasticity theory according to Stoney's formula[14].

The experimental set up for curvature measurements is a simple adaption of the laser beam technique to measure curvature. Stress σ_f is calculated for the case of the bending of a rigid beam, no slippage, and each lamella is elastically strained and the strain is istotropic.

This analysis given by:

$$\sigma_f = \frac{1}{6} \cdot \frac{E}{(1-v)} \cdot \frac{t_s^2}{t_f^2} \cdot \left(\frac{1}{R_f} - \frac{1}{R_0}\right) \qquad (3)$$

where E and γ are Young's modules and Poisson's ratio for the silicon substrate, t_s and t_f are the thickness of the substrate and film respectively while R_o and R_f are the values of the radius of curvature of the substrate before and after film growth or deposition.[15]

In FIGURE 3 we can see plotted oxide stress against concentrations of NF_3 at 800, 900 and 1000°C. It can be seen that there is a sharp reduction in the value of stress with NF_3 concentration between 100 and 200 ppm. Profile analysis of samples using secondary ion mass spectroscopy (SIMS) is consistent with this data. The hypothesis that fluorine incorporation results in interface strain relaxation and the suppression of interface state generation according to the bond strain gradient model[16] is also consistent with these findings.

The mechanism which causes this strain relaxation effect is thought to be related to the fluorine induced density reduction and opening of the oxide structure because the Si-F bond can break into Si-O-Si rings. Published x-ray photoemission spectroscopy (XPS) data indicate that fluorine atoms primarily bond to silicon and not to oxygen[17] thus resulting in a more open structure for the fluorinated oxides. This correlates with the increase in the oxide parabolic growth rate. Another view of the mechanism for strain relief is that the fluorine decreases the oxide viscosity caused by structural modification to the

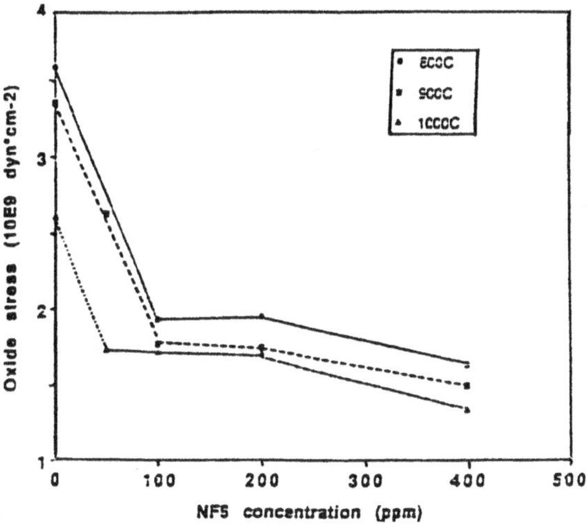

Figure 2. Oxide stress against NF_3 concentration with oxidation temperature as parameter.

SiO_2 network and allows viscoelastic relaxation to occur easier. The fluorine effect is particularly apparent when carrying out so-called 2D oxidation of trenches or pillars.

C. Stacking Fault Shrinkage

The growth and harmful effects of oxidation induced stacking faults (OSF) on devices are well known. Various methods have been developed to eliminate these OSF's during thermal heat treatments.[18,19]

It was found that OSF's anneal very rapidly even in an oxidation process with the additions of fluorine in contrast to the N_2 or nitridation processes usually carried out. It was also found that no new OSF's could be nucleated and generated even on mechanically damaged substrates.

FIGURE 3 shows OSF shrinkage rates for various thermal treatments as a function of temperature.[22]

It was proposed that the fast shrinkage rate can be explained as follows. It is generally believed that the behavior of OSF's in silicon is dependent upon the deviation from equilibrium of point defects during thermal heat treatment. It has been demonstrated that the local dynamic condition is a the vacancy supersaturation coupled with the undersaturation of interstitials during the N_2 annealing,[22] nitridation,[23] and also during O_2/NF_3 oxidation.[24] Thus, in this situation, OSF's should shrink by absorbing vacancies (or by emitting self-interstitials) at the dislocation that bounds OSF's.

We have observed a nonlinear shrinkage with time for O_2/NF_3 oxidation. We propose that the initial rapid shrinkage is due to the reaction of fluorine at the Si/SiO_2 interface giving rise to an increase of vacancies and their subsequent capture at the partial dislocations bounding the stacking fault.

In general, OSF shrinkage studies have been made using N_2 and other ambients which have little influence on the activation energy for shrinkage. Nishi and Antoniadas[20] suggested the existence of a fast transient shrinkage behavior during N_2 heat treatment is dominated by

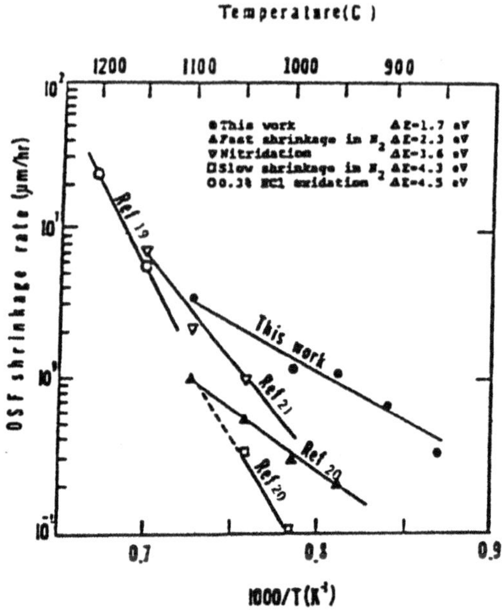

Figure 3. OSF shrinkage rate versus temperature.

self-interstitial capture at the Si/SiO_2 interface. In conclusion, we show the profound effect of a small amount of fluorine addition to dry O_2 oxidation on the OSF shrinkage, which lowered activation energy from almost 4eV to 1.7 eV.

D. Oxidation Enhanced and Retarded Diffusion

The thermal oxidation process is known to effect the concentration and flux of silicon point defects and then this means not only the behavior of OSF's as dealt with above but also the substitutional dopant diffusion in silicon.[26,27] The enhanced diffusion of B, P and As as well as the growth of OSF's are observed while the retarded diffusion of Sb is noted during dry oxidation in the temperature range of 900-1200°C.[28,29]

These oxidation related phenomena stem from a non-equilibrium concentration of point defects (V. and I.) which occur during oxidation[30] and their contribution to the diffusion process. In general it has been postulated that all substitutional dopants in Si diffuse by both the interstitialcy (I) and vacancy (V) mechanisms, which can be represented by a simple expression as follows:

$$\langle D\rangle/D^* = f_i\langle C_i\rangle/C_i^* + f_v\langle C_v\rangle/C_v^* \quad (4)$$

the fractional interstitialcy and vacancy mechanism is defined as follows $f_i = D_i^*/D^*$ and $f_v = D_v^*/D^*$. D_i^* and D_v^* are the interstitialcy and vacancy-motivated diffusivities. The diffusion rate of each of these mechanisms is proportional to the concentration of vacancies or interstitials as well as the fractional interstitialcy or vacancy mechanisms.

The processes of dry (and steam oxidation) favors the enhancing of the interstitial concentration over vacancies. On the other hand, the addition of chlorinated compounds such as HCl favor enhancing the vacancy concentration, retarding P diffusion[31]. The nitridation of SiO_2 capped and bare silicon in an NH_3 ambient enhances the diffusion of B and P but retards the diffusion of Sb in the SiO_2 capped silicon and the opposite effect is found when nitridation is performed on a bare silicon surface.[32] The effect of the various thermal heat treatments on dopant diffusion and OSF's length are summarized in TABLE III.

Table III. Effect of various heat treatment - dopant diffusion/OSF length

	Oxidation	Oxynitridation	Nitridation	Cl Oxidation
B,P Diffusion	Enhanced Diffusion	Enhanced Diffusion	Retarded Diffusion	Retarded Diffusion
Sb Diffusion	Retarded Diffusion	Retarded Diffusion	Enhanced Diffusion	Enhanced Diffusion
OSF Length	Growth	Growth	Shrinkage	Shrinkage

We have reported on the effect of fluorinated oxidation on the diffusion of boron.[24] Subsequently, the effect of the fluorinated oxidation effect on diffusion of phosphorous, arsenic and autimony were also investigated.[33] The detailed procedures for these studies are discussed[23,23] and the results for the various dopants are included in Reference 33.

From these studies we have calculated the point defect concentration after both a dry and O_2/NF_3 oxidation. The data for 1100°C and O_2/NF_3 oxidation is shown in FIGURE 4.

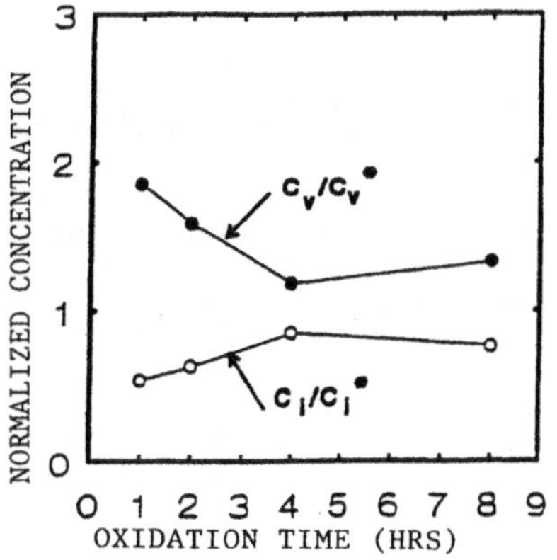

Figure 4. Change of $\langle C_i\rangle/C_i^*$ and $\langle C_v\rangle/C_v^*$ vs. oxidation time in silicon at 1100°C after O_2/NF_3 oxidation.

In contrast to the dry oxidation case, it is seen that the Si interstitial concentration is lower than the concentration of vacancies and thus injection of vacancies is favored during O_2/NF_3 oxidation. It is noted that with time as in the dry oxidation case, both defect concentrations tend to merge. We have further determined the estimated fractional interstialcy for these diffusions.[33]

From these and similar studies of diffusion under various "capping" layers we can summarize the diffusion results in TABLE IV.

TABLE IV. Summary of the junction motion under various covering layers with respect to the control region (SiO_2/Si_3N_4 layer) at 1100°C for up to 8 hrs. (B:bare silicon region, O:oxide covered region, N:nitride covered region, OED:Oxidation Enhanced Diffusion, ORD:Oxidation Retarded Diffusion, RD:Retarded Diffusion, OD:Out Diffusion, and NE:No effect)

Ambients	Area	P	As	Sb
dry oxidation	Bare	OED	OD	OD
	Oxide	OED	OED	ORD
	Nitride	RD(NE)	RD	RD
O_2/NF_3 oxidation	Bare	ORD	OD	OD
	Oxide	ORD	OED	OED
	Nitride	RD(NE)	RD	RD

The role of fluorine on point defect balance under various capping layers indicates that excess vacancies are generated during O_2/NF_3 oxidation. This leads to retardation of B and P and enhancement of As and Sb; just the opposite as that occurring under non-fluorine oxidations.

E. Electrical Characterization

The effect of fluorine on hot electron induced generations of interface traps[34,35] and improved hot electron immunity[36,37,38], dielectric breakdown[39] as well as radiation response[40] have already been reported. In our work we have investigated the influence of fluorine on as grown interface traps using MOS capacitors as test structures. In these studies we also performed SIMS analysis as a means of investigating the incorporation of fluorine into the oxide with varying oxidation processing parameters as well as effect on as grown interface trap density.

High frequency capacitance-voltage (C-V) testing was performed for all the MOS capacitor wafers used in these experiments.[41] With the oxide thickness given by ellipsometer measurements, the application of this method yielded not only C-V curves but also the values of quantities like the gate area, the flatband capacitance, the doping density and the depletion region width. The gate area as evaluated from C-V measurements was used, along with the surface potential versus gate voltage data derived from Q-V measurements, for the evaluation of the interface trap density for our MOS capacitors. High-frequency C-V curves were compared for dry and fluorinated oxides grown at the same temperature.

Charge-voltage (Q-V) characterization[42,43] was performed for the MOS capacitor wafers to extract surface potential against gate voltage data. The surface potential versus gate bias curves was obtained for oxides grown at 900°C for both dry and fluorinated (with 30 ppm NF_3) oxidation;

shift between the two curves. These data allow one to make the evaluation of interface trap density as a function of surface potential (FIGURE 5). The U-shaped curve is clearly wider for the fluorinated oxide case and thus indicating a lower value of interface state density for a given value of surface potential.

We believe that this fluorine-induced reduction of the interface trap density can be attributed to passivation of dangling bonds at the $Si-SiO_2$ interface as well as to removal of Si-Si and Si-O weak bonds, which are all defects that introduce energy levels within the silicon band gap[44]. Furthermore, fluorine bonding and incorporation in the

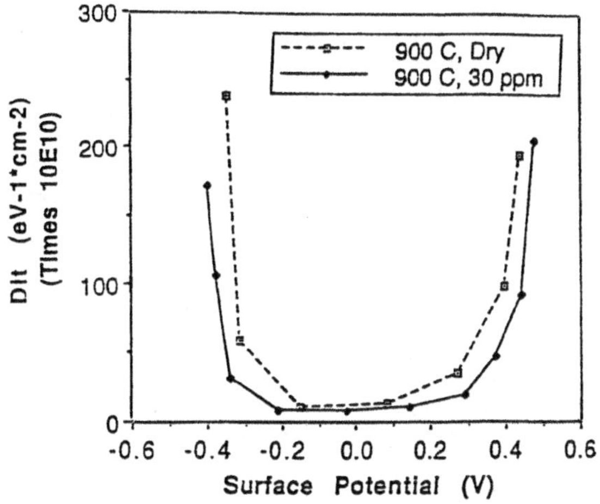

Figure 5. Interface trap density against surface potential, 900°C, one-step oxidation.

oxide network in the area of the oxide near the $Si-SiO_2$ interface would promote interfacial strain relaxation; this would reduce the migration towards the interface of non-bridging oxygen defects generating interface traps according to the bond strain gradient model.[40,45] Such a mechanism, which would be much more important in the presence of radiation or hot electron injection, is consistent with the observed fluorine-induced oxide stress relaxation.

This work was supported in part by the Army Research Office under grant DAAL 0388-K-0094 and Defense Nuclear Agency grant DNA001-88-C-0122-pl.

REFERENCES

1. R.J. Kriegler, Y.G. Cheng, and D.R. Colton, J. **Electrochem. Soc.**, 119:388 (1979); R.J. Kriegler, Y.G. Cheng, and D.R. Colton, U.S. Pat. 3,692,571 (1972).
2. J. Monkowski, **Solid State Technol.**, 58 (1979); ibid., 113 (1979).
3. Proposal to the Army Research Office, December, 1982.
4. P.F. Schmidt, R.J. Jaccodine, C.H. Wolowodiuk, and T. Kook, **Mat. Lett.**, 3:235 (1985).
5. M. Morita, T. Kub, T. Ishihara, and M. Hirose, **IEDM Tech. Dig.**, 144 (1984).
6. G. Erikssoon, **Acta Chem. Scand.**, 26:2651 (SOLGAS Program) (1971).

7. D.W. Hess and B.E. Deal, ibid., 124:735 (1977).
8. B.E. Deal, D. Hess, J.D. Plummer, C.P. Ho, ibid., 125:339 (1978).
9. B.E. Deal and A.S. Grove, J. **Appl. Phys.**, 36:3770 (1965).
10. A. Reisman, E.H. Nicollian, C.K. Williams, and C.J. Merz, J. **Electron. Mater.**, 16:45 (1987).
11. U.S. Kim, C. Wolowodiuk, R. Jaccodine, F. Stevie and P. Kahora, J. **Elec. Soc.**, 137:7 (1990).
12. J. Blanc, **Philos. Mag.**, 55 (1987).
13. P.D. Lazay and A.D. Pearson, **IEEE J. Quantum Electron**, QE-18:504 (1982).
14. G. Stoney, **Proc. of Royal Society**, London, Vol. A82, A553:1909.
15. J. Adamczewslea and T. Bredzynski,**Thin Solid Films**, 113:4 (1984).
16. E.F. DaSilva, Y. Nishioka and T.P. Ma, **IEEE Trans. Nuc. Sci.**, NS34:1190 (1987).
17. M. Morita, T. Kubo, T. Ishihara and M. Hirose, **App. Phys. Lett.**, 45:1312 (1984).
18. H. Shiraki, **Jpn. J. Appl. Phys.**, 14:747 (1975).
19. C.L. Claeys, G.J. Declerck, and R.J.V. Overstraeten, **Appl. Phys. Lett**, 35:794 (1979).
20. K. Nishi and D.A. Antoniadis, **Appl. Phys. Lett.**, 46:516 (1985).
21. Y. Hayafuji, K. Kajiwara, and S. Usui, J. **Appl. Phys.**, 53:8639 (1982).
22. T. Kook and R.J. Jaccodine, **Appl. Phys. Lett.** 49:1201 (1986).
23. P. Fahey, R. Dutton, M. Moslehi, **App. Phys. Lett.**, 43:683 (1983).
24. U.S. Kim, T. Kook, R. Jaccodine, J. **Electrochem. Soc.**, 1351:270 (1988).
25. S.M. Hu, J. **Appl. Phys.**, 45:1635 (1974).
26. A.M. Lin, R.W. Dutton, D.A. Antoniadis and W.A. Tiller, J. **Electrochem. Soc.**, 128:1121 (1981).
27. K. Taniguchi, K. Kurosawa and M. Kashiwagi, J. **Electrochem. Soc.**, 127:2243 (1980).
28. S. Mizuo and H. Higuchi, **Jap. J. Appl. Phys.**, 20:739 (1981).
29. D.A. Antoniadia and I. Moskowitz, J. **Appl. Phys.**, 53:6788 (1988).
30. W.A. Tiller, J. **Electrochem. Soc.**, 127:625 (1980).
31. Y. Nabeta, T. Uno, S. Kubo, and H. Tsukamoto, J. **Electrochem. Soc.**, 123:1416 (1976).
32. S.M. Hu, J. **Appl. Phys.**, 57:1069 (1985).
33. U.S. Kim, Ph.D. Thesis, Lehigh University, June, 1990.
34. Y. Nishioka, E.F. da Silva, Y. Wang and T.-P. Ma, **IEEE Electron Device Lett.**, 9:38 (1988).
35. K. Ohyu, T. Itoga, Y. Nishioka and N. Natsuaki, **Japan J. Appl. Phys.**, 28:1041 (1989).
36. P.J. Wright, N. Kasai, S. Inoue and K.C. Saraswat, **IEEE Electron Device Lett.**, 10:347 (1989).
37. P.J. Wright and K.C. Saraswat, **IEEE Trans. Electron Devices**, ED-36:879 (1989).
38. K.P. MacWilliams, L.F. Halle and T.C. Zietlow, **IEEE Electron Device Lett.**, 11:3 (1990).
39. Y. Nishioka, K. Ohyu, Y. Ohji, N. Natuaki, K. Mukai and T.-P. Ma, **IEEE Electron Device Lett.**, 10:141 (1989).
40. E.F. da Silva, Y. Nishioka and T.-P. Ma, **IEEE Trans. Nucl. Sci.**, NS-34:1190 (1987).
41. D. Kouvatsos, R.J.Jaccodine, F.A. Stevie, to be published.
42. P.D. Tanner and J.G. Simmons, **Review Sci. Inst.** 51, 10:1378 (1980).
43. T.-C. Lin and D. Young, accepted by the J. **Appl. Phys.**, April, 1992.
44. T. Sakurai and T. Sugano, J. **App. Phys.**, 52:2889 (1981).
45. F.J. Grunthaner, P.J. Grunthaner and J. Maserjian, **IEEE Trans. Nuc. Sci.**, NS-29:1462 (1982).

KINETICS OF OXIDATION OF SILICON BY ELECTRON CYCLOTRON RESONANCE PLASMAS

J. Joseph, Y.Z. Hu* and E.A. Irene*

Department of Physics and Chemistry
Ecole Centrale de Lyon, 69130 Ecully, France

*Department of Chemistry
University of North Carolina
Chapel-Hill, NC 27599-3290

ABSTRACT

In-situ ellipsometry, both single wavelength and spectroscopic, has been used to study the electron cyclotron resonance plasma oxidation of Si. Spectroscopic ellipsometry has been used to establish that the best fit optical model for the oxidation is a two layer model where the interface layer forms early and stabilizes and the outer layer is SiO_2. The interface layer is modeled well as a mixture of a-Si and oxide. The kinetics of film growth were followed using single wavelength ellipsometry at a temperature insensitive wavelength, and the results were in agreement with the Cabrera-Mott theory.

INTRODUCTION

Low temperature processing is required for the development of silicon technology. For this purpose Electron Cyclotron Resonance (ECR) plasma appears to be a promising tool. The most interesting feature of this method is that the high ionization ratio of the plasma is associated with low energy ionic species. Therefore, it is possible to obtain high quality films without the damage associated with impinging ionic species. The use of plasmas to grow SiO_2 films dates from the beginning of Si technology[1,2,3], but the process usually included elevated temperatures either during and/or after plasma treatment. A number of studies on ECR plasma oxidation are available[4,5,6,7,8]. In the latest of these studies[8], among the many experimental parameters examined, applied bias voltage effects and electronics quality of the oxides was studied, and these issues overlap with the present study and will be discussed further.

Since the ECR plasma technique enables the fabrication of thin oxides at low temperatures, the present study is aimed at the understanding of the oxidation mechanism

The Physics and Chemistry of SiO_2 and the Si-SiO_2 Interface 2
Edited by C.R. Helms and B.E. Deal, Plenum Press, New York, 1993

of Si between about 80°C and 400°C. In order to obtain a reliable description of the kinetics of the oxidation, in-situ during process ellipsometric measurements were used in two complementary ways: static spectroscopic and dynamic real time at a fixed wavelength. From the resulting experimental data the main features of the oxide layer are extracted using a two layer optical model[9,10].

EXPERIMENTAL PROCEDURES

A specially designed high precision spectroscopic ellipsometer was constructed with an independent process chamber[11,12]. The ellipsometer was a rotating analyzer system with a Xenon lamp allowing a spectral range between 2.5 eV and 4.5 eV. This range was chosen because it contains the main features of the dielectric function of silicon, viz. the two interband transitions at 3.4 eV and 4.2 eV. One of the main problems with performing in-situ ellipsometry is the unavailability of reliable optical properties versus temperature data. In order to circumvent this difficulty, we obtained the spectrum at room temperature after cooling the sample, and at the oxidation temperature we performed the real time kinetics measurements at a specific wavelength. These latter measurements were carried out using 340 nm (3.65 eV) light. This energy was chosen because we found that the spectra measured at different temperatures are nearly coincident at this energy[9]. Thus, at 340 nm we are able to ignore the temperature changes in the optical properties of Si and obtain the changing thickness of the oxide layer. Moreover, for the thin oxides on Si at this wavelength the accuracy of the rotating analyzer system is good.

The process chamber was equipped with a home-made electron cyclotron resonance source operating at 300 W and at 2.45 GHz. For this study the oxygen pressure was $5X10^{-4}$ torr and the distance between the mouth of the microwave cavity and the sample was 20 cm. The sample was heated by a light source located at the backside of the sample stage, and temperature was controlled by a thermocouple. The sample holder was electrically isolated from the chamber and a bias voltage between the sample and the plasma cavity was controlled by an independent power supply.

OPTICAL MODELLING

The physical parameters of the oxide layers are extracted both from the single wavelength and spectroscopic ellipsometric measurements using optical models. In order to interpret the single wavelength measurements in terms of SiO_2 thickness versus oxidation time, a simple one layer model was used. From a comparison of the theoretical trajectory corresponding to the growth of a layer with the experimental points, a thickness was obtained for each experimental point. For the spectra consisting of 41 data points between 2.5 and 4.5 eV, a stratified two layer model was used. Each layer is either a medium for which the optical properties are known, such as SiO_2, or where the properties can be described by a mixture of components. For this later case, the optical properties are determined from the dielectric functions of the components and the volume fraction using the Bruggeman effective medium approximation, BEMA[13]. To limit the number of parameters to be determined to four, we assumed that the entire oxide layer can be described by no more than two layers and that each layer was composed of no more than two components. The layers were assumed to be composed of one or two of the following components with well known dielectric functions: c-Si[14], a-Si[15], SiO_2[16] and voids. Four reasonable models shown in Figure 1 were evaluated and the results are summarized in Table 1 for a typical sample. From column 2 which shows the quality of the fit in term of an unbiased estimator[9], δ, model 4 yields the best fit. This model yields an a-Si volume fraction, f_v, of near 50% for the interface layer, and the relevant parameters are the thicknesses of the top SiO_2 film, L_{ox}, and the interface layer, L_{int}.

EXPERIMENTAL RESULTS

The ECR plasma oxidation of Si was studied at different sample bias and temperature conditions. We found that an applied sample bias of -30 V halted the film growth, and the plasma became unstable above 60 V. Thus our investigation was between these values.

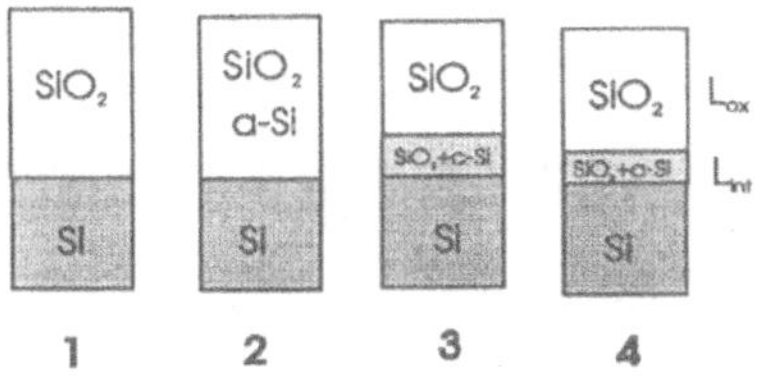

Figure 1. Optical models considered.

Table 1. Comparison of the optical models.

MODEL	$\delta X10^4$	L_{ox} (nm)	L_{int} (nm)	f_v (c-Si or a-Si) %
1	167	10.6 ± 0.5	\	\
2	104	9.6 ± 0.4	\	4.6 ± 1.5
3	115	7.6 ± 1.5	2.0 ± 0.7	39 ± 60
4	48	7.8 ± 0.3	2.2 ± 0.2	54 ± 4

Figure 2 shows the evolution of the ellipsometric parameters, Δ and Ψ, during the first 20 min of room temperature oxidation. In this Figure the solid line corresponds to a calculation performed with a one layer model (model 1 in Figure 1). From these results along with spectroscopic measurements, it appears that two different oxidation regimes occur during plasma oxidation. The first regime corresponds to about the first five degrees shift of Δ downwards, or about 2 nm in oxide thickness. The general behavior of the trajectories suggests that the ion bombardment is the main effect. The second oxidation regime corresponds to the growth of the SiO_2 film. During this period the experimental points are close to the theoretical line for an SiO_2 on Si layer. The different trajectories used for fitting were calculated using different refractive indexes for the SiO_2 layer. Spectroscopic measurements were performed during this second regime and were analyzed using Model 4 yielding L_{ox} and L_{int} values. The evolution of these parameters is shown in Figure 3. The interface layer was always found to be composed of nearly 50 % a-Si and 50 % SiO_2. These results show that the interface layer forms during the first stage of

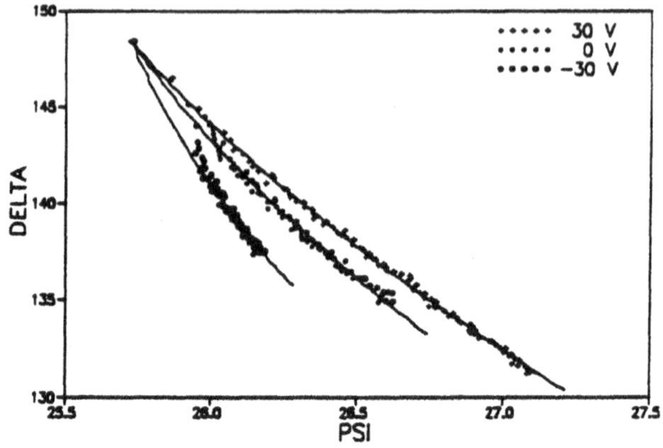

Figure 2. Evolution of the Δ,Ψ trajectories with oxidation.

oxidation and remains almost constant during the growth of the oxide layer.

Studies of oxidation were also done at four substrate temperatures (50°C, 200°C, 300°C, 400°C) at a bias of 30 V. The shape of the film thickness versus time data is similar for all the temperatures. However, for the longest times or the thicker oxides the unbiased estimator of the fitting procedure increases showing the limitation of the optical model.

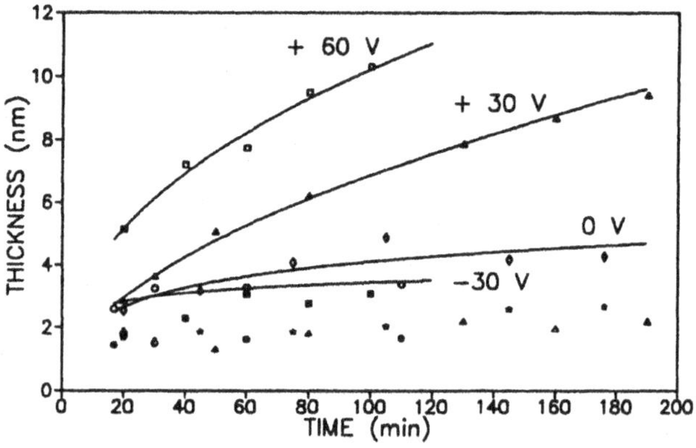

Figure 3. Results of two layer modelling of spectroscopic measurements. Open symbols represent oxide thickness, solid symbols represent interface thickness.

DISCUSSION OF RESULTS

For thermal oxidation of Si there exits significant evidence that the linear parabolic oxidation model is a reasonable approximation to the kinetics[17]. In the ECR plasma oxidation at low temperature the moving species are most likely different from those for thermal oxidation and the transport of these species is not only due to diffusion, but also and more importantly to the drift of charged species in the electric field. Many of these kind of models also yield a linear parabolic rate law[18]. In order to check the consistency of such an approach, the oxide thickness versus the square root of the time for the growth of one of our thicker oxides is plotted in Figure 4.

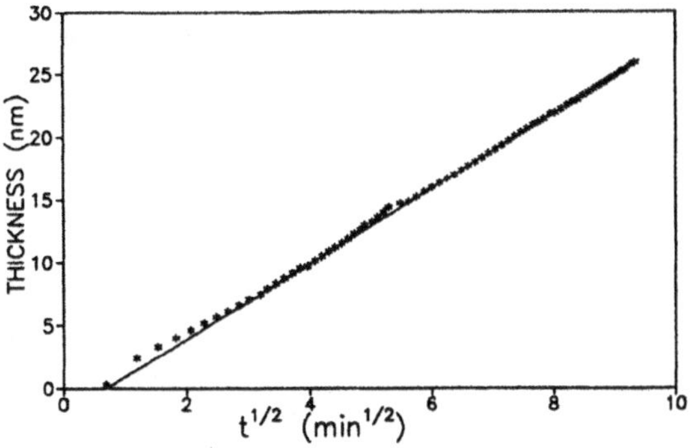

Figure 4. Thickness of the oxide layer as a function of the square root of ECR oxidation time obtained at 300°C for 30 V applied bias.

After the first few data points the results fit the $t^{1/2}$ dependance. Thus our results are interpreted with the simple relationship:

$$L^2 = B \cdot t + C$$

where L and t are the oxide thickness and the oxidation time, respectively, and B and C are constants. Parabolic behavior was also reported in a previous ECR oxidation study[8] for the early regime. All the data fit to this equation gave insignificant values for C. Thus only B was used for comparison with the model. Therefore, using the total thickness $L_t = L_{ox} + L_{int}$ as the thickness of the layer, Figures 5 and 6 show the kinetics data and the linear fits which provide the parabolic coefficient B.

Using the Cabrera-Mott theory[17] for oxidation by charged species in the limit of the low field, the growth rate law is given by:

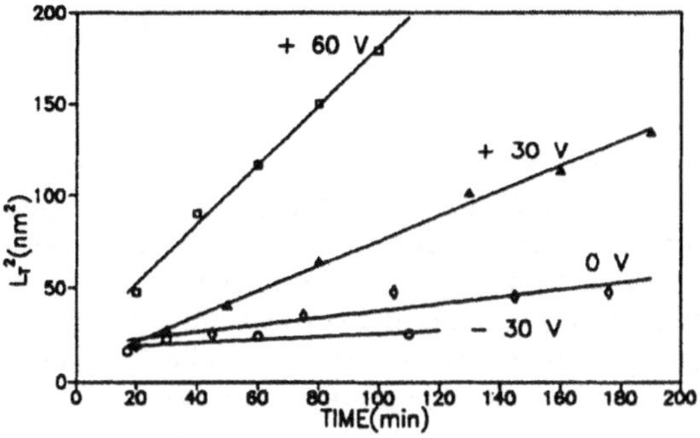

Figure 5. Growth law for four applied bias voltages.

$$L^2 = C_1 \exp\left(-\frac{E_a}{kT}\right) \cdot \frac{V_{ox} t}{kT} + C_2$$

where C_1 and C_2 are constants, V_{ox} is the potential drop across the oxide film, E_a is the activation energy for diffusion, t is the oxidation time, T is the temperature, and k is the Boltzman constant. In order to compare our results to this theory, the dependance of our experimental parabolic coefficient as a function of temperature and bias voltage is plotted in Figures 7 and 8.

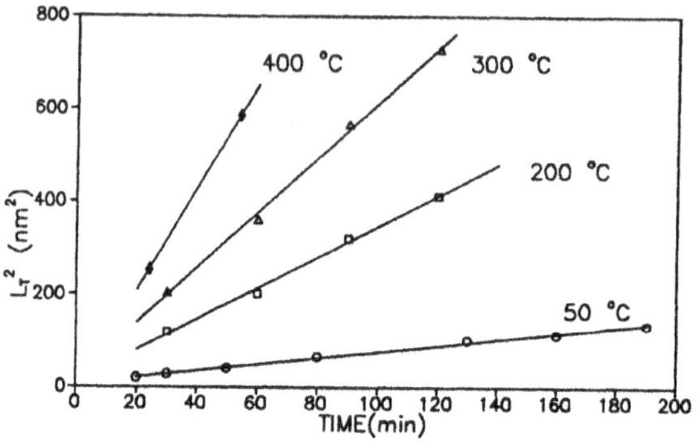

Figure 6. Growth law for four temperatures.

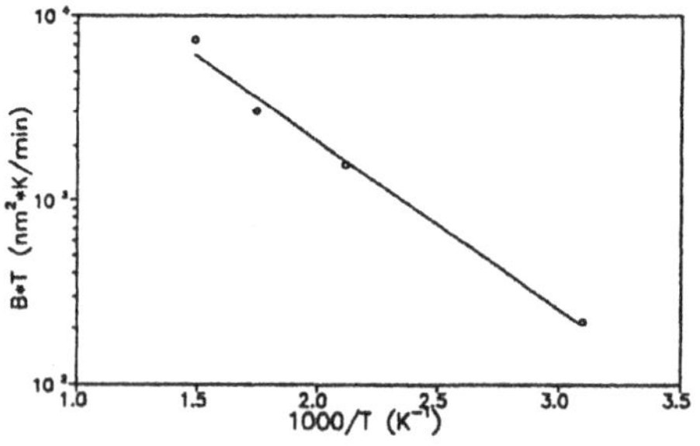

Figure 7. B vs. T.

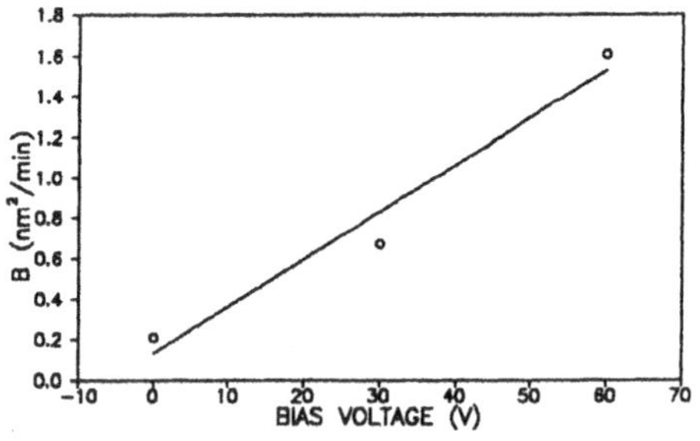

Figure 8. B vs. V.

The linear fit shown in Figure 7 yields an activation energy of 0.18 eV, which is notably lower than the activation energy for thermal oxidation (more than 1 eV[16]), and indicates a very different oxidation mechanism. The linear increase of the growth rate with the bias reveals that the oxidizing species are negative. Considering that in the ECR plasma the density of electrons is high, and consequently, the density of negative ions is low, a multistep process can be envisioned as was similarly reported for ECR formation of thicker ECR grown oxides[8]. For the present situation we consider first that electrons in the plasma attach to oxygen producing the molecular ion, O_2^-. This species is less stable than O_2, and will readily decompose to the atomic species O and O^-. The O^- can readily migrate through the oxide with a positive sample bias.

The best optical description of the interface layer was as a mixture of oxide and amorphous silicon. This description simulates the result of several phenomena: interface roughness, suboxides, and strained oxide. From our measurement we cannot decide which ones are dominant, but we know for sure that this interface layer is very different from a pure oxide and that its optical behavior is almost constant during plasma oxidation, hence this layer can be used as a marker. The immutability of this layer during the growth strongly suggests that the oxidation occurs between this layer and the outer surface. In this way oxidation occurs from the movement of both the Si outward as a cation and the oxidant species inward as O^-. In this hypothesis the interface layer is partly the result of the migration of Si^+ through the first layer of the oxide. Our previous study of the nature of this interface layer[9] has shown that with positive or zero bias this layer is optically indistinguishable from the interface layer seen for thermally grown SiO_2, and the recent Carl et al work[8] has shown similar electrical quality for the positive and zero biased ECR oxide films with thermal oxides.

ACKNOWLEDGEMENTS

This work was supported in part by the Office of Naval Research, ONR, and the NSF Engineering Research Center at NC State Univ. One of us (J.J.) is grateful to NATO for a travel fellowship.

REFERENCES

1. J.R.Ligenza, J.Appl.Phys., 36, 2703 (1965).
2. V.Q. Ho and T. Sugano, Jpn.J.Appl.Phys., 19, 103 (1980).
3. R.P. Chang, C.C. Chang, and S. Darack, Appl.Phys.Lett. 36, 399 (1980).
4. S. Kimura, E. Muraki, K. Miyake, T. Warabisako, H. Sunami and T. Tokuyama., J.Electrochem.Soc., 132, 1460 (1985).
5. T. Ropel, D.K. Reinhard, and J. Asmussen., J.Vac.Sci. Technol. B, 4, 295 (1986).
6. C. Vinckier, P. Coeckelberghs, G. Stevens, M. Heyns, and De Jaegere., J.Appl.Phys., 62, 1450 (1987).
7. D.A. Carl, D.W. Hess, and M.A. Lieberman., J.Vac.Sci.Technol. A, 8, 2924 (1990).
8. D.A. Carl, D.W. Hess, M.A. Liberman, T.D. Nguyen and R. Gronsky, J. Appl. Phys., 70, 3301 (1991).
9. Y.Z. Hu, J. Joseph , and E.A.Irene, Appl.Phys.Lett., 59, 1353 (1991).
10. J. Joseph, Y.Z. Hu, and E.A. Irene, J.Vac.Sci.Technol.B, 10, Mar/Apr (1992).
11. J.W. Andrews, Y.Z. Hu, and E.A. Irene, SPIE Proc., 1118, 62 (1990).
12. J.W. Andrews, Ph.D Thesis , University of North Carolina. Chapel Hill, NC 1990.
13. D.E. Aspnes, Thin Solid Films, 89, 249 (1982).

14. D.E. Aspnes, and A.A. Studna., Phys. Rev. B, 27, 985 (1983).

15. D.E. Aspnes, A.A. Studna, and E. Kinsbron., Phys.Rev.B, 29, 768 (1984).

16. I.H. Malitson, J.Opt.Soc.Am., 55, 1205 (1965).

17. E.A. Irene, CRC Crit.Rev.Sol.State Mat. Sci., 14, 175 (1988).

18. N. Cabrera and N.F. Mott, Rep.Prog.Phys., 12, 163 (1948).

MECHANISMS OF OXIDATION RATE ENHANCEMENT IN NEGATIVE-POINT OXYGEN CORONA DISCHARGE PROCESSING OF SiO_2 FILMS ON Si

L.M. Landsberger

Concordia University
Department of Electrical and Computer Engineering
1455 deMaisonneuve Blvd. West
Montreal, P.Qué., Canada H3G 1M8

ABSTRACT

Negative-point oxygen corona discharge processing at 600°C-900°C dramatically enhances the oxidation rate, while creating SiO_2 films with the refractive indices and oxygen transport characteristics normally found in films dry-thermally-grown at 1000°C-1200°C. By an analysis of the atomic mechanisms affecting the corona-treated region, features of the film thickness enhancement profile are quantitatively explained by additive components: relaxed oxide density differential, field-aided O^- ion flux, and enhanced oxygen diffusion.

INTRODUCTION

As integrated circuit fabrication technology continues to move toward smaller device geometries, low-temperature processing continues to become more important. Considerable recent work [1-8] has investigated novel methods of creating SiO_2 films for the thin gate dielectrics critical to MOS technology. Since low-temperature processing has been associated with increased film stress levels, the study of stress in SiO_2 films continues to be important. The negative-point oxygen corona-discharge technique has been discussed in several publications [9-12] as an alternate method for achieving strain-relaxed, high-quality SiO_2 layers on Si at 600°C - 900°C.

REVIEW OF THE CORONA-RELAXATION TECHNIQUE

The corona technique has been shown to be effective in significantly enhancing the oxidation rate, at the same time producing uncharacteristically relaxed films [9-12]. While one would expect [13-14] a more rapid oxidation rate to aggravate the stresses already present during low-temperature growth, the resulting films instead have the refractive index and oxygen transport properties [13-16] of films dry-thermally-grown at much higher temperatures (1100°C - 1200°C), indicative of a substantially more relaxed film.

The Physics and Chemistry of SiO_2 and the Si-SiO_2 Interface 2
Edited by C.R. Helms and B.E. Deal, Plenum Press, New York, 1993

Figure 1 shows the thickness and refractive index profiles of a film grown by a long corona-enhanced oxidation. Note that in the control (corona-unaffected) regions, the oxide grew to the thickness normal [17-18] for a 900°C, 9 hour oxidation, while the refractive index likewise reflected the 1.469 value characteristic of 900°C growth. By contrast, in the thickness ranges of the corona-enhanced region where the refractive index was readily measurable, the refractive index dropped and saturated at 1.461, a value characteristic of a high-temperature-grown film. While Figure 1 shows results for an oxide film grown by corona from bare Si, the phenomena of enhanced oxidation and relaxed refractive index are also obtained when an existing strained (for example, 800°C-grown) SiO_2 layer is subjected to a short corona treatment [9-11].

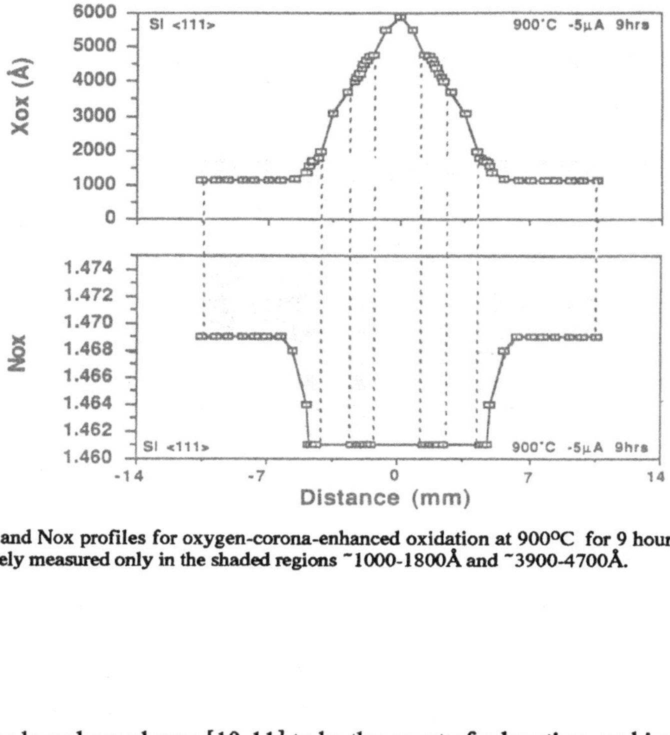

Figure 1. Xox and Nox profiles for oxygen-corona-enhanced oxidation at 900°C for 9 hours at -5µA. Nox could be accurately measured only in the shaded regions ~1000-1800Å and ~3900-4700Å.

O^- ions have been shown [10-11] to be the agent of relaxation, and isotope oxygen tracer studies have shown [10] that the corona causes bulk relaxation in the film network. Furthermore, studies on 2-D SiO_2-Si structures [12],where the stresses are expected to be much greater, have shown that very effective viscosity reduction must occur during the corona oxidation. This paper will explore further the atomic mechanisms present during corona-oxidation by analyzing components of oxide thickening on a corona-treated wafer.

The corona apparatus used and the protocols followed for these experiments have been described in earlier papers [9,10,19], and will only be briefly summarized here. The corona apparatus, constructed exclusively from quartz, sapphire, silicon and platinum, is operated in a standard oxidation furnace at atmospheric pressure. The negative point-to-plane corona is turned on and off while the oxygen is flowing. The wafers used were lightly-doped p-type Si<111> and <100>. Bare Si wafers were cleaned by the Reverse-RCA procedure, while for wafers covered with oxide films, the HF dip was omitted.[19].

THE MECHANISMS

(1) Refs. [13-16,20] show that the refractive index and density of thermally-grown SiO_2 films vary systematically with the growth temperature. A film dry-grown at 800°C (N_{ox} = 1.472) is ~3% denser than one dry-grown at ~1150°C (N_{ox} = 1.460). Moreover, such dense films grown at the lower temperatures actually thicken (swell) by the appropriate percentage [20] when annealed in inert ambients. This thickness enhancement is found to be in effect during corona processing as well. This component shall be called:

$$\Delta X_s = \frac{\{(\text{Initial } N_{ox}) - (\text{Final } N_{ox})\}}{(1.472 - 1.460)} (0.03)\, X_i$$

where X_i is the initial oxide thickness. For example, for an 800°C-grown 1100Å oxide, relaxed at 600°C, the final refractive index is 1.464 (Ref [10] provides data on the refractive index obtained as a function of treatment temperature), and

$$\Delta X_s = \frac{(1.472 - 1.464)}{0.012} (0.03)(1100\text{Å}) = {}^{\sim}22\text{Å}$$

(2) Previous work on corona-discharge processing [9-11] has shown that each negative charge (e-) emitted from the high-voltage needle reaches the Si sample (the plane in the point-to-plane discharge) as an O^- ion. These O^- ions are driven by the electric field through the existing film to the Si/SiO_2 interface, providing an extra oxidant flux in addition to the normal flux of diffusing oxidant (assumed neutral). This component shall be called:

$$\Delta X_{O-} = (J)(t)(q)(M_{ox}) = \Delta X_{O\text{-}post} + \Delta X_{O\text{-}dose}$$

where J is the beam current density, t is the corona treatment time, q is the electronic charge, and M_{ox} is the number of O atoms per cm^3 of SiO_2. As indicated in the above equation, this component can be further separated into $\Delta X_{O\text{-}dose}$, representing the extra thickness incurred directly from the dose of O^- required to completely relax the oxide, and $\Delta X_{O\text{-}post}$, representing the extra oxide due to continued ionic flux after complete relaxation.

Previous work [10-11] showed that $\Delta X_{O\text{-}dose}$ is proportional to {(Initial N_{ox}) - (Final N_{ox})}. For example, for dry 800°C-grown SiO_2 on Si<111>, it is ~1% of the initial oxide thickness X_i, while for dry 800°C-grown SiO_2 on Si<100>, it is ~1.5%.

(3) Other recent research [10,16] has shown that, for oxidation at low temperatures (700°C-1000°C), normal oxidant diffusion is significantly enhanced if the existing oxide layer has the refractive index (density) of an oxide grown at a higher temperature. Since a short corona-discharge treatment can render the treated region of oxide to be as relaxed as a high-temperature-grown film, the diffusivity of normal oxidant during the remainder of the corona processing will be substantially greater than that normal for the treatment temperature. For example, if the corona-discharge oxidation or relaxation is occurring at 800°C, the parabolic rate constant [17,18] for the component of oxide growth due to the flux of normal (O_2) oxidant will be enhanced by approximately x10. The oxidation growth law [17,18] would then indicate that, for an oxide ~1100Å thick, the component of oxide growth due to this flux of normal oxidant will be enhanced by about x2.5. This component shall be called:

$$\Delta X_{th} = \Delta X_{enh\text{-}diff} - \Delta X_{std} = \Delta X_{th*} + \Delta X_{thpost}$$

where $\Delta X_{enh\text{-}diff}$ is the oxide grown by enhanced oxidant diffusion, and ΔX_{std} is the oxide that grows in the control (corona-unaffected) region for the same temperature and time. As indicated in the above equation, this component can be, at least conceptually, separated into ΔX_{th*} (some partially-enhanced oxidation occurring before full relaxation has been achieved) and ΔX_{thpost} (after full relaxation has been achieved).

The above 3 mechanisms are all expected to be in effect during corona processing, and are used in this paper to account for the total oxide thickness enhancement in the corona-treated region. Therefore,

$$\Delta X_{ox} = \Delta X_{O-} + \Delta X_{th} + \Delta X_s$$

Figure 2 depicts schematically the case of corona-relaxation of a film pre-grown at a low temperature (for example 800°C). The thickness of the oxide in the treated region follows an approximately bell-shaped profile. Outside the treated region the refractive index is at the level normal for the processing temperature. Moving from the edges of the treated region toward the peak, the refractive index decreases and saturates at the relaxed level. At the outer edge of the treated region, the refractive index falls rapidly while the thickness rises. Of particular interest is the thickness at which the refractive index reaches the fully-relaxed value (the "just-relaxed thickness"). In terms of the above-described quantities, this just-relaxed thickness is:

$$\Delta X^* = \Delta X_{O\text{-dose}} + \Delta X_{th^*} + \Delta X_s$$

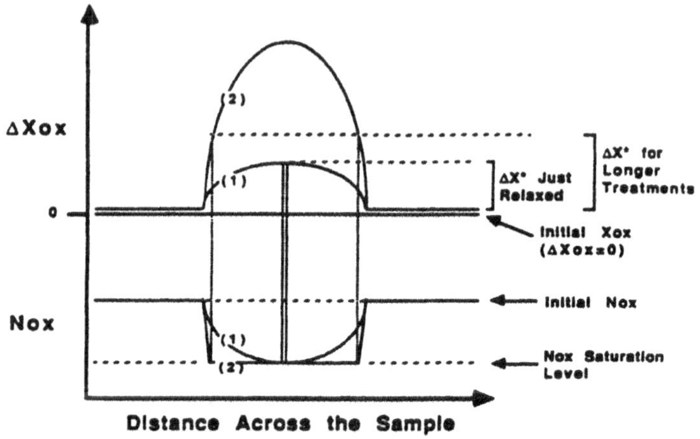

Figure 2: Schematic showing ΔX*, the "just-relaxed thickness", for two generic cases: (1) the film is barely relaxed at the center, (2) the saturation region has widened as the corona process has continued.

EXPERIMENTAL RESULTS

Analysis of the Just-Relaxed Thickness, ΔX*

Table 1 shows all the numerical information for a variety of experimental cases. Uncertainty in the observed ΔX* (the rightmost column) represents both the variation in observed ΔX* for repeated samples and the uncertainty in determining the location of the exact edge of the saturation region for any given sample. The saturation region edges are particularly uncertain for higher treatment temperatures.

For the predicted ΔX*, the lower bound assumes that no enhanced thermal oxidation begins until the oxide reaches the saturation level (this is the condition: $\Delta X_{th^*}=0$), while the upper bound assumes that it begins as soon as the corona is active and proceeds at its max. rate for the duration of the treatment. Any realistic case will be between these extreme limits.

By comparing the predicted and actual ranges in Table 1, one can see overlap between theory and experiment for all of these experimental cases. Figure 3 shows data for constant 20-minute treatments, while Figure 4 plots data for different treatment times at 800°C and 900°C. These two figures demonstrate neatly the following trends: (1) ΔX* increases with decreasing initial oxide growth temperature, indicating the effect of the larger swelling and dose components. (2) ΔX* increases with increasing corona treatment temperature, indicating primarily the effect of the increasing the ΔX_{th^*} component. (3) ΔX* increases with increasing corona treatment time, indicating again the effect of the increasing ΔX_{th^*} component. (4) Enhanced thermal oxidation (due to the O_2 oxidant flux) in the treated region begins substantially before the saturation level is reached, because theoretical values near the upper end of the predicted ranges are frequently necessary to match the experimental data. Thus $\Delta X_{th^*} \neq 0$ (except perhaps at 600°C and 700°C corona treatment temperatures).

Table 1. Combined Theoretical and Experimental Results for the Components of Oxide Thickening.

Si Orientation	Initial Growth Temp.	Initial N_{ox}	Initial X_{ox}	Corona Treatment Temp.	Final N_{ox}	Predicted % Swelling	Predicted ΔX_i	Predicted % Dose	Predicted ΔX_o-dose	Cor. Treatment time (min)	Enh. Ox. Rate (Å/min)	Norm. Ox. Rate (Å/min)	Diff. Ox. Rate (Å/min)	Predicted ΔX_{th}* (Max.)	Predicted ΔX^* Range	Actual Experimental Data Range
<111>	800C	1.472	1100Å	900C	1.461	2.75%	30Å	1%	11Å	10	3.5	1.4	2.1	21Å	41/62	48/76
<111>	800C	1.472	1100A	900C	1.461	2.75%	30A	1%	11Å	20	3.5	1.4	2.1	42Å	41/83	68/128
<111>	800C	1.472	1100Å	900C	1.461	2.75%	30Å	1%	11Å	40	3.5	1.4	2.1	84Å	41/125	100/134
<111>	800C	1.472	1100Å	800C	1.462	2.5%	27Å	1%	11Å	10	0.9	0.3	0.6	6Å	38/44	35/49
<111>	800C	1.472	1100Å	800C	1.462	2.5%	27Å	1%	11Å	20	0.9	0.3	0.6	12Å	38/50	40/60
<111>	800C	1.472	1100Å	800C	1.462	2.5%	27Å	1%	11Å	40	0.9	0.3	0.6	24Å	38/62	59/85
<111>	800C	1.472	1100Å	700C	1.463	2.25%	25Å	1%	11Å	20	0.15	0.0	0.15	3Å	36/39	25/39
<111>	800C	1.472	1100Å	600C	1.464	2.0%	22Å	1%	11Å	20	0.0	0.0	0.0	0Å	33	23/33
<111>	800C	1.472	1100Å	500C	1.464/5	1.75%	19Å	1%	11Å	90	0.0	0.0	0.0	0Å	30	20/40
<111>	800C	1.472	1750Å	800C	1.462	2.5%	44Å	1%	18Å	20	0.6	0.2	0.4	8Å	61/69	54/72
<100>	800C	1.476	1000Å	800C	1.462	3.8%	38Å	1.5%	15Å	20	0.3	0.1	0.2	4Å	53/57	42/58
<111>	1000C	1.466	1000Å	600C	1.464	0.5%	5Å	<0.5%	<5Å	20	0.0	0.0	0.0	0Å	5/10	5/11
<111>	1100C	1.463	1000Å	600C	1.464	0.0%	0Å	<0.3%	<3Å	20	0.0	0.0	0.0	0Å	0/3	0/5
<111>	1150C	1.461	1000Å	800C	1.461	0.0%	0Å	<0.1%	<1Å	20	0.9	0.7	0.2	4Å	0/5	0/5

Predicted Instrinsic Swelling Component: Predicted ΔXi

Predicted Dose of O- Ions Required to Reach Saturation Level: Predicted ΔXo- dose

Predicted Enhanced Thermal Oxidation Level in the Treated Region: Predicted ΔXth

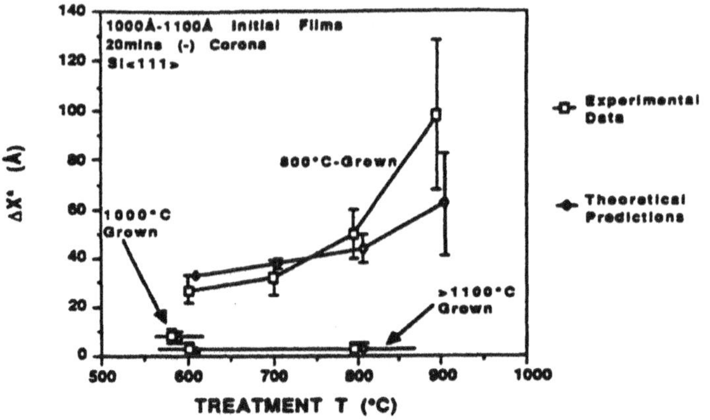

Figure 3. Comparison from Table 1 of theoretically modelled ΔX^* with observed experimental values for constant 20-minute corona treatments.

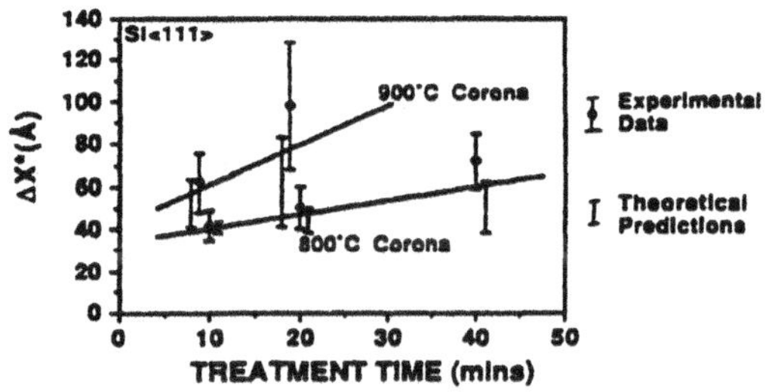

Figure 4. Comparison from Table 1 of theoretically modelled ΔX^* with observed experimental values for different treatment times at 800°C and 900°C.

In addition, other data from Table 1 demonstrate good matches between theory and experiment for thicker (1750Å) initial films and for <100>-grown initial films. These strongly support the trends relating swelling and dose components to the initial conditions of the film. The data for corona treatments at 500°C again agrees with the trends, although the uncertainty in this data is larger.

It should be noted that, if the predicted % swelling, ΔX_s, were chosen to be dependent only on the Initial N_{ox} (instead of being also dependent on the Final N_{ox}) the theoretical data would not overlap with the experimental values at low temperatures (500°C - 700°C). This is shown in Figure 5 for several of the cases from Figure 3. Using the predicted $\Delta X_{O\text{-}dose}$ and ΔX_{th}^* from Table 1, and a ΔX_s of $(0.03)X_i$ for initial oxides grown at 800°C, a predicted ΔX^* was calculated for comparison with the actual experimental data. As Figure 5 shows, these modified theoretical values are too high at 600°C and 700°C. Since scatter in the actual data is smallest at these lower temperatures, the use of X_i dependent on the Final N_{ox} is supported.

Oxidation Enhancement After Full Relaxation

As explained in Ref [10,11], after achievement of the saturation level in the center of the corona-treated region, the film continues to grow as the saturation region widens. The thickness in the relaxed region increases w. time due to ΔX_{th}^{post} and $\Delta X_{O\text{-}post}$ components. In order to compare with theory, the volumes of extra oxide grown in the treated regions

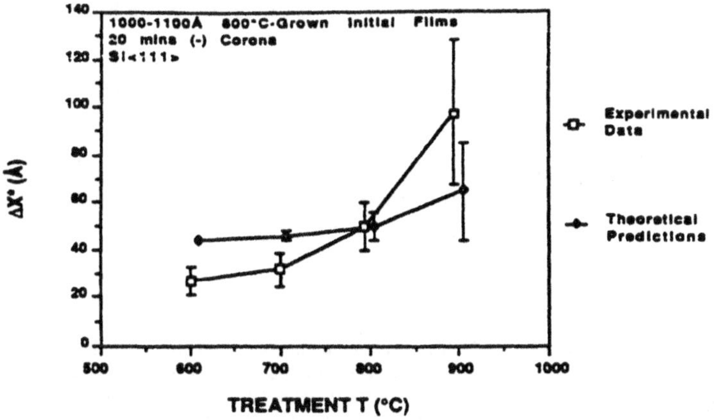

Figure 5. Comparison of observed experimental values for ΔX* with a modified theoretical model: instead of ΔXi, a constant 3% of the initial oxide thickness was used for 800°C, dry <111>-grown oxides, along with the other components from Table 1.

were estimated from $\Delta X_{ox} - \Delta X^*$, and compared with the extra volume calculated from the number of oxygen atoms expected from the ionic flux $\Delta X_{O\text{-}post}$. Each electronic charge q in the corona dose was assumed to yield one extra oxygen atom in the resulting film.

Although the results were subject to considerable error in estimation of volumes, in no case did the estimated extra volume exceed that expected from the ionic flux $\Delta X_{O\text{-}post}$. It can be concluded that, although the corona-treated oxide was exposed to significant concentrations of neutral O atoms in the corona space-charge region as well as O- ions, only the O- ions caused extra oxide growth.

Table 2 compiles a detailed comparison between measured and expected quantities for a small subset of the experimental data. In order to minimize uncertainty, only a selection of samples with large expected $\Delta X_{O\text{-}post}$ is studied. ΔX_{ox} profiles were approximated by cones of peak height ($\Delta X_{ox} - \Delta X^*$), and diameter equal to the observed treated region width. The implied number of O atoms in the volume of extra growth best correlated with O^- ion dose at 800°C, where it approached to within ~20%. At lower temperatures, the extra volume fell short of predicted values by as much as 50%. This reduction can be explained by considering the effects of electronic and ionic conductivity in SiO_2.

Table 2. Expected and Actual Oxygen Atoms in the Saturated Region

Corona Treat. Temp.	Corona Treatment Parameters	Expect. Extra Oxygen Atoms/cm^3	ΔX_{ox}peak - ΔX^*.	Obs'vd. Treated Region Radius	Actual Extra Oxygen Atoms/cm^3	Actual % of Expect.	% ionic Conductivity in SiO_2 from [21]
900°C	-4μA 20mins	2.25x10^{16}	(80Å)	6mm	1.35x10^{16}	(60%)	~80-90%
800°C	-4μA 20mins	2.25x10^{16}	90Å	6.5mm	1.8x10^{16}	80%	~90%
700°C	-4μA 20mins	2.25x10^{16}	65Å	7mm	1.5x10^{16}	67%	~80-90%
650°C							~50-70%
600°C	-4μA 20mins	2.25x10^{16}	50Å	7.5mm	1.3x10^{16}	58%	-
550°C							~30-60%
500°C	-4μA 20mins	2.25x10^{16}	70Å	7.5mm	1.85x10^{16}	(50%)	-

Actual extra O atoms are calculated by modelling post-relaxation corona profiles as cones of volume $(1/3)(\pi r^2)(\Delta X_{O\text{-}post})$, where $\Delta X_{O\text{-}post} = \Delta X_{O-} - \Delta X_{O\text{-}dose}$.

Due to the large uncertainties inherent in both the observed thicknesses and the modelling of the extra volume as a cone, this table is intended only to indicate a trend, not for quantitative use.

Work by Srivastava et al [21] showed that the partitioning between electronic and ionic conductivity in thermal SiO_2 varies with temperature. In particular, the ionic component reaches a maximum near 90% at 800°C, dropping slightly at higher temperatures and falling off sharply at lower temperatures. Their results for SiO_2 on p-type Si are also tabulated in the rightmost column of Table 2.

This agreement can be understood by considering separation of the corona beam ionic flux into ionic and electronic conduction components. A significant component of the incoming O^- must dissociate into O atoms and conduction electrons (e^-). The extra oxide growth is accounted for primarily by the ionic O^- flux, while the neutral O atoms produced are likely to return to the gas phase.

IMPLICATIONS FOR CREATING UNIFORM SiO_2 FILMS BY CORONA

If the corona treatment technique is ever to approach viability for modern Si processing, the uniform treatment of full wafers must be achieved. At present this is not possible, because the bell-shaped region in which oxidation is enhanced is an inherent property of the point-to-plane geometry. Even using a grid of points (instead of a single point) suspended above the wafer, uniformity promises to be a formidable task. However, if one uses an understanding of the mechanisms detailed above, one can choose processing conditions which minimize the effect of mechanisms which lead to the most non-uniformity, while maintaining the relaxed property of the resulting oxide.

For example, ΔX_{th} can be minimized, perhaps by keeping the corona treatment / oxidation temperature as low as possible (Table 1 shows $\Delta X_{th} \sim 0$ at 600°C -700°C). Also, since the swelling component is upper-bounded by $(0.03)X_i$, uniformity will be favored by full relaxation. Further, ΔX_{O^-} should be kept as low as possible, while maintaining its ability to fully relax the oxide strain. In other words, it should be held at ΔX_{O^-dose}.

The above conditions can be satisfied by producing thin oxides by corona-enhanced oxidation of bare Si at a low temperature (e.g. 700°C-800°C). If the corona current density were kept low, just high enough to relax the strain as the film grows, there would be the potential to achieve uniform, high-quality thin gate dielectrics.

REFERENCES

[1] G.Lucovsky, J.T.Fitch, D.V.Tsu. and S.S.Kim, *J. Vac.Sci.Technol.A*, **7**:1136 (1989).
[2] J.T.Fitch, S.S.Kim, and G.Lucovsky, *J. Vac.Sci.Technol.A*, **8**:1871 (1990).
[3] S.P.Tay, A.Kalnitsky, G.Kelly, J.P.Ellul, P.DeLalio, and E.A.Irene, *J.Electrochem.Soc.*, **137**:3579 (1990).
[4] S.Chongsawangvirod, E.A.Irene, A.Kalnitsky, S.P.Tay, and J.P.Ellul, *J.Electrochem. Soc.*, **137**:3536 (1990).
[5] K.Eljabaly, and A. Reisman, *J.Electrochem.Soc.*, **138**:1064 (1991).
[6] C.Vinckier and S.DeJaegere, *J.Electrochem.Soc.*, **137**:628 (1990).
[7] K.Fujino,Y.Nishimoto,N.Tokumasu,K.Maeda,*J.Electrochem.Soc.*,**137**:2883 (1990).
[8] D.Kouvatsos,J.G.Huang, and R.J.Jaccodine, *J.Electrochem.Soc.*, **138**:1752 (1991).
[9] D.N.Modlin and W.A.Tiller, *J.Electrochem. Soc.*, **132**:1163 (1985).
[10] L.M.Landsberger and W.A. Tiller, *J. Electrochem. Soc*, **139**:218 (1992).
[11] L.M.Landsberger and W.A.Tiller, Mat.Res. Soc. Symp. Proc. Vol 75, (MRS Meeting in Boston MA, Dec 1-6 1986), p. 803.
[12] L.M.Landsberger, D.B.Kao and W.A.Tiller,*J. Electrochem. Soc*, **135**:1766 (1988).
[13] E.A.Irene, E.Tierney and J.Angilello, *J. Electrochem. Soc.*, **129**:2544 (1982).
[14] E.A.Taft, *J.Electrochem. Soc.*, **125**:968 (1978).
[15] E.A.Taft, *J.Electrochem. Soc.*, **127**:993 (1980).
[16] L.M.Landsberger and W.A.Tiller, *J.Electrochem.Soc.*, **137**:2825 (1990).
[17] B.E.Deal and A.S.Grove, *J.Appl. Phys.*, **36**:3770 (1965).
[18] C.J.Han, and C.R.Helms, *J.Electrochem.Soc.*, **134**:1297 (1987).
[19] L.M.Landsberger, Ph.D. Thesis, Stanford Univ., E.E. Dep't, Stanford, CA, 1983.
[20] L.M.Landsberger and W.A.Tiller, *Appl. Phys. Lett.*,**51**:1416 (1987).
[21] J.K.Srivastava, M.Prasad, and J.B.Wagner Jr., *J.Electrochem.Soc.*, **132**:955 (1985).

HIGH PRESSURE OXIDATION FOR LOW TEMPERATURE PASSIVATION OF $Si_{1-x}Ge_x$ ALLOYS

C. Caragianis, Y. Shigesato, and D.C. Paine

Brown University, Division of Engineering
Box D, Division of Engineering Brown University
Providence

ABSTRACT

Thermal passivation of $Si_{1-x}Ge_x$ using high pressure (70 MPa) oxidation was studied for potential use in MOS-device applications. Alloys of CVD-grown $Si_{1-x}Ge_x$ (with x=10 and 15 at. %), 200 and 150-nm thick respectively, were oxidized using high purity dry oxygen at a pressure of 70 MPa and a temperature of 500°C. For comparative purposes, a second set of alloys were oxidized using conventional wet atmospheric pressure oxidation at 800°C. X-ray photoelectron spectroscopy (XPS), Raman spectroscopy, Transmission electron microscopy (TEM) and MOS C-V measurements were used to characterize the as-grown oxides. Chemical analysis by XPS confirmed that under high pressure conditions, compositionally congruent oxides are grown from these alloys. High resolution TEM and Raman spectroscopy show that the as-grown oxide/semiconductor interface is planar and free of Ge enrichment on a scale of 1-2 monolayers. A midgap interface state density for both the 10 and 15 at. % samples of 1×10^{12} $cm^{-2}eV^{-1}$ was estimated based on 1 MHz C-V measurements.

INTRODUCTION

Strain-layer heterostructures of $Si_{1-x}Ge_x$ grown on <001> Si substrates have been used in the fabrication of heterojunction bipolar transistors with exceptional performance [1]. The next level of technological advancement in this materials system will probably take advantage of higher hole mobilities [2] in $Si_{1-x}Ge_x$ to form channel regions in applications such as CMOS and BiCMOS. In order to fully exploit this opportunity, a high quality surface passivation must be developed for $Si_{1-x}Ge_x$. High dielectric breakdown fields, low density of interface and oxide states, thermal and chemical stability, and compatibility with standard device processing are all essential requirements for such a passivation material.

The Physics and Chemistry of SiO_2 and the Si-SiO_2 Interface 2
Edited by C.R. Helms and B.E. Deal, Plenum Press, New York, 1993

Conventional Si atmospheric oxidation is not useful for the $Si_{1-x}Ge_x$ system because, under these conditions, Si is selectively removed from the alloy such that Ge accumulates at the oxide/alloy interface [3,4]. The formation of the resulting $Si_{1-x}Ge_x/Ge/SiO_2$ sandwich has been shown [5] to be due to the much more negative free energy of formation of SiO_2 compared to GeO_2. The presence of Si sets the equilibrium oxygen potential at the reaction interface far below that at which Ge is oxidized and, consequently, the near-interface region of the alloy is denuded of Si. The breadth of the denuded region depends upon the rate of Si diffusion through Ge. High pressure techniques allow increased oxidation rates at low temperatures so that diffusional transport of underlying Si to the reaction interface is retarded.

In previous work we reported [3,5] the use of high pressure dry oxygen (>70 MPa) for the passivation of alloys of $Si_{1-x}Ge_x$ at low process temperature (<550°C). With this approach, compositionally congruent oxides were produced from $Si_{1-x}Ge_x$ so that macroscopic enrichment at the oxide/alloy interface was eliminated. Although high quality $Si_{1-x}Ge_x$ passivation has been produced using MBE alloy growth and subsequent *in situ* deposition of Si_3N_4 in an UHV chamber [e.g. 6], we believe that the high pressure approach offers a practical alternative. In both the MBE and the HPO techniques the insulator/semiconductor interface is never exposed to the atmosphere. In MBE this is done using multiple UHV chambers while in HPO the parent material is consumed during oxidation such that the final oxide/alloy interface is formed from pure, unexposed material.

In this study we have examined the feasibility of using dry high pressure oxidation for the production of MOS quality passivation for alloys of $Si_{1-x}Ge_x$ (x=10 and 15 at. %) grown by CVD. The oxides were characterized chemically, structurally, and electronically using many of the techniques that have been previously applied to the SiO_2/Si system. These techniques include X-Ray Photoelectron Spectroscopy (XPS) for chemical and compositional analysis, Raman Spectroscopy for the detection of interfacial Ge, and Transmission Electron Microscopy (TEM) for structural analysis of the oxide/semiconductor interface. In addition, MOS devices were fabricated to allow Capacitance-Voltage (C-V) measurements which were used for the evaluation of the electronic performance of the oxide/semiconductor system. These measurements are of obvious importance for many device applications since interface state charges are in direct electrical communication with the underlying semiconductor and can lead to a wide variety of degrading electronic effects.

EXPERIMENTAL

Alloys of $Si_{1-x}Ge_x$ with two compositions, x=10 and 15 at.%, were grown on 20-25 Ω·cm <100> Si by CVD [1] to a thickness of 200 and 150 nm respectively. The alloys were doped during growth with boron to roughly match the substrate. Subsequent depth profiling by spreading resistance measurements showed that the alloys (and underlying substrate) were p-type in the range $0.5\text{-}1\times10^{15}/cm^3$. The alloys were oxidized using ultra high purity dry oxygen at a pressure of 70 MPa and a temperature of 500°C using our high pressure oxidation (HPO) apparatus. Prior to HPO, the wafers were cleaned in hot electronic grade trichloroethylene, followed by acetone and methanol baths, a rinse in deionized water, and finally a 2 minute 10% $HF:H_2O$ dip. After a final rinse in deionized water the samples were blown dry and immediately loaded into the high pressure oxidation chamber. For comparative purposes, a second set of alloys were oxidized using conventional wet atmospheric pressure oxidation at 800°C.

Our HPO system is described elsewhere [5] and is based on a similar system used by others [7] for the oxidation of pure Ge. Briefly, samples were loaded into a quartz lined pressure vessel which was repeatedly flushed with N_2 followed by evacuation to a pressure of ~6.7 Pa. Under vacuum the furnace was ramped to the oxidation temperature of 500°C. The furnace tube was then filled with dry oxygen to a working pressure of 70 MPa which initiated the oxidation run. Oxidation was terminated by moving the furnace away from the reactor tube. After the reactor and contents had cooled to 200°C, the high pressure O_2 was bled-off and the samples removed. An oxidation time was selected based on kinetics studies (which will be presented elsewhere) to produce oxides 40 to 55 nm thick (depending on Ge concentration) suitable for MOS C-V measurements.

After each run the thickness and refractive index of the HPO oxides were determined using ellipsometry. The thickness measurements were later confirmed using cross-sectional TEM. Detailed XPS studies were used to evaluate the composition of the oxide and to provide insight into the chemical state of the Ge and Si incorporated in it. For these XPS studies a Phi 5500 multitechnique system, equipped with an aluminum anode, non-monochromatic K-alpha source, was operated at an energy of 1486.6 eV. The procedure was to acquire integrated peak spectra, linearly subtract the background, and apply previously determined sensitivity factors to determine the concentrations of the Ge in the as-grown oxides.

Raman spectroscopy was used as a sensitive and non-destructive probe [8] for the presence of Ge enrichment at the oxide/alloy interface. A double monochrometer and excitation by the 514.5 nm line of an Ar ion laser were used. To complement the Raman studies, selected samples were examined using a JEOL 2010 TEM operating at 200 keV for high resolution imaging of the oxide/alloy interface. Standard techniques were applied to prepare cross-sectional TEM samples for studying the HPO/alloy interface.

Filament evaporation of Al (99.99% pure) through a shadow mask was employed to form an array of MOS capacitor dots, 0.10 mm^2 in area, which were deposited on 40-55 nm thick oxides. Prior to C-V analysis, these MOS structures were annealed in an 80/20 N_2/H_2 forming gas mixture for 20 minutes at 450°C. A PAR 410 automatic C-V plotting system was used for high frequency (1 MHz) C-V measurements .

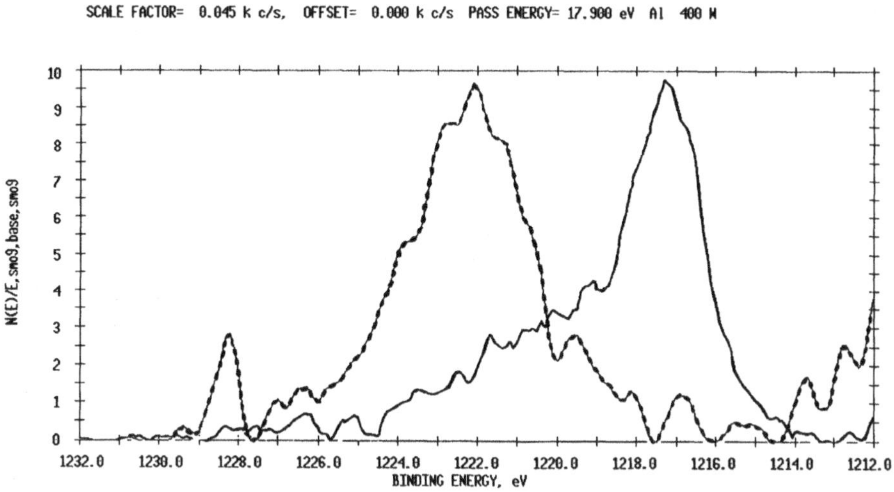

Figure 1. XPS spectra of the $Ge2p_3$ peak from a 50 nm thick HPO oxide (dry O_2, 500°C and 70 MPa) grown on $Si_{85}Ge_{15}$, under two conditions: a. (dotted curve) as oxidized, b. (solid curve) after an anneal at 800°C for 1 hr. A peak at the elemental Ge binding energy (1217.20 eV) appears in (b) but not (a).

RESULTS AND DISCUSSION

In previous research we established growth conditions for the synthesis of compositionally congruent oxides from $Si_{1-x}Ge_x$ alloys. In this paper we report on the characterization of 40 and 55 nm thick oxides grown by HPO on 10 and 15 at.% Ge alloys as part of an evaluation of this technology for potential MOS applications. Analysis began with the use of XPS for the determination of the composition and chemical state of the oxides. The chemical state of the Ge in the oxide was of particular concern since Ge in a substoichiometric (e.g. Ge vs GeO vs GeO_2) state can lead to electron traps and/or oxide leakage. The structural quality of the oxide/alloy interface with respect to Ge enrichment and interfacial smoothness were evaluated in the as-grown condition using high resolution TEM and Raman Spectroscopy. In addition, annealing studies at 700 and 800°C were carried-out in order to establish the thermal stability of the oxide and its compatibility with potential post-growth device processing. Finally, the oxides were evaluated using C-V measurements to determine the electronic performance of the HPO/alloy interface.

XPS and Raman Spectroscopies

In previous work [3,4,5] it has been shown that conventional wet atmospheric oxides grown at 800°C from alloys of $Si_{1-x}Ge_x$ consist of pure SiO_2 and are free of Ge. We have found that this is not the case when the oxidation temperature is lowered to 500°C and pure oxygen at a pressure of 70 MPa is used for oxidation. Detailed compositional analysis from XPS data (from both Ar ion sputter cleaned and uncleaned samples) show that the

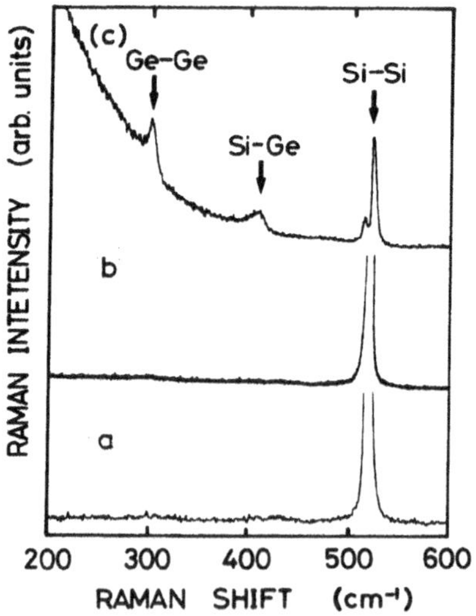

Figure 2. Raman Spectra of $Si_{85}Ge_{15}$: (a) before oxidation, (b) after HPO in pure O_2 at 70 MPa, 500°C, (c) after conventional atmospheric wet oxidation. A Raman peak corresponding to the elemental Ge-Ge excitation mode is present in (c) but not in (a) or (b).

relative Si-Ge composition of the oxide is identical to that of the underlying alloy for both the 10 and 15 at. % samples. Of equal importance is the chemical state of the Ge in the oxide which was analyzed using a Gaussian curve fitting routine to analyze the Ge_{2p3}-peak in the XPS spectra. An unoxidized $Si_{1-x}Ge_x$-alloy sample was used to provide a calibration for the position of the elemental Ge XPS peak. In Fig. 1(a) an XPS spectrum obtained from a 50 nm thick HPO oxide (dry O_2, 500°C and 70 MPa) grown on a 150 nm thick $Si_{85}Ge_{15}$ alloy film is presented. Gaussian analysis shows that the Ge_{2p3} peak is consistent with a bonding configuration for GeO_2 (1220.40 eV) and, to a lesser extent, GeO (1221.83 eV) [9]. Importantly, there is no detectable elemental Ge peak (1217.20 eV) in the XPS spectra which shows that Ge was chemically incorporated into the oxide. Similar results were obtained from oxides grown from $Si_{90}Ge_{10}$ alloys.

By annealing these oxides above 800°C for 1 hour, elemental Ge precipitates form in the oxide and the resulting XPS spectra (Fig. 1 (b)) shows a clear shift in the Ge_{2p3} peak to the lower binding energy level characteristic of elemental Ge. The precipitation phenomenon has been discussed elsewhere [10] and is a result of the chemical metastability of these $Si_{1-x}Ge_xO_2$ oxides when in contact with elemental Si. Our annealing studies have shown that a similar anneal for 1 hour at 700°C does not result in either a chemical or structural change in the oxide that is detectable by either Raman, XPS, or TEM. Based on these studies the HPO oxides are certainly stable at device operation temperatures and are sufficiently stable for many (but not all) device processing steps.

Conventional XPS studies of the oxide/alloy interface are not feasible for the 40-55 nm thick oxides that form the basis for this report. Argon ion sputtering is required in order to expose the interface and this is known to cause preferential removal of one or more species from the oxide. As an alternative, Raman spectroscopy was used to non-destructively evaluate the oxide/alloy interface for the presence of elemental Ge. This tool is highly sensitive to interface accumulation of Ge due to strong absorption of the Ar+ (514.5 nm) laser light and has been used previously [8] to analyze 1 to 6 monolayers of Ge on a Si substrate surface. Spectra are presented in Fig. 2 for the $Si_{85}Ge_{15}$ alloy in (a) the unoxidized state and (b) after oxidation in pure O_2 at 70 MPa and 500°C. Figure 2(c) shows the $Si_{85}Ge_{15}$ after conventional atmospheric steam oxidation at 800°C. In all three cases the Si-Si vibration mode due to the substrate can be seen at 510 cm^{-1} but only in the conventional steam oxidized sample is there a peak at 302 cm^{-1} which corresponds to the elemental Ge-Ge vibration mode. In addition, the deterioration of the interface due to conventional oxidation can be seen in Fig. 2(c) where interfacial roughening has caused increased scattering of the Ar+ laser light and consequently a sharp increase in the level of background noise.

High Resolution Transmission Electron Microscopy

The importance of interface morphology has been demonstrated in the Si/SiO_2 system where the presence of interfacial hillocks or asperities in thin gate oxides used in MOS devices results in premature dielectric breakdown, a reduction in channel carrier mobility, and an increase in interface states. We have used high resolution TEM to evaluate the interface planarity of the $Si_{1-x}Ge_xO_2/Si_{1-x}Ge_x$ interface formed by HPO. A high resolution TEM image of an HPO oxide/$Si_{85}Ge_{15}$ alloy interface oriented so that the substrate <110>

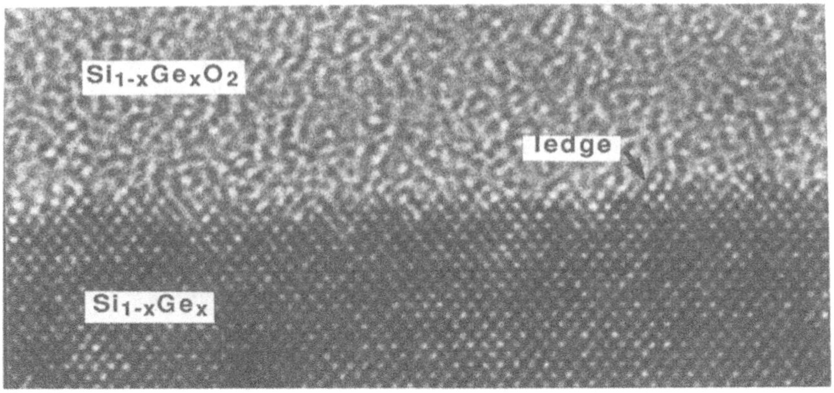

Figure 3. A high resolution TEM image of an HPO oxide/$Si_{85}Ge_{15}$ interface oriented so that the substrate <110> direction is parallel to the incident electron beam. There is an abrupt transition from amorphous to crystalline contrast at this oxide/$Si_{85}Ge_{15}$ interface.

direction is parallel to the incident electron beam is shown in Fig. 3. This figure shows that there is an abrupt transition from amorphous to crystalline contrast at the oxide/alloy interface. Variations in the planarity of the interface appear to be no greater than 1-2 monolayers in amplitude and the interface morphology is qualitatively comparable to that which is reported [11,12] for Si/SiO_2 interfaces used in MOS devices.

It should be noted, however, that the oxide and the alloy/oxide interface in all of the HPO oxide/alloy samples studied were susceptible to electron beam damage after relatively short exposures (<3 minutes) to the 200 keV beam. This damage is probably due to electron beam stimulated reaction between the substrate and the oxide resulting in the production of elemental Ge and pure SiO_2. For this reason a minimum dose technique was applied in order to obtain the image presented in Fig. 3.

Capacitance-Voltage Electronic Characterization

High frequency 1 MHz C-V curves were measured from Al metallized structures fabricated from alloys of 10 and 15 at.% Ge oxidized using the HPO conditions previously described. A representative example of the measured C-V data (solid line) is presented in Fig. 4 for $Si_{1-x}Ge_x$ with x=15 at.%. There is no observable hysteresis displayed for this measured curve or for any other curves acquired from samples of either Ge concentration. These curves were, as compared to the ideal (dotted line) curve, elongated along the voltage axis corresponding to a continuous distribution of states in the band gap. Past studies have revealed that if an MOS structure contains a relatively large number of states the C-V curve will exhibit shape distortions [13]. Broad slope variations are generally attributed to the presence of a continuous distribution of states in the bandgap while specific breaks in the curve are likely due to single levels in the gap. The location of these distortions can be related to the position of the states in the gap. The Terman method [14] was used to estimate the number of states associated with the observed elongation .

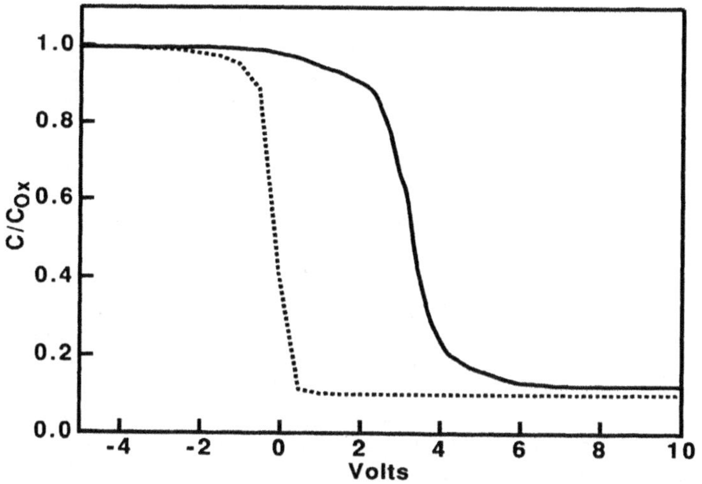

Figure 4. Example high frequency C-V curves for a MOS device fabricated from a $Si_{85}Ge_{15}$ alloy after HPO. Solid line: measured curve, dotted line: ideal (no interface traps) calculated curve. The measured curve is elongated along the voltage axis as compared to the ideal curve.

In order to apply this method, high frequency curves were obtained from Goetzberger's calculations [15] for the ideal (no surface states) case while assuming the same dielectric constant and bandgap as Si. These assumptions were made because the thickness of the $Si_{1-x}Ge_x$ layer is an order of magnitude thinner than the maximum depletion depth [16] calculated given the matched doping of the $Si_{1-x}Ge_x$ and Si substrate. In addition, our ellipsometric measurements of the refractive index of the $Si_{1-x}Ge_xO_2$ are very close to that of SiO_2 (n~1.46). With these assumptions in mind, the shape of the measured $Si_{1-x}Ge_x$ high frequency (1 MHz) C-V curves were compared with calculated ideal curves. Based on the Terman method, the midgap interface state density was found to be about $1x10^{12}$ $cm^{-2}eV^{-1}$ for both the 10 and 15 at.% samples.

The total charge per unit area induced in the semiconductor by charges in the dielectric layer, (Q_s'), was estimated by a charge analysis method which is described in detail elsewhere [17]. Using this approach, the flatband voltage (V_{FB}) shift of the experimentally measured C-V curve was determined and the induced charge, Q_s', was estimated using:

$$Q_s' = (-V_{FB} + \phi_{ms}) * C_0$$

where ϕ_{ms} is the metal-semiconductor work function difference, and C_0 is the capacitance per unit area of the dielectric layer. Analysis of C-V curves from the two alloys reveals a V_{FB} shift which can be attributed to Q_s'/q, in the 10 at.% Ge case, of positive $+1x10^{11}$ cm^{-2} (negative V_{FB}) and, in the 15 at.% case, of negative $-1x10^{12}$ cm^{-2} (positive V_{FB}). The origin of these large differences is difficult to unambiguously determine from the present data. However, previous studies [4,18] of conventionally oxidized, dilute $Si_{1-x}Ge_x$ alloys have shown large positive V_{FB} shifts in C-V curves similar to that seen in the 15 at.% sample. In the previous work these shifts were largely attributed to the existence of a large negative Q_f (oxide fixed charge) resulting from intermediate Si oxidation states and elemental Ge at the Si/SiO_2 interface. It is also known that GeO_2 rather readily becomes oxygen deficient [19] which can result in the formation of deep level electron traps. Such electron traps would result in a high level of Q_{ot} (Oxide trapped charge) and, if these charges are in electrical communication with the underlying semiconductor, an increase in D_{it} as well. We speculate that the positive shift that was seen in the 15 at.% case but not the 10 at. % was because of a slight accumulation of elemental Ge at the interface. Note that the positive shift was only seen in the 15 at. % samples and not in the dilute 10 at.% Ge case. Although Ge accumulation was not experimentally observed by either Raman or TEM it is consistent with previous reports of a negative Q_s' associated with interfacial enrichment. More careful Raman and TEM studies are underway to resolve this question.

The positive Q_s' obtained in the 10 at. % Ge sample is typical of that which is seen after conventional oxidation of pure Si [16]. The Q_s' observed in conventionally grown SiO_2 oxides has been associated with Q_f produced from excess Si ions in a narrow region at the oxide/Si interface and has been determined to be a strong function of both oxidizing and annealing conditions. In this study, the forming gas anneals which were employed did result in a decrease in both Q_s' and D_{it} in all measured oxides. Thus, HPO of $Si_{1-x}Ge_x$ for x=10 at. % does indeed produce a compositionally congruent oxide with positive Q_s' like that observed for Si under similar conditions. It is possible that by altering the HPO and annealing conditions, specifically temperature and gas (O_2, N_2, H_2) purity, similar results may be obtained for the 15 at. % Ge sample. Additionally, a more thorough investigation of the Ge bonding as a function of Ge concentration, oxidizing conditions, as well as depth in the oxide will be necessary to examine their possible effects upon Q_s' and possibly even D_{it}. Cooling rate and ambient constitution particularly moisture content, all interact and contribute to the oxide charge densities [20]. Similarly, it is quite possible that the value of D_{it} obtained in this investigation may reflect contributions due to metallic ions, specifically incorporated in these non-gettered structures. Continued process modifications will lead to a more thorough understanding of the above effects.

CONCLUSION

High pressure oxidation of $Si_{1-x}Ge_x$ alloys containing 10 and 15 at. % Ge has been investigated for potential MOS applications. Chemical analysis by XPS confirms that compositionally congruent oxides are grown from these alloys when oxidation proceeds in pure oxygen at a pressure of 70 MPa and a temperature of 500°C. High resolution TEM and Raman spectroscopy show that the as-grown oxide/semiconductor interface is planar and free of Ge enrichment on a scale of 1-2 monolayers. Although the value of D_{it} of 1×10^{12} $cm^{-2}eV^{-1}$ which was estimated from 1 MHz C-V analysis is probably unacceptably high for MOSFET applications, the process has not yet been optimized.

ACKNOWLEDGMENTS

This work was funded by the Office of Naval Research under Grant No. N00014-91-J-1837 and the National Science Foundation under equipment grant No. ECS-9112378. We are grateful to B. Meyerson of IBM for providing some of the material used in this research.

REFERENCES

1. S.S.Iyer, G.L. Patton, J.M.C. Stork, B.S. Meyerson, D.L. Harame, IEEE Trans Electron Dev. **36**, 2043(1989).
2. P.M. Garone, V. Venkataraman, J.C. Sturm, IEEE Electron Device Let. **134**, 56(1992).
3. C. Caragianis, D.C. Paine, C. Roberts, and E. Crisman, High Pressure Oxidation of Strained $Si_{1-x}Ge_x$ Alloys, in: "Chemical Perspectives of Microelectronic Materials II", MRS, Pittsburgh(1991).
4. F.K.LeGoues, R. Rosenberg, T. Nguyen, F. Himpsel, and B.S. Meyerson, J. Appl. Phys., **65**(4), 1724(1989).
5. D.C. Paine, C. Caragianis, and A. F. Schwartzman, J. Appl. Phys.,70(9), 5076(1991).
6. J. Reed, D.S.L. Mui, W. Jiang, and H. Morkoc, Electronics Letters, **27**,1826,(1991).
7. E.E. Crisman, Y.M. Ercil, J.J. Loferski, and P.J. Stiles, J. Electrochem. Soc. **129**,1845(1982).
8. J.C. Tsang, S.S.Iyer, and S.L. Delage, Appl. Phys. Lett., **51**(21), 1732(1987).
9. C.D. Wagner et. al., "Phi Handbook of Photoelectron Spectroscopy", Perkin-Elmer Corp., Eden Prairie, Mn (1979).
10. D.C. Paine, C. Caragianis, Y. Shigesato, Appl. Phys. Lett., **60**(23), 8(1992).
11. S.M. Goodnick, D.K. Ferry, C.W. Wilmsen, Z. Liliental, D. Fathy, and O.L. Krivanek, Phys. Rev. B, **32**, 8171(1985).
12. S.M. Goodnick, R.G. Gann, D.K. Ferry, C.W. Wilmsen, and O.L. Krivanek, Surf. Sci.**113**,233(1982).
13. B.E.Deal, E. L. Mackenna, P.L. Castro, J. Electrochem. Soc., **116**, 997(1969).
14. L.M. Terman, Solid State Electronics, **5**, 285 (1962).
15. A. Goetzberger., Bell Syst. Tech. J., **45**, 1097(1966).
16. G.W. Neudeck, R.F. Pierret, Field Effect Devices, Addison-Wesley Pub. Co., Reading, MA (1990).
17. A.S. Grove, B.E. Deal, E. H. Snow, and C.T. Sah, Solid State Electronics, **8**, 145(1965).
18. S.Margalit, A. Bar-lev, A.B. Kuper, H.Aharoni, and A. Neugroschel, J. of Crys. Growth, **17**:288(1972).
19. P. Balk, J. Electrochem. Soc., **118**(3), 494(1971)9.
20. R.R. Razouk, B.E. Deal, J. Electrochem. Soc., **126**(9), 1573(1979).

A NEW ELLIPSOMETRY TECHNIQUE FOR INTERFACE ANALYSIS: APPLICATION TO $Si\text{-}SiO_2$

E.A. Irene and V.A. Yakovlev*

Department of Chemistry CB# 3290
University of North Carolina
Chapel Hill, NC 27599-3290

* Institute of Crystallography, Acad. of Sci., Leninsky pr. 59 Moscow 117333 Russia

ABSTRACT

In this paper we report a new spectroscopic ellipsometry technique that overcomes much of the ambiguity associated with measuring an interface under a film. For this technique we match the refractive index of the overlayer with an immersion liquid and then perform spectroscopic ellipsometry at several angles of incidence. Essentially, the overlayer is optically (not physically) removed, thereby rendering the ellipsometric measurement sensitive to the interfacial layer which is often known to be optically and chemically different than either substrate or film. The $Si\text{-}SiO_2$ interface resulting from thermal oxidation of Si, and the evolution of the interface with annealing is studied using the new technique.

INTRODUCTION

It is apparent that the interface region between a semiconductor surface and a film, particularly dielectric films for MOSFET devices, is of crucial importance. Consequentially, the interface region of electronically relevant films on semiconductors has been widely studied by a variety of techniques (see for example ref 1 and refs 1-29 in ref 1). These techniques fall into a few broad classes that will be briefly discussed with the use of Fig. 1. First, there are the optical techniques that can access the optical response of the interface through an optically transparent overlayer, as seen in Figure 1a. While many successful studies have been done with the optimization of this technique, particularly ellipsometry studies, the key drawback is that the optical signals from the top interface must somehow be taken care of (minimized or subtracted etc.), in order to extract purely interface information. Second, there are techniques that can observe a surface with the overlayer somehow removed. Typically, the overlayer can be chemically etched or

The Physics and Chemistry of SiO_2 and the $Si\text{-}SiO_2$ Interface 2
Edited by C.R. Helms and B.E. Deal, Plenum Press, New York, 1993

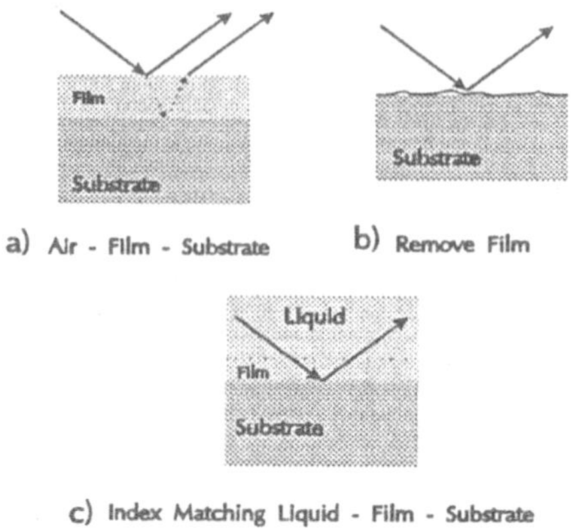

Figure 1. Comparison of interface analysis methods.

physically sputtered or a combination. Figure 1b indicates that the interface region can be altered by these aggressive processes, and that the information obtained about a previous interface contains an ambiguity. There are intermediate techniques such as cross-sectional transmission electron microscopy, XTEM, which neither removes the overlayer nor is overly sensitive to extraneous sources of signal. However, this technique requires considerable sample preparation, in particular the sample thinning, which can alter the interface. Notwithstanding the limitations of each technique, and taken in total, these techniques have contributed significantly to our understanding of the important interfaces.

The present paper deals with the development and application of a novel in-situ ellipsometry technique that can access the interface region for an optically transparent film on a surface[1,2]. The principle of operation of the technique is to immerse the film covered substrate sample in a liquid that refractive index matches to the overlayer film as illustrated in Figure 1c. In this way, the optical response consists only of reflections from the film-substrate interface region thereby removing the above mentioned ambiguity. It is assumed that the interface region is optically distinct from either the bulk film and substrate, an assumption borne out in many studies. The enhanced sensitivity of the technique will be demonstrated. By way of application, the Si-SiO_2 interface is chosen and the nature and evolution of this interface is examined as a function of thermal annealing.

THE TECHNIQUE

Index Matching. The ideal immersion liquid should be non reactive and transparent over a wide spectral range. Both the refractive index and the dispersion must correspond to the average refractive index of the overlayer. The refractive indices for the immersion liquids were calculated using a three term Cauchy dispersion formula, taking into account temperature and using literature values for the various constants[3,4]. Using liquid mixtures it is possible to adjust the refractive indexes of the ambient to that for SiO_2. Figure 2 shows the spectral dependencies of the refractive index of pure carbon tetrachloride, CCl_4, benzene, C_6H_6, and mixtures[3]. Both CCl_4 and C_6H_6 are nonpolar organic liquids that do not interact with SiO_2. Also in Figure 2 is shown the calculated spectral dependencies of the refractive index of bulk and thin film SiO_2 calculated using a single term Sellmeier approximation after Jellison[5]. Since n for thermally grown SiO_2 films is dependent on the thickness[6,7], the use of immersion liquids with different refractive indexes corresponding to the average refractive indexes of the films under investigation renders accuracy improvements possible.

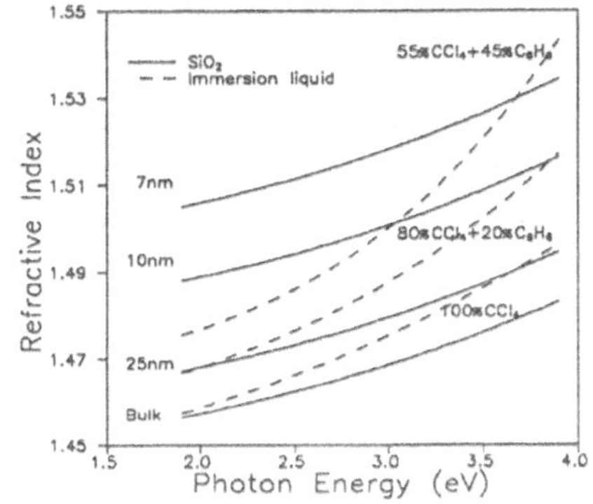

Figure 2. The refractive index versus photon energy for pure carbon tetrachloride (CCl_4), and mixtures with benzene (C_6H_6), along with bulk and thin-film SiO_2.

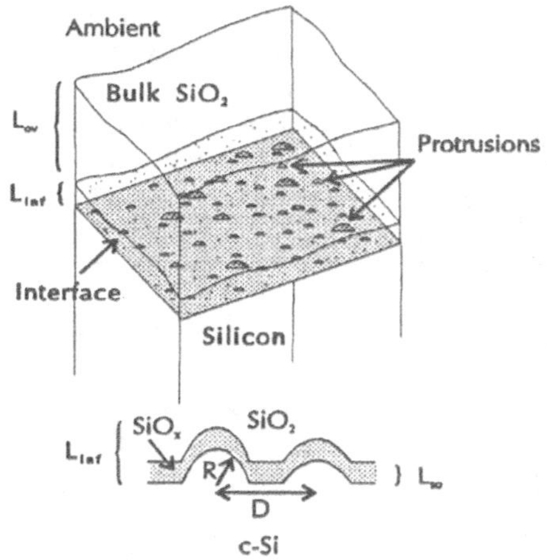

Figure 3. Model for the substrate-interface-overlayer-ambient system.

Analysis. In order to assess the sensitivity of the technique, the complicated structure shown in Figure 3 and thought to represent the Si-thermally grown SiO_2 system is simplified to a two film system with interface and overlayer thicknesses L_{inf} and L_{ov}, respectively. Thus, at this juncture the interface region in Figure 3 is considered to be a homogenous layer with an effective dielectric function, ϵ_{inf}. The overlayer or bulk SiO_2 film is represented with an average refractive index n_{ov}. The complex reflection coefficient, ρ, for the system is given by:

$$\rho = \tan\Psi \exp(i\Delta) = \rho(\lambda,\phi,n_o,n_{ov},n_s,k_s,e_{inf},L_{ov},L_{inf}) \qquad (1)$$

where Ψ,Δ are the ellipsometric measurables and the terms in the parenthesis on the right are parameters, some known a priori and some to be determined. The maximum change in ρ, $\delta\rho/\rho$, is the condition of optimum sensitivity in the measurables, Δ,Ψ. It is desirable to determine the optimum sensitivity in terms of the controllable parameters: angle of incidence, ϕ, and wavelength of light, λ. From the analytical solution for optimized ellipsometric measurements of interfaces with thicknesses, $L_{inf} << \lambda/4$ in a thin film structure[8], it was shown that the condition of $\delta\rho/\rho$ divergence or maximum sensitivity is:

$$r_{01}^{p(ands)} + r_{12}^{p(ands)} \exp(-2i\beta) = 0 \quad (2)$$

where the r's are the Fresnel reflection coefficients with subscripts corresponding to the interface between the media with numbering starting from 0 with the ambient and β is given as:

$$\beta = \frac{2\pi L_{ov}}{\lambda}\sqrt{(n_{ov}^2 - n_o^2 \sin^2\phi)} \quad (3)$$

It should be observed that the properties of the interface do not influence the best sensitivity conditions.

If the refractive index of the ambient is close to the index of the overlayer, i.e. $n_o \approx n_{ov}$, then $r_{01} \approx 0$. The optimum sensitivity condition from eqn(2) becomes r_{12} or, taking into account $n_o \approx n_{ov}$, $r_{02} \approx 0$. For the p wave this condition is equal to a minimum of the ellipsometric angle, $\Psi(\phi,\lambda)$[9].

We determine the optimal spectral range and angles of incidence from a simulation of $\Psi(\phi,\lambda)$ and $\Delta(\phi,\lambda)$ dependencies using the model in Figure 3 for the Si-SiO_2 system with the following assumptions: a. crystalline Si substrate with known dielectric function[9]; b. interface microroughness (see Figure 3) with effective height of 0.2nm and composition of 50% c-Si and 50% suboxide SiO_x, with x = 0.4 and the Bruggeman effective medium approximation, BEMA, was used to calculate a dielectric function of the mixture; c. interface suboxide transition zone of 0.6nm composed of $SiO_{0.4}$; d. SiO_2 overlayer with an average refractive index $n_{ov}(L_{ov})$; e. an air or pure CCl_4 ambient with refractive index n_o.

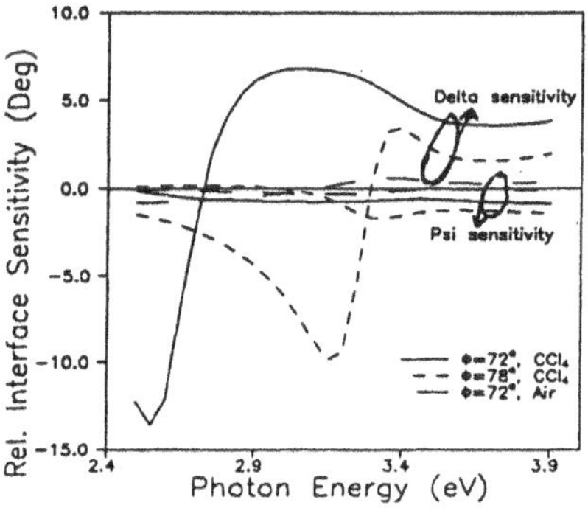

Figure 4. Relative interface sensitivity for Δ and Ψ for air and CCl_4 ambients.

Figure 4 shows a plot of the interface sensitivity for both Ψ and Δ in terms of a relative interface sensitivity function $\delta\Delta(E) = \Delta_0(E)-\Delta_{inf}(E)$ for air and CCl_4 ambients. Where $\Delta_0(E)$ and $\Delta_{inf}(E)$ were calculated for the cases without and with an interface layer, respectively. It is seen that the sensitivity of Δ to the presence of an interfacial layer is increased by more than an order of magnitude with the parameters at optimum sensitivity, but the Ψ sensitivity is low. However, changing ϕ to 78° for the same system enhances the Ψ sensitivity to the interfacial layer and this is also shown in Figure 4.

Calculation Scheme. In order to find unknown parameters of the modeled interface, we used the Marquardt non-linear best-fit algorithm which minimizes the value of the error function:

$$Q = \sum_{i,j} [(\Delta_{i,j}^{cal}(\phi_i, E_j, P) - \Delta_{i,j}^{exp})^2 + (\Psi_{i,j}^{cal}(\phi_i, E_j, P) - \Psi_{i,j}^{exp})^2] \quad (4)$$

where P is a vector of N unknown interface parameters and E_j is the photon energy, and the superscripts cal and exp refer to calculated and experimentally derived values. Δ^{cal} and Ψ^{cal} are the values obtained using the vector P, the Fresnel formulas[10] and a matrix algorithm for the multilayer system complex reflection coefficient. Optical parameters for the immersion liquid and SiO_2 overlayer included in the fitting procedure were calculated from dispersion equations mentioned above. BEMA was used to calculate the effective dielectric function for a mixture of constituents with known optical properties. Volume fractions of the constituents are treated as unknown parameters.

The calculation returns the vector P when an initial guess, P^o, is input, i.e. the program returns a value of P at the local minimum of Q, near the initial value. In addition, correlations between the fitting parameters are expressed in terms of the correlation matrix of the derivatives $\partial\Psi/\partial p_n$, $\partial\Delta/\partial p_n$ and a 95% confidence limit was used to calculate errors in the fitting parameters.

EXPERIMENTAL PROCEDURES

A commercially available vertical ellipsometer bench was modified to become rotating analyzer spectroscopic ellipsometer with the details previously described[11] and calibrated according to a published procedure[12] and the details described[1]. Calibration was performed on a sample with $\Delta \approx 90^\circ$ (Si wafer with thermally grown SiO_2 film) both in air and in the immersion cell in liquid.

A fused silica immersion cell, as shown in Figure 5, has been designed for the variable angle of incidence and spectroscopic measurements. The most significant feature of the cell is that the two optically flat and annealed fused silica plates, serving as the entrance and exit windows, are connected rigidly to the polarizer and analyzer arms, but not to the cell. The windows were adjusted to be exactly orthogonal to the incident light at the straight-through position of the ellipsometer ($\phi = 90^\circ$), in order to avoid any deviation of the incident light beam when it passes through adjacent media with different indexes. Two fitted metal tubes in each arm permit some lateral movement of the window position without a change of the window tilt. The cell is rigidly attached to a stage and connected to the tubes using chemically inert flexible tubing. A change of ϕ in the range of 67°-90° is possible while maintaining the window alignment precision during the immersion measurements.

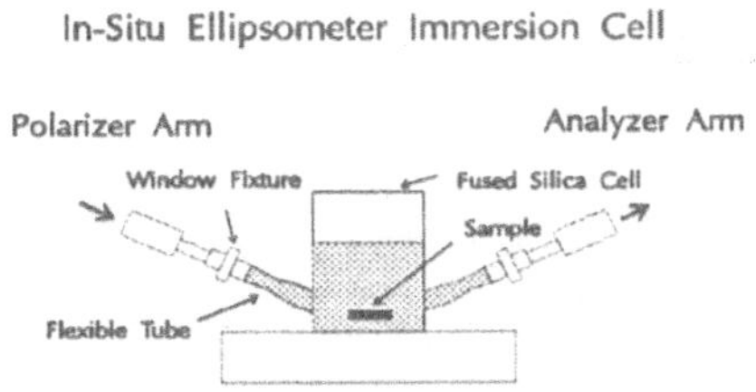

Figure 5. Sketch of the variable angle of incidence immersion cell.

Single-crystal (100) oriented 2 Ωcm p-type silicon wafers were cleaned prior to oxidation using a slightly modified RCA cleaning procedure[13]. The samples were thermally oxidized to about 20nm at 800° C in a fused silica tube furnace in clean dry oxygen which yielded MOS quality SiO_2 films on Si. For the annealing studies, after a particular oxidation, one sample was removed without annealing as a control, and the others were annealed in a clean nitrogen atmosphere for annealing temperatures and times in the range 750-1100°C and 1-120 min, respectively.

In order to use the convergence routines, an estimation of the overlayer thickness is required. One layer (substrate-overlayer) and two layer (substrate-interface-overlayer) models have been used to analyze the spectroscopic ellipsometry scans of the about 25nm SiO_2 films on Si samples in air. From these measurements in air, the difference in the minimum error function, Q, for one and two layer models was negligible, which indicates low interface sensitivity. However, these measurements yield a good estimate of the overlayer thickness to a precision of better than 0.5%. The refractive index of the film was calculated from dispersion relations. Immersion measurements were performed in pure CCl_4 at 20°C. Carbon tetrachloride becomes nontransparent at energies higher than ~4 eV which then defines the upper limit of the spectral range to be 4eV. The precision of rotating analyzer ellipsometry without using an achromatic compensator falls significantly when Δ approaches 180° or 0° [14]. To avoid this situation, we have used 2.5 eV as the lower energy limit where Δ is more than 20°. Therefore, Δ is restricted to $20° < \Delta < 160°$ in the spectral range 2.5-4.0 eV for the measurements in the immersion liquid at the angles of incidence ϕ = 70°-80°.

RESULTS: Evolution of the Si-SiO_2 Interface During Annealing

Model Independent Results. A series of annealing experiments were performed on SiO_2 covered Si samples, and the results are first displayed without recourse to a model, so that an objective view of the interface evolution can be obtained. Figures 6 and 7 display experimental results in terms of the interface parameter $\delta\Delta_{inf}(T_{an},t_{an})$ defined as:

$$\delta\Delta_{inf}(T_{an},t_{an}) = \Delta^{exp}(T_{an},t_{an}) - \Delta_o^{exp} - \delta\Delta_{ov}^{cal}(T_{an},t_{an}) \qquad (5)$$

where $\Delta^{exp}(T_{an}, t_{an})$ is the experimental ellipsometric angle Δ at a specific annealing temperature and time, Δ_o is the ellipsometric angle for a non-annealed sample and the term $\delta\Delta_{ov}{}^{cal}(T_{an}, t_{an})$ is the overlayer relaxation correction. This correction term represents the difference in Δ for the single-film (without interface) system of a non-annealed and

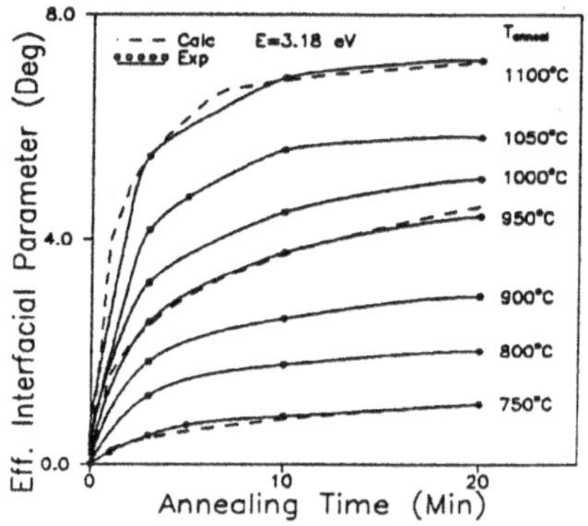

Figure 6. The experimental dependence of the effective interface parameter, Δ_{inf}, on annealing time at a number of annealing temperatures. A simulated dependency is shown with the dashed curve.

annealed overlayer with the refractive index calculated from a consideration of stress relaxation[6,15,16,17]; and represents only a few percent of $\delta\Delta_{inf}$. The chosen photon energy E=3.18 eV is in the range of the maximum interface and the minimum overlayer sensitivity. Figure 6 shows the values for $\delta\Delta_{inf}$ measured at 3.18 eV versus anneal time at a number of temperatures, and versus anneal temperature at two times in Figure 7. Other information relating to modelling is also included in Figure 6, but discussion of these items will be deferred. It is seen in Figure 6 that at all annealing temperatures two

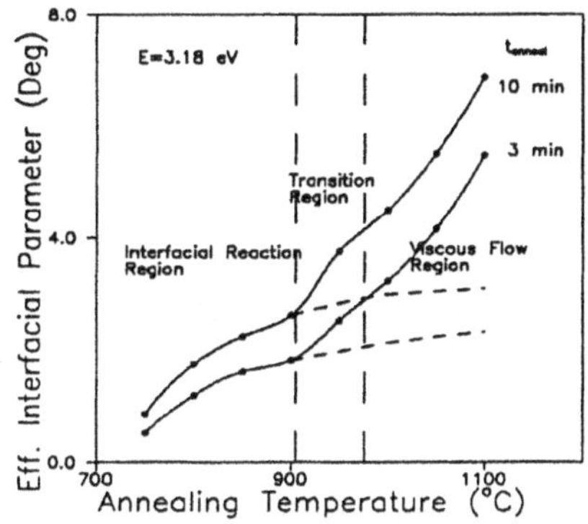

Figure 7. The experimental dependence of the effective interface parameter, Δ_{inf}, on annealing temperature for two annealing times.

temporal regions of behavior are present. Initially, the anneals yield a relatively fast increase in $\delta\Delta_{inf}$ that slows after 5 min and then nearly saturates. The data in Figure 7 is from Figure 6 at two anneal times, one before and one near saturation, and this plot versus anneal temperature reveals a distinct break in the 900°-970°C range, which corresponds to the viscous flow range, i.e. a temperature above which viscous flow of the oxide is fast[18]. Thus, we consider that the viscous relaxation dominates at the higher anneal temperatures, but at lower temperatures where a large part of the change in $\delta\Delta_{inf}$ occurs, we need to consider other possible mechanisms.

If the interface region is treated as a single homogenous film, then the extent of the interface is observed to decrease with both annealing time and temperature as evidenced by the increasing Δ_{inf}. With both a decrease in the interface region for short and long times and low and high temperatures, the model that is chosen for the different modes of behavior must show the observed functionality.

An Interface Model. Previous models for the Si-SiO_2 interface have agreed that the interfacial region is structurally and chemically different from either Si or SiO_2. Components of the interfacial region include roughness and a suboxide, SiO_x, as is shown in Figure 3. The parameters that quantify this working model, and then the experimental data $\Delta^{exp}(E,\phi,T_{an},t_{an})$, $\Psi^{exp}(E,\phi,T_{an},t_{an})$ are reduced to yield values for the model parameters which are compared with other studies about the interface region. We describe the roughness by crystalline silicon protrusions as hemispheres with an average radius R, which form a hexagonal network with an average distance D between centers. The protrusions and the region between them are covered by a layer of suboxide, SiO_x, with $0<x<2$, and with an average thickness L_{so}. The effective thickness of the complex transition layer is given as:

$$L_{inf} = R + L_{so} \qquad (6)$$

and an effective dielectric function ϵ_{inf}, which represents a mixture of crystalline silicon c-Si, silicon suboxide SiO_x and the SiO_2 overlayer and written as:

$$\epsilon_{inf} = \epsilon_{inf}(\epsilon_{c-Si}, f_{c-Si}, \epsilon_{SiO_x}, f_{SiO_x}, \epsilon_{SiO_2}, f_{SiO_2}) \qquad (7)$$

ϵ_{inf} was calculated using the BEMA, where the dielectric properties, ϵ, and relative volume fractions, f, of all of the interfacial layer constituents are known a priori. The dielectric function of SiO_x was calculated[1,2] using the BEMA, and by considering that SiO_x is a mixture of a-Si, and SiO_2[19].

In order to model the evolution of the interface during annealing, we use a power law to describe the reduction of both the protrusions and the chemical transition layer with the powers p and g respectively are taken to be p=g=0.5 which implies a diffusion model and is justified below.

The minimization of the error function (eqn.(4)) for the sets of experimental data for the non-annealed wafer at $\phi=72°$ and 75° gives the average distance between the centers of protrusions D = 44±4 Å, initial radius of the protrusions R° = 9.8±0.3 Å and initial thickness of suboxide, SiO, L_{so} = 3.4±0.2 Å. These results are in agreement with the interface geometry sensitive TEM study of the SiO_2-Si, which shows that distances between protrusions at non-annealed interface are distributed in the range 40-50 Å and heights are 9-15 Å[20]. Also, photoelectron spectroscopy revealed a chemical transition layer with a thickness of 2.4-4 Å[21].

A simulation of the kinetic dependencies, $\delta\Delta_{inf}(T_{an}, t_{an})$, with p=g=0.5 yields close agreement (dashed lines in Figure 6) with the experimental plots of $\delta\Delta_{inf}(t_{an})$.

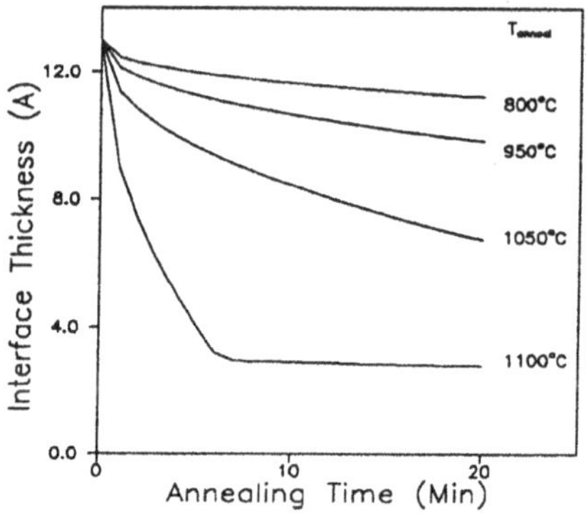

Figure 8. The dependence of the effective interface thickness on annealing time at a number of annealing temperatures.

Figure 8 displays the decrease of the extent of the interface, L_{inf}, i.e. effective thickness from eqn.(6) with a minimum thickness of about 3 Å realized at the highest anneal temperatures. The interface effective refractive index vs t_{an} at several T_{an} is shown in Figure 9.

Initially both the index and thickness of the interface layer decrease due to the shrinking of the silicon protrusions. This is followed by the slower decrease of the SiO_x transition layer which becomes dominant after considerable reduction of the height of the protrusions. The sharp increase in the index seen in Figure 9 for the 1100°C annealed sample is indicative of the index returning quickly to the SiO value as the fraction of the Si as protrusions in the interface layer rapidly goes to zero. The same effect would occur for the other anneal temperatures but more slowly. A slower rise in index is seen for the 1050°C annealed sample after 10 min and the time is too short to see the rise for the lower temperature anneals. The overall shape of the results in Figure 9 is typical of and dictated by the BEMA model used to interpret the data.

The results show two distinct modes for the evolution of the Si-SiO_2 interface upon annealing. The low temperature mode corresponds to a reduction in the suboxide and protrusions. The coefficients, p and g that scale this evolution yield an excellent fit to the data when a value of 0.5 is used in modelling. This suggests that Si atom diffusion is operative for the smoothening reaction and inward trace oxidant diffusion for the suboxide disappearance. At high temperatures these reactions occur but the dominant mode is viscous relaxation of the oxide overlayer. These model dependent assertions are consistent

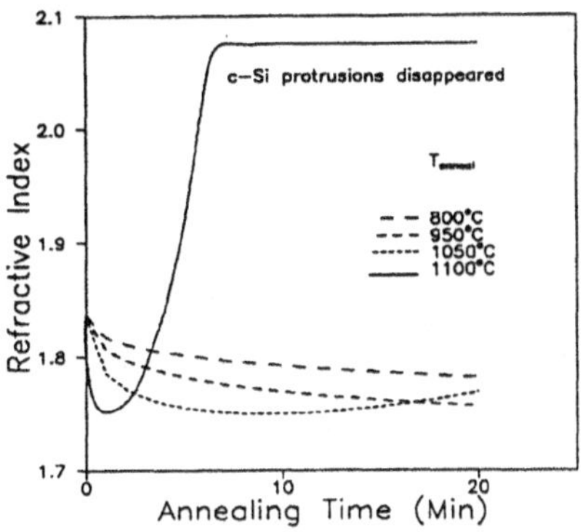

Figure 9. The dependency of the effective interface refractive index on annealing time at a number of annealing temperatures.

with other studies and are shown to be accessible by the novel ellipsometric technique herein described.

ACKNOWLEDGEMENTS

This research was supported in part by the Office of Naval Research, ONR.

REFERENCES

1. V.A. Yakovlev and E.A. Irene, J. Electrochem. Soc.,139, 1450 (1992).

2. V.A. Yakovlev, Q. Liu and E.A. Irene, J. Vac. Sci. and Tech. A, 10(3), May/June (1992).

3. "Techniques of Chemistry", Ed. A. Weissberger, Vol.II Organic Solvents, Ed. J.A. Riddick and W.B. Bunger, Wiley-Interscience, NY, 1970.

4. P.Perez, T.E.Block, and C.M.Knobler, J. Chem. and Engineering Data, 16, 333 (1971).

5. G.E.Jellison, Jr., J. Appl. Phys., 69, 7627 (1991).

6. A.Kalnitsky, S.P.Tay, J.P.Ellul, S.Chongsawangvirod, J.W.Andrews, and E.A.Irene, J. Electrochem. Soc., 137, 234 (1990).

7. S.Chongsawangvirod, E.A.Irene, A.Kalnitsky, S.P.Tay, and J.P.Ellul, J. Electrochem. Soc., 137, 3536 (1990).

8. S.A.Alterovitz, G.H.Bu-Abbud, and J.A.Woollam, Thin Solid Films, 128, 183 (1985).

9. D.E.Aspnes, A.A.Studna, Phys. Rev. B: Condensed Matter, 27, 985 (1983).

10. R.M.A.Azzam and N.M.Bashara, "Ellipsometry and Polarized Light", North-Holland, Amsterdam, 1977.

11. X.Liu, J.W.Andrews, and E.A.Irene, J. Electrochem. Soc., 138, 1106 (1991).

12. J.M.M. de Nijs, A.H.M.Holtslag, A.Hoeksta, and A. van Silfhout, J. Opt. Soc. Am. A, 5, 1466 (1988).

13. W.Kern and D.A.Puotinen, RCA Rev., 31, 187 (1970).

14. D.E.Aspnes, J. Opt. Soc. Am., 64, 639 (1974).

15. E.A.Irene, E.Tierney, and J.Angillelo, J. Electrochem. Soc., 129, 2594 (1982).

16. L.M.Landsberger and W.A.Tiller, Appl. Phys. Lett., 51, 1416 (1987).

17. K.Taniguchi, M.Tanaka, C. Hamaguchi, and K.Imai, J. Appl. Phys., 67, 2195 (1990).

18. E.P.EerNisse, Appl. Phys. Lett, 35, 8 (1979).

19. G.Zuther, Phys. St. Sol.(a), 59, K109 (1980).

20. H.Akatsu, Y. Sumi, and I. Ohdomari, Phys. Rev. B, 44, 1616(1991).

21. T.Hattori and T.Suzuki, Appl. Phys. Lett., 43, 470 (1983).

OBSERVATION OF THIN SiO_2 FILMS USING IR-RAS

Shuzo Fujimura, Kenji Ishikawa, and Haruhisa Mori

Process Development Division 1-1, C523
FUJITSU LIMITED
1015 Kamikodanaka Nakahara-ku, Kawasaki 211, Japan

ABSTRACTS

Thin oxide films thinner than 200Å on the common silicon wafers with rough back surface were observed by IR-RAS (Infrared reflection absorption spectroscopy). The IR-RAS spectra calculated using Grosse's oscillator model[1] fitted with the measured spectra fairly well, but the peak of the resonance of transverse optical phonon polariton (TO phonon) in the calculated spectra was rather sharper than that in the measured spectra. In addition, the model calculation did not explain the dependence of the peak height of the resonance of longitudinal optical phonon polariton (LO phonon) and the TO phonon on SiO_2 film thickness and thermally oxidation temperature. However, the LO and TO phonon calculated from a new model in which a thin crystal SiO_2 layer existed between the amorphous SiO_2 and the silicon crystal showed same dependence on the film thickness and the oxidation temperature as those of the measured spectra.

INTRODUCTION

Requests to the SiO_2 films as the insulators of high density semiconductor devices become diversified and strict. Flash memories need the gate insulator which has high breakdown voltage and controlled leakage current, and DRAM capacitors need high breakdown voltage and high dielectric constant for their insulation layers. To satisfy these requests, we have to control electrical properties of SiO_2 layers through oxidation processes. However, the relations between oxidation process parameters and film qualities have not been clarified systematically. The relations between process parameters and film qualities can be understood only through the physics of the SiO_2 layer and the oxidation process mechanism. The physics of the SiO_2 films shows the relations between the chemical structure of the SiO_2 layer and its electrical properties and the oxidation process mechanism explains the relation between the chemical structure change in SiO_2 films and process parameters. First of all, therefore, we have to find out an observation method of the SiO_2 chemical structure. In practice, measurement of electrical characteristics of the devices fabricated using the SiO_2 layers whose chemical structure has already analyzed is one of the most effective method to probe the relation between the chemical structure and electrical properties of SiO_2 films. Nondestructive and untouched analytical methods, which can apply to SiO_2 layers on the normal Si wafer, are preferable for this purpose. Many analytical methods to the chemical structure of thin oxide films on the silicon substrate have already reported such as XPS[2-3], EELS[4], IR-GIR[5,6,7], etc.. From the above reason, however, we chose the infrared reflection absorption spectroscopy (IR- RAS).

Grosse et al. observed the thin SiO_2 films on the silicon substrate using IR-RAS and showed that the oscillator model explained IR-RAS spectra fairly well[1]. In the paper, they assumed that the thermally grown SiO_2 layer was amorphous, which had an isotropic uniform structure. However, Ourmard et al. insists that there is crystalline structure at the Si-SiO_2 interface[8], moreover Lucovsky et al. reported that TO phonon peak of the thermally grown SiO_2 on the IR spectra shifted its position depending on the intrinsic stress in the SiO_2 layer induced by the oxidation process[9]. These discussions suggest that the thermally grown SiO_2 is not isotropic and uniform amorphous but anisotropic and nonuniform condensed matter. Thus the oscillator model should be altered so as to include the stress effects.

Our purposes in this paper are two: one is application of the IR- RAS measurement to 4-inch wafers, and another is the comparison of the dependence of IR peak height on the oxide thickness and oxidation temperature between the spectra experimentally measured and those calculated by the Grosse's oscillator model. If the structure of thermally grown SiO_2 layers is not uniform and depends on the oxidation process, calculated IR spectra can not follow the spectra change caused by the thickness of oxide layer or oxidation temperature.

EXPERIMENTAL

Silicon wafers used for this experiment were P-type, (100) surface, 10±1Ω-cm, 4-inch, and rough back wafers. After SC1 cleaning (NH_4OH/H_2O_2, H_2SO_4/H_2O_2, HCl/H_2O_2), thin SiO_2 films were formed on wafers by the thermal oxidation in a dry O_2 atmosphere or a plasma CVD. Thermal oxidation temperature was 800°C, 900°C, and 1000°C. A parallel electrode reactor was used for plasma CVD. Its operating condition was as follows: gas condition was SiH_4 8cc/min, NO_2 400cc/min, and N_2 105cc/min, pressure was 1.0Torr, applied RF frequency was 200kHz, RF power was 35W, wafer stage temperature was 350°C, and distance between two electrodes was 20mm. To avoid the thermal oxide growth during the plasma CVD, the CVD films deposited on the native oxide layer formed by the SC1 cleaning. In addition, we removed back side oxide of all wafers before IR-RAS observation. This was because a little back side oxide influenced to the IR-RAS spectra.

IR-RAS measurements have been carried out by means of Fourier transform spectrometer (JEOL JIR-6500) and a mirror unit for RAS measurement (JEOL IR-RSC120). Measurement condition was as follows: the angle of incidence on the sample was 80°, the spectra were recorded at 8cm^{-1} resolution, and scan time was 200. Only P-polarized light was used. A calibration spectrum was obtained from each sample wafers after the oxide layers were removed by the 5% HF solution.

We measured the thickness of oxide layers with the ellipsometer. Then the refractive index of the oxide layers was fixed on 1.462.

MODEL CALUCULATIONS

For the dielectric function a model has been used which is based on oscillator terms of the following structure

$$\varepsilon(\omega) = (n+ik)^2 = \varepsilon_\infty + \sum_j \frac{\Omega_{pj}^2}{\Omega_{oj}^2 - \omega^2 - i\Omega_{\tau j}\omega} \tag{1}$$

Here Ωoj, Ωpj, $\Omega \tau j$ are the resonant wavenumbers, the plasma wavenumber and the damping wavenumber of the jth vibration mode. ε is due to the electronic polarizability. Table 1 shows the each resonant wavenumber Ω of the oscillator mode of the thermally grown SiO_2. These wavenumbers are obtained by Grosse et al.[1] as the optimal fit parameters for the dielectric functions, which reproduced their experimental spectra of a 8400Å SiO_2 layer grown in 1050°C dry O_2 atomsphere. However some differences, particularly concerning to the peak height of the spectra, have still remained.

Table 1. Parameter of the dielectric functions of a quartz and a thermally grown oxide[1].

Mode number		-	1	-	2	3	4	ε_∞
Thermally	Ωo	-	445	-	800	1075	1190	2.4
grown SiO_2	Ωp	-	422	-	240	867	266	
	$\Omega \tau$	-	30	-	70	20	70	
Quartz E ⊥ C	Ωo	393	449	698	789	1068	1162	2.36
	Ωp	278	421	99	299	894	164	
	$\Omega \tau$	6	6	11	12	9	20	
Quartz E ∥ C	Ωo	365	494	-	778	1073	-	2.38
	Ωp	301	450	-	270	911	-	
	$\Omega \tau$	3	6	-	11	8	-	
Quartz glass	Ωo	-	460	-	795	1090	1195	2.2
	Ωp	-	374	-	308	808	293	
	$\Omega \tau$	-	18	-	70	22	95	

Frequencies [cm^{-1}]

The generalized Fresnel formula of the p-polarized light for a film (material 2) between two half space (front half space material 1 and back half space material 3)are given by

$$r=\frac{\cos(\delta N_2)\left(\frac{N_3}{\varepsilon_3}-\frac{\cos\alpha}{N_1}\right)-i\sin(\delta N_2)\left(\frac{N_2}{\varepsilon_2}-\frac{\varepsilon_2 N_3\cos\alpha}{N_2\varepsilon_3 N_1}\right)}{\cos(\delta N_2)\left(\frac{N_3}{\varepsilon_3}+\frac{\cos\alpha}{N_1}\right)-i\sin(\delta N_2)\left(\frac{N_2}{\varepsilon_2}+\frac{\varepsilon_2 N_3\cos\alpha}{N_2\varepsilon_3 N_1}\right)} \quad (2)$$

r is the amplitude reflection. (α: angle of incidence, $\delta=(\omega/c)d=2\pi d/\lambda$: relative thickness of the film, $N_3=\sqrt{\varepsilon_3-\varepsilon_1\sin^2\alpha}$: generalized index of refraction). Then the reflectance R is given by

$$R=rr^* \quad (3)$$

Throughout the spectral region in the vicinity of $\varepsilon(\omega)=0$, and for films thin compared to the wavelength, the approximations $\cos(\delta N_2)\cong 1$ and $\sin(\delta N_2)\cong \delta N_2$ are varied. Thus, eq.(2) become

$$r=\frac{\dfrac{N_3}{\varepsilon_3}-\dfrac{\cos\alpha}{N_1}-i\delta\left(1-\dfrac{\varepsilon_1}{\varepsilon_2}\sin^2\alpha-\dfrac{\varepsilon_2 N_3\cos\alpha}{\varepsilon_3 N_1}\right)}{\dfrac{N_3}{\varepsilon_3}+\dfrac{\cos\alpha}{N_1}-i\delta\left(1-\dfrac{\varepsilon_1}{\varepsilon_2}\sin^2\alpha+\dfrac{\varepsilon_2 N_3\cos\alpha}{\varepsilon_3 N_1}\right)} \tag{4}$$

To calculate the spectra which are measured in our experiment, we need to know the value of R_0, T_0, A, R', R_{-1}, and T_{-1} which are shown in Fig. 1. R_0 is the reflectance at the thin films from air to air. T_0 is the transmittance through the thin SiO_2 layer from air into the silicon substrate. A is the attenuation of intensity by the substrate for one path. R' is the internal reflectance at the back side rough surface. R_{-1} is the reflectance at the thin SiO_2 layer from the silicon balk to the silicon balk. T_{-1} is the transmittance through the oxide film from Si substrate to the air. Total reflectance which we measured is

$$\begin{aligned} R_{total} &= R_0 + T_0A^2R'T_{-1} + T_0(A^2R')^2R_{-1}T_{-1} + T_0(A^2R')^3R_{-1}{}^2T_{-1} + \cdots\cdots \\ &= R_0 + \frac{aT_0T_{-1}}{1-aR_{-1}} \end{aligned} \tag{5}$$

$$(a = A^2R')$$

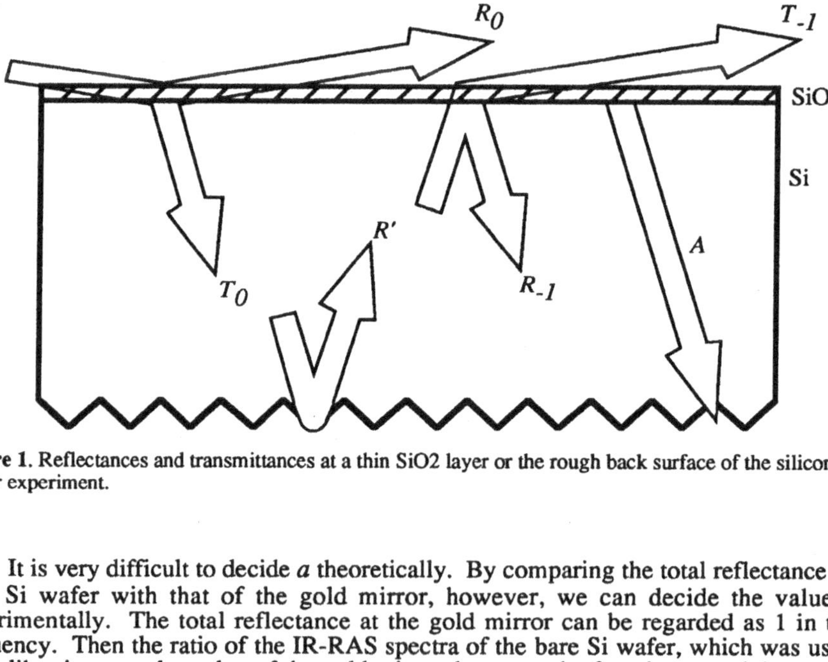

Figure 1. Reflectances and transmittances at a thin SiO2 layer or the rough back surface of the silicon wafer in our experiment.

It is very difficult to decide a theoretically. By comparing the total reflectance of the bare Si wafer with that of the gold mirror, however, we can decide the value of a experimentally. The total reflectance at the gold mirror can be regarded as 1 in the IR frequency. Then the ratio of the IR-RAS spectra of the bare Si wafer, which was used for the calibration sample to that of the gold mirror, becomes the function containing only one variable a.

$$\frac{IR_{Si-bare}}{IR_{Au-mirror}} = R_{0-bare} + \frac{aT_{0-bare}T_{-1-bare}}{1-aR_{-1-bare}} \tag{6}$$

("I" is the intensity of incident wave. $R_{0\text{-}bare}$, $T_{0\text{-}bare}$, $R_{-1\text{-}bare}$, and $T_{-1\text{-}bare}$ are values of R_0, T_0, R_{-1}, T_{-1} at d=0)

Then, for example, a is 0.0866 at ω=1254cm^{-1} and 0.0814 at ω=1060cm^{-1}.

RESULTS AND DISCUSSION

Figure 2 shows the spectra of thermally grown SiO_2 layers (1000°C, 800°C) and a plasma CVD SiO_2. Thickness of both films was 100Å. Both of them have a downward peak at about 1050cm^{-1} and an upward peak at about 1250cm^{-1}. This downward peak at about 1050cm^{-1} is assigned as the resonance of the transverse optical phonon polariton (TO phonon) and the upward peak at about 1250cm^{-1} was assigned as the resonance of the longitudinal optical phonon polariton (LO phonon)[1,10,11]. The LO and the TO phonon signal of the plasma CVD oxide layer appeared at smaller wavenumber than those of the thermally grown SiO_2. This difference of peak frequency between the plasma CVD oxide and the thermally grown oxide has been reported by Koller et al.[11], and agreed with their result.

Figure 3 shows the measured spectra of a 100Å SiO_2 grown in 1000°C dry O_2 and two calculated spectra at d=100. The peak frequency of TO phonon was a little different from the calculated spectra using Ω of the thermally grown oxide (TG Ω). The shape of the LO phonon in the calculated spectra with the TG Ω agreed with that in the measured spectra fairly well. However, the TO phonon peak in the spectra calculated from the TGΩ was sharper than that in the measured spectra. The peak frequency of the TO phonon calculated from the TG Ω is also somewhat different from that of the measured TO phonon.

Figure 4 shows the dependence of the peak height of the LO and TO phonon on film thickness. Concerning to the LO phonon peak, the peak height calculated from the TG Ω as the function of film thickness increased more steeply than the measured peak height. Moreover, the peak height of the TO phonon calculated from TG Ω increased also more steeply than that of the thermally grown oxide as film thickness becomes larger. Among the measured spectra, the increase in the peaks hight of the LO and TO phonon with film thickness became speep with the oxidation temperature rise.

These differences between the spectra calculated from the TG Ω and the measured spectra show obviously that Grosse's model is not perfect. In particular, to explain the deference of the peak height change with oxidation temperature, every Ω should be a function of film thickness and oxidation temperature. Lucovsky et al. showed that the average intrinsic stress in the SiO_2 was a function of the film thickness and the oxidation temperature. Thus Ω is the function of the stress in the SiO_2. Lucovsky et al., moreover, showed that the intrinsic stress in the SiO_2 increased approximately linearly with the film thickness to about 200Å, in addition, the stress was almost saturated in the thicker oxide than 500Å [9].

Lucovsky's results suggest that the stress existed in the oxide part in the vicinity of the Si-SiO_2 interface and the thickness of such oxide part increased almost linearly. Thus we calculated IR-RAS spectra assuming a quartz layer which relaxed the stress existed between the crystal silicon and amorphous SiO_2. Then we assumed the Ω of quartz glass in Table 1 as Ω of amorphous oxide. In addition, we calculatred the dielectric function of the quartz part ε_c as $\varepsilon_c = \varepsilon_\perp \sin 60° + \varepsilon_\parallel \cos 60°$, taking account of the angle of the criystallographic axis to the electric field[12]. The spectra at d=100 calculated from this new model is also shown in Fig. 3. The peak frequencies of the LO and TO phonon is still different from the measured spectra, but somewhat broader that the calculated spectra from the TG Ω. As shown in Fig.4, however, our new model explain the dependence of the LO peak height on the film thickness and oxidation temperature. The TO phonon calculated from the new model is complicated. In the new model, TO signal was divided two peaks: 1090cm^{-1} peak and 1070cm^{-1} peak. Between the two peaks, peal height of the 1070cm^{-1} peak showed same dependence on the film thickness and the oxidation temperature as the measured TO phonon.

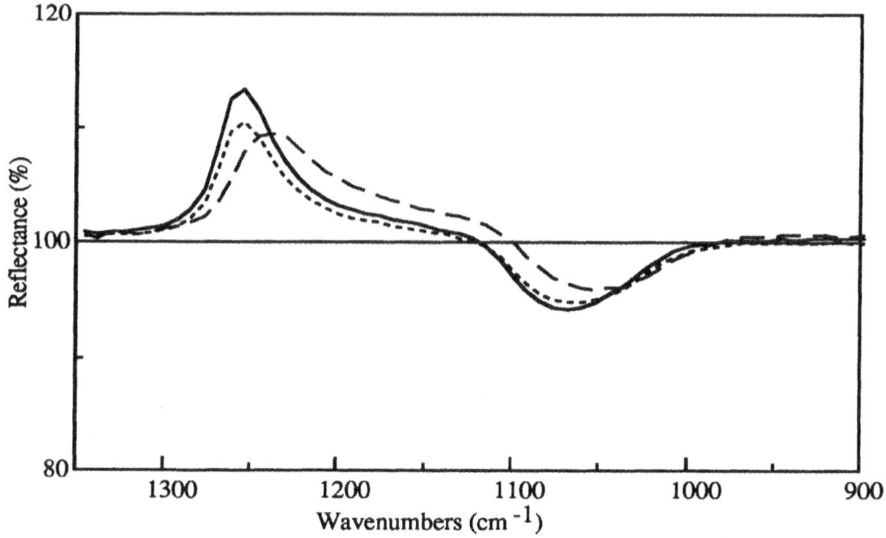

Figure 2. The IR-RAS spectra of the oxide films. Solid line is the spectra of the 1000°C thermally grown oxide of 100Å. Dotted line is the spectra of the 800°C thermally grown oxide of 100Å. Dashed line is the specta of the CVD oxide of 120Å.

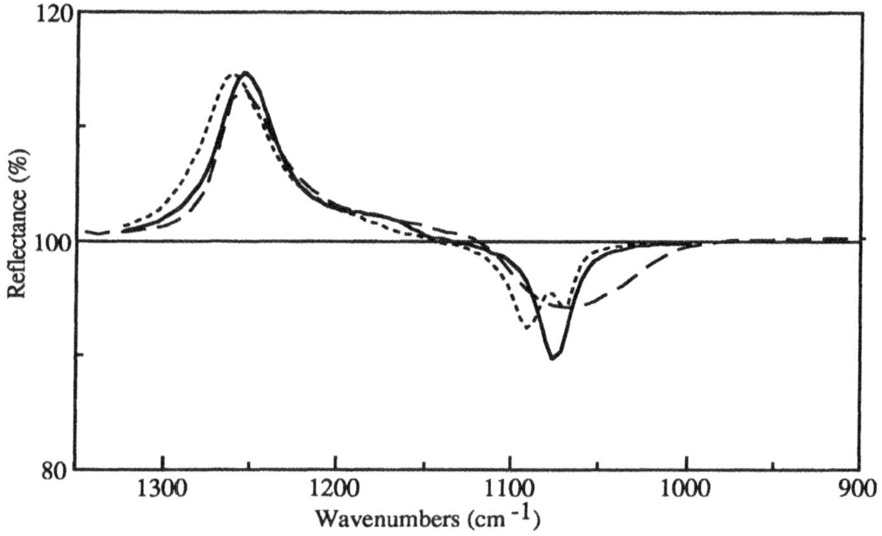

Figure 3. IR-RAS signal of the LO and TO phonon. Solid line shows the spectra calculated using TG Ω. Dotted line shows the spectra calculated the new model (20% crystral SiO_2). Dashed line shows the measured spectra.

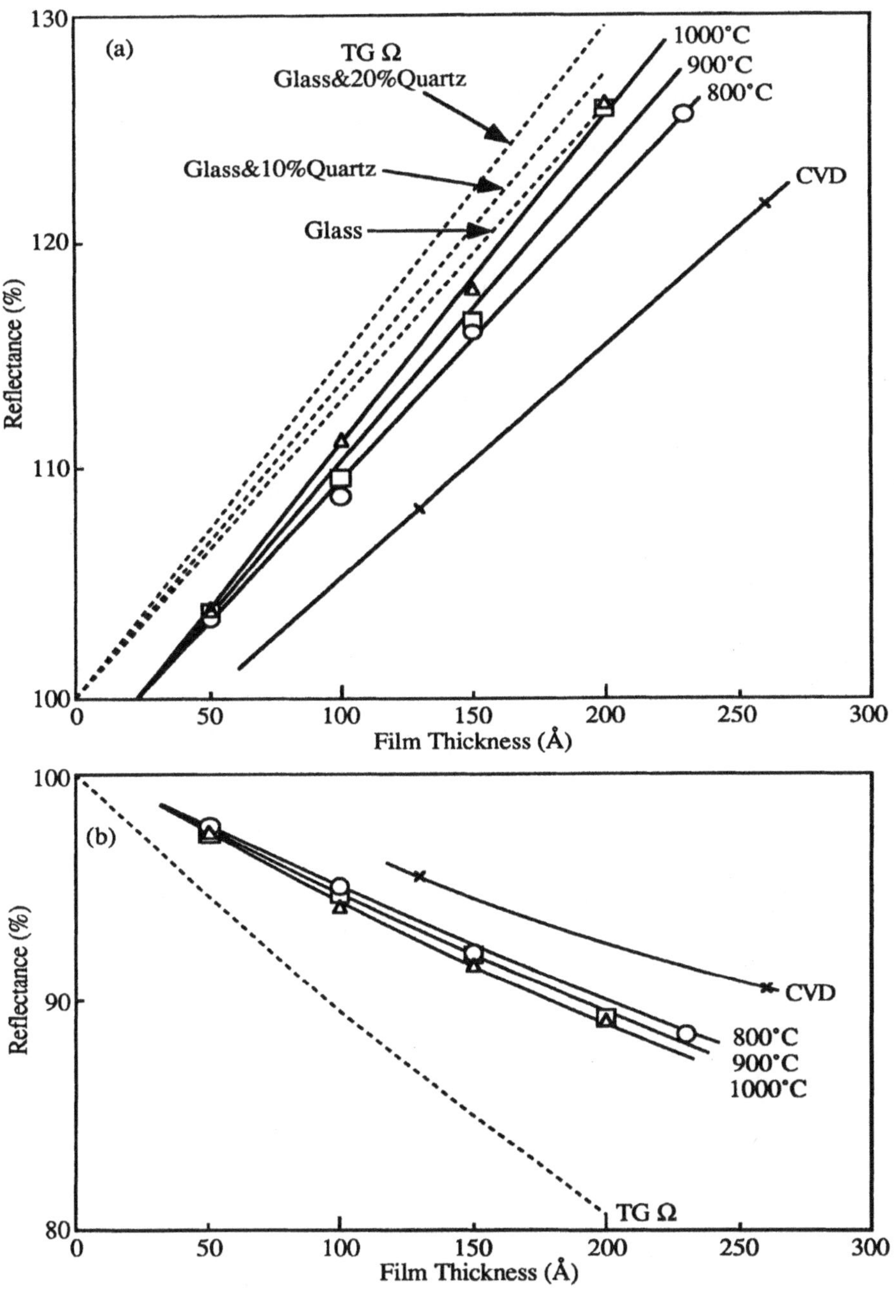

Figure 4. The dependence of the peak height of the LO and TO phonon on the SiO2 film thickness and the oxidation temperature. (a) shows the LO peak height. In this graph, the line showng the LO peak calculated from TG Ω and that from a new model including 20% crystal SiO2 overlap each other. (b) shows the TO peak height.

CONCLUSION

We observed thin SiO_2 layers on the common silicon wafers by the IR-RAS method. On the analysis of the spectra, the influence of the attenuation of IR intensity by the silicon substrate and the IR scattering at the rough back surface of the Si wafers was eliminated by comparing IR-RAS spectra between a bare silicon wafer and a gold mirror.

The IR-RAS spectra calculated using Grosse's oscillator model followed measured spectra fairly well, but it did not explain the dependence of peak height of the LO and TO phonon on the SiO_2 film thickness and the oxidation temperature. By assuming that a thin quartz layer grows between the amorphous SiO_2 and the crystal silicon substrate relaxs the stress in the SiO_2 layer, however, we can explain the dependence of peak height of the LO and TO phonon on the film thickness and the oxidation temperature.

REFERENCES

1. P.Grosse, B.Harbecke, B.Heinz, R.Meyer, and M.Offenberg, *Appl. Phys.*, A39, 257 (1986).
2. F. Rochet, S.Rigo, M.Froment, C.d'Anterroches, C.Maillot, H.Roulet, and G.Dufour, *Adv. Phys.*, 35, 237 (1986).
3. F.J.Grunthaner and P.J.Grunthaner, *Material Sci.Rep.*, 1, 65 (1986).
4. D.Graf, M.Grundner, R.Schuly, and L.Muhlhoff, *J.Appl.Phys.*, 68, 5155 (1990).
5. R.Brendel, *Appl. Phys.* A 50, 587 (1990).
6. I.E.Olsen and F.Shimura, *J. Appl. Phys.*, 66, 1353 (1989).
7. Y.Matsui, Y.Miyagawa, J.Izumitani, M.Okuyama, and Y.Hamakawa, *Jpn. J. Appl. Phys.*, 31, 369 (1992).
8. A.Ourmazd, D.W.Taylor, and J.A.Rentschler, *Phys. Rev. Lett.*, 59, 213 (1987).
9. G.Lucovsky, J.T.Fitch, E.Kobeda, and E.A.Irene, "The physics and chemistry of SiO2 and the Si-SiO2 interface" ed. by C.R.Helms and B.E.Deal, pp.139, PLENUM Press, 1988.
10. B.Harbecke, B.Heinz, and P.Grosse, *Appl. Phys.* A 38, 263 (1985).
11. K.B.Koller, W.A.Schmid, and J.E.Butler, *J. Appl. Phys.*, 64, 4704 (1988).
12. D.A.Kleinman and W.G.Spitzer, *Phys. Rev.*, 125, 16 (1962).

DECONVOLUTION OF THICKNESS-AVERAGED STRUCTURAL AND OPTICAL PROPERTIES OF THERMALLY GROWN AND RPECVD SiO_2 FILMS

C. E. Shearon Jr., C. H. Bjorkman, and G. Lucovsky

Departments of Physics, and Materials Science and Engineering
North Carolina State University
Raleigh, NC 27695-8202

ABSTRACT

This study compares the optical characteristics of SiO_2 thin films, grown by high temperature dry thermal oxidation of crystalline Si, or deposited by low-temperature remote plasma enhanced chemical vapor deposition (RPECVD), and subjected to rapid thermal annealing (RTA). Infrared (IR) spectrophotometry is used to determine the frequency of the Si-O-Si bond-stretching vibration for films of varying thickness, both before and after RTA. The thickness dependence of the stretching frequency as a function of distance from the SiO_2/Si interface is determined from analysis of these measurements. The resulting profiles combined with a previous study of the index of refraction, n, for thermally grown SiO_2 films is used to estimate the variation of n as a function of the distance from the SiO_2/Si interface.

INTRODUCTION

The present trend in dielectric research is towards thinner films grown or deposited at low-temperatures, ≤ 850°C, and therefore below the traditional oxidation and or annealing temperatures of 950°C or more. It would be desirable to utilize lower oxidation and annealing temperatures, but at the same time maintain the SiO_2 film and Si/SiO_2 interface quality that are characteristic of oxide layers processed at the higher temperatures. Previous studies have shown that the structural and optical properties of thermally grown SiO_2 films vary as a function of the film thickness, and as a function of processing

The Physics and Chemistry of SiO_2 and the Si-SiO_2 Interface 2
Edited by C.R. Helms and B.E. Deal, Plenum Press, New York, 1993

temperatures, including both growth and post-oxidation annealing.[2-15] This study is a continuation of the previous ones, focusing on the frequency of the Si-O-Si bond-stretching vibration, ν_s, which provides a measure of the bond angle at the oxygen atom sites, and the index of refraction, n, which provides information relative to the density, and the atomic polarizabilities of the Si and O-atoms. The study has been expanded to include SiO_2 films deposited by Remote Plasma-Enhanced Chemical-Vapor Deposition (RPECVD). The IR optical properties were examined by etching the films back in 100 Å increments, and measuring the IR absorption spectrum at each thickness. A method of deconvolution [11] was applied to the thickness dependent data, initially to confirm the results of the earlier studies, and then to provide a more continuous range of thicknesses for the SiO_2 films, complementing the thickness-averaged values that were previously measured. In addition to the IR measurements, the thickness dependence of the index of refraction was estimated by combining the present IR studies with the results of previous research, [1] which demonstrated a scaling relationship between ν_s and n.

EXPERIMENTAL PROCEDURES

The SiO_2 films were grown on device-quality p-type (100) oriented silicon wafers with resistivities in the range of 10-30 Ω-cm. The samples used for the IR spectroscopy measurements were etched in a CP-4A solution to reduce scattering losses from the unpolished backside of the wafers. The thermal oxidation runs were performed in a dry oxygen (H_2O < 1ppm) ambient at atmospheric pressure and at temperatures of 850°C and 950°C. The oxidation times were adjusted to give a final oxide thickness of approximately 1000 Å at each of these temperatures. In addition, SiO_2 films of approximately the same thickness were deposited using RPECVD at 300°C. Half of the samples were then subjected to a rapid thermal anneal (RTA) at 1100 °C for 100 seconds using an AG model 410 Heatpulse in an argon ambient to prevent any additional oxidation. Following this, all samples were wet-chemically etched back in 100Å increments using a buffered oxide etch (BOE); this provided films ranging in thickness from approximately 900 Å to 100 Å. IR spectroscopy measurements, were then performed at each thickness between 1000 Å and 100 Å, using a Perkin Elmer 983 Spectrophotometer. The spectral peak frequency of the dominant bond-stretching mode at ~1075 cm^{-1}, ν_s, was determined from several averaged spectra to an uncertainty of $\pm 1 cm^{-1}$. We have previously shown that this frequency provides a convenient measure of the Si-O-Si bond angle, 2θ, i.e., $\nu_s = \nu_o \sin\theta$, where ν_o is a constant, determined from the properties of a thermally-relaxed oxide film. ν_s also provides a convenient measure of the local, and the macroscopic strain in the SiO_2 film.[9-11]

EXPERIMENTAL RESULTS

The results of the IR measurements described above are shown in Figs. 1-9. Figure 1 displays the frequency of the IR active Si-O bond stretching vibration, ν_s, as a function of the oxide thickness for films grown at 850°C. The as-measured curve of Fig. 1 yields a thickness-averaged bond stretching frequency. The difference-curve illustrates how the IR spectroscopic data can be analyzed to yield the average stretching frequency in 100Å layers by determining frequency differences between successive spectra of the etched-back films, i.e., $a(\delta) = a(t+\delta) - a(t)$, where a is the absorbance, t is the remaining film thickness and δ is the amount of oxide removed by etching, ~100Å per step. The dotted curve is a deconvolution of the thickness-averaged values using the self-consistent deconvolution method described in Ref. 11. Since the thickness dependent data reported in this paper are all averages over the film thickness at which they were measured, to deconvolve the data we divided the SiO_2 film into 50 Å thick incremental layers and calculated the bond stretching frequency for each individual layer by requiring that the average stretching frequency for the first n layers is equal to the thickness-averaged value at n times 50Å. Figure 2 shows a similar analysis for films grown at 950°C. In Figs. 1 and 2, we have established that the deconvolution methods of Ref. 11 is consistent with incremental measurements of the IR bond-stretching frequency. The combination of these data show that the bond stretching frequency decreases as the distance between the incremental layer of oxide and the Si/SiO_2 interface decreases, indicating a highly strained SiO_2 region at that interface.

In addition, films annealed at 1100°C have higher bond-stretching frequencies than the as-grown films throughout any thickness range explored (compare data in Figs. 3 and 4, and 5 and 6). These data then indicate increased bulk relaxation in the annealed films with respect to the films grown at the lower oxidation temperatures of 850°C and 950°C. In all of the data presented below, we then present the IR data, and the analysis of that data using the deconvolution procedure. Figs. 3 and 4 present data, respectively, for a film grown at 850°C, and grown at 850°C, and then subjected to an 1100°C RTA for 100s.

Figs. 5 and 6 present similar data for a film grown at 950°C. The following observations apply: i) prior to the RTA, the vibration frequencies were higher at all oxide thicknesses in the film grown at 950°C; ii) after the RTA, the frequencies at every oxide thickness in both films increased, with corresponding increases in the strain gradients as the distance to the SiO_2/Si interface decreased; and iii) the vibrational frequencies as a function of film thickness in the films grown at 850°C and 950°C after RTA were about the same. This means that an RTA at 1100°C relaxed film stress to the same degree.

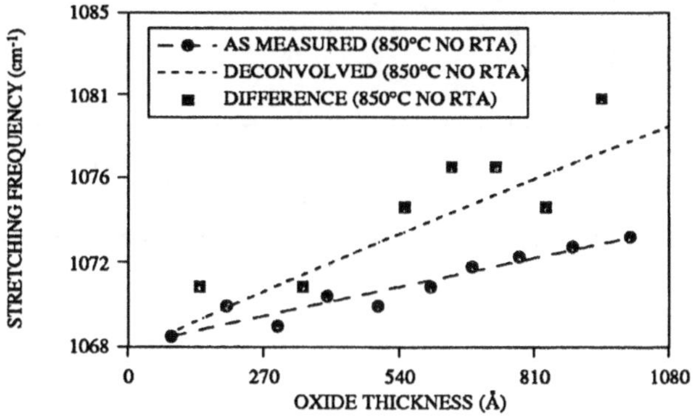

Figure 1. Bond stretching frequency as a function of SiO_2 thickness for films grown at 850°C with no RTA.

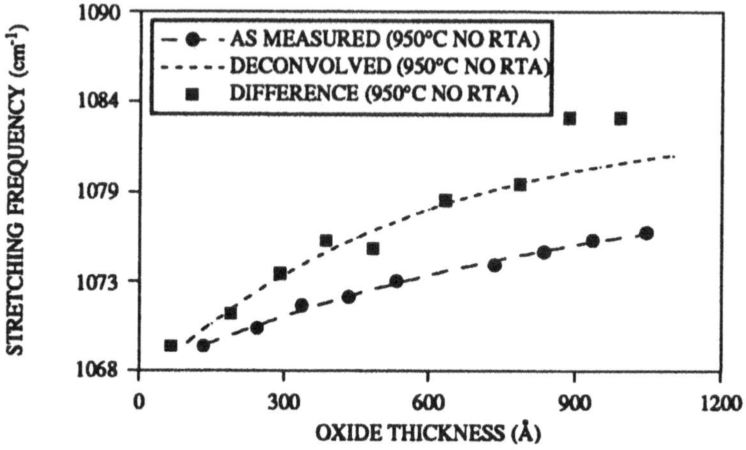

Figure 2. Bond stretching frequency as a function of SiO_2 thickness for films grown at 950°C with no RTA.

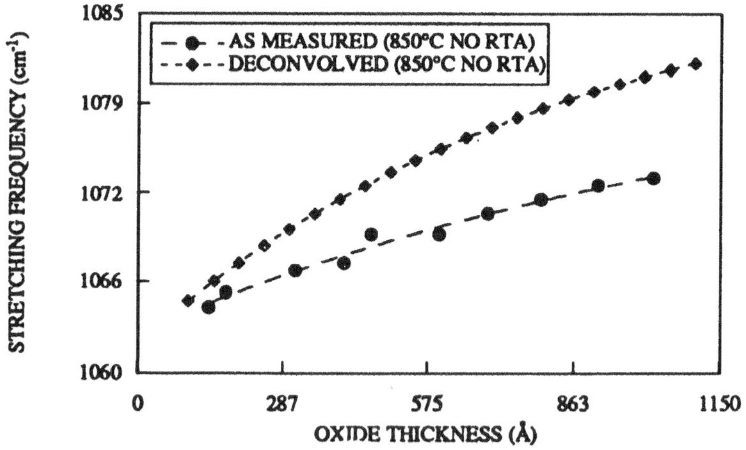

Figure 3. Bond stretching frequency as a function of SiO_2 thickness for films grown at 850°C with no RTA.

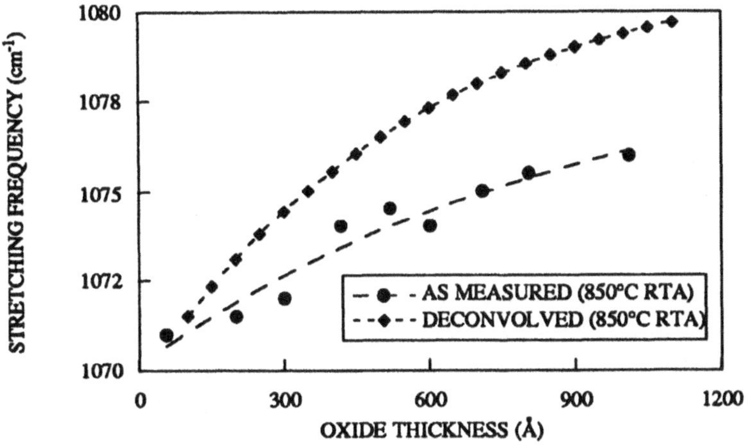

Figure 4. Bond stretching frequency as a function of SiO_2 thickness for films grown at 850°C with an RTA.

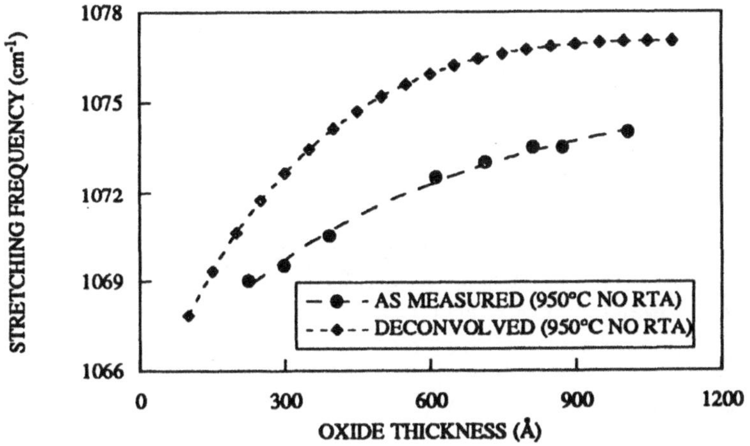

Figure 5. Bond stretching frequency as a function of SiO_2 thickness for films grown at 950°C with no RTA.

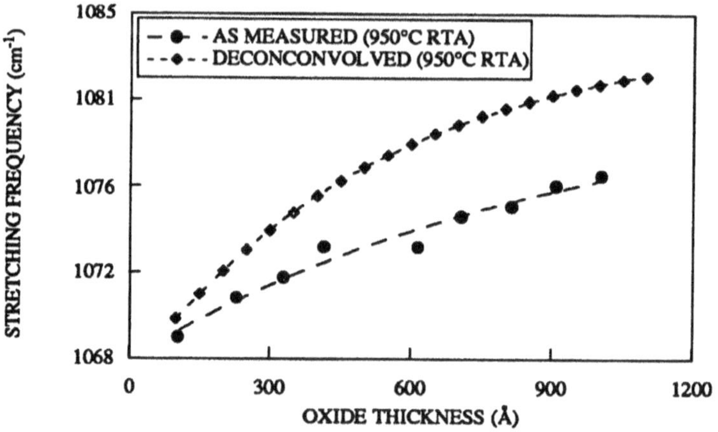

Figure 6. Bond stretching frequency as a function of SiO_2 thickness for films grown at 950°C with an RTA.

The IR data in Fig. 7 are for a 1000Å SiO_2 film deposited at 300°C by RPECVD. Note that in RPECVD, the film is deposited onto the Si substrate, and the substrate is not consumed to generate the oxide. The high levels of strain in thermally grown oxides in the immediate vicinity of the SiO_2/Si interface result from consumption of the Si to generate the oxide,[5] and reflect the molar volume mismatch between the Si substrate and the grown oxide. The data in Fig. 7 indicate relatively small changes in ν_s as a function of oxide thickness, ~1 to 2 cm^{-1} at most over a thickness range of 1000 Å.
Since these variations of ν_s fall within the same range as the experimental uncertainty, ±1 cm^{-1}, in the determination of ν_s, we did not attempt to deconvolve these data. Figure 8 presents ν_s as a function of film thickness for the RPECVD film after the RTA at 1100°C for 100 s. The deconvolved profile is essentially the same as that of the thermal oxides subjected to the same RTA. This is demonstrated in Fig. 9 which compares both experimental data, and deconvolution profiles for the RPECVD film subjected to RTA, and for the oxide film grown at 850°C, also subjected to RTA.

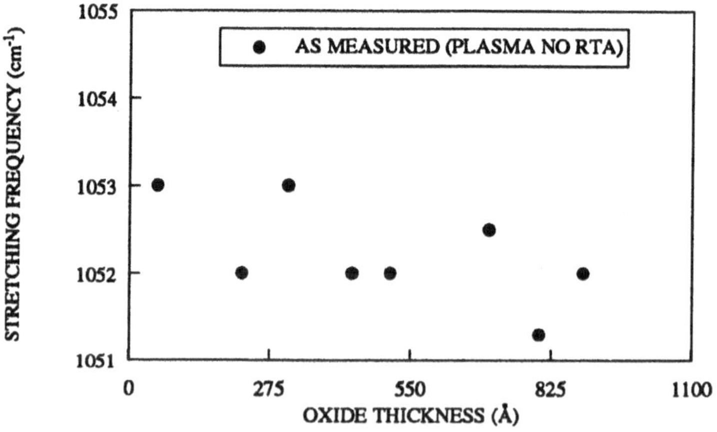

Figure 7. Bond stretching frequency as a function of SiO_2 thickness for plasma grown films with no RTA.

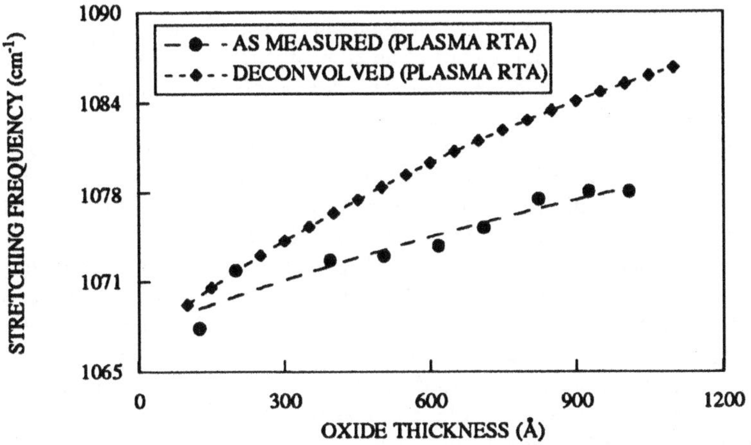

Figure 8. Bond stretching frequency as a function of SiO_2 thickness for plasma grown films with an RTA.

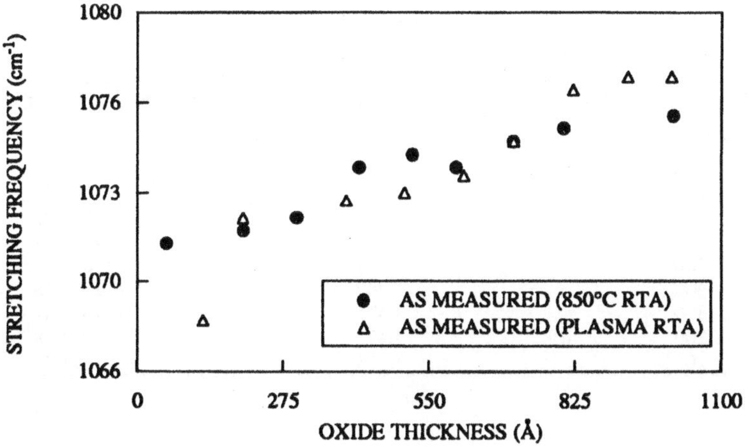

Figure 9. Bond stretching frequency as a function of SiO_2 thickness for plasma grown films and films grown at 850°C both with an RTA.

In addition, in a previous study, a relationship between ν_s and the index of refraction, n was established [1]

$$\frac{n^2 eff\,(1)}{n^2 eff\,(2)} = \frac{\nu_s\,(2)^{3/2}}{\nu_s\,(1)^{3/2}} \qquad (1)$$

The studies performed in Ref. 1 focused on how the stretching frequency and the index of refraction varied as a function of growth temperature and oxygen pressure. The oxide in this study was grown to about one-half of an ellipsometric period to obtain reliable values for the refractive index. The uncertainties for index of refraction were ±0.003. The studies presented in Ref. 1 demonstrated that the index of refraction varied linearly with the stretching frequency for thermally grown oxides over a temperature range from about 700 to 1100°C. For the restricted range of values of ν_s and n, explored in Ref. 1, this linear variation is consistent with the result of the model calculation, i.e., Eqn. (1). From what we have discussed above these measurements should now be interpreted as being representative of thickness-averaged values of ν_s and n. For these measurements, the index of refraction ranged from 1.465 to 1.48, and the average vibration frequency varied between about 1080 cm^{-1} and 1065 cm^{-1}. This range of ν_s is about the same as we find for the variation of n with position from the SiO_2/Si interface, see Figs. 1-6 and 8 and 9. If we assume that the empirical relationship between ν_s and n presented in Ref. 1, and validated in terms of a model calculation presented in that paper, applies to thickness-dependent as well as thickness-averaged data, we can then calculate a variation of n with position from the SiO_2/Si interface from the IR data. This is correlation is shown in Table 1.

Table 1. Deconvolved stretching frequencies of 850°C sample with a rapid thermal anneal and estimated index of refraction data.

Oxide Thickness (Å)	Index of refraction (632.8 nm)	IR stretching frequency (cm^{-1})
	(±0.003)	(±1)
1000	1.464	1079.39
900	1.464	1079.01
800	1.465	1078.55
700	1.466	1077.99
600	1.467	1077.31
500	1.468	1076.51
400	1.469	1075.55
300	1.471	1074.42
200	1.473	1073.08
100	1.475	1071.51

DISCUSSION

As can be seen from the data in Figs. 1 and 2, the stretching frequencies obtained by deconvolving the thickness-averaged data are in very good agreement with the stretching frequencies for the 100 Å layers obtained by analyzing the differences in IR frequencies between successive spectra for films of different thicknesses. The same method of deconvolution has previously been used to obtain stress profiles for the SiO_2 films [11] and we infer that it can be used for any thickness-averaged property, even though IR spectroscopy is the only experimental technique which allows us to readily confirm the validity of the deconvolution method by a direct difference measurement. It should be noted that, the deconvolution of the stress data, obtained by measuring wafer deformation by optical techniques, is consistent with strain profiles that have been generated from the deconvolved IR data. As a continuation of this work we have used this deconvolution method to obtain additional vibrational frequency profiles, as well as predicted variations of the index of refraction based on the measured stretching frequency and index of refraction in Ref. 1.

For the as-grown thermal oxides, and the etched-back thermal and RPECVD oxides subjected to a RTA, there is a relatively steep gradient in ν_s in the vicinity of the SiO_2/Si interface. Coupled with the data presented in Ref. 1, this predicts an increase in the index of refraction, n, in the same region. In both cases, the increase in n, and the decrease in ν_s, is due to increased compressive strain in the oxide in the vicinity of the SiO_2/Si interface. In Ref. 15 it was shown that when SiO_2 films are etched-back the BOE has a tendency to enlarge pores and voids on the surface of the oxide and lower the overall index of refraction. Since the films on which n was measured were not etched back, this has not affected the results shown here.

We have used equation (1) relating ν_s and n, along with the IR data, determined from the studies presented in this paper, to estimate the variation of the index of refraction with film thickness.[1] We have then compared this estimated variation of n as a function of distance from the SiO_2/Si interface with calculations of n, as a function of stress and film thickness, and with recent ellipsometric measurements. [13 - 15] In Ref. 13 a first order compressibility relationship calculating the index of refraction based on a stress gradient in the film, shown here by Eqn. (2), was developed.

$$n = n_0 + \frac{\Delta n}{\Delta \sigma} * \sigma \qquad (2)$$

The profile and the magnitude of the increase of the index of refraction is comparable with the estimated values we found from our data. In Ref. 13 there was an increase of approximately 0.96 percent in n at the vicinity of the SiO_2/Si interface as compared to the value calculated at 1000Å for a film grown at 900°C. From our data presented in Table 1, we calculated a 0.75 percent increase, at 100Å, in the index of refraction from that which was calculated at 1000Å. However, experimental results of index of refraction measurements made by ellipsometry found an increase of 14 percent, at 100Å, from that which was measured at 1000Å.[14] This difference between calculated and/or estimated values of index of refraction and actual ellipsometric measurements identifies a need for additional studies.

CONCLUSIONS

In conclusion, we have studied the IR frequency, ν_s, as a function of distance from the SiO_2/Si interface. These measurements indicated considerable strain gradients in: i) as-grown oxides; ii) both as-grown oxide and RPECVD oxides subjected to high temperature

RTA, 1100°C at 100s; but iii) no evidence for significant strain gradients in as-deposited RPECVD films. Moreover, the variation of ν_s with distance from the SiO_2/Si interface was essentially the same in thermal and RPECVD oxides subjected to high temperature RTA's. These properties were examined by incrementally etching back the SiO_2 films and performing IR spectroscopy measurements at each thickness. The thickness-averaged values measured were then deconvolved to provide continuous values of the frequency of the dominant bond stretching mode for thin layers of SiO_2, as a function of the distance from the SiO_2/Si interface. We have experimentally verified the deconvolution method by comparing it to determination of the bond stretching frequency obtained from difference measurements using IR spectra from successively etched back films. Finally, we have estimated the variation of n with position relative to the SiO_2/Si interface from the experimentally determined variation of ν_s. We have found that the changes in n estimated in this way are much smaller than what has been measured using ellipsometry[14] but are similar to what has been predicted in Ref 13.

ACKNOWLEDGMENTS

This work in part supported by ONR and the NC State University NSF Engineering Research Center for Advanced Electronic Materials Processing.

REFERENCES

1. G. Lucovsky, M. J. Mantini, J. K. Srivastava, E. A. Irene, J. Vac. Sci. Technol. **B5**, 530 (1987).
2. R.J. Jaccodine and W. A. Schlegel, J. Appl. Phys. **37**, 2429 (1966).
3. E. P. EerNisse, Appl. Phys. Lett. **30**, 290 (1977); **35**, 8 (1979).
4. A. K. Sinha, H.J. Levinstein, and T. E. Smith, J. Appl. Phys. **49**, 2433 (1978).
5. E. A. Irene, E. Tierney, and J. Angilello, J. Electrochem. Soc. **129**, 2594 (1982).
6. E. Kobeda and E. A. Irene, J. Vac. Sci. Technol. **B4**, 720 (1986); **B5**, 15 (1987); **B6**, 574 (1988).
7. J. T. Fitch and G. Lucovsky, MRS Symp. Proc. **92**, 89 (1987); MRS Symp. Proc. **105**, 151 (1988); AIP Conf. Proc. **167**, 124 (1988).
8. G. Lucovsky, J. T. Fitch, E. Kobeda, and E. A. Irene, in "The Physics and Chemistry of SiO_2 and the Si-SiO_2 Interface", ed. by C. R. Helms and B. E. Deal (Plenum Press, New York, 1988), p. 139.
9. J. T. Fitch, E. Kobeda, G. Lucovsky, and E. A. Irene, J. Vac. Sci. Technol. **B7**, 153 (1989).
10. J. T. Fitch, C. H. Bjorkman, J. J. Sumakeris, and G. Lucovsky, MRS Symp. Proc. **130**, 289 (1988).
11. C. H. Bjorkman, J. T. Fitch, and G. Lucovsky, MRS Symp. Proc. 146, 197 (1989).
12. L. M. Mack, A. Reisman, P. K. Bhattacharya, J. Electrochem. Soc. **136**, 3433 (1989).
13. A. Fargeix and G. Ghibaudo, J. Phys. D: Appl. Phys., **17**, 2331 (1984).
14. A. Kalnitsky, S. P. Tay, J. P. Ellul, S. Chongsawangvirod, J. W. Andrews, and E. A. Irene, J. Electrochem. Soc., **137**, 234 (1990).
15. S. Chongsawangvirod, E. A. Irene, A. Kalnitsky, S. P. Tay, and J. P. Ellul, J. Electrochem. Soc., **137**, 3536 (1990).

TEM INVESTIGATIONS OF THE OXIDATION KINETICS OF AMORPHOUS SILICON FILMS

Manfred Reiche

Max-Planck-Institut für Mikrostrukturphysik
Weinberg 2
D - 0 4050 Halle/S., Germany

INTRODUCTION

Stacked layers of polycrystalline silicon (polysilicon) are widely applied to VLSI-processing, such as dynamic RAM´s, EPROM´s, etc. In these structures, the polysilicon layers are isolated from each other by thin oxide (SiO_2) films. In general, the oxide is thermally grown on the bottom polysilicon layer.

The kinetics of the oxidation of thin polysilicion layers has been investigated under different conditions such as oxidation temperature and time, respectively, doping, and atmosphere (Saraswat and Singh, 1982; Sze,1985; Deal, 1988; Wang et al., 1991). All the investigations show that the oxidation kinetics is complex, differing from that of single-crystalline silicon. It was assumed that there is an intergranular oxidation at the grain boundaries of the polysilicon layer. In addition, the oxidation at grain boundaries is also the reason of the higher roughness of the interface making it delicate to produce ultra-thin oxide layers on polysilicon. On the other hand, applying amorphous deposited silicon layers (a-Si) instead of polycrystalline deposited ones results in more homogeneous SiO_2 films even at thicknesses lower than 20 nm. Furthermore, their interface roughness is much lower, yielding better values of the breakdown characteristics (Hendriks and Mavero, 1991). Both the homogeneity and the electrical properties of the oxide films refer to differences in the oxidation process of polycrystalline and amorphous silicon films.

The present paper deals with the oxidation of a-Si thin films. Differences are discussed as to polycrystalline deposited ones, especially with respect to the oxidation kinetics.

EXPERIMENTAL

P- (boron-doped) and n-type (phosphorus-doped) wafers (diameter 4 in., {100} and {111} orientation, respectively) were applied. Oxide and silicon layers were prepared on them by the following procedures:

1) A 100 nm thick SiO_2 layer was produced by oxidation at 960°C (dry oxygen; 3% HCl was added).

The Physics and Chemistry of SiO_2 and the Si-SiO_2 Interface 2
Edited by C.R. Helms and B.E. Deal, Plenum Press, New York, 1993

2) Silicon layers having thicknesses of 300 to 400 nm were prepared on it by LPCVD. Amorphous, semicrystalline, and crystalline layers were obtained by varying the deposition temperature between 560 and 625°C.
3) The layers were doped with phosphorus by ion implantation using doses of $1 \cdot 10^{15}$, $5 \cdot 10^{15}$, and $1 \cdot 10^{16}$ cm^{-2}, respectively. Undoped films were regarded, too.
4) The silicon layers were oxidized in dry oxygen (3% HCl was added) at 960°C. In order to measure the kinetics of the process, the oxidation time was varied between 5 and 100 minutes.
5) The grown oxide films were protected by additional polysilicon layers (thickness 400 nm). They were deposited by LPCVD at 625°C.

Plan-view and cross-sectional samples were prepared to be investigated in a high-voltage transmission electron microscope (HVEM, accelerating voltage 1 MV). Film thickness d and roughness Δd of the individual layers were measured directly from electron microscope images on the basis of a computer-aided interpretation of TEM images. Film thicknesses, reported here, are mean values of 300 or more individual measurements. The error of the thickness measurements was calculated to be lower than 7% (d > 50 nm), and 10 - 15% for d < 50 nm, respectively. The roughness Δd was defined to be the sum of the differences of individual thickness measurements d_i and the mean thickness d_m, divided by the number of measurements n, i.e.

$$\Delta d = \frac{1}{n} \cdot \sum_{i=1}^{n} (d_i - d_m) \quad . \tag{1}$$

The results of the measurements of d and Δd have been compared to spectroscopic ellipsometry for several samples. The differences between TEM and ellipsometric measurements range from 0.8 to 10% (for the oxide film grown on the silicon layer). They were strongly affected by the interface roughness. The differences were smallest for smooth interfaces (amorphous deposited silicon layers), increasing, however, with increasing roughness.

RESULTS

Crystallization Process of the Amorphous Silicon Layers

The amorphous silicon layers crystallize during thermal oxidation. This process influences the oxidation. Therefore it is important to know the crystallization velocity and the mechanism of the amorphous/crystalline phase transition.

TEM *in situ* experiments on plan-view and cross-sectional samples have been performed to analyze the crystallization process of a-Si layers used here (Reiche and Hopfe, 1990; Reiche, 1991). Comparisons have been made to furnace annealed samples, too. The investigations proved that surface-induced crystallization is the dominating process. Here, interfacial stresses are the main reason for nucleation. The crystallization starts at the interface to the subjacent SiO_2 layer, i. e. on the side opposite to which the oxidation process begins (Figure 1). Furthermore, a needle-like shape of the grains occurs (needle axes <111>). Twins, which can frequently be observed in the grains, were generated during further stages of growth (coalescence). Moreover, statistical measurements of the grain size have shown that the crystallization of the films is characterized by a $t^{1/2}$ dependence which is independent of the doping and of the type of the subjacent layer (SiO_2 and Si_3N_4 layers, respectively, were regarded).

Dynamics of the Oxidation

The oxidation kinetics of amorphous deposited silicon layers was studied for dry oxidation at 960°C and for oxidation times up to 100 minutes. The results were compared to polycrystalline layers.

Figure 2 shows typical TEM cross-sectional images of the oxidation sequences of undoped and heavily doped silicon layers, respectively. Even after short times thin homogeneous SiO_2 layers have grown. Their thickness increases continuously with increasing time. For instance, oxidation of an undoped a-Si layer for 5 minutes results in a thickness of 13.2 nm of the oxide film. For medium-doped layers, the oxide thickness is nearly equal (d_{ox} = 12.6nm), whereas higher values were measured for heavily doped a-Si layers (d_{ox} = 15.5 nm). The differences in the thickness of the oxide films grown on undoped and medium-doped a-Si, on the one hand, and on heavily doped layers, on the other, increase with increasing oxidation time. They amount to about 40 nm after an oxidation time of 100 minutes (see Fig. 3).

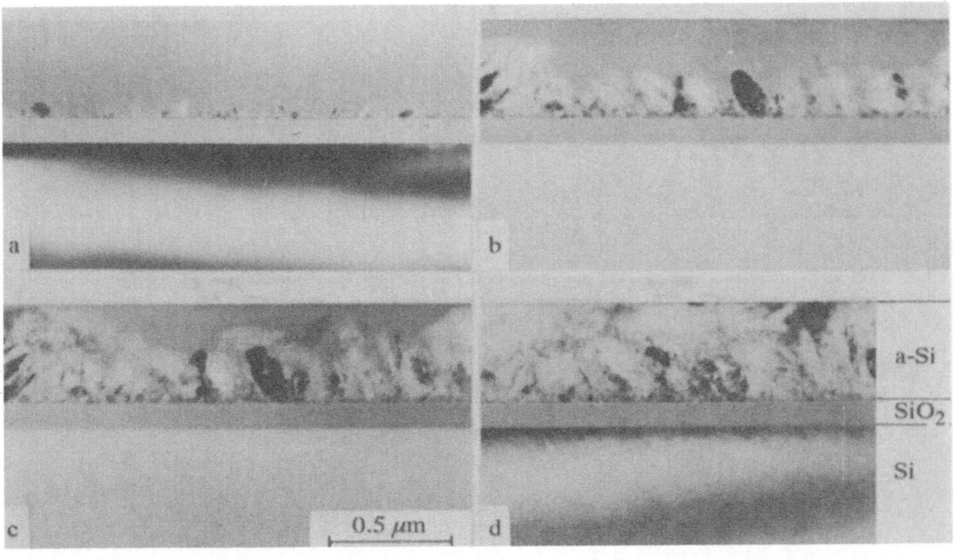

Figure 1. ***In situ*** **crystallization of an amorphous, undoped Si layer deposited on a 100nm thick oxide film.** ***As deposited*** **at 560°C (a), and after annealing at 690°C for 9 (b), 12 (c), and 60 minutes (d). Note that the micrographs show the same object region.**

The differences of the thicknesses of the grown oxide films induced by different doping of the silicon layer are found to be most significant. On the other hand, there is no strong influence of the deposition temperature on the Si layers for $d_{ox} > 30$ nm (assuming an oxidation temperature of 960°C for these experiments). In other words, the oxidation kinetics should not differ between amorphous and polycrystalline deposited silicon layers after reaching oxide thicknesses of about 30 nm. The differences found for lower values of d_{ox}, however, refer to different kinetics induced by different oxidation mechanisms. In contrast to the homogeneous oxide films grown on amorphous deposited Si, on poly-crystalline deposited ones only SiO_2 islands are detected at the first stages (Fig. 4). It is important to note that the lateral dimensions of the islands are up to 130 nm, i.e. they are significantly larger than the mean grain size (by more than a factor of 2). Moreover, first continuous films occurring after oxidation of heavily doped Si layers at 960°C for about 5 minutes show thicknesses of more than 20% lower than for amorphous deposited silicon (Fig. 3). In addition, there are great differences also in the roughnesses measured of the interfaces.

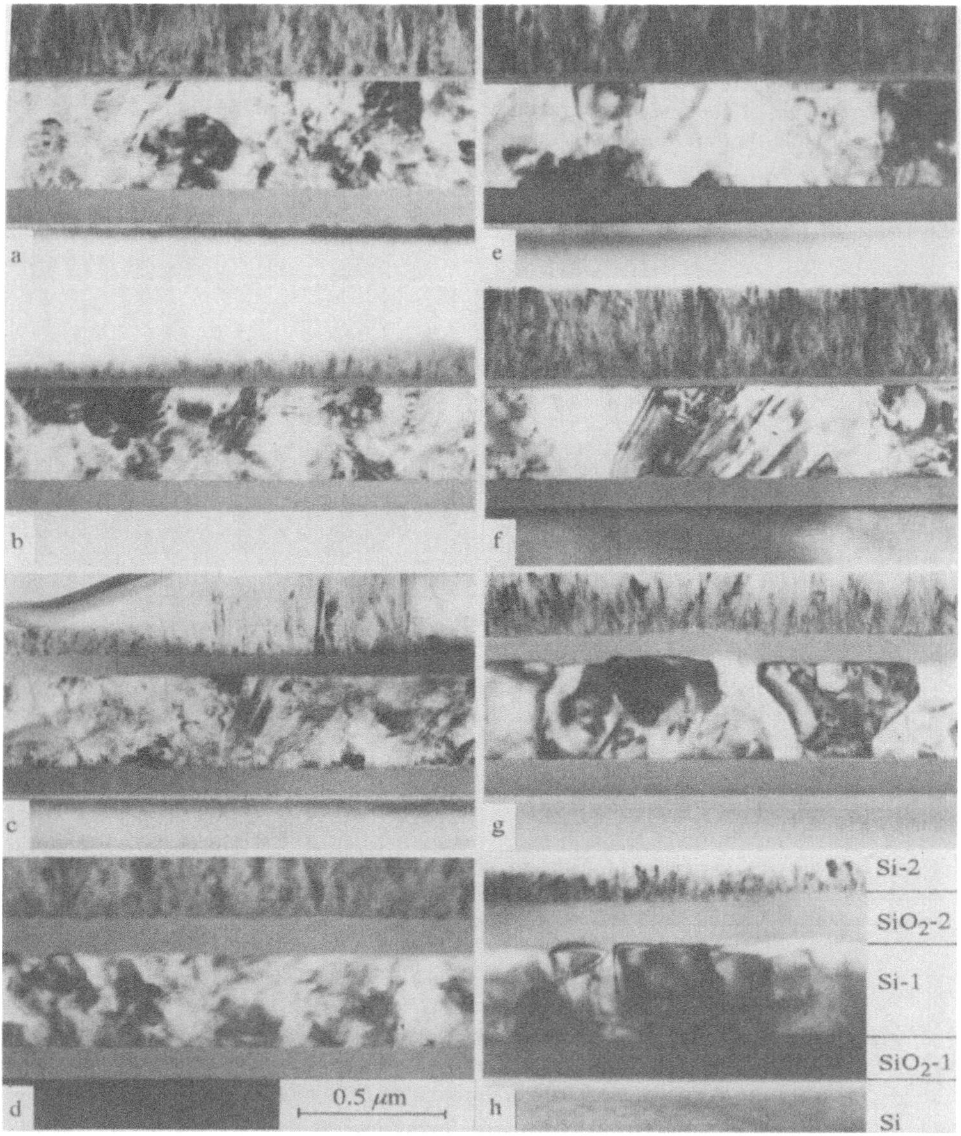

Figure 2. TEM cross-sectional images of undoped (a - d) and heavily doped silicon layers (e - f) after oxidation times of 5 (a,e), 10 (b,f), 40 (c,g), and 100 minutes (d,h). Amorphous deposited layers are used. SiO_2-2 denotes the oxide film grown on the silicon layer Si-1.

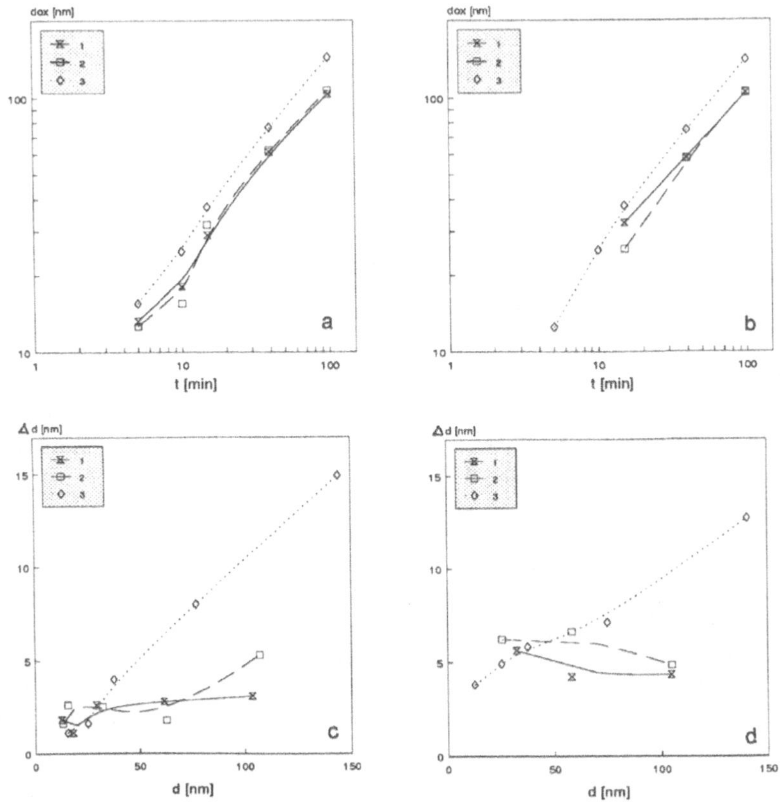

Figure 3. **Dependence of the oxide thickness d_{ox} (a,b) and the interface roughness Δd (c, d), resp., on the oxidation time. (a,c) Amorphous deposited silicon layers. (b,d) Polycrystalline deposited layers. Data for undoped (1), medium-doped (2) and heavily doped samples (3) are shown.**

Besides the dependence of the oxidation kinetics on the doping of the Si layers and the deposition temperature, also the influence has been investigated of additional postannealing after ion implantation (drive-in annealing). The subsequent oxidation process was shown not to vary (with respect to the thickness and the roughness of the grown SiO_2 film) by applying amorphous deposited layers, whereas the use of polycrystalline deposited ones effected an increasing roughness. This is not only true of furnace annealing but also of RTA processes between 1000 and 1200°C. As it is proved by TEM analysis, the reason is the full maintenance of the primary interface flatness by the formation of large grains during crystallization after drive-in annealing. Size distribution measurements of the grains reveal only small deviations from the mean size. On the other hand, recrystallization of the polycrystalline deposited layers during this process results in an increasing interface roughness associated with a larger scattering of the grain sizes in the polycrystalline deposited silicon layer.

Interface Roughness

Increasing the oxidation time increases not only the thickness of the growing SiO_2 film but also varies their roughness, which is defined by eq. (1). Therefore, analogous to

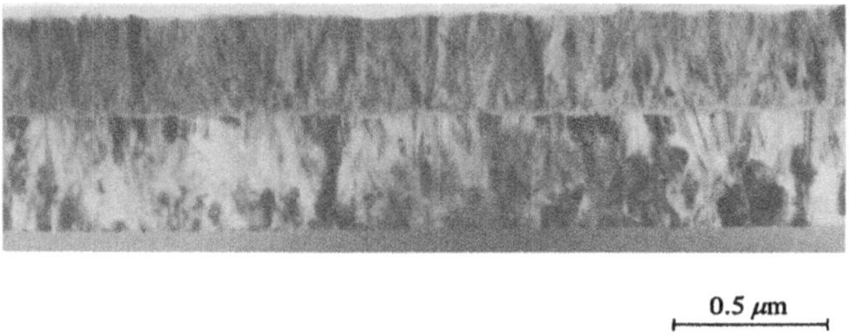

Figure 4. Oxidation of a polycrystalline deposited layer for short times (5 minutes) does not result in a closed SiO_2 film (cf. Fig. 2 for the same oxidation of an amorphous deposited Si layer).

the dependence of d_{ox}, the doping of the silicon layer is the most important factor influencing the roughness of the interface to the oxide film (Fig. 3). Applying heavily doped Si-layers causes the interfacial roughness Δd to continuously increase with increasing oxide thickness. This correlation is described by

$$\Delta d \sim 0.1 \cdot d_{ox} \quad [\text{nm}] \quad . \tag{2}$$

On the other hand, Δd is nearly constant or increases only slightly by oxidation of undoped and medium-doped a-Si layers, respectively (Fig. 3b). For instance, for the oxidation of undoped silicon layers, Δd values of about 3 nm were measured for oxide thicknesses up to 100nm. They are about 5 times smaller than those observed for oxide layers grown on heavily doped a-Si (here a value of Δd = 14.9 nm was measured for d_{ox} = 100nm). An analogous correlation between Δd and d_{ox} as a function of the doping was also proved for oxide films grown on polycrystalline deposited Si (Fig. 3). There are, however, higher absolute values of Δd even for medium- and undoped layers, respectively. The differences are larger by a factor of 5 especially for very thin oxide films (thicknesses smaller than 20 nm). In addition, increasing the oxidation time results in a slight decrease of Δd which is probably caused by the transition to the diffusion-controlled oxidation and the reduction of stress (Carim and Sinclair, 1987).

Furthermore, there also is a correlation between Δd and the grain size in the crystallized silicon films. The experiments revealed that

$$\Delta d \sim 0.28 \cdot D_{Si} - 12 \quad [\text{nm}] \tag{3}$$

where D_{Si} denotes the grain diameter. There is no anaologous correlation for polycrystalline deposited silicon.

DISCUSSION AND CONCLUSIONS

The investigations have proved a nearly linear dependence between the thickness of the oxide layer and the oxidation time (cf. Fig. 3). This is observed not only for the oxidation of amorphous deposited silicon layers but also for polycrystalline deposited ones. Based on this result, the oxidation kinetics is described by applying the model of Deal and Grove (1965), which was evolved for the oxidation of single-crystalline material. In this model, for the linear growth region (i.e. short times) the dependence of d_{ox} on the oxidation time t is given by

$$d_{ox} = \frac{B}{A} \cdot (t + \tau) \tag{4}$$

where B/A denotes the linear rate constant, and τ a time constant. For the calculations, values of B/A of $7.2 \cdot 10^{-2}$ μmhr^{-1} (for {100} oriented Si) and $1.0 \cdot 10^{-1}$ μmhr^{-1} (for {111} oriented Si), resp., were used, which were given by Sze (1985) for dry oxidation (O_2 - 3%HCl mixtures). An initial value of d_{ox} = 1.5 μm (native oxide) was also assumed. Hence it follows that the oxidation rate of undoped and medium-doped silicon layers is higher than for {100} oriented Si. For oxide thicknesses of more than 15 nm, however, the oxidation rate is equivalent to {111} oriented material. This is fulfilled for amorphous and polycrystalline deposited layers. Furthermore, at a characteristic time $A^2/4B$ = 24 min., corresponding to d_{ox} = 42.8 nm, the oxidation changes from the linear region into the parabolic one.

Of heavily doped silicon layers the oxidation is faster than of undoped and medium-doped ones, respectively. By modifying B/A, the experimental data can be fitted to the model of Deal and Grove, too. From the calculations it follows that the correlation is best for B/A = $1.3 \cdot 10^{-1}$ μmhr^{-1} and for the thickness of 2 μm of the native oxide (Fig. 5). This correlation ensues, however, only for oxidation times up to about 30 minutes. At prolonged times, the transition into the parabolic growth region causes oxide thicknesses lower than expected from the linear growth model. This is true of the oxidation of both amorphous and polycrystalline deposited layers. On the other hand, at very short oxidation times ($t <$ 10 min), the linear growth model can also be applied to the polycrystalline layers, whereas the higher oxidation rate of amorphous deposited layers, however, results in deviations from the calculated thickness (Fig. 5). They are caused by the different oxidation mechanisms of both types of layers even at the first stages. It can be assumed that atomic processes of incorporating oxygen into the amorphous silicon and on the anisotropic, crystalline silicon surface are most important (breaking of Si-Si bonds and the formation of Si-O-Si

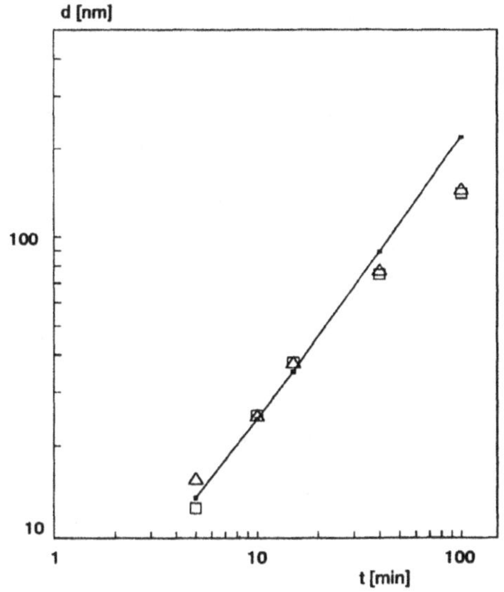

Figure 5. The experimental data are fitted to the model of Deal and Grove (1965) by applying B/A = $1.3 \cdot 10^{-1}$ μmhr^{-1} (linear growth region). Data for amorphous (Δ) and polycrystalline deposited silicon layers (□), both heavily doped with phosphorus.

bonds). They are different as long as the crystallization front, starting simultaneously at the opposite interface of the a-Si layer, crosses the oxidation front. In addition, as it is shown by HREM, effects of dopant segregation or the influence of stresses discussed e.g. by Wang et al. (1991) can be avoided (oxide layers grown at 960°C should almost be free of stress (cf. e.g. Claeys et al., 1988a)).

Furthermore, the enhancement effect of doping on oxidation is generally known also of polycrystalline layers. There are, however, differences in the influence of this effect on the present investigations, which is referred to in the literature. They show that doping the silicon layers strongly varies the influence of this effect on the oxidation. While ion implantation was used for the present investigations, the literature mostly anoted doping from $POCl_3$ sources. For instance, Claeys et al. (1988b) reported B/A variations of more than 1 order of magnitude by increasing the phosphorus concentration from 1 to $5 \cdot 10^{20}$ cm^{-3}. On the other hand, the medium-doped samples described here have the same B/A as the undoped ones; the phosphorus concentration calculated from the implantation dose is, however, $9 \cdot 10^{20} cm^{-3}$. Moreover, the value of $1.3 \cdot 10^{-1}$ μmhr^{-1} of the linear rate constant found for the heavily doped samples (i.e. a phosphorus concentration of $4 \cdot 10^{21} cm^{-3}$) is more than one order of magnitude lower than that extrapolated from the data of Claeys et al. (1988b).

Acknowledgements

The author is especially indebted to Mr. B. Pappe (Erfurt, Dortmund) for preparing the samples. The financial support by the Deutsche Forschungsgemeinschaft (DFG) is gratefully acknowledged.

References

Carim, A.H. and Sinclair, R., 1987, The evolution of the Si/SiO_2 interface roughness, *J. Electrochem. Soc.* 134:741.

Claeys, C.L., De Keersmeacker, R.F., and Declerck, G.J., 1988a, The oxidation process and Si-SiO_2 system properties, in:"The Si-SiO_2 System," P.Balk, ed., Elsevier Science Publishers, Amsterdam.

Claeys, C.L., De Keersmeacker, R.F., and Declerck, G.J., 1988b, Technology and kinetics of SiO_2 growth, in:"The Si-SiO_2 System," P.Balk, ed., Elsevier Science Publishers, Amsterdam.

Deal, B.E., 1988, The thermal oxidation of silicon and other semiconductor materials, in:"Semiconductor Materials and Process Technology Handbook," G.E.McGuire, ed., Noyes Publications, Park Ridge.

Deal, B.E. and Grove, A.S., 1965, General relationship for the thermal oxidation of silicon, *J. Appl. Phys.* 36:3770.

Hendriks, M. and Mavero, C., 1991, Phosphorus doped polysilicon for double poly structures, *J. Electrochem. Soc.* 138:1466.

Reiche, M., 1991, TEM *in situ* investigations of the crystallization of a-Si thin films, in: "Microscopy of Semiconducting Materials 1991," A.G.Cullis and N.J.Long, eds., Institute of Physics Conference Series Number 117, Institute of Physics, Bristol.

Reiche, M. and Hopfe, S., 1990, TEM *in situ* investigations of the crystallization behaviour of amorphous silicon thin films, *Ultramicroscopy* 33:41.

Saraswat, K.C., and Singh, H., 1982, Thermal oxidation of heavily phosphorus-doped thin films of polycrystalline silicon, *J. Electrochem. Soc.* 129:2321.

Sze, S.M., 1985, "Semiconductor Devices - Physics and Technology," John Wiley and Sons, New York.

Wang, Y., Tao, J., Tong, S., Sun, T., Zhang A., and Feng, S., 1991, The oxidation kinetics of thin polycrystalline silicon films, *J. Electrochem. Soc.* 138:214.

III. DEPOSITION AND PROPERTIES OF SiO_2

INTRODUCTION

H. Z. Massoud
Dept. of Electrical Engineering
Duke University, Durham, NC 27706

S. I. Raider
IBM T. J. Watson Research Center
Yorktown Heights, NY 10598

In this chapter, there are seven papers covering the deposition and properties of silicon dioxide layers. The chapter starts with a review of the thermal and X-ray production of point defects in vitreous SiO_2 by Galeener. In this invited paper, the three main spin-active defect centers observed in vitreous SiO_2 are reviewed. These are the E' electron center, the nonbridging oxygen hole center (NBOHC), and the peroxyl radical oxygen hole center (PROHC). Defect generation is examined as a result of thermal and mechanical shock, particle irradiation, neutron bombardment, and above-band-gap electromagnetic radiation, especially X- and γ-irradiation. A model involving creation, activation, annihilation, and deactivation is found to fit the highly nonlinear defect concentration vs. dose curves for electron spin resonance and luminescence. Photon energy studies reveal that electrons ejected by photoabsorbed X-rays cause the spin-active defects.

Paramagnetic defects in α-quartz as characterized in detail by electron paramagnetic resonance spectroscopy are reviewed by Weil, McEachern, and Mombourquette. The discussion covers intrinsic or impurity-free defects and extrinsic defects. The dominant intrinsic defect is the E'_1 center, an O^- vacancy of special importance in the fields of glasses and MOS devices. The vast majority of point defects in α-quartz are extrinsic impurity-related defects, with the primary E'_2 centers being the $E'_2(H)$, $E'_2(Li)$, and $E'_4(H)$ single oxygen vacancies. This paper concludes with a discussion of recent work on generating powder/glass EPR lineshapes from the single-crystal parameters.

The remaining four papers in this chapter deal with the processing conditions and properties of SiO_2 layers deposited either by plasma-assisted or non-plasma-assisted chemical vapor deposition.

Lucovsky, Ma, Yasuda, and Habermehl describe the properties of SiO_2 deposited at low temperatures in the 200-300°C range by a PECVD process that allows the separate control of the interface formation and oxide deposition processes. This technique includes the oxidation by remotely generated O atoms, following an RCA clean that concludes with a rinse in dilute HF, and deposition of an oxide or a multilayer dielectric (NO or ONO) by remote plasma-enhanced CVD.

The dependence of the physical, optical, and electric properties of deposited SiO_2 on the deposition conditions at low temperatures (<120°C) in a distributed electron cyclotron resonance (DECR) plasma-enhanced chemical vapor deposition (PECVD) is presented by Agius et al. They report that high-density films with near-SiO_2 stoichiometry and low hydrogen concentration were obtained by optimal choice of pressure, microwave power, total glass flow, and oxygen-to silane flow ratio.

In the study of Chakravarthy, Levy, and Grow, SiO_2 layers were deposited in the 375-475°C range using diethylsilane (DES) by low-pressure chemical vapor deposition (LPCVD). The deposition rate was found to follow an Arrhenius behavior with an activation energy of 10 Kcal/mole, increased at higher pressures, and varied with the DES flow rate and O_2/DES ratio. The oxide thus grown has a density of 2.2 g/cm^3, IR spectra peaks at 1060, 810, and 440 cm^{-1}, a refraction index of 1.46, and an etch rate of 900 Å/min at 25°C. These properties allow its use as an intermetal dielectric or as a top-layer passivation coating.

The electrical characteristics of these low-temperature deposited oxides generally indicate that there is room for improvement in the concentration of interface traps, leakage current, process window, reproducibility, and reliability. They also indicate the great potential of these depositions technologies for incorporation into future standard manufacturing practices.

THERMAL AND X-RAY PRODUCTION OF POINT DEFECTS IN VITREOUS SiO_2

Frank L. Galeener

Department of Physics
Colorado State University
Fort Collins, CO 80523

INTRODUCTION

In this review, I will outline our recent work on the thermal and x-ray production of spin active point defects in vitreous (v-) SiO_2. Experimental details and more extensive comments on the nature of the defects and the relevant work of others will be found in Refs. 1-15.

There are three main spin active defect centers seen at room temperature in pure v-SiO_2: the E′ electron center, the nonbridging oxygen hole center (NBOHC) and the peroxyl radical oxygen hole center (PROHC). The nominal structures for these point defects are shown in Fig. 1. The E′ electron spin resonance (ESR) signal is due to the unpaired spin of an electron trapped on a Si atom that is bonded to only three oxygen atoms.[1] The H_W or NBOHC signal is due to a hole on an oxygen atom singly bonded to Si.[2] The H_D or PROHC signal is due to a hole on an oxygen atom singly bonded to O.[3] Many aspects of these and other defects in v-SiO_2 are discussed by Griscom[4-6] and in references given by him.

Each structure shown in Fig. 1 represents the "spin activated" form of the defect, i.e., the structure has an unpaired $S = 1/2$ spin which is seen in our ESR experiment. Upon trapping or untrapping of an appropriate charge, such structures may *continue to exist but exhibit no net spin*; they are then said to be "deactivated," and are not seen in ESR. Upon much more energetic rearrangements of local structure the proper SiO_2 bonds may be reestablished and the defect is said to be "annihilated," or "recombined" into the network.

When an *undefected* piece of SiO_2 network is permanently broken, e.g., as a result of thermal agitation or irradiation, the resultant defect is said to have been created. If this newly created defect has *no* unpaired spin, it will *not* be seen in ESR. By "activation" we mean the subsequent untrapping or trapping of an appropriate charge which produces an unpaired spin and renders the structure observable by ESR. We assume that "activation" requires relatively little energy, perhaps less than 1 eV, and therefore *probably* corresponds to a minor change in the geometry of the structure.

The Physics and Chemistry of SiO_2 and the Si-SiO_2 Interface 2
Edited by C.R. Helms and B.E. Deal, Plenum Press, New York, 1993

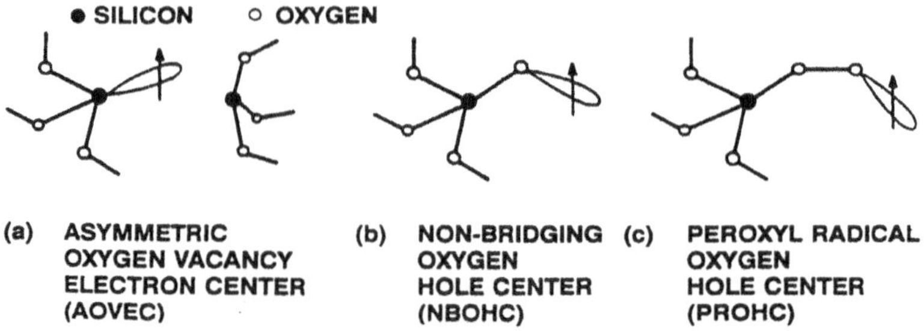

Figure 1. Schematic representation of the spin active defect structures in $v-SiO_2$. Each ↑ represents the unpaired spin of a charge trapped in the orbital shown. These structures have been assigned[1,2,3] to the E′ (a), H_W (b) and H_D (c) ESR lines.

The spin−*inactive* form of the defect is sometimes referred to as the "precursor" of the structure that exhibits ESR.[4−6] Thus the "precursor" of the E′ defect is said to be a structure like that in Fig 1(a) but having no unpaired spin.

In the context of our experiments involving radiation induced defects we refer to the spin−inactive form as the "preexisting form" of the defect, or more simply the "preexisting" defect. We use this language to distinguish between two kinds of sources (or precursors) of spins that are observed after irradiation. When the spin−active defect was made by breaking the bonds in an undefected piece of network (as source, or precursor), we say the ESR observed defect has been "created" (from the undefected network). When the observed spin−active defect is made merely by radiation induced transfer of an appropriate charge, we say it was "activated" from a "preexisting" (form of the) defect.

At the present time we therefore distinguish four processes: (1) "creation" of a spin−active defect from an undefected network, and its converse (2) "annihilation" or "recombination" then (3) "activation" of a spin−active defect from its spin−inactive preexisting form or precursor, and its converse (4) "deactivation."

A sample of $v-SiO_2$ quenched from the melt at normal rates shows no ESR signals at room temperature. Defects are seen by ESR only after exposure to various additional disturbances, including thermal shock, mechanical shock, particle irradiation, neutron bombardment and above band gap electromagnetic radiation. Our research is aimed at a better understanding of the mechanisms of generation of spin active defects in SiO_2 under X− and γ−irradiation (as well as electron bombardment) for samples with different thermal histories and trace impurity concentrations.

X−RAY DOSE DEPENDENCE

Figure 2, taken from Ref. 7, exhibits the monotonic increase of E′ and H_W spin signal strength S with X−ray exposure time t in otherwise high purity "wet" Suprasil−1, which contains large amounts of hydroxyl impurity, [OH] ~ 1200 ppm. Figure 3, also from Ref. 7, shows the increase of S versus t for E′ and H_D spin signals in high purity "dry" Suprasil−W1, in which there is very little of the normally ubiquitous hydroxyl, [OH] < 2 ppm. The samples were annealed to structural equilibrium at temperatures of 1350, 1000 and 700°C, then rapidly air quenched to room temperature, hence they have the indicated fictive temperatures $T_F \simeq T_A$. The 3 mm × 6 mm × 1 mm samples were irradiated at room temperature from one flat side by the X−rays from a Cu−target tube,

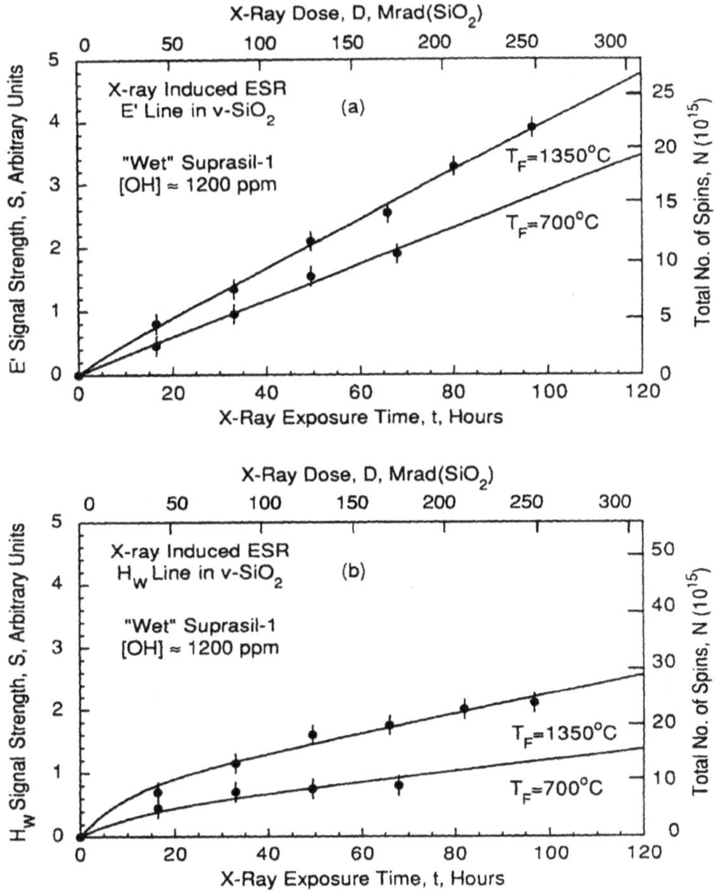

Figure 2. The total number of E′ spins and H_W spins induced in "wet" v−SiO_2 vs. total Cu−tube X−ray dose, for two different fictive temperatures T_F. The solid curves are fits using the parameters in Table 1.

with about 70% of the output power in the Cu K_α and K_β lines at about 8 keV. The strengths of spin signal and exposure times were calibrated to yield the alternative scales in Figs. 2 and 3, which are labeled Total Number of Spins, N, and X−ray Dose, D in Mrad (SiO_2).

INTERPRETATION IN TERMS OF NET CREATION AND NET ACTIVATION

The data points are fit quite well by a model which assumes that for all our defects only two processes occur: That is, the X−ray bombardment leads to (1) creation of new spin−active defects, by the rupturing of bonds and (2) the spin−activation of similar preexisting defects, by trapping of electrons or holes onto the preexisting structures. According to this model,

$$N(D) = M_C D + N_0\left[1 - \exp(-m_A D)\right] , \qquad (1)$$

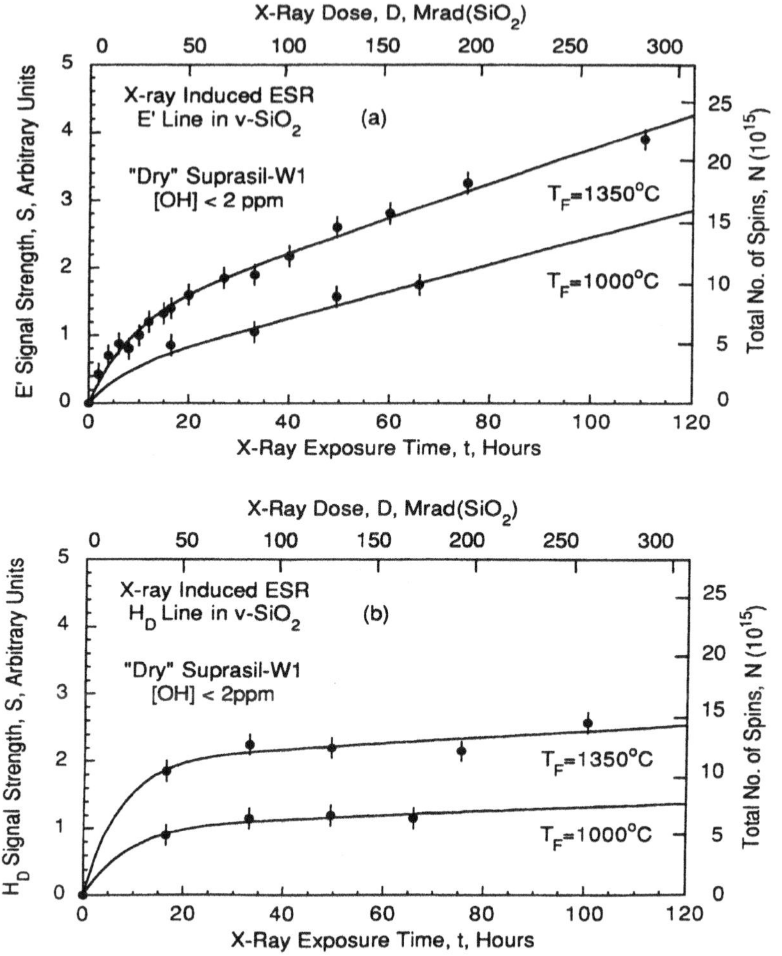

Figure 3. The total number of E′ spins and H_D spins induced in "dry" $v-SiO_2$ vs. total Cu−tube X−ray dose, for two different fictive temperatures T_F. The solid curves are fits using the parameters in Table 1.

where N(D) is the total number of spins of a given type induced after accumulated dose D, M_C is the net rate of creation of the defects giving that type of spin, N_O is the number of preexisting defects of the specified type, and $M_A \equiv m_A N_O$ is the initial net rate of activation of those defects, at low D. *Net* creation in this expression is the combined result of both the "creation" and "annihilation" processes discussed above. Similarly, net activation is the combined result of "activation" and "deactivation."

The first term in equation (1) assumes that spin signals due to newly created defects will increase linearly with X−ray dose. The second term in equation (1) states that the spin signals due to activation of preexisting defects will be zero initially and will rise exponentially to a saturation value of N_O, when all the preexisting defects have been activated.

The solid lines in Figs. 2 and 3 are our best fits of equation (1) to the data points. The resultant parameters N_O, M_C and M_A are given in Table 1 and discussed extensively in Refs. 7 and 8.

Table 1. Parameters that provide a fit of Eq. (1) to the data, giving the solid line curves shown in Figs. 2 and 3.

[OH]	Spin Type	T_F (°C)	$N_O \times 10^{-15}$ (spins)	$M_C \times 10^{-13}$ ($Mrad^{-1}$)	$M_A \times 10^{-13}$ ($Mrad^{-1}$)
"Wet"	E′	1350	0.7	8.6	~3
(S−1)		700	0.1	6.4	~1
	H_W	1350	7.5	6.9	~30
		700	3.7	3.8	~30
"Dry"	E′	1350	6.8	5.8	31
(S−W1)		1000	2.5	5.8	~14
	H_D	1350	11.0	0.9	~61
		1000	5.6	0.6	~26

In general,[7−9] we find that the concentration N of spins for each type of defect increases from zero *nonlinearly* with radiation dose D. These N versus D dose curves are strongly affected by both hydroxyl concentration [OH] and fictive temperature T_F.

Many conclusions regarding creation and activation are drawn in Refs. 7 and 8 from the [OH] and T_F dependencies of N_O and M_C. For example, from N_O we learn about the concentrations of E′, NBOHC and PROHC defects in the *unirradiated* material. Thus, N_O is very small (or zero) in samples with high [OH] ~ 1200 ppm, but is large ($\geq 10^{16}$ cm^{-3}) for low [OH] ≥ 2 ppm. In other words, *"wet" Suprasil−1 contains far fewer preexisting E′ defects than "dry" Suprasil−W1.* From values of M_C we learn that *a given material is made more radiation hard for all three traps at all exposures by annealing it to an equilibrium structure at the lowest practicable temperature* T_F. Also from M_C we learn that *achievement of enhanced radiation hardening by raising or lower [OH] is predicted, but the desired direction depends on the type of trap and the anticipated level of exposure.* M_C also reveals that *it is very difficult to make PROHC defects by X−ray bombardment.*

It also follows that *at low dose* creation dominates in low [OH] or "dry" samples, while activation dominates in high [OH] or "wet" samples. We have taken advantage of this to learn more about the creation and activation processes, individually, as follows.

EVIDENCE THAT *ELECTRON* DAMAGE PRODUCES THE OBSERVED SPINS

In one study of the *creation* process[10], two identical "wet" samples were subjected at the Stanford Synchrotron Radiation Laboratory to equal energy exposure of narrow band X−radiation, one at photon energy just above the Si−K_α absorption edge and the other just below, as illustrated in Fig. 4, reproduced from Ref. 10. Approximately four times as many X−rays were absorbed on Si atoms in the higher photon energy case as in the lower, yet the numbers of spin active E′ defects created in the samples were the same (±10%). Clearly the number of defects created is *not* proportional to the number of ionized Si atoms (or O atoms) produced in the initial X−ray photoabsorption events. We concluded that the majority of *created* defects must be produced by the energetic photo and Auger electrons ejected by the photoabsorption process. Damage caused by these electrons will be independent of whether they were ejected initially from Si or O photoabsorption

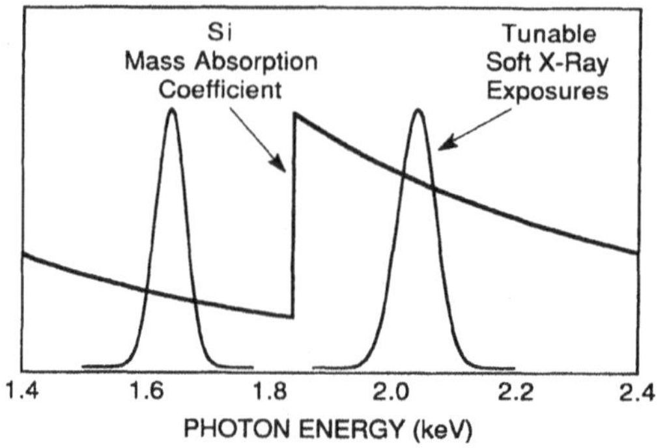

Figure 4. A schematic illustration of the bandwidths of the two monochromatic exposures along with a schematic illustration of the Si mass absorption coefficient. For the 1.64−keV exposure most of the ionization occurs on O atoms, while for the 2.04−keV exposure most of the ionization occurs on Si. For equal energy exposures the measured number of E′ spins was unchanged.

events although the extent of damage would be expected to depend on the initial energy E_O of the ejected electrons. Thus we find that X−irradiation is a convenient means for studying *electron* damage processes in v−SiO_2!

It is reasonable to infer from this study that *activation* of spins in our samples is also due overwhelmingly to electron damage. Likewise, we infer that other point defects such as NBOHC and PROHC are also created and activated primarily by the energetic electrons resulting from X−ray absorption.

THE ENERGY DEPENDENCE OF E′ SPIN PRODUCTION BY X−RAYS

In another study of the creation and activation processes[11] "wet" and "dry" samples were subjected at the National Synchrotron Light Source to narrow band tunable X−rays from 5 to 20 keV. The total doses were kept well below 5 MRad (SiO_2) so that, as mentioned before, creation dominated in the "wet" sample while activation dominated in the "dry" one.

Figure 5 shows the results, as reported in Ref. 11. The E′ signal cross section per photon for "wet" material (creation) increased almost linearly with X−ray photon energy, while that for "dry" material (activation) was highly nonlinear. The data was reanalyzed in terms of the energies of photo and Auger electrons ejected by photoabsorption of the X−rays into Si− and O−K_α shells, since these electrons would be the dominant contributors to electron damage. The model assumed a power law E_O, and n was the only unknown in the model. The expression used is given by Eqs. (2)−(4):

$$S = S_{Si} + S_O \quad , \tag{2}$$

where

$$S_{Si} = p(h\nu)\left[(h\nu - 1.84)^n + (1 - \gamma_{Si})1.62^n\right] \tag{3}$$

and

$$S_O = [1 - p(h\nu)]\left[(h\nu - 0.53)^n + (1 - \gamma_O)0.51^n\right] \quad . \tag{4}$$

The fraction of the photons of energy hv which photoionize Si and O atoms is p(hv) and 1 − p(hv), respectively. These are easily calculated from the published values of the mass absorption coefficients of Si and O. The fluorescence yields, γ_{Si} and γ_O, can also be

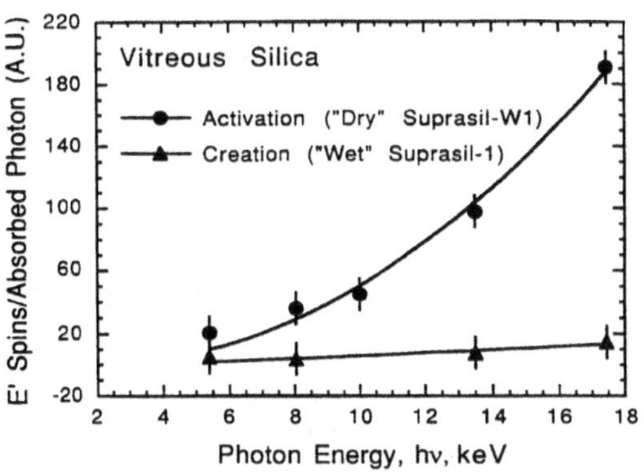

Figure 5. The E′ spin production efficiency per *absorbed* photon in arbitrary units (a.u.) for vitreous silica as a function of *photon* energy. The dry Suprasil−W1 approximately follows a square law in the ejected photo− and Auger *electron* energies, while the wet Suprasil−1 is nearly linear in the ejected electron energies. The linear and square law behaviors are ascribed to the creation of new E′ defect centers and the activation of preexisting E′ defects, respectively. The solid lines are Eqs. (2)−(4) with $n_D = 2.1$ and $n_W = 1.3$ for dry and wet Suprasil, respectively.

found in the literature, and represent the probability that an ionized atom emits an X−ray photon. The probability that the ionized atom will emit an Auger electron instead of an X−ray photon is then $1 - \gamma$. The K−edge energies for both Si and O are 1.84 and 0.53 keV, and the KLL Auger electron energies are 1.62 and 0.51 keV, respectively. These appear explicitly in Eqs. (2)−(4). Aside from a normalization constant, n is the only unknown parameter.

The best fit of the dry Suprasil−W1 data by Eqs. (2)−(4) results in an exponent $n_D = 2.1 \pm 0.2$, suspiciously close to the integer 2. The best fit for the wet Suprasil−1 data results in an exponent $n_W = 1.3 \pm 0.4$, which is fairly close to linear. The solid lines in Fig. 5 are the fits of Eqs. (2)−(4) to the data using these exponents. Because neither pure creation nor pure activation occurs completely alone in either type of silica sample, we inferred that the true exponents are nearer to the following *idealized* values: $n_W = 1$ (creation) and $n_D = 2$ (activation).

These idealized empirical exponents were then shown to be consistent with collision models in which a large energy ~30 eV is taken from a photo or Auger electron to create a defect ($n_C = 1$) while a much smaller energy (<1 eV) is taken to activate a preexisting defect ($n_A = 2$). In the latter case we argued that the number of defects which are *activated* is proportional to the length of the damage path of the decelerating electron, which is in turn proportional to E_O^2.

Figure 6 is a schematic picture of our microscopic view of the general case where both creation and activation occur along the electron path. An incoming photon produces an energetic electron by photo− or Auger ejection, and this electron transfers energy to the v−SiO_2 network in numerous inelastic scattering events with other electrons. Some of the scattering events result in newly created spin active defects, and others in the spin activation of preexisting defects; however, it seems likely that many scattering events will produce no stable observable spins of the type studied−−rather, the energy lost in scattering will be dissipated as heat, luminescence, or in other ways.

The very different dependencies of creation and activation on X−ray (or photoejected electron) energies supports the basic notion that our nonlinear dose curves are evidence for two importantly different processes in the production of spin−active defects.

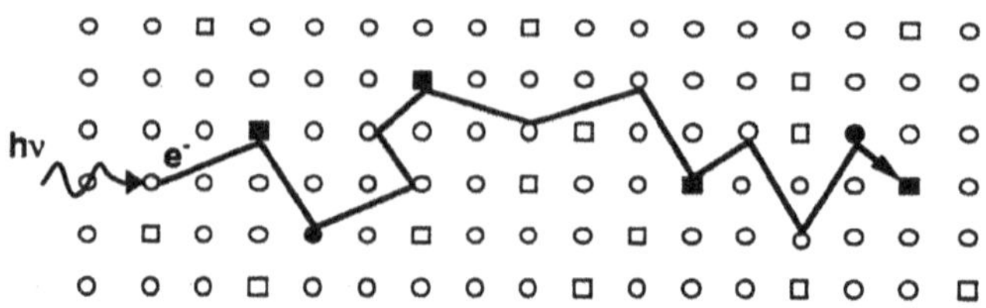

Figure 6. A highly schematic picture of X−ray (hν) absorption producing an energetic photo− or Auger electron (e^-) which is multiply scattered by other electrons in a dry Suprasil−W1 silica network. (The actual network is *not* periodic.) The ○ are undefected network sites, while the □ are preexisting defect sites. An $e^- - e^-$ scattering event near a preexisting defect site activates the preexisting defect (■). Occasionally, a scattering event can create a new defect (●) which is automatically spin activated.

1.5−3.5 eV X−RADIOLUMINESCENCE VERSUS X−RAY DOSE

Kerwin et al.[12] have subjected "dry" Suprasil−W1 and "wet" Suprasil−1 samples to X−radiation and measured the optical frequency luminescence emitted *during irradiation*. In Ref. 13 we refer to this X−ray driven optical frequency luminescence as "X−radioluminescence" (XRL). Figure 7, taken from Ref. 13, shows the integrated intensity of XRL from about 1.5 eV to about 3.5 eV as a function of time in front of the X−ray tube, i.e., vs. dose, for both "dry" (a) and "wet" (b) material. This band spans a broad luminescence line which peaks at about 2.7 eV. Although the dose curves are very strong functions of T_F, they are almost independent of [OH]. In particular the straight line projections of the high dose data to zero dose, marked X in Fig. 7, are independent of [OH] within experimental error for a given T_F. This means that the concentration of XRL centers in the unirradiated sample is *not* dependent on [OH], in marked contrast to the strong [OH] dependencies of the N_O of preexisting E′, H_D and H_W centers seen in Table 1.

Since the XRL and ESR dependencies of defect concentration on [OH] do not correlate, we conclude[13] that the dominant component of the 1.5−3.5 eV XRL band centered at about 2.7 eV does not involve an AOVEC, PROHC or NBOHC as initial or final state in the transition. This also rules out the self−trapped exciton recombination model of Itoh et al.[14] for our data. Since this particular component of XRL does not involve the dominant point defects or the dominant impurity (OH), its origin is most likely intrinsic to the structure of SiO_2, an hypothesis which we are further investigating.

EVIDENCE THAT "ACTIVATION" IS ENHANCED UNDER γ−IRRADIATION

Galeener[9] has compared previously published curves showing electron spin resonance spin count versus ionizing radiation dose in bulk vitreous SiO_2 for the two cases of ~1.25 MeV γ−rays and ~8 keV X−rays. The data for E′ spins are repeated in Figs. 8 and 9. Figure 8 is for "wet" and "dry" silica with T_F = 1350°C, and is an overlay of curves from Figs. 2(a) and 3(a), respectively. Recall that the nonlinear behavior of the "dry" silica is caused by the exponentially saturating "activation" of preexisting defects, while the essentially linear behavior of the "wet" silica is due to the *absence* of preexisting E′ defects, accompanied by the linear "creation" of new ones. Since the "creation" rate in "wet" material is greater than that in "dry," the curves eventually cross over, after saturation of preexisting defects is completed. This crossover is at about 175 Mrad (SiO_2) exposure from a Cu−target X−ray tube, whose major output is at ~8 keV. Because both samples were exposed to X−rays and ESR spectroscopy identically, the existence of this

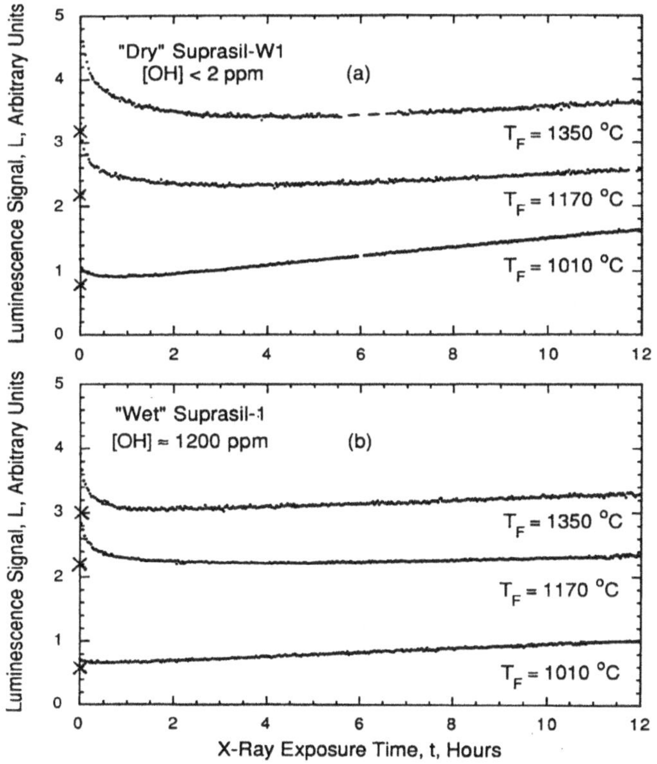

Figure 7. The 1.5–3.5 eV band X–radioluminescence intensity versus X–ray dose curves for "dry" (a) and "wet" (b) v–SiO_2. The zero dose projections (x) are *independent* of [OH], in marked contrast to the projections for the point defects in Figs. 2–3, hence the values of N_D given in Table 1.

crossover is an empirical fact not effected by any experimental uncertainties concerning calibration of the absolute values of spin count or dose (given in Ref. 9).

Figure 9 is also for "wet" and "dry" Suprasil, but now exposed to ~1.25 MeV ^{60}Co γ–rays, as reported by Devine and Arndt,[15] and reproduced from Ref. 9. The similarity between γ–ray and X–ray dose curves in Figs. 9 and 8 is striking: essentially linear dose dependence for "wet" material and highly nonlinear for "dry" material, accompanied by a crossover of the two curves. For γ–rays the existence of the crossover (at D ~ 8 Mrad) is also independent of any possible uncertainty in the absolute calibration of spin count or γ–ray dose.

After consideration of possible calibration errors, it is concluded in Ref. 9 that the existence of the γ–ray crossover at a much lower dose (~8 Mrad) than the X–ray crossover (~175 Mrad) means that *the γ–rays are at least one order of magnitude more efficient per unit of dose in activating all the preexisting E' defects*. This is consistent with our earlier observation (see Fig. 5) that up to say 20 keV "activation" increases with ejected electron energy, E_O, as E_O^2; that is, much more rapidly than "creation," which increases as E_O. Work is underway to see if the E_O^2 dependence actually holds at the much higher electron energies encountered with 1.25 MeV γ–rays.

Other dose curves are compared in Ref. 9 to show that for activation of *NBOHC* defects γ–rays are also *at least* one order of magnitude more efficient per unit of dose. I presume that something similar will also be true for PROHC defects.

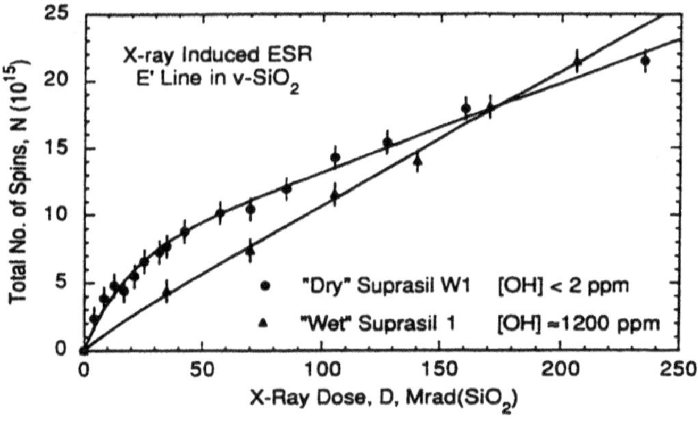

Figure 8. The dependence of the number of E′ spins on the dose of predominantly 8 keV X−rays, taken from Ref. 9. These are merely an overlay of the T_F = 1350°C curves from Figs. 2(a) and 3(a), for "wet" and "dry" silica, respectively. As before, the solid lines are least squares fits to Eq. (1) using parameters given in Table 1. Note the crossover at ~175 Mrad (SiO_2).

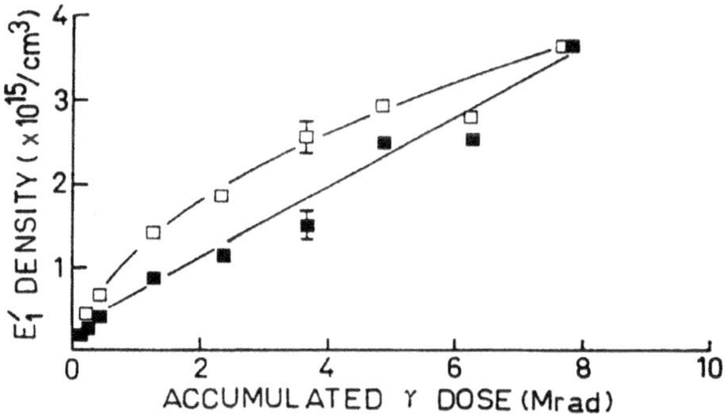

Figure 9. The dependence of the concentration of E′ spins on the dose of ~1.25 MeV ^{60}Co γ−rays, taken by Devine and Arndt[15] and reproduced from Ref. 9, for "wet" Suprasil−1 (■) and "dry" Suprasil−W1 (□). Here the solid lines were drawn to aid the eye. Note the very strong similarity of these curves to those in Fig. 8 for ~8 keV X−irradiation. Crossover at ~8 Mrad rather than ~175 Mrad suggests that the γ−rays are much more efficient in activating E′ defects than are the X−rays, per unit of dose.

CONCLUSIONS

Suffice it to conclude that studies of dose and X−ray photon energy dependencies of ESR and luminescence signals reveal much about the thermal and energetic electron production of spin active point defects in v−SiO_2. We are extending studies to higher irradiation energies (γ−rays) and significantly higher doses [Grads (SiO_2)].

ACKNOWLEDGEMENTS

We are deeply grateful for support of this work by Colorado State University (start up funds) and the U.S. Navy Office of Naval Research under contract N00014−91−J−1607 (G. B. Wright).

REFERENCES

1. R.A. Weeks, *J. Appl. Phys.* **27**, 1376 (1956); and F.J. Feigl, W.B. Fowler and K.L. Yip, *Solid State Commun.* **14**, 225 (1974).
2. M. Stapelbroek, D.L. Griscom, E.J. Friebele and G.H. Sigel, Jr., *J. Non−Crystal. Solids* **32**, 313 (1979).
3. E.J. Friebele, D.L. Griscom, M. Stapelbroek and R.A. Weeks, *Phys. Rev. Lett.* **42**, 1346 (1979).
4. See, e.g., D.L. Griscom, in *Glass: Science and Technology*, (Edited by D.R. Uhlmann and N.J. Kreidel), Vol. **4B**, p. 151ff, Academic, Boston (1990).
5. D.L. Griscom, *J. Ceram. Soc. Japan* **99**, 923 (1991).
6. D.L. Griscom, *Nucl. Inst. and Methods B* **46**, 12 (1990).
7. F.L. Galeener, D.B. Kerwin, A.J. Miller and J.C. Mikkelsen, Jr., *Solid State Commun.* **82**, 271 (1992).
8. F.L. Galeener, D.B. Kerwin, A.J. Miller and J.C. Mikkelsen, Jr., *Phys. Rev. B*, in press.
9. F.L. Galeener, *J. Non−Crystall. Solids*, in press.
10. D.B. Kerwin and F.L. Galeener, *Appl. Phys. Lett.* **59**, 2959 (1991).
11. D.B. Kerwin and F.L. Galeener, *Phys. Rev. Lett.* **68**, 3208 (1992).
12. D.B. Kerwin, A.J. Miller and F.L. Galeener, in preparation.
13. F.L. Galeener, D.B. Kerwin and A.J. Miller in "The Physics of Non−Crystalline Solids," ed. by L.D. Pye, W.C. LaCourse and H.J. Stevens (Taylor and Francis, London, 1992), p. 597.
14. C. Itoh, K. Tanimura and N. Itoh, *J. Phys. C: Solid State Phys.* **21**, 4693 (1988).
15. R.A.B. Devine and J. Arndt, *Phys. Rev. B* **39**, 5132 (1989).

A REVIEW OF THE EPR SPECTROSCOPY OF THE POINT DEFECTS IN α-QUARTZ: THE DECADE 1982-1992

John A. Weil

Department of Chemistry
University of Saskatchewan
Saskatoon, SK, S7N 0W0, Canada

ABSTRACT A review (126 references) is presented of progress made, primarily in the decade 1982-1992, in the field of electron paramagnetic resonance spectroscopy of 'point' defects occurring in crystalline silicon dioxide (α-quartz).

INTRODUCTION

This review article has been designed to summarize the field of paramagnetic 'point' defects in crystalline silicon dioxide, emphasizing the developments of the last decade. Thus it follows and builds on the contents of our previous article [W84]. Both are written for the non-specialist in electron paramagnetic resonance (EPR) spectroscopy, which is by far the most rewarding and productive technique in deriving detailed information about the defect centres of interest, i.e., those with unpaired electrons.

Paramagnetic species in α-quartz can occur due to radiation damage, including photo-ionization, due to electrolytic treatment, and of course due to incorporation of paramagnetic ions. The field of study by EPR of defects in α-quartz was initiated in the 1950s [W84] with discovery of donor centres by Griffiths, Owen and Ward, of oxygen-vacancy centres by Weeks, and of acceptor centres by Anderson and Weil. So far, relatively few of the elements constituting the ~109 members of the periodic table have been found to occur either substitutionally or interstitially in the quartz structure. There may well be potent restrictions for incorporation, based on the charge and size of the ion. More work on this question is called for. Table 1 summarizes the elements that have been found in α-quartz EPR, excluding those that are likely to exist only in macroscopic flaws (dislocations, bubbles, cracks, etc.) or at the surface.

Nevertheless, a large number (> 50) of unpaired-electron defects in α-quartz are now known. It is probable that many more await to be produced and discovered.

Table 1. Chemical Elements observed by EPR in Point Defects existing in Crystalline Quartz

Si,O / H, Li, Na, Ag / Al, Ga / Ti, Mn, Fe, Ni, Cu / P, Ge

The Physics and Chemistry of SiO_2 and the Si-SiO_2 Interface 2
Edited by C.R. Helms and B.E. Deal, Plenum Press, New York, 1993

QUARTZ STRUCTURE

Crystalline SiO_2 in its α-quartz form is one of nature's purest chemicals. However, man in crystallizing it hydrothermally is able to improve on this. This purity is crucial in the primary use of α-quartz, as the piezoelectric oscillator material of choice [B85]. For many of our purposes, α-quartz can be considered as a particularly well-defined solvent. Thus we'll not focus on heavily damaged crystals. However, even though we know the time-average positions of all Si and O atoms from accurate x-ray diffraction studies as a function of temperature and applied pressure, it is nevertheless a formidable problem to characterize the local distortions associated with each point defect. Also, to understand the resulting dynamic effects (i.e., acoustic and piezoelectric aspects) is still a future goal.

Formally, one can speak of SiO_4 tetrahedra, linked together to form the quartz structure, occurring in three distinct orientations (see Fig. 1). There are (at most) six chemically equivalent sites, inter-related by the symmetry operations of the proper point group, D_3, of the crystal. In reality, each tetrahedron in pure SiO_2 is irregular, distorted so that only one two-fold rotational symmetry axis remains. These axes a_i ($i = 1, 2, 3$) pierce the Si^{4+} ions and interchange the O^{2-} ions pairwise. Thus each basic unit $[SiO_4]^0$ contains two different pairs of Si-O entities ("short-bond" and "long-bond"; at 94 K: 1.6101 and 1.6145 Å [LCG80]). In the above, we speak of 4+ and 2– ions for book-keeping purposes only; in practice, there is considerable covalency [P80, SWD80]. We ascribe only one minus charge from each oxide ion to each basic unit, which thus is neutral.

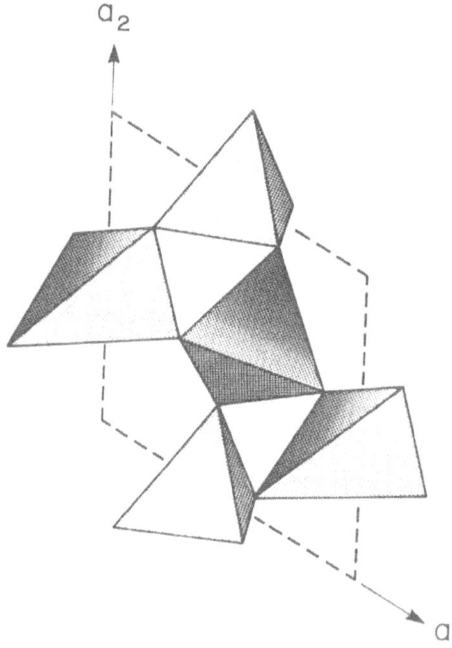

Fig. 1. Projected view along the optic axis **c** of α-quartz of a smallest unit cell (dashed lines), showing the three different orientations of the quasi-tetrahedra $[SiO_4]^0$. Two of the three two-fold rotational axes are indicated.

Substitution by some atom X for Si is described by symbol $[XO_4]^q$, where q is the total charge number. Notation $[XO_4/M_1M_2\ ...]^q$ allows consideration of one or more nearby interstitial ions $M^{q'}$. The latter are thought to be able to occur in and diffuse along the open c-axis channels running parallel to the optic axis **c** (three-fold screw) of the crystal. The two-fold axes $\mathbf{a}_i$ are perpendicular to **c**, and at 120° to each other.

EPR SPECTROSCOPY

Due to the excellent sensitivity (allowing detection of fewer than 10^{-6} spins/Si) and the high resolution (allowing simultaneous study of various unpaired-electron species; see Fig. 2), EPR continues to be the tool of choice for the study of the point defects in SiO_2.

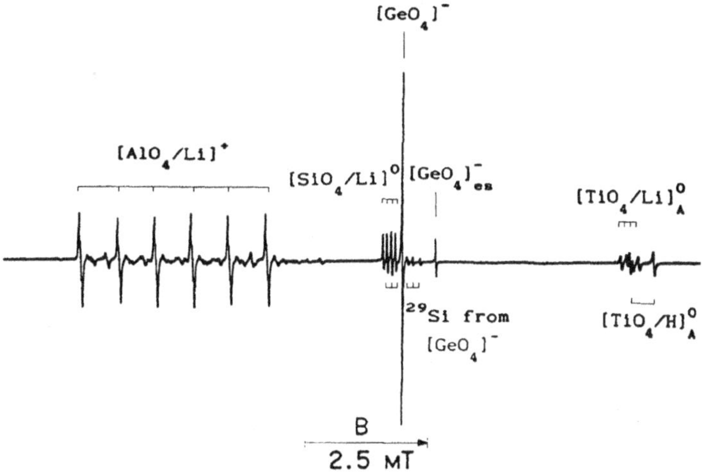

Fig. 2. EPR spectrum (1st derivative) of a rose-quartz crystal at 35 K showing various species (see text) simultaneously present after x-irradiation at 77 K, measured at ν = 9.957 GHz with **B** ∥ **c**. From [BW91].

EPR spectroscopy depends on the presence of one or more (i.e., S = 1/2, 1, 3/2,...) unpaired electron spins per defect. This net unpaired spin is generally a multi-electron phenomenon, and it is spread over many atoms. However the centres do not interact very appreciably in SiO_2. One important task of the experiment is to ascertain the unpaired-electron distribution, in sundry orbitals on various atoms. Hyperfine splittings (usually 2I+1 lines) due to nuclear magnetic moments (spins I) are especially useful in this. Another task, of course, is to measure the relative (or absolute) concentrations of the centres present. The temperature dependence of each EPR spectrum often is very informative. Knowledge of the spin relaxation times too can be helpful.

The line-position information derived from EPR is summarized for each centre by a set of spin-hamiltonian parameters, typically as 3 x 3 matrices. Such matrices are utilized since they are able to quantify the anisotropy of the spectrum, i.e., exactly how the relative orientation of the quartz crystal and the externally applied magnetic field **B** affects the spectrum. Each matrix can be diagonalized, furnishing both geometric and quantum-mechanical information. Additional insight often is obtained from the EPR line widths and relative intensities.

It cannot be overemphasized that quantum-mechanical modelling of the defect centres is absolutely essential in quantitatively interpreting EPR spectra. Happily, dramatic progress continues in our capability to do so, both with mathematical techniques and with computer power. We cannot review this field here. We content ourselves by citing a few recent examples: excitons [SS90, RMW91]; oxygen vacancies [GM88, EF90, F90, SCDW91]; AlO_4 [MW86, SCDW91]; hydrogarnet defects [PJHOC92].

We shall first examine those defects which do not directly depend on the presence of foreign atoms, and then turn to the vast majority which are impurity linked, i.e., extrinsic.

Excitons

Since SiO_2 in its ground state is diamagnetic, excitation and consequent spin unpairing is required to see EPR of intrinsic species. In the last decade, great progress has been made in this area, with the discovery and characterization of a photo-excited self-trapped exciton in α-quartz [HKSWD84, FHS90, H90]. With x-ray excitation at 1.7 K, the optical emission intensity caused by recombination of the charge-separated species was observably modulated by chopping (at 185 Hz) the microwave frequency used for EPR (at B ~ 0.3 T). This optical detection of magnetic resonance (ODMR) allowed identification of an unstable triplet spin (S=1) system. The two spins are thought to reside on Si and O at the "ends" of a temporarily broken bond, with especially the oxygen ion displaced from its normal location. No second exciton, based on the other type of Si–O bond, was reported.

E' Centres

The E' centres by definition all contain oxygen vacancies. The primary centre, E'_1, does not appear to contain any impurity atom, whereas the others do (e.g., H^+, Li^+) [W84]. It is thought to have the structure $[O_3Si. SiO_3]^+$, with the unpaired spin density mostly on the silicon ion of the "short-bond" type with respect to the departed O^- ion. (The parent neutral centre, E_1, is of course diamagnetic and not amenable to EPR study.)

Summaries of the various known E' centres (E'_1, $E'_2(M)$, E'_4 (= $[O_3Si{:}H.SiO_3]^0$) and several triplet-state related species) are in the literature [HJB84, W84]. In the last decade, disappointingly, very few new EPR studies of E' centres in α-quartz appear to have been published, despite the importance of these defects. One exception is a careful EPR and ^{29}Si ENDOR study [JBH83] of the E'_1 centre, which demonstrated that one dominant and two peripheral silicon nuclei are observable. On the other hand, numerous theoretical papers on the topic have appeared, but cannot be reviewed here (however, see [F86, F90, SCDW91]).

The E'_1 centres in α-quartz are useful for investigating basic EPR phenomena. For instance, signal oscillations caused by sudden magnetic-field changes have been reported [FR82, KY85]. The decay of the electron spin echo signal from E'_1 in powdered quartz yielded an early example of Fourier-transform EPR [HF86]. In recent years, use of the E'_1 EPR signal as a high-level radiation dosimeter has been suggested [WR90]. Doses (γ-rays) from 100 Gy to 50 MGy can be covered, at temperatures up to 600 K.

Trapped-Atom Centres

X-irradiation of selected quartz crystals below 100 K produces hydrogen atoms H^0. The occurrence and nature of hydrogen in α-quartz has been discussed [AR84, AKR84, H84]. Analysis of the anisotropic EPR spectrum from H^0 discloses that it carries the complete spin population, i.e., is hardly bonded to other atoms and has a compressed ground-state wavefunction [W84]. The atoms are trapped in sites off the main c-axis channels. Warming above 100 K detraps them, whereupon they react chemically in various ways (e.g., to produce $[TiO_4/H]^0_{A,B}$ and $[GeO_4/H\ Li_a\ Li_b]^0$).

The analogous short-lived atom muonium (i.e., μ^+ with e^-) has been studied in α-quartz by muon spin rotation spectroscopy [BHLDEV83, W84, BGGKKMSBKR85, BGGKKMSBKR86]. One source of interest has been the electronic quadrupole moment of this hydrogenic atom [W81, BKR83, BGGKKMSBKR85]. Another aspect has been observation of the muonium signal both in trap sites and after thermal detrapping, i.e., as it diffuses along c-axis channels [W84]).

Among the alkalis, only atomic lithium has been observed in crystalline SiO_2 [JHH86, WWR86, BW91]. The atom Li^0 appears to be trapped between silica tetrahedra in the major c-axis channels, interacting preferentially with one of these to reduce its electron spin density remarkably and to alter its charge. This centre has been associated with a thermoluminescence peak occurring at 190 K [HJH86]. Trapped atoms Cu^0 and Ag^0 too are known in α-quartz [W84], and may experience similar effects. (Paramagnetic ions Ni^+, Cu^{++} and Ag^{++} also have been studied [SMS74, SM78]).

Al Hole Centres

Considerable research activity has continued in studies and applications of the aluminum-based centres in α-quartz. The diamagnetic precursors, $[AlO_4/M]^0$ on irradiation act as donors of electrons and of ions M^+ (= H^+, Li^+, Na^+, Ag^+), and in some instances of atoms M^0, to yield paramagnetic centres. Each such species $[AlO_4/M]^+$ and $[AlO_4]^0$ contains a single unpaired electron (i.e., a hole) occurring mostly in a non-bonding orbital on one of the four bridging oxygen ions. Some of these centres can be generated by ultraviolet radiation; others require higher energy [LWW85]. The species $[AlO_4]^+$ is known and contains two such electrons (triplet state) on symmetry-related oxygens [W84].

The centre $[AlO_4/Na]^+$ has recently been characterized by EPR [DW90]. Only a single type of such a species was observed, so that the Na^+ position adjacent to the Al^{3+} ion appears to be unique. Decay, i.e. release of Na^+, becomes evident on warming to 150 K.

An electron-nuclear double resonance (ENDOR) study [BH86] of $[AlO_4]^0$ reinforced the earlier EPR spin-hamiltonian data [NW81], and contributed discussion regarding the mechanism giving rise to the ^{27}Al hyperfine effects. The latter were interpreted [AJS85] in terms of an exchange polarization mechanism which (evaluated semi-empirically) yielded reasonable values of the coupling parameters as well as an estimate of the Al-O distance.

Studies of aluminum hole centres as a function of irradiation dose (1.7 MeV electrons) have been made available [MCH88]. The concentration dependence of $[AlO_4]^0$ and of $[AlO_4/H]^+$ formed, as measured by EPR, is sample dependent and not simple, largely as a result of competition between the amounts of various $[AlO_4/M]^0$ (i.e., alkali ions and protons) present. The dose dependence under γ-irradiation for centres $[AlO_4]^0$ and H^0 was measured and discussed in terms of hole/exciton migration [MAYF88].

A series of further hole centres, some based on presence of aluminum, were discovered in natural citrines and rose-colored crystals [ML83a,b]. One of these species features an oxygenic hole in the bond sequence Al–O–P.

A whole series of EPR papers by Brik et alii [BMI80, BMLS80, B82a,b, BIMO82, RBIML84, B85, BMTK86, BV86, RBG88] reported investigations of properties (new magneto-electric phenomenon, hole tunneling, spin-lattice relaxation, spin cooling) of $[AlO_4]^0$ affected by externally applied electric fields (dc and ac). They follow up on the pioneering work of Taylor and Farnell [TF64] and of Schnadt and Schneider [SS70]. Dielectric relaxation measurements at low temperatures disclose presence of single-phonon-assisted tunneling [TLN88]. The Al^{3+} and hole-bearing oxygen form a local electric dipole which is transferable, since the hole can jump between neighboring oxygens. The optical band (smoky quartz color) associated with this process has been delineated by means of optically detected magnetic resonance [MLSW84].

A decade of research into the very low-temperature dielectric and heat-capacity aspects of $[AlO_4]^0$ has been summarized by Saint-Paul and Nava (SN84). For instance, hole jumping has been investigated via electric-field pulsed echo spectroscopy.

Evidence has been produced [KELCH87], in part from EPR measurements of $[AlO_4]^0$ concentrations, that vacuum-swept crystals contain the diamagnetic uncompensated species $[AlO_4]^-$, which is isoelectronic with $[SiO_4]^0$.

Electrolysis at high temperatures (1100 K) has led to discovery [VEL89] via EPR of a new defect analogous to $[AlO_4]^0$. It has its hole localized, presumably because of some nearby structural disturbance (oxygen vacancy?).

Hydrogenic Hole Centres

Multi-proton centres $[(OH)_4]^0$ (alias hydrogarnet-type defects), in which four H^+ are thought to replace a single Si^{4+}, have been reported [W84]. Irradiation can produce EPR-sensitive centres by removing an electron, presumably from one oxygen ion, to produce $[(OH)_4]^+$ and when a proton absents itself: $[O(OH)_3]^0$. The ^1H hyperfine matrices have now been obtained, but interpretation is not yet complete [W92]. These centres have been implicated in recombination luminescence (see below) and as possible agents in hydrolytic dissociation of rocks [MCHT83, H84, P86].

Ti Centres

The oxidation state Ti^{3+}, created by irradiation, is detectable in rose quartz by EPR, and occurs in a number of distinct centres. It is formed from Ti^{4+} (i.e., $[TiO_4]^0$) which is an electron acceptor.

Species $[TiO_4]^-$ constitutes a rare example of a $3d^1$ ion in tetrahedral surroundings. Recent EPR work [BPSSW92] has revealed from $^{47,49}Ti$ hyperfine and relaxation (τ_1) data that this centre is indeed close to tetrahedral in symmetry, relaxing via an Orbach process, and has close to 100% of its spin density on the titanium ion (Fig. 3). Centre $[TiO_4]^-$ is unstable above ca. 120 K, relative to alkali-compensated $[TiO_4/M]^0$ formed from it by thermally activated diffusion of an alkali ion M^+. By comparison, $[TiO_4/H]^0$ appears [RW72] to be formed primarily by direct diffusion of atomic hydrogen H^0 to $[TiO_4]^0$.

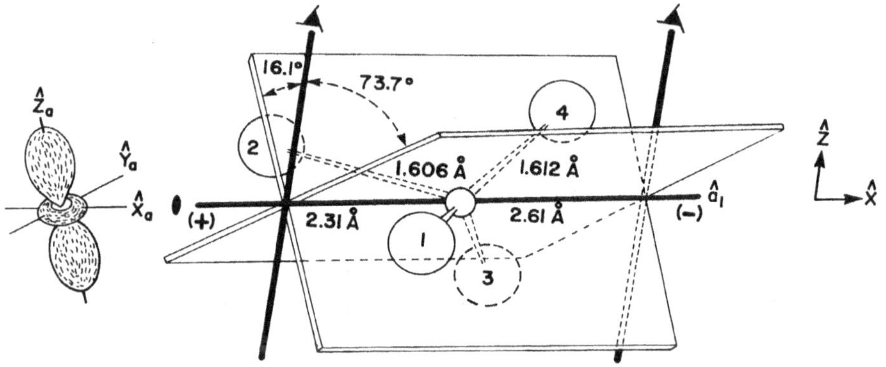

Fig. 3. A SiO_4 tetrahedron in crystalline quartz at 300 K. With Ti^{3+} in place of Si^{4+}, yielding the $S = 1/2$ center $[TiO_4]^-$, the unpaired electron is distributed in a Ti 3d orbital as shown at the left. From [BPSSW92].

Good progress has been made in characterizing the centers $[TiO_4/Li]^0$ and $[TiO_4/Na]^0$. With Li, several such Ti^{3+} species (labelled A, B and γ) are now known. Centres $[TiO_4/Li]^0_{A,B}$ exhibit two-fold symmetry ($\geq$ 35 K) consistent with Ti substituted for Si, with differing positions of the Li^+ ion [IBNW83, ITW88, BW92b]. Centre $[TiO_4/Na]^0_A$ has only C_1 symmetry, but its 9 GHz EPR spectrum shows a remarkable Na^+ dynamic effect, leading to effective C_2 symmetry above ca. 60 K [BW92a]. A second Na centre (labelled β) also exhibits C_1 symmetry [BW92b].

The variety of Ti^{3+} centres (occurring in rose quartz) has its counterpart with Ge^{3+} and Fe^{3+}, as will be seen below. Much more interpretative work, in terms of ion positions and local electric fields, is needed.

Fe Centres

Iron is known potentially to occur in numerous oxidation states, including +1, +2, ..., +6. EPR has, so far, only disclosed presence of the +3 and +4 ions in α-quartz.

The +2 state appears not to have been detected by EPR in SiO_2. Some optical evidence for presence of Fe^{2+} in α-quartz does exist [CH74, KPZ76]. However, consistent with its integral electron spin (typically S=2), its spin-lattice relaxation time τ_1 will normally tend to be so short that EPR detection would require very low temperatures, and its energy-level spacings may well be too large for absorption to occur at usual frequencies. More careful work is called for.

The +3 state is abundant in many natural and cultured quartz crystals. The EPR characteristics of the uncompensated centre $[FeO_4]^-$ have been well characterized [MTW86], but whether the ferric ion here occurs substitutionally for Si^{4+} or interstitially is still the subject of some debate [SL89a]. Basically, the controversy arises because no evidence exists that ions M^+ diffuse up to $[FeO_4]^-$ to neutralize it. However symmetry analysis [MTW86, MW87] of its surroundings does suggest that the Fe^{3+} ion is located in a silicon site.

In recent years, EPR studies of the charge-compensated ferric centres $[FeO_4/M]^0$ with M = H, Li, Na have been numerous and quantitative, largely due to the development of suitable computer programs capable of analysis of the complicated EPR patterns (Fig. 4). The central ion is of the high-spin (S = 5/2) type, with the unpaired-electron distribution held rather closely, as is evident from ^{57}Fe hyperfine measurements [BL80, SL89a, MW89]. Here the hyperfine splitting is very anisotropic, even though its parameter matrix **A** suggests isotropy, due to the influence of the electronic quadrupole effect (matrix **D**). The latter is sensitive to the compensator ion; for instance, ^6Li and ^7Li show distinct EPR lines (Fig. 5) because of this [SL89a]. It has now become evident that very numerous centres $[FeO_4/M]^0$ occur: With lithium, at least five such species occur and have recently been studied by EPR [HC87, HHMMW89, CC89, MWM89, SL89a, HC89, CW90]. Presumably several Li^+ positions at each Fe^{3+} exist, possibly with occurrence of other still–unknown structural differences. The sodium-compensated centres, prepared by exchange electrolysis, have also been investigated [KLNH87, SL89b, MMW89]. The sodium ion, although larger than the lithium ion, shows much greater dynamic effects, presumably due to its off-axis location and decreased bonding to oxygen ions. The primary proton-compensated centre, $[FeO_4/H]^0$, too has been characterized by EPR; here hydrogen-bonding to one neighboring (long-bonded) oxygen ion is evident [MMHWH89].

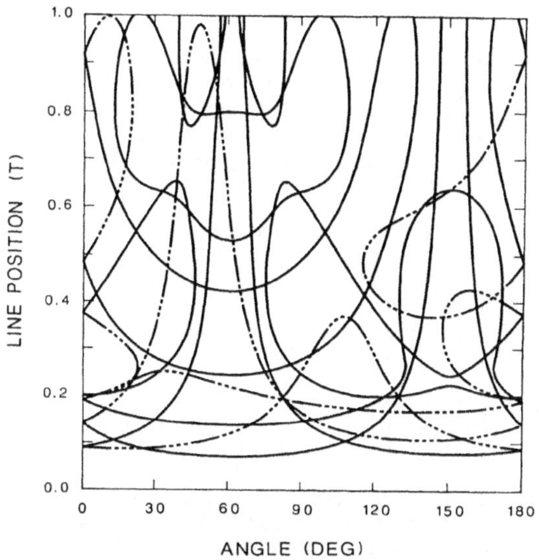

Fig. 4. EPR line positions of $[FeO_4/Li]^0{}_A$ at 9.915 GHz and T = 20 K as a function of crystal rotation about a crystal axis a_i. All symmetry-related sites are included: solid curves - site 1, dot-dash curves - sites 2 and 3. The ^7Li quartet hyperfine structure is not discernible, for the field scale used. Angle 0° is at **B** ‖ **c**. From [HHMMW89].

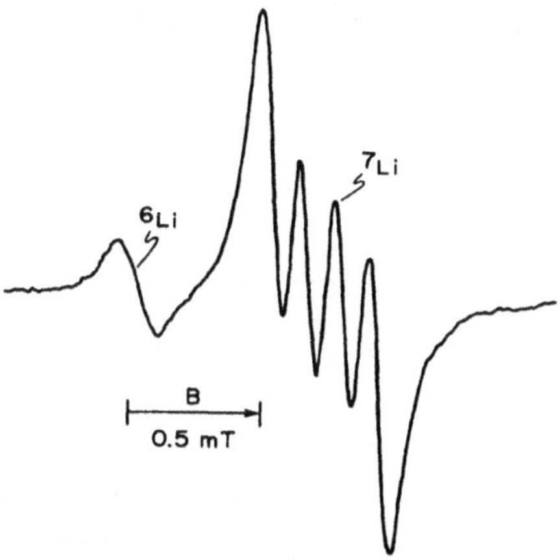

Fig. 5. EPR spectrum of $[FeO_4/Li]^0{}_A$ at 9.52 GHz and 50 K, showing the large isotope shift between ^6Li (I = 1) and ^7Li (I = 3/2) for a fine-structure transition of one site, with **B** ‖ **a**. The ^6Li hyperfine triplet is not resolved. From [SL89a].

The EPR linewidths can give a measure of the concentration of faults in quartz crystals, as was demonstrated with the signals from various Fe^{3+} centres [NSLS86]. This may have implications for quality-control testing.

Irradiation, using an Van de Graaff accelerator at 1.75 MeV in the electron mode, disclosed that M = Li and Na are replaced by H in $[FeO_4/M]^0$, when it is carried out above ca. 200 K [HHH90]. Presumably initial exciton formation causes oxidation to Fe^{4+} with release of the ions M^+, possibly followed by H-atom incorporation.

EPR of the +4 state has been difficult to study because of rather large electronic quadrupole (zero-field) splittings, but has been reported [C76]. Its presence is linked to amethystine color.

Ge Centres

A germanium analog of the pure-quartz self-trapped exciton discussed earlier was discovered and characterized by ODMR [HJ88]. Here the triplet state is based on a distorted Ge—O pair, with recombination luminescence 10% lower in energy than that observed in undoped quartz.

As discussed below, it is well known that Ge^{4+} can trap an electron to form anionic centres $[GeO_4]^-$ containing Ge^{3+}. A theory for this effect has been made available [ERD84]. A surprising development in the last decade has been the discovery [HJ85, HJ86, JKSH87] that $[GeO_4]^0$ can also trap holes, i.e., it is amphoteric. This occurs on x-irradiation at 4 K. Presumably the hole (missing electron) is trapped in a non-bonding O adjacent to Ge. Three unstable centres were reported, and are thought to contain: a hole on long-bond O, a hole on short-bond O, and a hole on O between two Ge ions. These await study using an ^{17}O-enriched quartz crystal, i.e., to investigate the resulting hyperfine splittings.

Such ^{17}O experiments have recently been reported [MWS92] for centres $[GeO_4]^-_{I,II}$, where II is a thermally excited (≥ 50 K) version of ground-state species I. The EPR study of these confirms that most of the previously missing spin population occurs on the oxygen ions (Ge^{3+} bears only ca. 50%). In Ge(I), no two adjacent oxygens are equivalent; no such distortion is observable on Ge(II), which shows normal two-fold rotational symmetry about $\mathbf{a}_i$ (at least in the absence of ^{17}O). Similar studies on the neutral centres $[GeO_4/Li]^0_{A,C}$, obtained from $[GeO_4]^-$ via Li^+ self-diffusion on warming, are now complete [MW92]. A study of spin cross-relaxation between the Ge^{3+} centres and $[AlO_4]^0$ has been published [BLM90].

Detailed EPR measurements on $[GeO_4/Na]^0_{A,C}$ have now been carried out [DWD91]. Both centres exhibit reversible change in symmetry from C_1 to C_2 on warming from 75 K (Fig. 6). The change is associated with the defect jumping between configurations related by axis $\mathbf{a}_i$, with energy barriers 0.223(6) eV for centre A and 0.178(4) eV for C. Above 270 K, centres A and C are observably in equilibrium. They are thus deemed to be isomeric species. Note that here too, as with Ti and Fe, compensating ion Na^+ is more dynamic than is Li^+ (or H^+) in its EPR behavior.

Little progress has been made in the study of the Ge^{2+} substitutional species $[GeO_4/M_1M_2]^0$ which can incorporate atomic hydrogen to form paramagnetic Ge^{3+} centres [W84]. Similarly the multi-alkali-ion germanium centres reported by Lorenze and Feigl [LF73] remain quite obscure.

Germanium centres exhibit room-temperature EPR signals in many natural quartz-type polycrystalline samples [RMKM86]. Attempts to use these and other species in geological aspects (e.g., in prospecting for rare-metal mineralization [RMKM86, MBLDKGS89, VCRK90]), abound.

Miscellaneous Centres

No progress appears to have been made in elucidating the structure of the potentially important defect reported by Baker and Robinson [BR83]. This was suggested to be an O_2^- ion, incorporated somehow into the quartz structure. Measurement of ^{17}O hyperfine structure is much needed.

Attempts to find any paramagnetic centre featuring carbon substituted for silicon and tetrahedrally bonded, gave negative results [RWW89]. A search for E' centres containing carbon should continue. Such centres with germanium are known [W84].

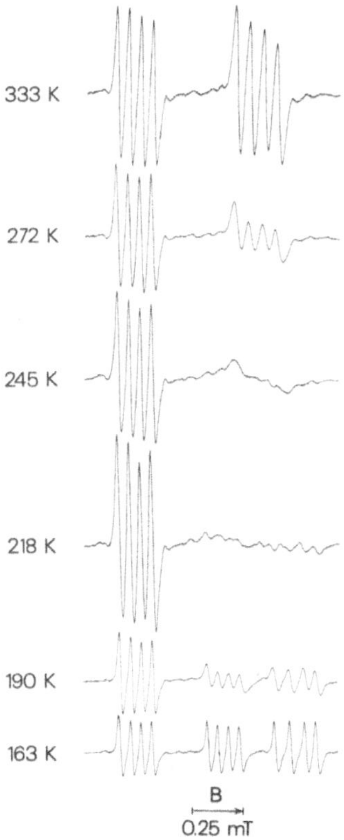

Fig. 6. Temperature dependence of the EPR spectrum of $[GeO_4/Na]^0_A$, showing the effect (site averaging) of the dynamic motion of the ^{23}Na (I = 3/2) ion. Here ν = 9.98 GHz, and **B** is 20° from c. From [DWD91].

The thermoluminescence observed on heating crystalline quartz (383 K) has been shown [YM90, YM88, MCH85] via EPR measurements to be caused by electron transfer from excess-electron centres (e.g., $[GeO_4]^-$ and $[GeO_4/M]^0$) to hole centres (e.g., $[AlO_4]^0$ and $[O(OH)_3]^0$). An EPR study of the behavior of various centres in quartz under thermal annealing (575-1225 K) has been recently published [R89]. These various radiation-damage annealing results are relevant to the dating and dosimetry techniques being developed (for a review, see [M84]).

Various applications utilizing EPR of α-quartz centres are being explored in the realm of geochronometry (dating of mineral, rock and sediment formation) and geothermometry (formation temperatures). Examples may be found in the recent literature [YFQ85, LCCF91, FYM91, RO91, TI91, ABDHP91].

A comparative study of the effect of fast-neutron irradiation on crystalline and vitreous SiO_2 has been published [CGE88]. The powder E' EPR spectrum shows considerable broadening as a function of increasing dose. Similarly, EPR and static magnetic susceptibility measurements [VVS86] reveal a complicated radiation-damage situation. Work by Jani and Halliburton [JH84] on quartz single crystals irradiated with fast neutrons revealed presence of E'_1 and $[AlO_4]^0$, as well as of two previously unreported double-hole (S=1) centres. Two other new centres, each containing a proton, have been observed in neutron-irradiated natural quartz [ML84].

Neutron irradiation can transmute silicon to phosphorus, with paramagnetic centres $[PO_n]^q$ formed. These have been observed by EPR [CZ84]. The species $[PO_4]^0_{I,II}$ in α-quartz doped with ordinary phosphorus (^{31}P) have been known for some time and studied by EPR [W84].

SUMMARY

Herein, we have demonstrated the power of the EPR technique, and the complexity of the crystalline quartz system in the realm of paramagnetic defects. Many major questions remain to be resolved. For instance, we note the virtual absence of knowledge about ions X^{2+} and X^{5+} substituted for Si^{4+}, about ions replacing O^{2-}, or interstitial ions M^{2+}. Much more work is needed to elucidate the structure of many of the centres seen by EPR and listed in this review. We have of necessity had to ignore closely related areas: e.g., surface phenomena including SiO_2/Si interfaces, and fused-quartz defects. These two topics are amply reviewed elsewhere (e.g., [B89, G90]).

Clearly, there is much further effort needed in the study of the point defects in α-quartz. Promising EPR areas include: short-lived species observed by pulsed techniques (e.g., Fourier-transform EPR), optically detected EPR, as well as high-temperature and high-pressure EPR. Incorporation of chemical elements other than those listed in Table 1 also is called for. We trust that interest in and support for such research will materialize.

ACKNOWLEDGMENTS

This manuscript was produced under the auspices of the Natural Sciences and Engineering Research Council of Canada. We would like to take this opportunity to thank Sawyer Research Products (Eastlake, OH), and in particular Dr. Baldwin Sawyer, for their interest and assistance over the years.

REFERENCES

Adrian F.J., Jette A.N., and Spaeth J.-M., 1985, Theory of indirect hyperfine interactions of oxygen-aluminum defects in ionic crystals, Phys. Rev. B31:3923.

Agel A., Bershov L.V., Dusausoy Y., Hafner S.S., and Petrov A., 1991, Paramagnetische defekte in quarz und deren thermische und radiologische abhängigkeit, KTP Report 91:224 (Germ.).

Aines R.D. and Rossman G.R., 1984, Water in minerals? A peak in the infrared, J. Geophys. Res. 89:4059.

Aines R.D., Kirby S.H., and Rossman G.R., 1984, Hydrogen speciation in synthetic quartz, Phys. Chem. Minerals 11:204.

Bailey P. and Weil, J.A., 1991, EPR study of the $[SiO_4/Li]^0$ centre in α-quartz, J. Chem. Soc. Faraday Trans. 87:3143.

Bailey P., Pawlik T., Söthe H., Spaeth J.-M., and Weil J.A., 1992, $[TiO_4]^-$ in α-quartz studied by low-temperature electron paramagnetic resonance, J. Phys.: Condens. Matter 4:1.

Bailey P. and Weil J.A., 1992, The EPR spectral parameters and dynamic properties of the centre $[TiO_4/Na]^0_A$ in x-irradiated α-quartz, J. Phys. Chem. Solids 53:309.

Bailey P. and Weil J.A., 1992, EPR of $[TiO_4/Li]^0_B$ and related centres in x-irradiated α-quartz, J. Phys. Chem. Solids 53:601.

Baker J.M. and Robinson P.T., 1983, EPR of a new defect in natural quartz: possibly O_2^-, Solid State Commun. 48:551.

Barsov S.G., Getalov A.L., Gordeev V.A., Kruglov S.P., Kuzmin L.A., Mikirtychyants S.M., Shcherbakov G.V., Baryshevskii V.G., Kuten S.A., and Rapoport V.I., 1985, Investigation of muonium quadrupole interaction in α-quartz in the temperature regions 75-120 K, Phys. Lett. 110A:225.

Barsov S.G., Getalov A.L., Gordeev V.A., Kruglov S.P., Kuzmin L.A., Mikirtychyants S.M., Shcherbakov G.V., Baryshevskii V.G., Kuten S.A., and Rapoport V.I., 1986, Temperature-dependence of the quadrupole interaction and diffusion of muonium atoms in α-quartz, Hyperfine Interact. 32:631.

Baryshevskii V.G., Kuten S.A., and Rapoport V.I., 1983, Quadrupole interactions of muonium in α-quartz, J. Phys. C: Solid State Phys. 16:6651.

Becker W. and Lehmann G., 1980, Anomalous hyperfine splitting of ^{57}Fe in α-quartz, Solid State Commun. 35:367.

Bossoli R.B. and Halliburton, L.E., 1986, ^{27}Al hyperfine and quadrupole interactions for the $[AlO_4]^0$ center in quartz, Phys. Stat. Solidi (b) 136:709.
Brice J.C., 1985, Crystals for quartz resonators, Rev. Mod. Phys. 57:105.
Brik A.B., Matyash I.V., and Ishchenko S.S., 1980, Increase of magnetization of a paramagnet by an alternating electric field, Sov. Phys. JETP 52:960.
Brik A.B., Matyash I.V., Litovchenko A.S., and Samoilovich M.I., 1980, Electric field effect on paramagnetic $Al–O^-$ centers in quartz, Sov. Phys. Solid State 22:1849.
Brik A.B., 1982, Kinetics and mechanism of the relaxation magnetoelectric effect, Sov. Phys. Solid State 24:282.
Brik A.B., 1982, Determination of internal electric fields in quartz using the relaxational magnetoelectric effect, Sov. Phys. Solid State 24:912.
Brik A.B., Ishchenko S.S., Matyash I.V., and Okulov S.M., 1982, Pressure-induced localization of paramagnetic centers in one of the structural positions, Sov. Phys. Solid State 24:1044.
Brik A.B., 1985, Anomalous relaxational magnetoelectric effect and its characteristics, Sov. Phys. Solid State 27:91.
Brik A.B., Matyash I.V., Takzei G.A., and Kostyshin A.M., 1986, Effect of an electric field on the static magnetization of $Al–O^-$ centers in quartz, Sov. Phys. Solid State 28:536.
Brik A.B. and Vikhnin V.S., 1986, Effect of an electric field on the spin-lattice relaxation of $Al–O^-$ centers in quartz, Sov. Phys. Solid State 28:662.
Brik A.B., Larikov A.L., and Matyash I.V., 1990, Steady-state electric-field cooling of paramagnetic Ge^{3+} centers in quartz, Sov. Phys. Solid State 32:757.
Brower K.L, 1989, Electron paramagnetic resonance studies of Si-SiO_2 interface defects, Semicond. Sci. Technol. 4:970.
Brown J.A., Heffner R.H., Leon M., Dodds S.A., Estle T.L., and Vanderwater D.A., 1983, Double electron-muon resonance experiments on muonium in α-quartz, Phys. Rev. B27:3980.
Chan S.L., Gladden L.F., and Elliott S.R., 1988, Magnetic resonance studies on neutron-irradiated quartz and amorphous silica, J. Non-cryst. Solids 106:413.
Cherlov G.B. and Zatsepin A.F., 1984, Formation of phosphorus centres during nuclear transmutation of α-quartz, Radiats. - Stimul. Yavleniya Tverd. Telakh 6:43 (Russ.).
Choi D. and Choh S.H., 1989, Temperature dependence and hyperfine structure of Fe^{3+} centres in synthetic quartz, J. Phys.: Condens. Matter 1:7661.
Choi D. and Weil J.A., 1990, EPR study of Fe^{3+} in α-quartz: further lithium-compensated centers, Phys. Rev. B42:9759.
Cohen A.J. and Hassan F., 1974, Ferrous and ferric ions in synthetic α-quartz and natural amethyst, Am. Mineral. 59:719.
Cox R.T., 1976, ESR of an S=2 center in amethyst quartz and its possible identification as the d^4 ion Fe^{4+}, J. Phys. C: Solid State Phys. 9:3355.
Dickson R.S. and Weil J.A., 1990, The magnetic properties of the oxygen-hole aluminum centres in crystalline SiO_2. IV. $[AlO_4/Na]^+$, Can. J. Phys. 68:630.
Dickson R.S., Weil J.A., and Davis P.H., 1991, The paramagnetic germanium-sodium impurity centres $[GeO_4/Na]^0_A$ and $[GeO_4/Na]^0_C$ in α-quartz, Can. J. Phys. 69:761.
Edwards A.H. and Fowler W.B., 1990, Relationship between hyperfine parameters and the geometry of defects in non-metallic solids, Phys. Rev. B41:10816.
Ekenberg U.A., Robertson J., and Dow J.D., 1984, Impurity states in SiO_2, Phys. Rev. B29:2216.
Falgueres C., Yokoyama Y., and Miallier D., 1991, Stability of some centers in quartz, Nucl. Tracks Radiat. Meas. 18:155.
Fedoruk G.G. and Rutkovskii I.Z., 1982, Free induction of E'_1 centres in crystalline quartz, Phys. Stat. Solidi (b) 112:453.
Fisher A.J., Hayes W., and Stoneham A.M., 1990, Structure of the self-trapped exciton in quartz, Phys. Rev. Lett. 64:2667.
Fowler W.B., 1986, Electronic properties of SiO_2 and models for its point defects, Struct. Bonding Noncryst. Solids (Int. Symp. 1983), Walrafen G.E. et alii - Editors, Plenum, New York, NY, pp. 157-183.
Fowler W.B., 1990, Theory of defects and defect processes in silicon dioxide, Rev. Solid State Sci. 4:551.
Georgiev M. and Manov A., 1988, Pseudo-Jahn-Teller and relaxation-rate analysis of a model for the hydride ion at the E'_4 center in alpha-quartz, Solid State Commun. 65:513.

Griscom D.L., 1990, Electron spin resonance, Glass Sci. Technol. 48:151.
Halperin A., Jani M.G., and Halliburton L.E., 1986, Correlated EPR and thermoluminescence study of the $[SiO_4/Li]^0$ center in quartz, Phys. Rev. B34:5702.
Halliburton L.E., Jani M.G., and Bossoli R.B., 1984, Electron spin resonance and optical studies of oxygen-vacancy centers in quartz, Nucl. Intrum. Methods Phys. Res. B1:192.
Halliburton L.E., Hantehzadeh M.R., Minge J., Mombourquette M.J., and Weil J.A., 1989, EPR study of Fe^{3+} in α-quartz: a reexamination of the lithium-compensated center, Phys. Rev. B40:2076.
Han C.S. and Choh S.H., 1987, EPR study of Korean natural amethyst, Bull. Magn. Reson. 9:91.
Han C.S. and Choh S.H., 1989, EPR study of iron centers in natural amethyst, J. Korean Phys. Soc. 22:241.
Hantehzadeh M.R., Han C.S., and Halliburton L.E., 1990, Radiation-induced mobility of interstitial alkali ions in iron-doped quartz, J. Phys. Chem. Solids 51:425.
Hayes W., Kane M.J., Salminen O., Wood R.L., and Doherty S.P., 1984, ODMR of recombination centres in crystalline quartz, J. Phys. C: Solid State Phys. 17:2943.
Hayes W. and Jenkin T.J.L., 1985, Paramagnetic hole centres produced in germanium-doped crystalline quartz by x-irradiation at 4 K, J. Phys. C: Solid State Phys. 18:L849.
Hayes W. and Jenkin T.J.L., 1986, Charge-trapping properties of germanium in crystalline quartz, J. Phys. C: Solid State Phys. 19:6211.
Hayes W. and Jenkin T.J.L., 1988, Optically detected magnetic resonance studies of exciton trapping by germanium in quartz, J. Phys. C: Solid State Phys. 21:2391.
Hayes W., 1990, The structure of the self-trapped exciton in quartz, Rev. Solid State Sci. 4:543.
Hobbs B.E, 1984, Point defect chemistry of minerals under a hydrothermal environment, J. Geophys. Res. 89:4026.
Hornak J.P. and Freed J.H., 1986, Spectral rotation in pulsed ESR spectroscopy, J. Magn. Reson. 67:501.
Isoya J., Bowman M.K., Norris J.R., and Weil J.A., 1983, An electron spin echo modulation study of lithium nuclear hyperfine and quadrupole coupling in the A(Ti-Li) center of alpha-quartz, J. Chem. Phys. 78:1735.
Isoya J., Tennant W.C., and Weil J.A., 1988, EPR of the TiO_4/Li center in crystalline quartz, J. Magn. Reson. 79:90.
Jani M.G., Bossoli R.B., and Halliburton L.E, 1983, Further characterization of the E'_1 center in crystalline SiO_2, Phys. Rev. B27:2285.
Jani M.G. and Halliburton L.E, 1984, Point defects in neutron-irradiated quartz, J. Appl. Phys. 56:942.
Jani M.G., Halliburton L.E., and Halperin A., 1986, Observation of a simple lithium-associated electron trap in crystalline SiO_2, Phys. Rev. Lett. 56:1392.
Jenkin T.J.L., Koppitz J., Schirmer O.F., and Hayes W., 1987, Bound small-polaron optical absorption in germanium-doped quartz, J. Phys. C: Solid State Phys. 20:L367.
Kahan A., Euler F.K., Lipson H.G., Chen C.Y., and Halliburton L.E., 1987, Radiation effects in vacuum-swept quartz, Proc. Annu. Freq. Control Symp. IEEE 41st: 216.
Keilson S., Ling S., Nowick A.S., and Halliburton L.E., 1987, Dielectric relaxation and epr in quartz crystals containing iron, Proc. Annu. Freq. Control Symp. IEEE 41st: 223.
Khalilov V. Kh., Pivovarov S.S., and Zakharov V.K., 1976, Coordination of iron(2+) in iron-doped crystalline and vitreous silicon dioxide, Fiz. Khim. Stekla 2:496 (Russ.).
Kuz'min V.S. and Yashin A.N., 1985, Transition mutation signals in inhomogeneously broadened media, Zh. Prikl. Spektrosk. 43:635 (Russ.).
LePage Y., Calvert L.D., and Gabe E.J., 1980, Parameter variation in low-quartz between 94 and 298 K, J. Phys. Chem. Solids 41:721.
Leyderman A., Weil J.A., and Williams J.A.S., 1985, Generation of paramagnetic centres in crystalline quartz by ultraviolet irradiation, J. Phys. Chem. Solids 46:519.
Liang X., Chen M., Chen J., and Feng J., 1991, Estimation of landslide deformation and failure time by α-quartz ESR technique, Hejishu 14:79.
Lorenze R.V. and Feigl F.J., 1973, Defects in crystalline quartz: electron paramagnetic resonance of multiple-alkali-compensated centers associated with germanium impurities. Phys. Rev. B8:4833.

Maschmeyer D. and Lehmann G., 1983, New hole centers in natural quartz, Phys. Chem. Miner. 10:84.
Maschmeyer D. and Lehmann G., 1983, A trapped-hole center causing rose coloration of natural quartz, Z. Kristallogr. 163:181.
Maschmeyer D. and Lehmann G., 1984, New electron centers in neutron-irradiated natural quartz, Solid State Commun. 50:1015.
Matyash I.V., Brik A.B., Larikov A.L., Derskii L.S., Kravchenko G.L., Gamarnik M. Ya., and Shvets D.I., 1989, The EPR characteristics of quartz of gold-ore mineralized ferruginous-silicious rocks in the southern Ukraine, Dokl. Akad. Nauk Ukr. SSR, Ser. B: Geol. Khim. Biol. Nauki (11):21 (Russ.).
McEachern R.J., Weil J.A., and Sawyer B., 1992, Distortion and ^{17}O hyperfine interaction in the centres $[GeO_4]^-_{I,II}$ in α-quartz, Solid State Commun. 81:207.
McEachern R.J. and Weil J.A., 1992, EPR studies of $[GeO_4/Li]^0_{A,C}$ in ^{17}O-enriched α-quartz, to be published.
McKeever S.W.S., 1984, Thermoluminescence in quartz and silica, Radiat. Protect. Dosim. 8:81.
McKeever S.W.S., Chen C.Y., and Halliburton L.E., 1985, Point defects and the pre-dose effect in natural quartz, Nucl. Tracks 10:489.
McLaren A.C., Cook R.F., Hyde S.T., and Tobin R.C., 1983, The mechanisms of the formation and growth of water bubbles and associated dislocation loops in synthetic quartz, Phys. Chem. Minerals 9:79.
Meyer B.K., Lohse F., Spaeth J.-M., and Weil, J.A., 1984, Optically detected magnetic resonance of the $[AlO_4]^0$ centre in crystalline quartz, J. Phys. C: Solid State Phys. 17:L31.
Minge J. and Weil J.A., 1989, ^{57}Fe hyperfine structure of $[FeO_4/Li]^0$ center in α-quartz, J. Phys. Chem. Solids 50:997.
Minge J., Weil J.A., and McGavin D.G., 1989, EPR study of Fe^{3+} in α -quartz: characterization of a new type of cation-compensated center, Phys. Rev. B40:6490.
Minge J., Mombourquette M.J., and Weil J.A., 1989, Dynamic interchange between $[FeO_4/Na]^0$ configurations in α-quartz, Phys. Rev. B40:6523.
Minge J., Mombourquette M.J., and Weil J.A., 1990, EPR study of Fe^{3+} in α-quartz: the sodium-compensated center, Phys. Rev. B42:33.
Miyazaki T., Azuma N., Yoshida S., and Fueki K., 1988, Comparison of long-range activation transfer in gamma-radiolysis of single crystalline quartz with amorphous silica, Radiat. Phys. Chem. 32:695.
Mombourquette M.J. and Weil, J.A., 1986, Structure determination of the AlO_4 centers in alpha-quartz by EPR and SCF MO, J. Magn. Reson. 66:105.
Mombourquette M.J., Tennant W.C., and Weil J.A., 1986, EPR study of Fe^{3+} in α-quartz: A reexamination of the so-called I center, J. Chem. Phys. 85:68.
Mombourquette M.J. and Weil J.A., 1987, Modeling Fe^{3+} in quasitetrahedral surroundings using the Newman superposition model, J. Chem. Phys. 87:3385.
Mombourquette M.J., Minge, J., Hantehzadeh M.R., Weil J.A., and Halliburton L.E., 1989, Electron paramagnetic resonance study of Fe^{3+} in α-quartz: hydrogen-compensated center, Phys. Rev. B39:4004.
Mondragon M.A., Chen C.Y., and Halliburton L.E., 1988, Observation of a dose-rate dependence in the production of point defects in quartz, J. Appl. Phys. 63:4937.
Nienhaus K., Stegger P., Lehmann G., and Schneider J.R., 1986, Assessment of quality of quartz crystals by EPR and γ-ray diffraction, J. Cryst. Growth 74:391.
Nuttall R.H.D. and Weil J.A., 1981, The magnetic properties of the oxygen-hole aluminum centers in crystalline SiO_2. I. $[AlO_4]^0$, Can. J. Phys. 59:1696.
Paterson M.S., 1986, The thermodynamics of water in quartz, Phys. Chem. Minerals 13:245.
Pauling L., 1980, The nature of silicon-oxygen bonds, Am. Mineral. 65:321.
Purton J., Jones R., Heggie M., Öberg S., and Catlow C.R.A., 1992, LDF pseudopotential calculations of the α-quartz structure and hydrogarnet defect, Phys. Chem. Minerals 18:389.
Rakov L.T., Milovidova N.D., Kuvshinova K.A., and Moiseyev B.M., 1986, An ESR study of Ge centers in natural polycrystalline quartz, Geochem. Int. 23:61.
Rakov L.T., 1989, Behavior of paramagnetic defects during thermal annealing of quartz, Sov. Phys. Crystallogr. 34:154.
Rao P.S., Weil J.A., and Williams J.A.S., 1989, An investigation of carbonaceous natural quartz single crystals, Can. Mineral. 27:219.

Rao P.S., McEachern R.J., and Weil J.A., 1991, On a proposed radiation-induced polaronic hole in silicon dioxide, J. Comput. Chem. 12:254.

Rink W.J. and Odom A.L., 1991, Natural alpha recoil particle radiation and ionizing radiation sensitivities in quartz detected with EPR: implications for geochronometry, Nucl. Tracks. Radiat. Meas. 18:163.

Rinneberg H. and Weil J.A., 1972, EPR studies of Ti^{3+}–H^{+} centers in x-irradiated alpha-quartz, J. Chem. Phys. 56:2019.

Roitsin A.B., Brik A.B., Ishchenko S.S., Matyash I.V., and Litovchenko A.S., 1984, Paraelectric resonance of localized Al–O^{-} electric dipoles in quartz, Sov. Phys. Solid State 26:1792.

Roitsin A.B., Brik A.B., and Gokhman V.L., 1988, Electric-field tunnel effects in EPR of paraelectric centers, Sov. Phys. JETP 67:975.

Saint-Paul M. and Nava R., 1984, Very low temperature properties of gamma-irradiated quartz, Ferroelectrics 51:193.

Schnadt R. and Schneider J., 1970, The electronic structure of the trapped-hole center in smoky quartz, Phys. Condens. Matter 11:19.

Shluger A. and Stefanovich E., 1990, Models of the self-trapped exciton and nearest-neighbor defect pair in SiO_2, Phys. Rev. B42:9664.

Sim F., Catlow C.R.A., Dupuis M., and Watts J.D., 1991, Ab initio self-consistent field molecular orbital calculations on defects associated with radiation damage in alpha-quartz, J. Chem. Phys. 95:4215.

Solntsev V.P., Mashkovtsev R.I., and Shcherbakova M.Ya., 1974, Copper and nickel centers in α-quartz, Sov. Phys. Solid State 16:1192.

Solntsev V.P. and Mashkovtsev R.I., 1978, Stabilization of silver and copper atoms in α-quartz, Sov. Phys. Solid State 20:471.

Stegger P. and Lehmann G., 1989, The structures of three centres of trivalent iron in α-quartz, Phys. Chem. Minerals 16:401.

Stegger P. and Lehmann G., 1989, Dynamic effects in a new substitutional center of trivalent iron in quartz, Phys. Stat. Solidi (b) 151:463.

Stewart R.F., Whitehead M.A., and Donnay G., 1980, The ionicity of the Si–O bond in quartz, Am. Mineral. 65:324.

Taylor A.L. and Farnell G.W., 1964, Spin-lattice experiments on color centers in quartz, Can. J. Phys. 42:595.

Toulouse J., Ling S., and Nowick A.S., 1988, Dielectric relaxation of the aluminum-hole center in α-quartz: an example of phonon-assisted tunneling, Phys. Rev. B37:7070.

Toyoda S., and Ikeya M., 1991, ESR dating of quartz and plagioclase from volcanic ashes using E'_1, aluminum and titanium centers, Nucl. Tracks Radiat. Meas. 18:179.

Van Den Bosch A., Vansummeren J., and Stesmans A., 1986, Magnetism of neutron-damaged alpha-quartz, Radiat. Effects 97:183.

Van Moort J.C., Cohen D.D., Russell D.W., and Katsaros A., 1990, Correlations betwen chemical composition as determined by PIXE and the paramagnetism of auriferous vein quartz, Nucl. Instrum. Methods Phys. Res. B49:323.

Vassilikou-Dova A.B., Eftaxias K., and Lehmann G., 1989, An electrolytically generated localized hole center in quartz, Z. Naturf. 44a:278.

Weil J.A., 1981, Electric field effects on the ground-state hyperfine structure of hydrogenic atoms, Can. J. Phys. 59:841.

Weil J.A., 1984, A review of electron spin spectroscopy and its application to the study of paramagnetic defects in crystalline quartz, Phys. Chem. Minerals 10:149.

Weil J.A., 1992, The hyperfine parameters of hole centers $[(OH)_4]^+$ and $[O(OH)_3]^0$ in irradiated α-quartz, to be published.

Wieser A. and Regulla D.F., 1990, Ultra–high–level dosimetry by ESR spectroscopy of crystalline quartz and fused silicate, Radiat. Prot. Dosim. 34:291.

Wilson T.M., Weil J.A., and Rao P.S., 1986, Electronic structure of the interstitial lithium-associated electron trap in crystalline quartz, Phys. Rev. B34:6053.

Yang X.H. and McKeever S.W.S., 1988, Characterization of the pre-dose effect using ESR and TL, Nucl. Tracks. Radiat. Meas. 14:75.

Yang X.H. and McKeever S.W.S., 1990, The pre-dose effect in crystalline quartz, J. Phys. D: Appl. Phys. 23:237.

Yokoyama Y., Falgueres C., and Quaegebeur J.P., 1985, ESR dating of quartz from Quaternary sediments, Nucl. Tracks Radiat. Meas. 10:921.

FORMATION OF Si/SiO_2 HETEROSTRUCTURES BY LOW-TEMPERATURE, PLASMA-ASSISTED OXIDATION AND DEPOSITION PROCESSES

G. Lucovsky, Yi Ma, T. Yasuda and S. Habermehl

Departments of Physics, and Materials Science and Engineering
North Carolina State University
Raleigh, North Carolina 27695-8202

ABSTRACT

A new approach to the low-temperature, < 500°C, formation of SiO_2/Si heterostructures formed by plasma-assisted processing is discussed. Following an ex-situ RCA clean, Si(100) wafers are exposed to O-atoms generated in a remote plasma. This: i) eliminates residual C-atom contamination; and ii) forms a thin oxide layer, ~0.5-0.6 nm. SiO_2/Si heterostructures are completed by remote PECVD deposition of an oxide film ~15 nm thick. The SiO_2/Si interface and bulk oxide electrical properties of MOS devices are similar to those of devices with thermal oxides grown at ~1000°C. This paper: i) compares the use of pre-deposition H-atom and O-atom treatments, prior to oxide deposition; and ii) discusses changes in electrical properties, when composite oxide/nitride dielectrics are prepared by sequential oxide and nitride remote PECVD depositions.

INTRODUCTION

There is considerable interest in the development of low-temperature, T < 500°C, processes for the formation of SiO_2/Si heterostructures that can be integrated into field effect transistors, FET's. This derives from a necessity to reduce thermal budgets, i.e., processing temperatures and/or times, to limit dopant atom diffusion, and thereby preserve sub-micron geometrical features, as for example, in sharp metallurgical p-n junctions. To understand the significance of these results, it is first necessary to identify the ways SiO_2/Si interfaces are formed during oxide growth or deposition. In the conventional high-temperature, T = 850-1050°C, thermal, or rapid thermal oxidation processes, the SiO_2/Si interface is continuously regenerated during the growth of the *bulk* oxide layer. The Si substrate is consumed to generate the oxide film, so that the SiO_2/Si interface is located in the bulk of the Si, below the original Si surface (Fig. 1(a)). In contrast, in an idealized low-temperature deposition process, e.g., photo- or plasma-assisted chemical-vapor deposition, CVD, the SiO_2/Si interface is expected to be at the original Si surface, as shown in Fig. 1(b). However, studies have shown,[1,2] that during oxide depositions, active oxygen species, generated either by photons, or in a plasma, can react at the Si surface during the initial stages of film deposition to grow a thin, ~0.5 to 2 nm, oxide film (see Fig. 1(c)). This means that the SiO_2 interface is located in the bulk Si, not at the original Si surface. Studies, in which the oxide deposition was done by remote plasma-enhanced CVD,[1-3] hereafter remote PECVD, have demonstrated that these *grown* oxide layers have a significant effect on the electrical properties of the SiO_2/Si structures. In addition, studies have shown that the electrical properties of the SiO_2/Si interface formed by this parasitic oxidation are sensitive to the pre-deposition Si processing, including ex-situ cleaning of the Si surface by wet-chemical or vapor techniques, and on-line or in-situ cleaning and/or passivation processes.[4,5] Therefore, attempts to form SiO_2/Si heterostructures by thin film deposition processes must also include *customized* pre-deposition ex-situ and/or on-line processing of the Si surface.

The formation of SiO_2/Si heterostructures by remote PECVD, has emphasized the importance of pre-deposition Si surface processing, and the necessity to *match* the in-situ surface processing to the plasma-assisted deposition process.[1,2,4-7] This paper focuses on a recently developed low-temperature process in which SiO_2/Si interface formation and oxide deposition have been separately controlled (see Fig. 1(d)), and also optimized.[5-7] To put

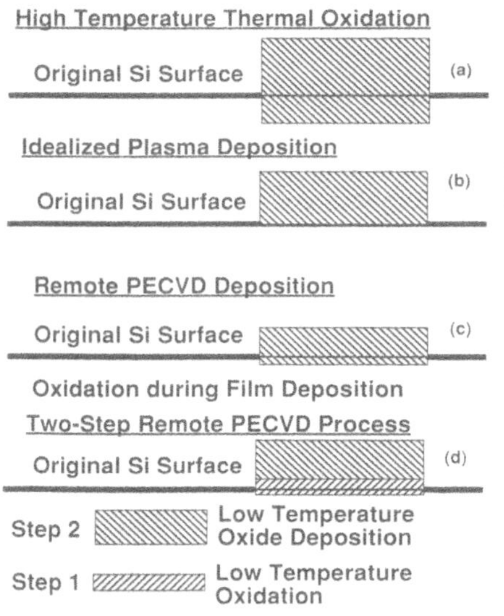

Fig. 1. Schematic representations of SiO_2/Si interfaces formed by: (a) thermal oxidation of Si (b) idealized PECVD; (c) remote PECVD; and (d) the new two-step oxidation/deposition process described in this paper.

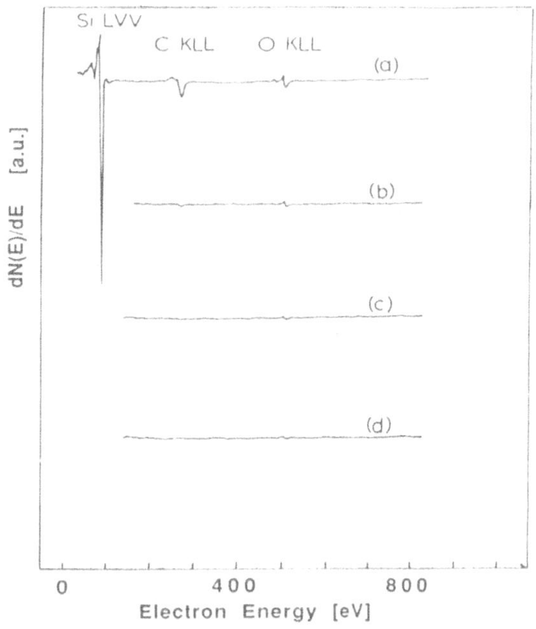

Fig. 2 Derivative AES spectra of the Si(100): (a) after an RCA clean and rinse in dilute (1:30) HF; and after: (b) 30 second; (c) 2 minute; and (d) 5 minute exposures to plasma-generated atomic-H. The strong feature at ~90 eV in (a) is the Si LVV feature; this has been deleted from traces (b) to (d). The features at ~280 eV and 500 eV are due respectively to C and O.

these new results into perspective, and identify the driving forces that have motivated the development of the *new* process (designated as Process B), we first discuss the *original* low-temperature remote PECVD process (designated as Process A). We then describe the *new* process and the way it remedies the deficiencies of the *original* process. The emphasis is on: i) in-situ and/or on-line pre-deposition Si surface processing prior to oxide deposition by remote PECVD; ii) interactions at the processed Si surface at the beginning of the oxide deposition; and iii) the effects i) and ii) on the electrical properties of the SiO_2/Si heterostructures, and in particular the SiO_2/Si interface.

THE *ORIGINAL* REMOTE PECVD PROCESS -- PROCESS A

There are several process steps involved in the formation of an SiO_2/Si heterostructure such as an MOS device that can be used to evaluate the electrical properties of the SiO_2/Si interface and the oxide film. These include: i) ex-situ processing of the Si surface; ii) insertion of the Si wafer into the deposition system, and in-situ processing prior to oxide deposition; iii) oxide deposition, and generally in-situ annealing; iii) metallization, patterning and post-metallization annealing. To identify the effects each step on the electrical properties of the SiO_2/Si heterostructure, it is necessary to characterize the structural and chemical properties of the processed surfaces, interfaces and films at intermediate stages of the process sequence. This has been done by a combination of on-line and off-line probes. Rather than tracing the evolutionary development of this approach, we discuss an *optimized* version of the *original* process, and identify where that process falls short of achieving sufficient performance and reliability for meeting future device fabrication requirements.

The majority of these studies have used epitaxial p/p^+ Si wafers with a 100 orientation. The resistivity of the p^+ region was ~0.5 Ω-cm, and was sufficiently low to limit the series resistance of MOS devices, with areas of $\sim 2\text{-}3\times10^{-3}$ cm^2, to ~10 Ω. The resistivity of the p-region was ~2-5 Ω-cm, and was sufficiently high to: i) provide a significant capacitance swing, ~400 pF, between accumulation and inversion; and thereby ii) allow for the detection of interfacial defect state densities, D_{it}, to the mid 10^9 $cm^{-2}\text{-}eV^{-1}$ range when using the *conventional* high-frequency and quasi-static capacitance-voltage, C-V, technique.

The initial process step consisted of an ex-situ RCA clean, followed by a rinse in dilute HF (1:30); leaving the Si surface hydrophobic. The wafer was then inserted into one of two functionally-equivalent UHV-compatible systems that contained: i) processing chambers for surface preparation and/or dielectric deposition; and ii) an on-line analysis chamber with either back-view low energy electron diffraction, LEED, or reflection high energy electron diffraction, RHEED, and Auger electron spectroscopy, AES.[3] The ex-situ processed Si(100) surfaces: i) displayed a 1x1 symmetry as determined by electron diffraction; and ii) showed as much as 10-20% surface oxygen and carbon as determined from the relative strengths of the Si, C and O derivative signals using AES (see Fig. 2). Other studies have shown that this surface is *fully* hydrogenated with the dominant bonding arrangement being an SiH_2 group that terminates the two dangling bonds on each surface atom.[8] In our initial studies, defined as Process A, *pre-deposition* in-situ cleaning and passivation of the Si(100) surfaces were accomplished by exposure to plasma-generated atomic-H.[1,2] During the hydrogen exposure, the Si substrate was held at 300°C, and the processing times were varied between 30 seconds and 5 minutes. This surface processing: i) left the Si(100) surfaces monohydride terminated; ii) removed C-atom contamination to at least the limit of detection by AES, ~0.5 at. %, by creating volatile hydride species: but iii) did not completely remove native oxides: the

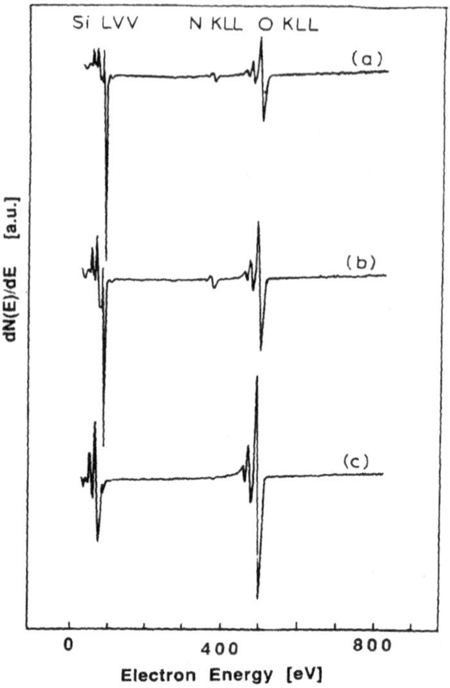

Fig. 3. Derivative AES spectra of Si(100) after RCA clean, rinse in dilute HF, and 2 min exposure to atomic-H: and after (a) 5 s; (b) 20 s; and (c) 2 min. of oxide deposition from N_2O. The strong features from ~50-100 eV are the Si LVV lines; the features at ~380 eV are N-atoms, and those at ~500 eV are O-atoms. Changes in the Si LVV lines indicate the transition from c-Si to a-SiO_2.

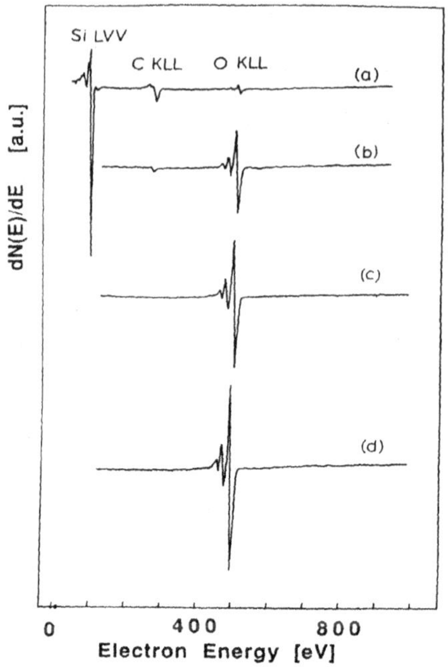

Fig. 5 Derivative AES spectra of Si(100) after: (a) RCA clean; and (b) and (c) 5 s and 15 s oxidations at 300°C. (d) is after a 5 s deposition of SiO_2 onto the surface in (c) using N2O as the O-atom source gas.

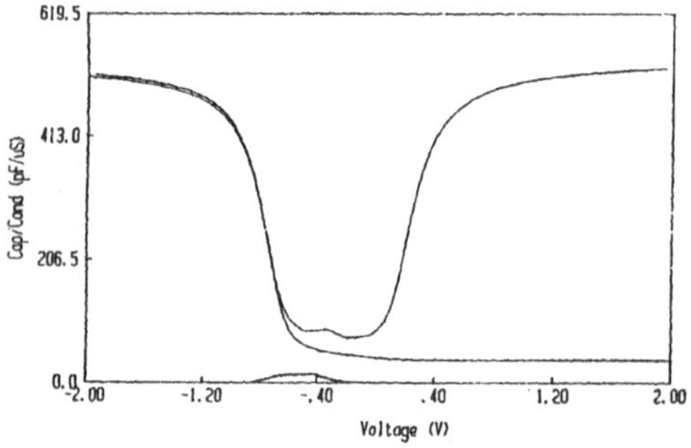

Fig. 4. Electrical characteristics of MOS device using Process A: high frequency (1 MHz) and quasi-static C-V traces.

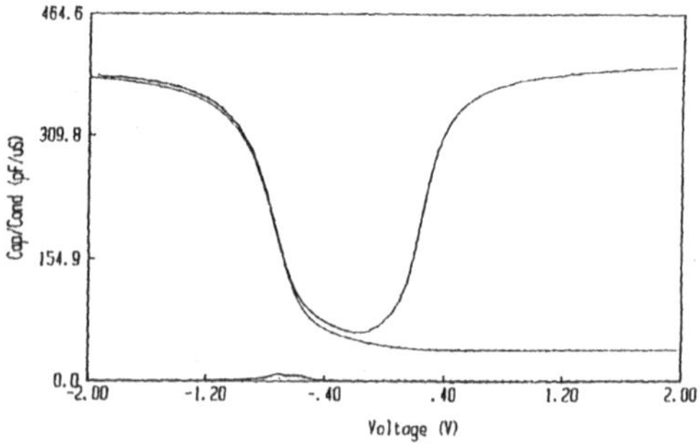

Fig. 6 Electrical characteristics of MOS device using Process B: high frequency (1 MHz) and quasi-static C-V traces.

residual O-coverage was ~1-5 at.%; (see Fig. 2 which shows the chemical composition of the Si surface as it evolved during the H-atom exposure). After a 30 s exposure, C and O-atom contamination were still readily detected. After a 2 minute exposure, the C-atom contamination is reduced to below the level of AES detection (< 0.5 at.%), but O-atom contamination was still evident. Finally, after a 5 minute exposure, there was still residual O-atom contamination (~1-2 at.%). Other studies have shown that after the exposure to atomic-H, the surface concentration of H drops, and the dihydride bonding configurations are replaced by monohydride groupings with a surface reconstruction that displays a 2x1 symmetry.[9] One indicator of the Si surface quality, that was used to evaluate the effectiveness of the H-atom cleaning/passivation process, was the relative strength of the 2x1 surface reconstruction as detected by on-line LEED or RHEED.[1,2] As the time of the H-atom exposure increased, the strength of the surface super-cell features in the electron diffraction patterns increased; however, contrary to the previous reports,[1,2] we have found no consistent correlation between the strength the super-cell features and improvements in the electrical properties of the SiO_2/Si interface.[6,7] A major problem in fixing a time for H-atom exposure was residual C-atom contamination. Variations in the initial C-atom concentrations could not be correlated with any particular aspects of ex-situ processing. Studies indicated that C-atom attachment could also occur in the load-lock chamber of the system where the typical pressure was in the low 10^{-7} to mid-10^{-8} Torr range. The longer the time in the load-lock, the more C-atom contamination, and the longer processing times required for C-atom removal.[10]

After a 1-2 minute pre-deposition exposure to atomic-H, an oxide film was deposited by remote PECVD using RF excitation of a He/N_2O or He/O_2 mixture, and downstream injection of He diluted SiH_4 (10:1). Following this deposition, the chamber pressure was reduced to a base-pressure of ~2×10^{-8} Torr, and the oxide layer was annealed in the deposition chamber at 400°C for 5 minutes. Following a cool down at the same base-pressure, the sample was removed from the multichamber system and the final steps of MOS device fabrication were completed. The back of the wafer was metallized by RF sputtering of Al, and dots of Al were produced on the front surface of the wafer by sputtering, followed by photo-resist patterning and selective etching. Typical device areas were ~$2\text{-}3\times10^{-3}$ cm^2. After metallization, the wafers were subjected to a post-metallization anneal, PMA, in a dry N_2 ambient, at 400°C for 30 minutes.

In Process A, the SiO_2/Si interface is formed by a *subcutaneous* oxidation of Si substrate that occurs at the onset of film deposition;[2,3] this generates ~0.5 to 1 nm of *grown* oxide (see Fig. 1(c)). When the temperature was raised above 300°C, and/or the deposited oxide thickness was increased above ~25 nm, the amount of subcutaneous oxidation increased significantly, and the electronic quality of the interface rapidly deteriorated.[1,2] These studies indicated that the maximum amount of oxidation that could be *tolerated* without increasing D_{it} significantly was of the order of 0.5 to 0.8 nm, or 2.-3 atomic layers.[2] In addition, even when subcutaneous oxidation was not excessive, process reproducibility was generally poor. Other studies[5-7] demonstrated that this could be correlated with either: i) a roughening of Si surfaces by exposure to atomic-H; ii) incomplete removal of C-atom contamination; and/or iii) N-atom attachment at the Si surface during initial stages of oxide deposition when using N_2O as the O-source gas. For oxide depositions using the N_2O source gas, N-atom bonding, as well as O-atom, bonding was detected at the Si-substrate at the initial stages of the oxide deposition (see Fig. 3). By interrupting the oxide deposition process, and by using etch-back techniques, it was established that N-atoms were bonded at the SiO_2/Si interface, and in the bulk oxide.[6,7] The H-atom pre-deposition process then contributed to two materials-related problems: i) damage to Si substrates; and ii) by creating a reactive Si surface immediately before oxide deposition, it also allowed bonding of N-atoms at the SiO_2/Si interface at the onset of oxide deposition. However, comparisons of electrical properties, using N_2O and O_2 source gases for the oxide depositions, showed that the hydrogen induced damage was the dominant contributor to high values of D_{it}.

The properties of the MOS devices have been evaluated by conventional C-V and I-V measurements, and representative data are displayed in Figs. 4(a) and 4(b), respectively. We have had considerable difficulty in using Process A to consistently fabricate *good* electrical-quality MOS devices; e.g., we have found that only 10 to 20% of the devices made using Process A have characteristics comparable to those shown in Figs. 4. The majority of the MOS devices have lower breakdown fields, $E_b < 5x10^6$ V/cm, and higher values of D_{it} at mid-gap, > $5x10^{10}$ cm^{-2}-eV^{-1}; a good fraction of these devices also display low field breakdowns, or the order of, or less than 10^6 V/cm.

Process B has been developed to remedy three problems with Process A; i) damage to the Si surface and bulk due to pre-deposition exposure to plasma-generated atomic-H; ii) bonding of N-atoms at the SiO_2/Si inteface during the initial stages of oxide deposition; and iii) formation of the SiO2/Si interface in Process A during the initial stages of filmd deposition via the parasitic, subcutaneous oxidation of the Si surface.

THE *NEW* REMOTE PECVD PROCESS -- PROCESS B

Process B

As in Process A, the Si wafers were initially subjected to ex-situ RCA cleaning followed by a rinse in a dilute HF solution.[5-7] We have found that dilute HF (pH ~ 2) could be replaced by HF (pH < 2), buffered HF (pH ~ 5) or NH_4F (pH ~ 8), and that samples could be rinsed in water following these treatments, and using Process B, without any detectable effects on electrical properties of the SiO_2/Si interface, the interfacial roughness (as determined by TEM), or the process reproducibility. Following the RCA clean and an *HF rinse*, the Si wafers were loaded into the three-inch multichamber system, in which they were processed in a dual-function surface processing/dielectric deposition chamber. Process B includes two plasma-assisted steps: i) a low-temperature oxidation; and ii) oxide deposition by remote PECVD. The oxidation step consists of an exposure of the Si surface to plasma generated O-atoms, and combines surface cleaning and oxide passivation. No attempt is made to produce an *atomically clean* Si surface as an intermediate step of the interface development. In process A, exposure to atomic-H was used in an attempt to remove both C-atom and O-atom contamination, and to leave the Si surface H-atom passivated prior to oxide deposition by remote PECVD. In Process A, C-atom contamination, mostly as hydrocarbons, was removed by creating volatile species such as H_nC-SiH_m, with $m,n \leq 3$. In contrast, in Process B, hydrocarbon contaminants are removed by oxidation, with the reaction products being, CO, CO_2, H_2O, etc., so that this mechanism does not promote removal of Si-atoms along with the C-atom contamination. The hydrocarbon oxidation and removal in Process B is accomplished at the same time a thin oxide layer is formed on the Si surface. This cleaning/oxidation/passivation step has been done at temperatures between 200 and 400°C with exposure times from 15 to 300 s. The other conditions are: i) a process pressure of 300 mTorr; ii) remote plasma excitation of a He/O_2 mixture with flow rates of 200 sccm for He, and 10 sccm for O_2; and iii) an RF power of 15 W at 13.56 MHz.

Conditions for optimum devices are a temperature of 200-300°C and a time of 15 s. The remote PECVD oxide deposition process parameters have been adjusted to give ~15 nm of oxide for deposition times of the order of 5-6 minutes. These parameters are: i) a process pressure of 300 mTorr; ii) a deposition temperature between 200°C and 400°C; iii) an RF power of 15 W at 13.56 MHz; iv) remote plasma excitation of a 10:1 He/N_2O mixture, at a combined flow rate of 200 sccm; and v) down-stream injection of 10% SiH_4 in He, with an effective SiH_4 flow rate of 0.1 to 0.5 sccm. For a substrate temperature of 300°C, and an SiH_4 flow rate of 0.2 sccm, the deposition rate was ~0.04 nm-s^{-1}. After deposition, the chamber pressure was reduced to ~2-4x10^{-8} Torr, and the SiO_2/Si structure was annealed at 400°C for 5 minutes. Previous studies showed that post-deposition annealing reduced OH concentrations in the oxide layer, with an accompanying decrease in pre-breakdown current.[2] After a cool-down at the base-pressure, the sample was removed from the multichamber system and MOS devices, with areas ~2-3x10^{-3} cm^2, were formed by applying sputtered Al electrodes to the back of the wafer, and to the top surface of the oxide. Following metallization and patterning, the Si wafers were subjected to a PMA in dry N_2 at 400°C for 30 minutes.

On-line Analysis

To follow changes in the chemical and structure character of the Si surfaces and the dielectric layers, on-line studies were performed by AES and LEED. The ex-situ wafer processing was the same as in Process A, i.e., an RCA clean and HF rinse. After this, the Si surfaces showed C- and O-atom contamination, sometimes as high as 10-20 at.%, as estimated from AES measurements (see Fig. 5). The Si(100) surfaces displayed a 1x1 reconstruction consistent with dihydride termination. Process B does not attempt to *clean* the Si surface, but instead promotes formation of SiO_2/Si interfaces by: i) oxidation and effective removal of C-atom contaminants (hydrocarbons); and ii) Si surface passivation by a thin oxide layer that becomes an integral part of the SiO_2/Si heterostructure. The AES data in Fig. 5 show that after a 15 s exposure to oxygen at 300°C: i) there is no evidence for C-atom contamination; and ii) the thickness of oxide was ~0.5-0.6 nm. This thickness was estimated from relative strengths of Si-Si and Si-O features in the Si LVV manifold using the previously determined values for the electron escape depth for electrons with 50 to 100 eV of kinetic energy. There were no crystalline features evident in the LEED pattern after this superficial oxidation. The thickness of oxide generated in 15 s was about ~0.5 to 0.6 nm for substrate temperatures between 200° and 400°C. At any oxidation temperature, the oxide thickness increased with increasing exposure time, with a five minute exposure typically yielding ~1.0 to 2.0 nm of oxide. Chemical bonding at the Si surfaces was studied by AES at the initial stages of oxide deposition using N_2O as the O-source gas. There was no evidence for incorporation of N-atoms in the SiO_2 film, or at the SiO_2 surface (see Fig. 5); however, studies on etched-back films showed that N-atoms can migrate through the oxide during deposition, and form Si-N bonds at the SiO_2/Si interface. However, when using N_2O as the O-atom source gas for remote PECVD, the concentration of N-atoms at this interface in Process B is significantly less than that found in Process A. Process B is therefore more effective that Process A in; i) eliminating C-atom contaminants from the Si surface (the shorter time period required), and ii) minimizing N-atom attachment at the SiO_2/Si interface. In addition, the elimination of any pre-deposition exposure to atomic-H reduces pre-deposition damage to the Si-surface and the Si-bulk that degrades electrical properties of MOS devices.

Electrical Evaluation of Device Structures

The properties of MOS devices fabricated using Process B have also been measured by standard C-V and I-V methods, and are compared with Process A in Figs. 4 and 6. Reduction of the C-V data yields: i) flatband voltages, V_{fb}; and ii) D_{it} as a function of energy in the Si band-gap. A typical D_{it} plot for an MOS device made by Process B is shown in Fig. 7(a). Analysis of data from a large number of similarly prepared MOS devices indicated: i) mid-gap interface trap densities, D_{it}, of ~1-3x10^{10} $cm^{-2}eV^{-1}$; ii) a spectral peak in D_{it} ~0.35 ± 0.05 eV above the valence band edge; and iii) flatband voltages of -0.7 ± 0.1 V, essentially the same as the calculated work-function difference of -0.7±0.1 eV between Al and 3-5 Ω-cm, p-type Si. I-V measurements yielded breakdown fields, E_b, of 9 to 12 MV/cm (see Fig. 6). Using Process B, there are then approximately 40% increases in E_b, and 50% reductions in D_{it}, with respect to Process A. Bias-temperature stress measurements performed at 150°C under application of a ±2 V bias, gave mobile charge densities, $Q_m <$ 2x10^{10} cm^{-2}. Samples subjected to pre-deposition oxidation reproducibly showed deep depletion effects (see Fig. 8), while samples prepared using Process A, did not. For example, C-V measurements on MOS structures formed with plasma-generated H-processing prior to oxide deposition did not show deep depletion behavior at ramping rates of 35 mV/s. Coupled with the data in Fig. 8, this indicated that minority carrier generation in bulk Si was lower in samples prepared with the pre-deposition O-process than with the H-process, by a factor of at least ten. Additional comparisons between the H- and O-pre-deposition steps indicated that: i) D_{it} increased with increasing H-atom exposure time; ii) combinations of H-exposure followed by O-exposure, and then by oxide deposition by remote PECVD yielded higher D_{it} values than O-exposure followed by oxide deposition; iii) breakdown fields were generally a factor of 1.3 to 1.5 lower with any pre-deposition exposure to H; and iv) exposure to H-atoms introduced additional spectral features in the D_{it} spectrum.

The decreased performance in MOS devices fabricated with H-atom pre-deposition processing has been correlated with roughening of Si-surfaces occurring during that exposure.[6,7] This was reinforced by showing that additional spectral features in MOS capacitors formed on Si(100) and exposed to H-atoms (as in Fig. 7(b)), also occur in MOS devices on Si(110) and Si(111) wafers, independent of the pre-deposition processing. High resolution transmission electron microscopic, HRTEM, images indicated surface roughening of Si(100) surfaces by exposure to H generates Si(111) facets, thereby establishing a probable origin for the additional features in D_{it} spectra.[6,7]

Returning to Process B, experiments in which D_{it} has been determined as function of the pre-deposition oxidation temperature (200-400°C) for fixed deposition temperature (300°C), and as function of deposition temperature (200-400°C) for a fixed oxidation temperature (300°C) (see Figs. 9(a) and 9(b)), have established that D_{it} was *determined* by the initial oxidation. This means that the first ~0.5 to 0.6 nm of oxide (~two atomic layers) is the important process step in establishing the electrical quality of the SiO_2/Si interfaces. For a given exposure time, oxidations at 200°C and 400°C yielded the same oxide thickness, so that the large differences in D_{it} for oxidation temperatures of 200°C and 400°C are not due to an increased oxide thicknesses. In previously reported studies using pre-deposition H-atom exposure, it was demonstrated that D_{it} increased as the amount of subcutaneous oxidation increased, and this was correlated with both higher deposition temperatures, and increased deposition times.[1,2] This difference is between the pre-deposition oxidation step of Process B, and the subcutaneous, parasitic oxidation of Process A is further illustrated by the fact that for an oxidation temperature of 200-300°C oxygen exposure times as long as 300 s in

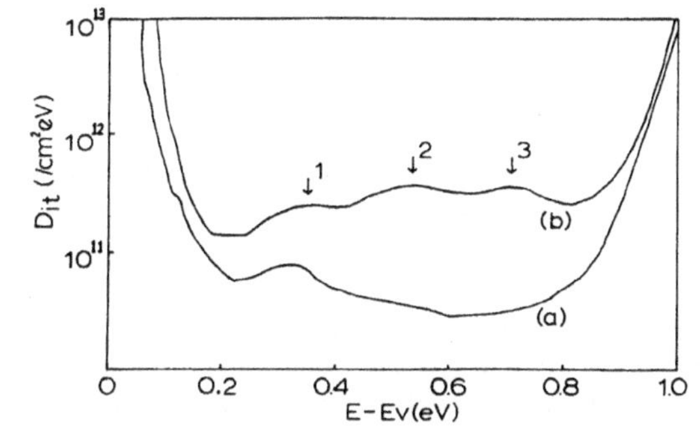

Fig. 7. Comparison between D_{it} spectrum of MOS capacitors with the pre-deposition being: (a) a 15 s exposure to plasma-generated O-atoms (lower trace), or (b) a 5 minute exposure to plasma-generated atomic-H (upper trace).

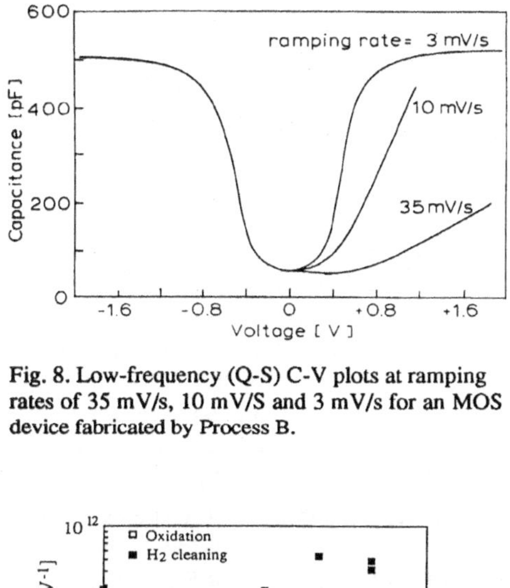

Fig. 8. Low-frequency (Q-S) C-V plots at ramping rates of 35 mV/s, 10 mV/S and 3 mV/s for an MOS device fabricated by Process B.

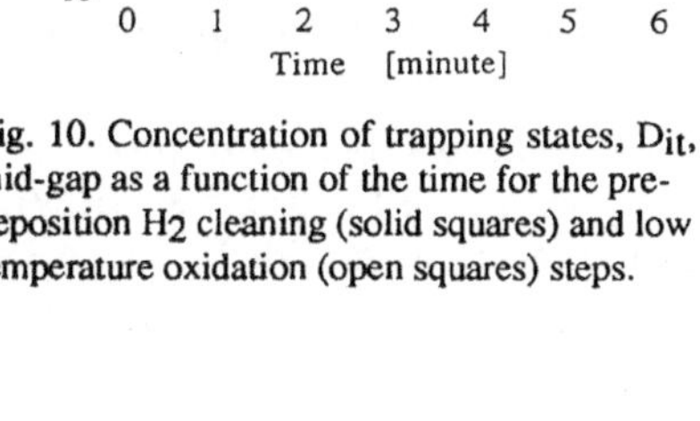

Fig. 10. Concentration of trapping states, D_{it}, at mid-gap as a function of the time for the pre-deposition H_2 cleaning (solid squares) and low temperature oxidation (open squares) steps.

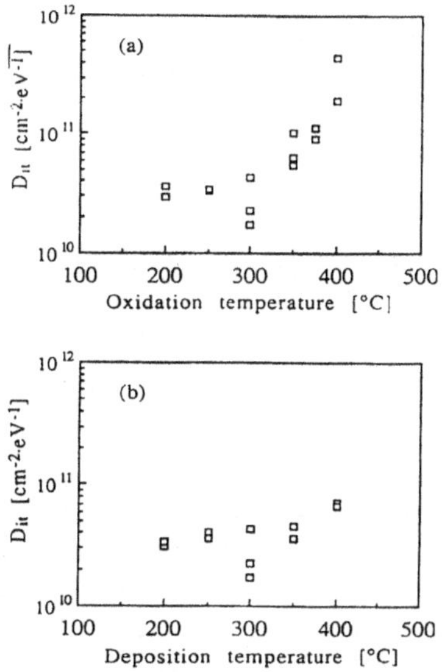

Fig. 9 (a) D_{it} as a function of the plasma-assisted oxidation process temerature. The deposition temperature 300°C. (b) D_{it} as a function of the remote PECVD deposition temperature. The pre-deposition oxidation was temerpature 300°C.

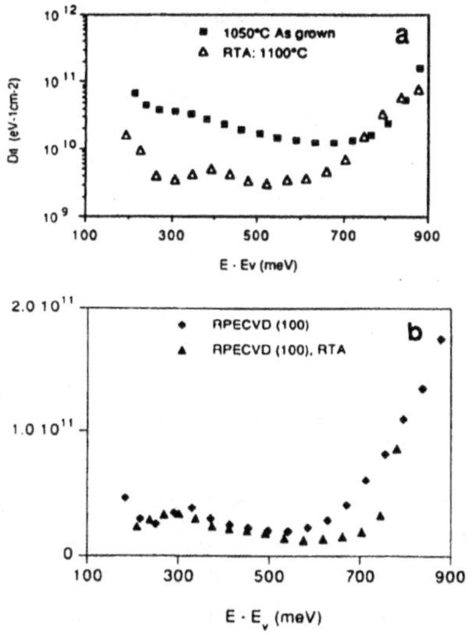

Fig. 11 D_{it} as function of position in the energy gap of Si: (a) thermally grown oxide (1050°C) and RTA at 1100° for 100s; and (b) remote PECVD oxide, and annealed in same way. The discrete feature at 0.35 eV above valence-band edge is the P_b center. D_{it} in the thermal oxide decreases after RTA, but does not in the plasma oxide.

Process B did not increase D_{it} (see Fig. 10). The thickness of oxide developed in the 300 s exposures was of the order of 1.5 to 2.0 nm, and was considerably greater that the thickness of subcutaneous oxide that could be generated in Process A without increasing D_{it}.

Post Deposition Thermal Processing

The formation of source and drain junctions in FET's generally requires high-temperature processing, > 900°C, and it remains to be established whether the thermal exposure of SiO_2/Si structures prepared by the new, low-temperature two-step process, that occurs during downstream high-temperature processing degrades the SiO_2/Si electrical properties. To address this, we have made comparisons of the distribution of interfacial defect states for thermal and remote PECVD oxides using doped microcrystalline Si gate electrodes (see Fig. 11).[11] We find that: i) the mid-gap D_{it} levels are essentially the same, D_{it} ~$2\text{-}3\times10^{10}$ cm^{-2}-eV^{-1}, for an interface formed by a 1050°C thermal oxidation, and formed by Process B; ii) the mid-gap D_{it} of the thermal oxide structure is reduced by a factor of about 5, when that oxide is subjected to a 100 s RTA at 1100°C; iii) the mid-gap Dit of remote PECVD oxide is not changed after the RTA; and iv) both the thermal oxide, subjected to the RTA and the remote PECVD oxide structures show a discrete interfacial defect state at ~0.35 eV above the valence band edge, the so-called P_b type center.[12] The reduction in D_{it} is associated with strain relief in the thermally grown oxide;[13] there is no comparable process of strain relief for the plasma deposited oxide. The higher mid-gap D_{it} in the plasma oxide derives from trapping states at the plasma process interface that are not strain-induced.

MULTILAYER DIELECTRIC STRUCTURES

MIS structures were formed by using combinations of oxide and nitride depositions, but always starting with oxide layers on the Si substrates. These studies were performed using Process B to form the SiO_2/Si interface and the first oxide layer. Composite dielectrics, with ON and ONO layers, used dielectric thicknesses from ~2 to 15 nm. Nitride depositions were done with: i) substrate temperatures of 250 to 300°C; ii) up-stream excitation of He/NH_3 mixtures with flow rates of 5 sccm of NH_3 and 200 sccm for He; iii) a plasma power of 50 W at 13.56 MHz; and iv) down-stream injection of 5 sccm of 10% SiH_4 in He. These conditions produced stoichiometric Si_3N_4 films, with < 5 at.% H in Si-NH arrangements.[3]

Studies of these structures have focused on chemical bonding at interfaces of the composite dielectrics.[14] As noted above, N-atoms diffuse through oxide layers, and accumulate at the SiO_2/Si interface during the remote PECVD deposition of SiO_2 from N_2O source gases. Two other experiments have also demonstrated that N-atoms can penetrate remote PECVD oxides and migrate to the SiO_2/Si interface: i) exposure of remote PECVD oxides to active N-species from remote plasma-excited He/NH_3 discharges; and ii) deposition of Si_3N_4 onto oxide layers using remote PECVD. In both cases, AES spectra showed that: i) N-atoms were not bonded at the oxide surface, or in the bulk oxide; but that ii) N-atoms were bonded at the SiO_2/Si interface in concentrations comparable to those observed for etch-back of oxide layers deposited from N_2O.

Qualitatively different results were obtained for oxide depositions onto nitrides.[14] Exposure of nitrides to remote plasma-generated O-species, as well as remote PECVD deposition of oxides onto nitrides, yielded evidence for interfacial oxy-nitride layers. AES studies showed Si-O bonding at Si_3N_4 surfaces after exposure of nitrides to plasma-generated O-species, and the IR studies showed oxy-nitride layers between remote PECVD nitride and oxide films. The IR data were for: i) 60 nm SiO_2 and 60 nm Si_3N_4 films, and ii) a six period structure consisting of alternating 10 nm films of SiO_2 and Si_3N_4. By subtracting IR absorbances of the 60 nm films from the six period composite, spectral features corresponding to oxynitride interfaces have been identified.[14] The IR data, combined with experiments in which SiO_2 and Si_3N_4 surfaces were respectively exposed to plasma-generated N and O-species, have established that oxy-nitride spectral features resulted from interfacial layers produced during deposition of SiO_2 onto Si_3N_4, and not from deposition of Si_3N_4 onto SiO_2.

The electrical properties of MIS devices with ON and ONO dielectrics, have not been studied as extensively as those of MOS devices with oxide layers. Experiments to date have indicated: i) increases in mid-gap D_{it} of the order of five with respect to MOS devices; ii) flatband voltage shifts in the negative direction of at least 0.2 V, and sometimes as much as 1-2 V; and iii) small decreases in the breakdown fields.[14] Post deposition annealing of the MIS devices produces improvements in the electrical characteristics (see Table 1).

Table 1 Comparison of Electrical Characteristics

	Oxide	ONO (before anneal)	ONO (after anneal)*
D_{it} ($\times 10^{10}$ cm^{-2}-eV^{-1})	1-5	5-10	0.9 - 2
E_b ($\times 10^6$ V-cm^{-1})	~10	~10	~10
V_{fb} (V)	-0.65±0.05	-1.9	-1.1
Q_f ($\times 10^{17}$ cm^{-3})**	<1	~8	~2

* 30 s at 900°C, ** fixed charge

There are still a number of important issues to resolve: i) the role of N-atoms at the SiO_2/Si interface; and ii) the origin of the fixed charge. Specifically, i) how do N-atom affect D_{it} ?, and ii) is the fixed charge in the ONO structures in the nitride or in the interfacial oxy-nitride region? These issues are currently under investigation.

SUMMARY

The results presented in this paper have established that it possible to independently control the electrical properties of the SiO_2/Si interface and the SiO_2 thin film by a combination of low-temperature, 200-400°C, plasma-assisted processes, specifically exposure of an RCA cleaned surface to atomic oxygen, followed by oxide deposition by remote PECVD. We have also demonstrated that the use of dual function chambers for accomplishing these two process steps, coupled with the availability of on-line characterization probes (LEED or RHEED, and AES) has provided important information relative to chemical and structural properties of the Si surface, the SiO_2/Si interface, the bulk SiO_2, and the interfaces between oxide and nitride films in composite ON and ONO dielectrics. The process has been extended to composite ONO dielectrics, and initial studies indicate that downstream high temperature processing does not degrade electrical properties.

Other low-temperature deposition processes have also yielded SiO_2/Si heterostructures with excellent electrical characteristics. These include the Batey-Tierney process,[15] and a variation of that process by Bright et al.[16] In these approaches, the process starts with a Si wafer with a native oxide. In the original process,[15] the SiO_2/Si heterostructure is formed by He-PECVD, a direct PECVD process in which the atomic source gases are SiH_4 and N_2O, but where the He dilution is very high, e.g., the He/SiH_4 ratio can be 10,000:1. In the modification of this process in Ref. 16, an in-situ oxidation step is inserted before the oxide deposition. In the Batey-Tierney process, it is likely that the SiO_2/Si interface is established by a subcutaneous oxidation process that occurs at the earlier stages of film deposition. In this respect, that process is qualitatively similar to the remote PECVD Process A. However, the Batey-Tierney process does not expose the Si surface to atomic-H, so that all of the detrimental aspects of that exposure are not present. On the other hand, the Bright et al. modification of the Batey-Tierney Process is similar to Process B, with the SiO_2/Si interface being formed by the oxidation step that has been inserted prior to the He-PECVD deposition.

ACKNOWLEDGMENTS

This work is supported by the Office of Naval Research, the National Science Foundation Engineering Research Center, ERC, for Advanced Electronic Materials Processing, and the North Carolina SEMATECH Center of Excellence, SCOE.

REFERENCES

1. G.G. Fountain, R.A. Rudder, S.V. Hattangady, R.J. Markunas, and P.S. Lindorme, J. Appl. Phys. **63**, 4744 (1988).
2. S.S. Kim, D.J. Stephens, G. Lucovsky, G.G. Fountain, and R.J. Markunas, J. Vac. Sci. Technol. **A8**, 2039 (1990).
3. G. Lucovsky, D.V. Tsu, R.A. Rudder and R.J. Markunas, *Formation of Inorganic Films by Remote Plasma-Enhanced Chemical-Vapor Deposition*, in Thin Film Processes II, Ed. by J.L. Vossen and W. Kern (Academic Press, New York, NY, 1991), p. 565.
4. T. Yasuda, Yi Ma, and G. Lucovsky, MRS Symp. Proc., **238** (1992), in press.
5. T. Yasuda, Yi Ma, S. Habermehl, and G. Lucovsky, Appl. Phys. Lett. **60**, 434 (1992).
6. Yi Ma, T. Yasuda, S. Habermehl, and G. Lucovsky, J. Vac. Sci. Technol **A10** (1992), in press.
7. T. Yasuda, Yi Ma, S. Habermehl, and G. Lucovsky, J. Vac. Sci. Technol **B10** (1992), in press.
8. F.J. Grunthaner and P.J. Grunthaner, Material Science Reports **1**, 65 (1986), and references therein.
9. B. Anthony, T. hsu, L. Breaux, S. Banerjee and A. Tasch, MRS Symp. Proc. **165**, 139 (1990).
10 T. Yasuda, Y. Ma, S. Habermehl, S.S. Kim, G. Lucovsky, T.P. Schneider, J. Cho, and R.J. Nemanich, MRS Symp. Proc., **202,** 395 (1991).
11. D.R. Lee, C.H. Bjorkman, C. Wang, and G. Lucovsky, MRS Symp. Proc. **219**, 395 (1991).
12. E.H. Poindexter, G.J. Geraldi, M.-E. Rueckel, P.J. Caplan, N.M. Johnson and D.K. Beigelsen, J. Appl. Phys. **56**, 2844 (1984), and E.H. Poindexter and P.J. Caplan, J. Vac. Sci. Technol. **A6**, 1352 (1990).
13. C.H. Bjorkman, J.T. Fitch and G. Lucovsky, Appl. Phys. Lett. **56**, 1983 (1990).
14. S.S. He, D.J. Stephens, Yi Ma, T. Yasuda, S. Habermehl and G. Lucovsky, MRS. Symp. Proc. (1992), to be published.
15. J. Batey and E. Tierney, J. Appl. Phys. **60**, 3136 (1986).
16. A.A. Bright, J. Batey, and E. Tierney, Appl. Phys. Lett. **58**, 619 (1991).

COMPARISON OF SiO_2 THIN FILM PROPERTIES DEPOSITED BY DISTRIBUTED ELECTRON CYCLOTRON RESONANCE PLASMA USING TWO DIFFERENT OXIDANT GASES: N_2O OR O_2

B. Agius,[1] M.C. Hugon,[1] N. Jiang,[1] F. Plais,[2]
D. Pribat[2] and T. Carriere[3]

[1]Institut Universitaire de Technologie, Université de Paris Sud (XI), Plateau du Moulon BP 127, 91403 Orsay Cedex, France

[2]Thomson-CSF LCR, Domaine de Corbeville, 91404 Orsay Cedex, France

[3]Matra Marconi Space, 37 avenue Louis Breguet, Velizy/Villacoublay 78140, France

ABSTRACT

Device quality SiO_2 thin films were reproducibly fabricated using distributed electron cyclotron resonance (DECR) microwave plasma without substrate heating (substrate temperature <100°C) and under floating potential. The purpose of this paper is to discuss the effects of the reactant gas mixture (O_2/SiH_4 or N_2O/SiH_4) on the physical, chemical and electrical properties of DECR SiO_2. Under optimum deposition conditions, the films show promising characteristics, comparable to those of thermal oxides grown at 850-1050°C, in terms of refractive index, atomic composition, p-etch rate, critical field and midgap trap density.

I INTRODUCTION

In recent years, there has been growing interest in thin film dielectrics that can be deposited at low substrate temperatures, in the range from 200 to 500°C. Numerous studies have shown that both films and interface quality are altered and generally degraded as the temperature is decreased below 1000°C [1].

Different chemical vapor deposition (CVD) technics have been developed for low temperature fabrications of SiO_2, e.g. low temperature CVD [2], plasma enhanced CVD (PECVD) [3], plasma anodization of silicon [4] and more recently electron cyclotron resonance (ECR) PECVD in its classical configuration [5] or in its distributed (DECR) one [6]. ECR plasma deposition has distinct advantages over conventional radio frequency (rf) plasma processes: the low process pressures ($5\ 10^{-4}$ - $5\ 10^{-3}$ mb) allow to use high reactive gases (SiH_4 and O_2), the low electronic temperature (2-3eV) induces a low floatting potential (20-30V) and the low ion bombardment energies (20-30eV) keep wafer temperature below 100°C at floating potential.

The Physics and Chemistry of SiO_2 and the Si-SiO_2 Interface 2
Edited by C.R. Helms and B.E. Deal, Plenum Press, New York, 1993

There are several reasons for the observed dependance of n on R_0: effects of film density and/or stoichiometry and/or contamination. A decrease in film density would be due to an increase degree of spontaneous gas-phase reactions. In the previous discussions we feeled (Fig. 1-b) that such mechanism is not dominant and it cannot account for the variation of n given in Fig. 2. A variation of film density and hence n may be also due to the effects of ion bombardment. But the value of the refractive index is not affected by the microwave power and then such decrease in n value cannot be attributed to ion bombardment.

Film stoichiometry was obtained over a wide range of oxidant ratios. This is illustrated in Fig. 3. Atomic densities of Si, O, N and H in the films are reported as function of the gas phase composition. These densities in at.cm^{-3} are deduced from NRA and ellipsometric measurements. We have reported on Fig. 3 the corresponding values (dotted lines) for oxygen and silicon density in thermal silicon oxide: N_{Si}(thermal SiO_2) = $2.2x10^{22}cm^{-3}$, N_O(thermal SiO_2) = $4.4x10^{22}cm^{-3}$. All O_2/SiH_4 oxides have a Si:O ratio close to that expected for stoichiometric SiO_2, whereas N_2O/SiH_4 films are stoichiometric only for $R_0 > 4$. The samples are slightly silicon rich.

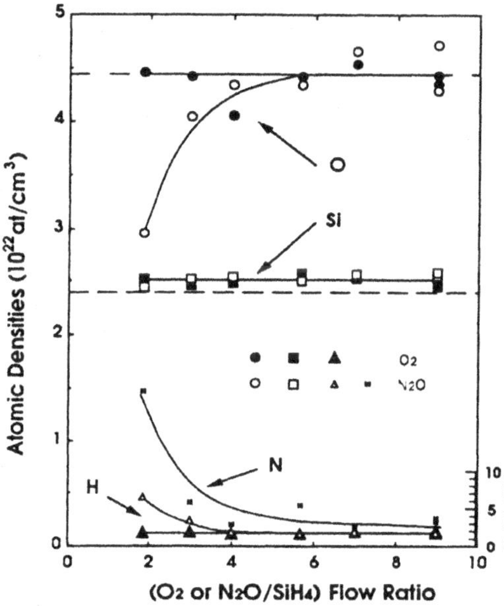

Figure 3. Atomic densities of Si, O, N and H in DECR silicon oxide as function of gas flow ratio (ratio of N_2O or O_2 flow to SiH_4 flow). The dotted lines indicate atomic densities of Si and O in thermal silicon oxide.

No impurities were detected in any of the oxides except nitrogen (3-4 at% in R_0>4 range) for N_2O/SiH_4 silicon oxides and hydrogen. These measurements (on 70 nm thick samples) gave a very low hydrogen concentration (2 at%), which could be incorporated as Si-OH or Si-H groups and then explained the low silicon excess. These data have to be compared with those obtained by ECR CVD technics (Fig. 4). In general, ECR films deposited at floating temperature and for similar conditions have much higher hydrogen contents (5-10 at%) [15]. In fact, at 300°C and with rf bias, the hydrogen concentration in ECR oxides is approaching the level observed in DECR oxides deposited without bias and at floating temperature. It should be noted that the amount of hydrogen observed in our floating temperature deposited films is near the upper limit (2-3 at%) of the hydrogen concentration range detected in thermal CVD SiO_x films deposited at 800°C [15].

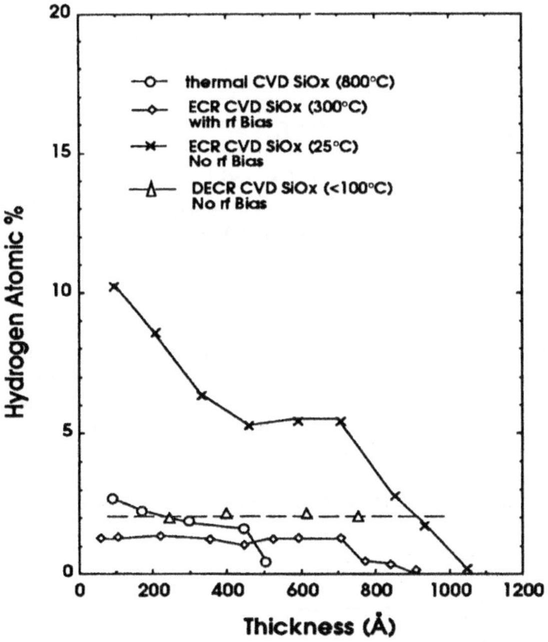

Figure 4. Hydrogen contamination of thermal CVD, ECR CVD (25°C with no rf bias or 300°C with rf bias) and DECR CVD SiO_2 as function of film thickness.

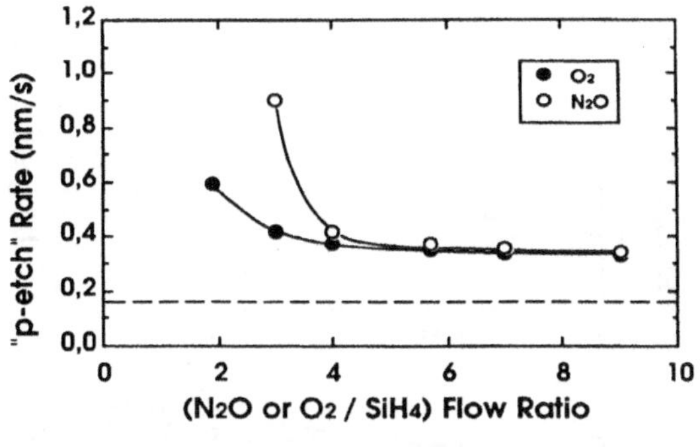

Figure 5. Etch rate of DECR oxides in "P-etch" solution[16] as function of gas flow ratio (ratio of N_2O or O_2 flow to SiH_4 flow). The dotted line indicates the value of thermal oxide etch rate.

3 *Etching rate*

Fig. 5 shows a strong correlation between the etch rate and the oxidant/silane flow ratio: the etch rate of the DECR oxides decreases when R_0 increases and it is lower for films deposited for $R_0 = 9$. The thermal oxide is etched very slowly (<1.5 Å/s) and the DECR floating temperature oxides deposited by our standart process ($R_0 = 9$) is etched only 2 times faster than the thermal oxide. This significant result seems to indicate that DECR and thermal oxide films are very similar in terms of film density or film porosity.

IV OXIDE ELECTRICAL PROPERTIES

The electrical properties of the DECR oxides were strong functions of processing conditions and post-processing treatment. In particularly, they depend on oxidant gas nature. In all cases, N_2O induces poor electrical properties. The results given below are only for oxygen reactant gas.

1 *Current-Voltage analysis*

We have measured by quasi-static ramp I-V technic, in accumulation regime, the resistivity of SiO_2 and the critical field (E_C) (leakage current 1 nA/cm^2). In Fig. 6, E_C versus R_0 for floating temperature and potential samples is shown. E_C increases with silane dilution from 4 MV/cm for $R_0 = 2$ to 6.5 MV/cm for $R_0 = 9$. Under optimum conditions ($R_0 = 9$), the DECR films exhibite the same E_C value than thermal oxide or rf PECVD one [17]. It must be pointed out that these last technics are performed at much higher temperature than our process. The film resistivity, determined as explained in the inset of Fig. 6, is close to 10^{16} Ω.cm (for $R_0 = 9$).

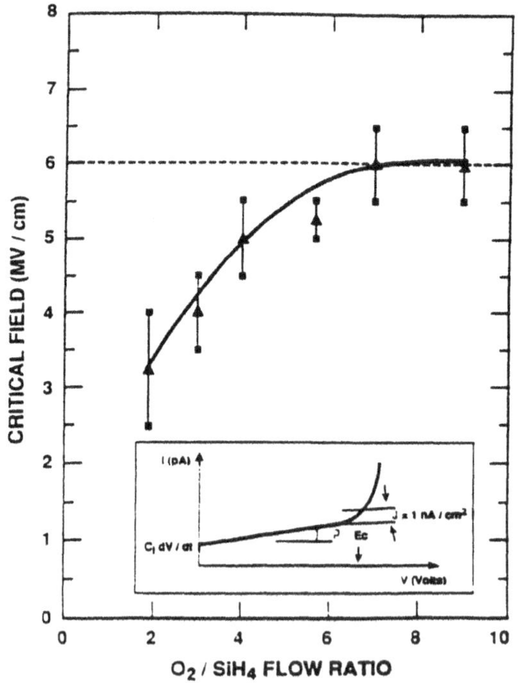

Figure 6. Critical fied (E_C) of MOS capacitors as function of gas flow ratio (O_2/SiH_4). The inset shows a typical quasi-static I(V) characteristic with ρ and E_C determination.

2 Capacitance-Voltage analysis

Fig. 7-a shows typical high frequency (HF) and quasi-static (QS) C-V curves. DECR oxides exhibite well defined accumulation, depletion and inversion regimes. They show no hysterisis upon sweeping from accumulation back into depletion (density of mobile ionic charges within the oxides < 10^{10} cm^{-2}). About this point, more investigations (bias and temperature stress) will be done. For all floating potential DECR films, the magnitude of flatband shifts ($\Delta V_{FB} \approx 0.1V$) is very low and not far from the uncertainty about the work function difference between gate and Si and nearly independent of the R_0 value. The density of Q_{ox} calculated using ΔV_{FB} is estimated to be about 2 10^{10} cm^{-2}. This low value indicates that UV photon and electron flux from the plasma do not induce any significant "damage" during the process.

The spectrum of active interface trapping state density (D_{it}) versus surface potential, calculated using QS and HF C-V curves is shown in Fig. 7-b. D_{it} at midgap is below 3 10^{10} cm^{-2} eV^{-1} and could be considered to be practically independent of the R_0. This low value is comparable to that one found for routinely thermally grown SiO_2/Si interfaces. It is interesting to note that we don't notice any peak in the Si band gap; this is also true for the thermal oxide control but not for some other PECVD process which indicates a peak (0.3eV above the valence band edge) associated with Si-atom dangling bonds at the SiO_2/Si interface[18].

(a)

(b)

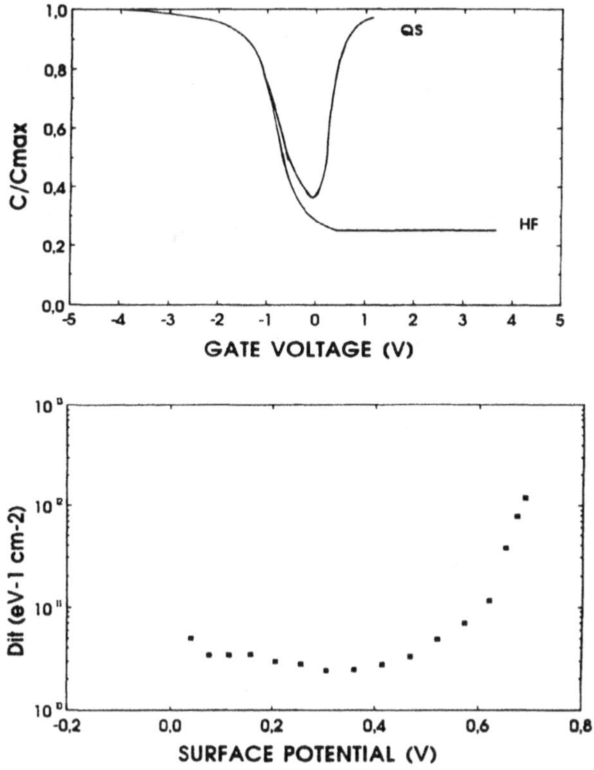

Figure 7. (a) high frequency and quasi-static C-V characteristics for floating temperature and potential DECR film deposited with a gas flow ratio of 9. (b) D_{it} spectrum corresponding to the C-V characteristics of (a).

CONCLUSION

In summary, high physical, chemical and electrical quality SiO_2 films can be deposited from DECR microwave plasma in O_2/SiH_4 or N_2O/SiH_4 reactant gas mixture without substrate heating and under floating potential. We can notice that the physical and chemical behavior of DECR films is practically insensitive to the O_2/SiH_4 flow ratio whereas it depends strongly on N_2O/SiH_4 one. Moreover, it was found that the critical field (6MV/cm) and the interface trapping density (3 10^{10} eV^{-1} cm^{-2}) are significantly better for oxygen reactant gas. Therefore, DECR-PECVD is a very hopeful method for high quality SiO_2 film deposition at low temperature below 100°C.

ACKNOWLEDGMENTS

This work was supported by CNRS (GDR 86) and by THOMSON-LCR.

REFERENCES

1- E. Kobeda and E.A. Irene, J. Vac. Sci. Technol. **B 4**, 722 (1986).
2- Y. Avigal, J. Bregman and Y. Shapira, J. Appl. Phys. **63**. 430 (1988).
3- U. Mackens and U. Merkt, Thin Solid Films **97**, 53 (1982).
4- S. Taylor, K.J. Barlow, W. Eccleston and A. Kiermasz, Electron. Lett. **23**. 309 (1987).
5- T. Hirao, K. Setsune, M. kitagawa, T. Kamada, K. Wasa, K. Tsukamoto and T. Izumi, Jpn. J. Appl. Phys. **27**, 30 (1988).
6- F. Plais, B. Agius, F. Abel, J. Siejka, M. Puech, G. Ravel, P. Alnot and N. Proust, J. Electrom. Soc.**139**, 5, 1489 (1992).
7- N. Proust, M. Petitjean, S. Cassette, A Huber, C. Grattepain, F. Plais, B. Agius and J. Perrin, Third International Conference on InP and Related Materials, Cardiff (GB) 543(1991).
8- F. Plais, B. Agius, N. Proust, S. Cassette, G. Ravel and M. Puech, Appl. Phys. Lett. **59**(7), 837 (1991).
9- W. Kern and D. A. Puotinen, RCA Rev. **31**, 187 (1970).
10- G. Amsel, J.P. Nadai, E. D'Artemare, D. David, E. Girard and J. Moulin, Nucl. Instr. and Methods **92**, 481 (1971).
11- I. C. Vickridge, Nucl. Instr. and Methods **B34**, 470 (1988).
12- T. Carriere, B. Agius, I.Vickridge, J. Siejka and P. Alnot, J. Electrochem. Soc. **137**, 1582 (1990).
13- J. L'Ecuyer, C. Brassard, C. Cardinal, J. Chabbal, L. Deschenes, J.P. Labrie, B. Terreault, J.G. Martel, R. St James, J. Appl. Phys. **47**, 381 (1976).
14- T. V. Herak, T. T. Chau, D. J. Thomson, S. R. Mejia, D. A. Buchanan and K. C. Kao, J. Appl. Phys. **65** (6), 2457 (1989).
15- S. V. Nguyen and K. Albaught, J. Electrochem. Soc. **136**, 2835 (1989).
16- W. A. Pliskin and H. S. Lehman, J. Electrochem. Soc. **112**, 1013 (1965).
17- J. Batey and E. Terney, J. Appl. Phys. **60**(9), 3136 (1986).
18- T. Yasuda, Y. Ma, S. Habermehl and G. Lucovsky, Appl. Phys. Lett. **60** (4) 434 (1992).

LOW TEMPERATURE SYNTHESIS AND CHARACTERIZATION OF SILICON DIOXIDE FILMS

G.S. Chakravarthy, R.A. Levy, J. M. Grow, and W. M. Attia*

New Jersey Institute of Technology
Newark, NJ 07102

ABSTRACT

Diethylsilane (DES) has been used as a precursor to produce silicon dioxide films by low pressure chemical vapor deposition. These films were synthesized in the temperature range of 350° to 475° C with the growth rate observed to follow an Arrhenius behavior with an apparent activation energy of 10 kcal/mol. The growth rate was seen to increase with higher pressure and to vary as a function of the square root of the O_2 flow rate and O_2/DES ratio. In both the pressure and the O_2/DES ratio studies conducted at 400° C, there were points of abrupt cessation in deposition. The density and index of refraction of the films were found to be 2.25 g/cm³ and 1.46 respectively independent of deposition conditions. The etch rate of the films in a 25° C P-etch solution was observed to decrease with higher deposition or annealing temperatures reflecting densification of the material. Despite severe aspect ratios, the films were seen to exhibit good step coverage.

INTRODUCTION

Low pressure chemical vapor deposition (LPCVD) of silicon dioxide is generally based on the reaction of SiH_4 and oxygen[1,2]. The feasibility of using organic liquid sources such as tetraethylorthosilicate (TEOS)[3-5], ethyltriethoxysilane (ETOS)[6-9], diacetoxyditertiarybutoxysilane (DADBS)[10,11], and tetramethylcyclotetrasiloxane (TMCTS)[12] to form silicon dioxide for microelectronic applications has been demonstrated by several researchers. Such liquid precursors offer numerous advantages over the use of silane including superior step coverage and increased safety.

Diethylsilane (DES) is a precursor capable of producing oxide films in the temperature range of 350° to 475° C thus allowing its use as an intermetal dielectric or as a top layer passivation coating. Using conventional hot wall LPCVD reactors, preliminary results[13,14] have shown films produced from DES to have several performance advantages over silane including superior conformality, low particulate formation, low stress and high crack resistance. Diethylsilane with a chemical formula of $(C_2H_5)_2SiH_2$ is a colorless precursor with a boiling point of 56° C and freezing point <-76° C at atmospheric pressure. It exhibits a vapor pressure of 207 Torr at 20° C and is commercially available from Schumacher Corporation with a 99.9999% chemical purity.

In this study, the growth kinetics of films produced from DES were investigated as a function of process parameters and the properties of resulting films characterized. The suitability of using this precursor for low temperature deposition of SiO_2 films is considered in context with potential microelectronic applications.

*Present address: Physics Department, Suez Canal University, Ismailia, Egypt

EXPERIMENTAL PROCEDURES

The synthesis of the SiO_2 films was carried out in a horizontal LPCVD reactor (Figure 1), consisting of a fused quartz insert tube 13.5 cm in diameter and 150 cm in length placed within another quartz tube 19.3 cm in diameter and 155 cm in length. Such a configuration was found essential to achieve uniformity with the 10 cm wafers used in this study. A five zone Lindberg furnace provided a uniform temperature profile across the reaction chamber. The back end of the reactor was connected to an Edwards vacuum station consisting of both a mechanical pump and a Roots blower. The reactor pressure was monitored with the use of an MKS baratron gauge and controlled with the use of an MKS exhaust valve. The reactor temperature profile was measured with the aid of an Omega type K thermocouple. Gas flows were regulated by Applied Materials model AFC 550 automatic N_2 mass flow controllers which were calibrated for DES and O_2.

P-type <100> silicon wafers 10 cm in diameter were placed vertically in a silica boat and spaced 0.9 cm apart. Typically, six wafers were loaded in each run. An average deposition rate was determined from the wafers located in slots 3, 4, and 5. A Perkin Elmer 580 IR spectrophotometer was used to monitor the vibrational modes of the deposits. The dielectric breakdown strength was determined with the aid of a Keithley 236 voltage source and the I-V characteristics plotted with a Keithley 251 software package. A Keithley 82 high frequency/quasi static system operating at 1 MHz was used for the C-V measurements.

RESULTS AND DISCUSSION

Film Synthesis

In the kinetics study, the process variables investigated were temperature, pressure, DES flow rate, and O_2/DES ratio. The deposition rate of the oxide films was determined as a function of temperature between 350° and 475° C while maintaining a constant pressure of 0.5 Torr, a DES flow rate of 50 sccm, and an O_2/DES of 2/1. As seen in Figure 2, a semilog plot of the average deposition rate versus 1000/T follows an Arrhenius behavior between 375 and 450° C which yields an apparent activation energy of 10 kcal/mol. This low value is similar to values obtained in cases of adsorption on the surface or gas phase diffusion to the surface. Above 450° C, the saturation normally associated with the mass transfer limited regime is observed.

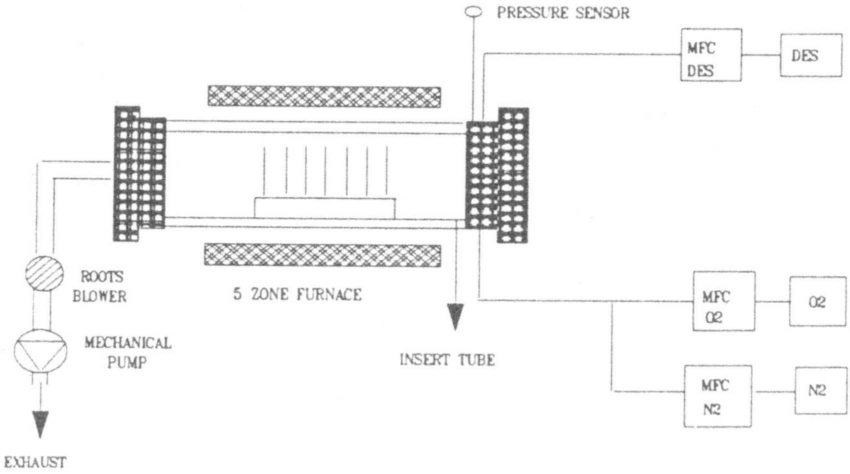

Figure 1. Schematic of LPCVD reactor.

The variation of deposition rate with chamber pressure is shown in Figure 3 for constant conditions of temperature (400° C), DES flow rate (50 sccm), and O_2/DES ratio (2/1). The deposition rate is observed to decrease monotonically from 0.75 Torr to 0.35 Torr where an abrupt cessation in deposition sets in.

A plot of deposition rate versus square root of O_2 flow rate is shown in Figure 4 for constant conditions of temperature (400° C), DES flow rate (50 sccm), and pressure (0.5 Torr). In the O_2 flow rate range of 25 to 117 sccm, a linear dependence is observed indicating that the reaction rate is controlled by the amount of O_2 reaching the surface. This linear region, corresponding to an O_2/DES ratio in the range of 0.5-2.35, is followed by an abrupt cessation in deposition at an O_2/DES ratio of 2.5. In both the pressure and flow rate studies where cessation occurs, the residence time is so short (<2 s) that the heat transfer is insufficient in raising the temperature of the input gases high enough for deposition to occur.

Similarly, a plot of deposition rate versus square root of O_2 flow rate for conditions where the O_2/DES ratio is kept at 2/1, the temperature at 450° C, and pressure at 0.5 Torr, yields a linear dependence for the investigated flow rate range of 12 to 150 sccm DES (Figure 5). For an O_2 flow rate of 300 sccm, the deposition rate is observed to be close to 160 Å/min, making such a process attractive for high throughput semiconductor manufacturing.

Film Characterization

The stoichiometric composition as well as the optical, chemical, electrical, and mechanical properties of the oxide deposits were ascertained. Figure 6 illustrates a typical RBS plot for an oxide deposited at 475° C. This plot reveals the absence of incorporated carbon in the film and an oxide stoichiometric composition of $SiO_{2.2}$. At 375° C, the RBS results indicated a stochiometry of $SiO_{2.5}$. This compositional variation is believed to be due to a higher incorporated hydrogen concentration in the films at lower temperatures which is in turn compensated by excess oxygen. Little change from the $SiO_{2.2}$ composition was seen when pressure, flow rate, and O_2/DES were varied. Energy recoil detection

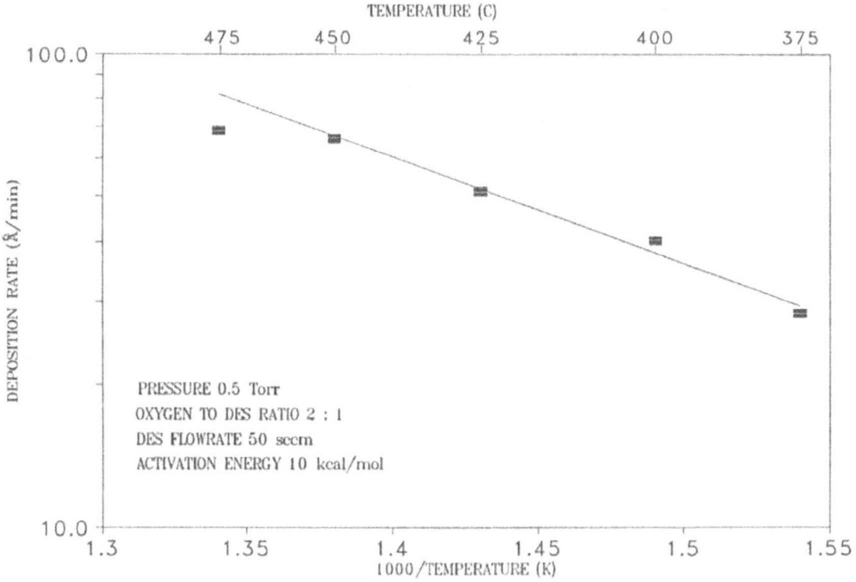

Figure 2. Variation of deposition rate as a function of temperature.

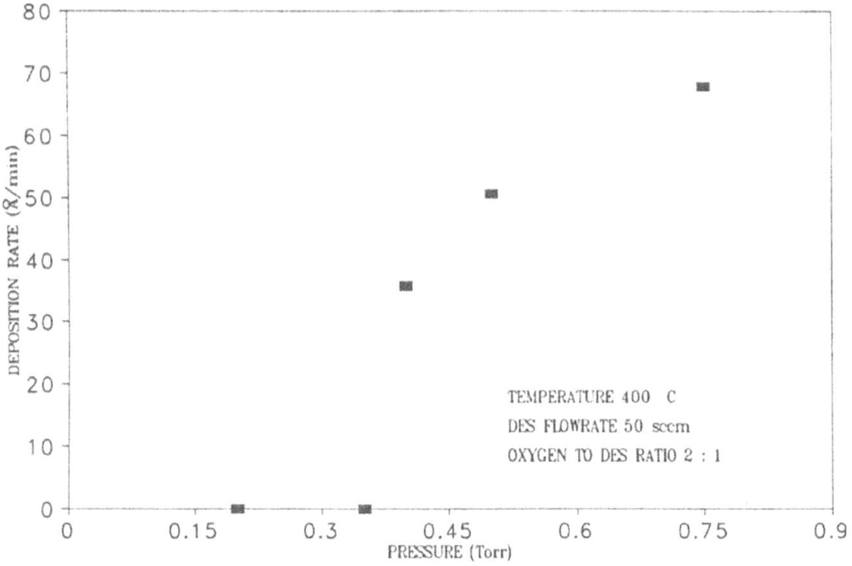

Figure 3. Variation of deposition rate as a function of pressure.

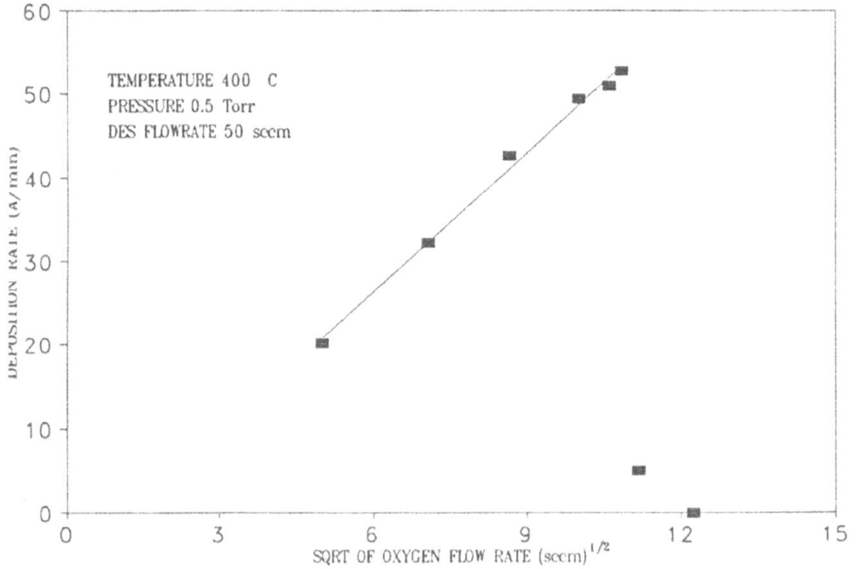

Figure 4. Variation of deposition rate versus square root of O_2 flow rate for constant DES flow rate.

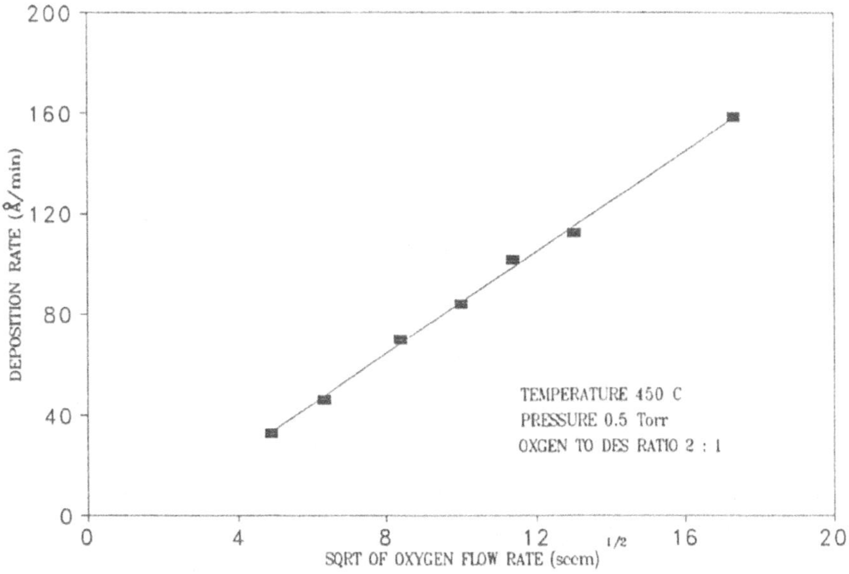

Figure 5. Variation of deposition rate versus square root of O_2 flow rate for O_2/DES ratio of 2/1.

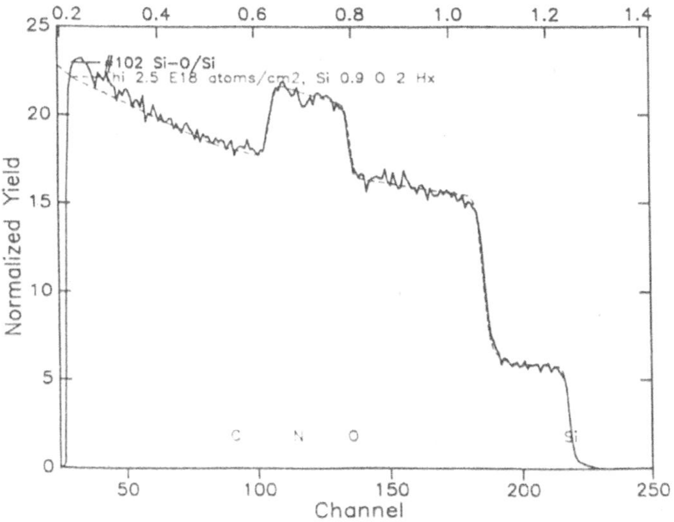

Figure 6. RBS plot for an oxide deposited at 475° C.

measurements taken on oxide films deposited at 375 and 400° C revealed a detectable level of incorporated hydrogen in agreement with the IR results of Figure 7. No hydrogen was detected in a thermal oxide control sample.

For all deposits, the Si-O stretching, bending, and rocking modes were identified at 1060, 800 and 440 cm^{-1} respectively by IR spectroscopy. Figure 7 illustrates an IR spectrum of an oxide film deposited at 400° C. Besides the standard peaks, a small peak at 880 cm^{-1} was also observed and identified as an $HSi\text{-}O_3$ moiety[15]. This peak normalized with respect to the 1060 cm^{-1} was found to decrease with higher temperatures (Figure 8), lower DES flow rates (Figure 9), and lower pressures (Figure 10) in agreement with the work of Huo et al[14].

The refractive index data, obtained by standard ellipsometry at the He-Ne laser wavelength of 632.8 nm, revealed a value of 1.46 independent of all deposition conditions.

The density of the deposited films was measured as a function of process parameters. A plot representing average film thickness as a function of mass for all process parameters is shown in Figure 11. The minimal deviation of the data from the linear least square fit indicates little dependence of density on process variables. From the slope of the curve and the known area of the wafer, the density was calculated to be 2.25 g/cm^3 which is in close agreement with the value of 2.27 g/cm^3 for thermal oxide.

The etch rates of the oxide films in a P-etch solution (15 parts Hydrofluoric acid (49%), 10 parts nitric acid (70%), and 300 parts of water) were determined by measuring the change in film thickness as a function of etch time. The room temperature etch rate was found to decrease with higher deposition (Figure 12) or higher annealing temperature (Figure 13) reflecting the gradual removal of the $HSi\text{-}O_3$ moiety and associated reduction in the break-up of the network forming structure. The etch rate measurements were found to be independent of flow rate variation at a constant deposition temperature.

The dielectric breakdown strength of a 450 nm oxide film deposited at 400° C (0.5 Torr, 50 sccm DES, and a O_2/DES ratio of 2/1) was found to be 2 MV/cm. This value is a factor of four lower than reported values for LPCVD oxides produced from the silane-based chemistry. From C-V measurements on the same film, the fixed oxide charge density (N_{ot}) was calculated from the difference between the ideal and measured flat band voltage values and found to be $\sim 6 \times 10^{11}$ cm^{-2}.

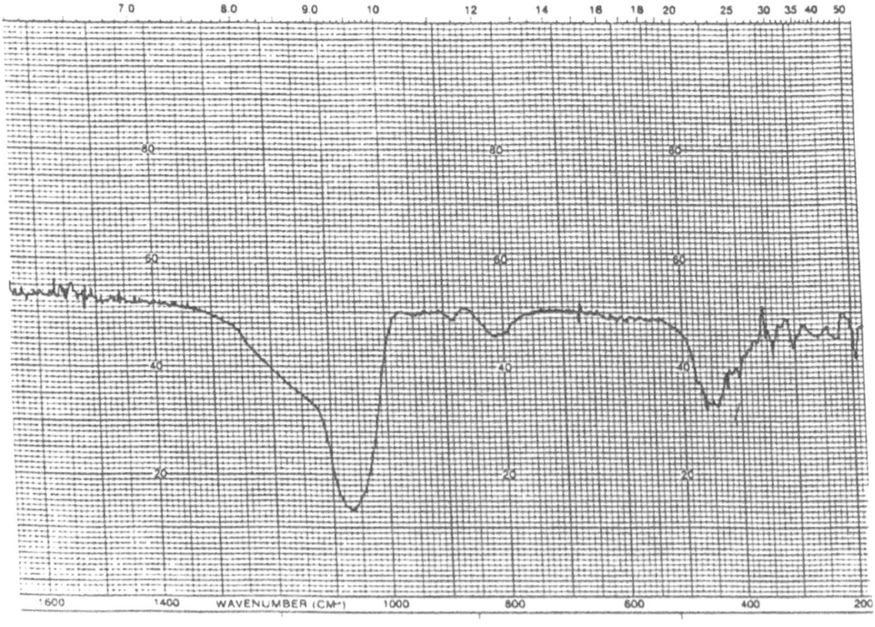

Figure 7. IR spectrum of oxide film deposited at 400° C.

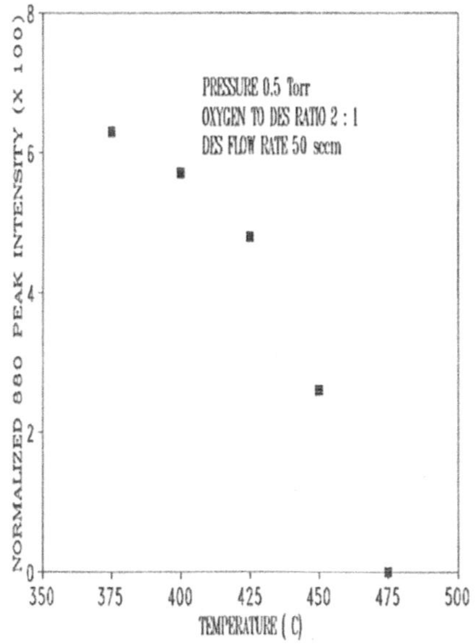

Figure 8. Normalized 880 cm^{-1} peak as a function of deposition temperature.

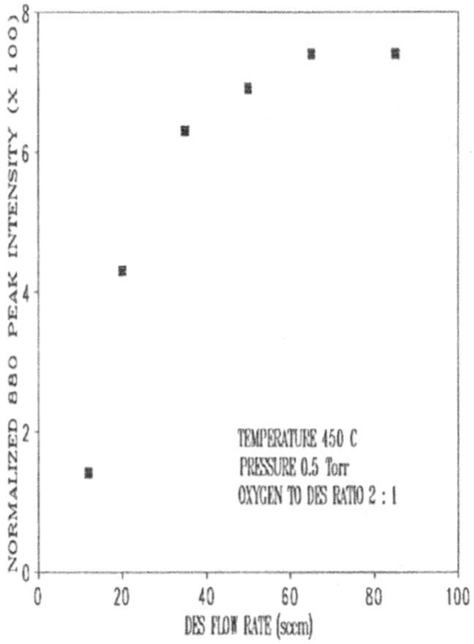

Figure 9. Normalized 880 cm^{-1} peak as a function of flow rate.

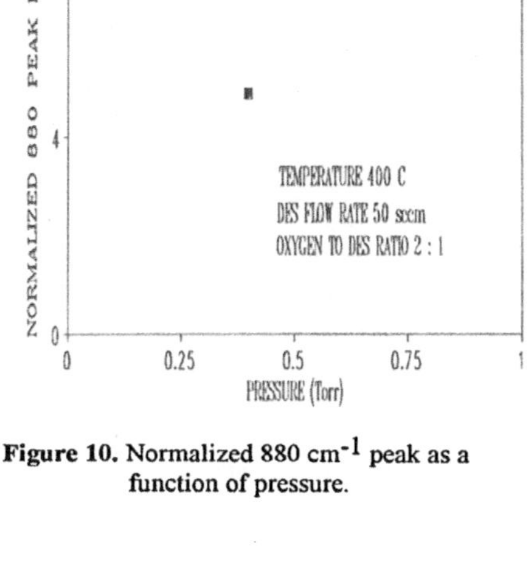

Figure 10. Normalized 880 cm^{-1} peak as a function of pressure.

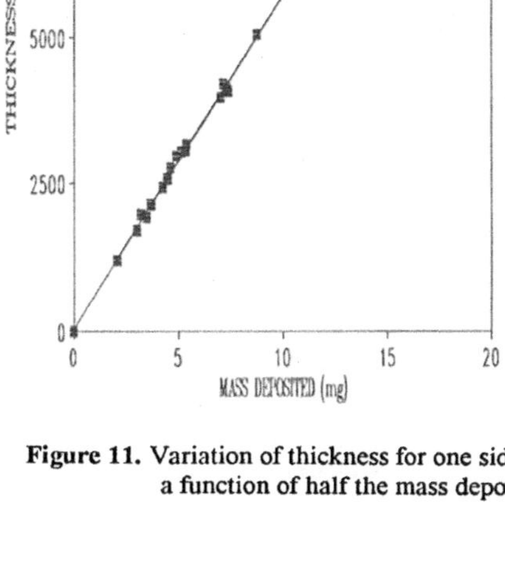

Figure 11. Variation of thickness for one side as a function of half the mass deposited.

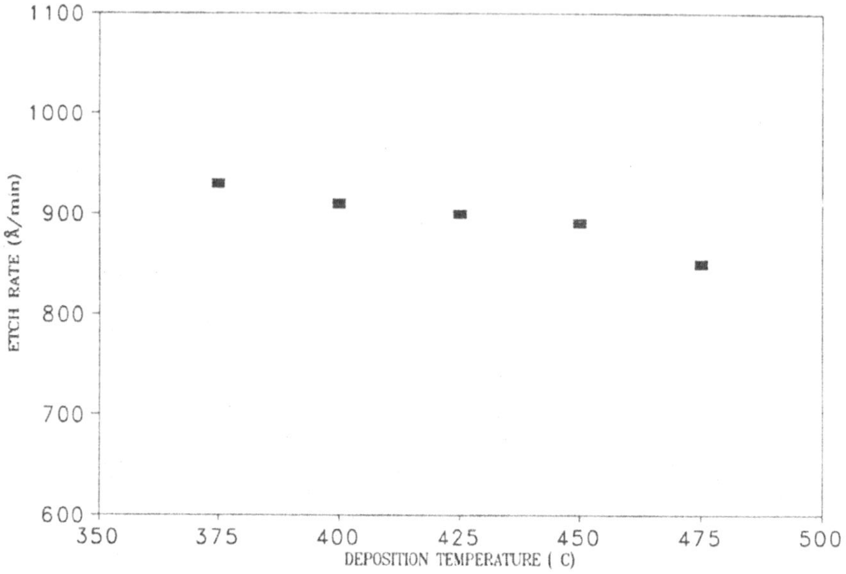

Figure 12. Variation in etch rate as a function of deposition temperature.

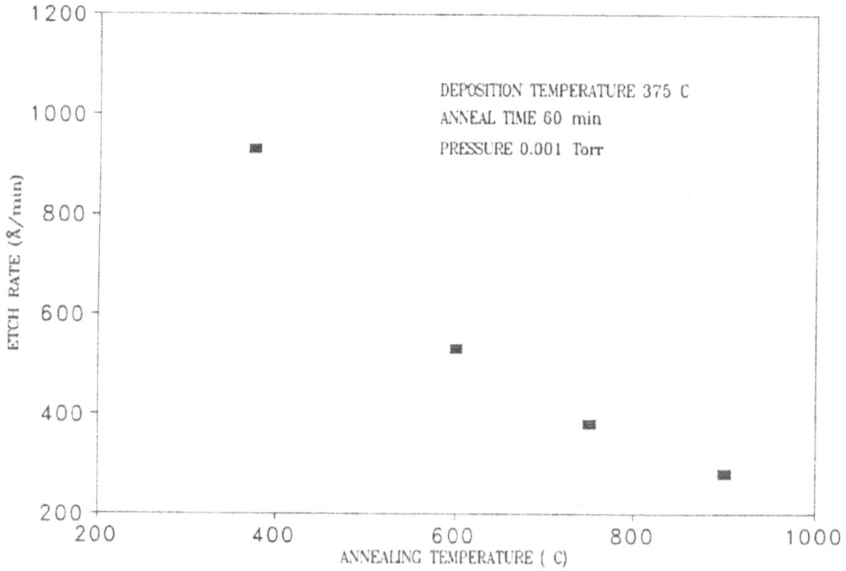

Figure 13. Variation in etch rate as a function of annealing temperature.

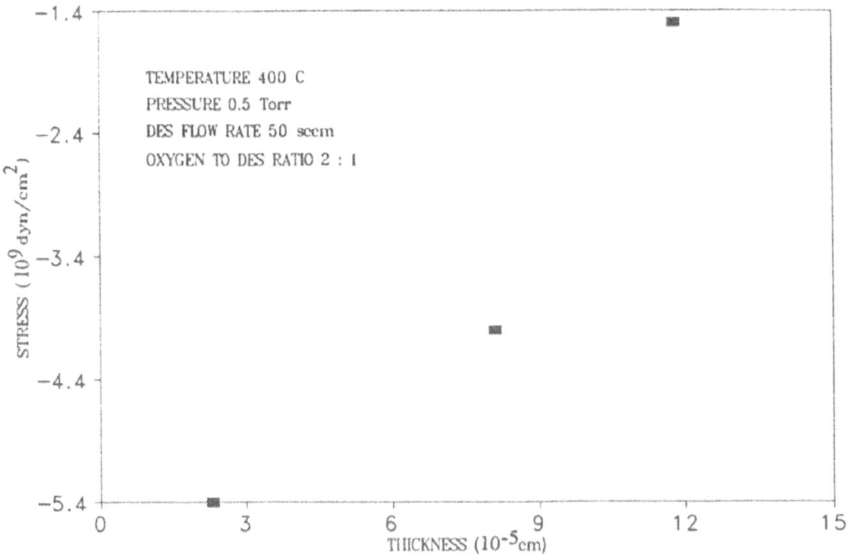

Figure 14. Variation in film stress as a function of film thickness for oxide films.

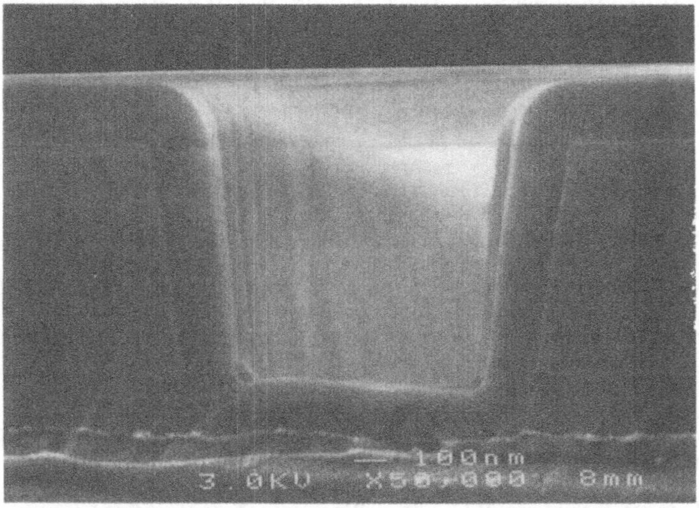

Figure 15. Step coverage of 200 nm thick oxide film deposited at 400° C, 0.5 Torr, 50 sccm DES, and 100 sccm O_2.

Film stress was measured as a function of thickness for process conditions of 400° C, 0.5 Torr, 50 sccm DES flow rate, and 100 sccm O_2. The stress was in all cases compressive and seen to increase with higher thicknesses as illustrated in Figure 14.

The step coverage of a 200 nm thick oxide film deposited at 400° C in a via ~1 μm deep and ~1.25 μm wide yielded, as shown in Figure 15, a better than 85% conformality. For a more severe aspect ratio exemplified in a via ~1 μm deep and ~0.75 μm wide, the conformality was found to be better than 55%.

CONCLUSIONS

In this study, we have demonstrated the use of diethylsilane for synthesizing silicon oxide films by low pressure chemical vapor deposition. The deposition rate as a function of temperature was found to follow an Arrhenius behavior yielding an apparent activation energy of 10 kcal/mole. The growth rate was observed to increase with higher pressure and to vary as a function of the square root of the O_2 flow rate. In both the pressure and the O_2/DES ratio studies, there were points of abrupt cessation in deposition. The films were determined to be carbon free and slightly oxygen rich. An additional peak was detected in the IR spectrum at 880 cm^{-1} reflecting the presence of $HSi\text{-}O_3$ which may be useful in monitoring hydrogen content. The refractive index and the density were observed to be close to 1.46 and 2.25 g/cm^3 respectively and in good agreement with corresponding values of thermal oxide. The dielectric breakdown values were found to be lower than expected and additional work is required to address this issue to improve applicability of the material to microelectronic applications. The etch rate of the films in a 25° C P-etch solution was observed to decrease with higher deposition or annealing temperatures reflecting densification of the material. Despite severe aspect ratios, the films were seen to exhibit good step coverage rendering the process attractive for submicron applications.

ACKNOWLEDGMENTS

The authors wish to thank A. Lagendijk at J. C. Schumacher for providing the DES and for the many fruitful discussions. Thanks are also due to B. Wilkens at Bellcore and C. C. Lee now at Motorola for the RBS and ERD measurements, as well as G. J. Campisi at NRL for the C-V measurements. The authors wish to acknowledge the Research Corporation for supporting this work.

REFERENCES

1. T. A. Foster, G. Hoeye and J. Goldman, J. Electrochem. Soc., **132**, 505 (1985).
2. A. J. Learn and B Baerg, Thin Solid Films, **130**, 103 (1985).
3. J. Orshonik and J. Kraitchman, J. Electrochem. Soc., **115**, 649 (1968).
4. A. C. Adams and C. D. Capio, J. Electrochem. Soc., **126**, 1042 (1979).
5. R. A. Levy, P. K. Gallagher, and F. Schrey, J. Electrochem. Soc., **134**, 430 (1987).
6. J. Klerer, J. Electrochem. Soc., **112**, 503 (1965).
7. J. M. Albella, A Criado, and E. Munoz Merino, Thin Solid Films, **36**, 479(1976).
8. J. Klerer, J. Electrochem. Soc., **108**, 1070 (1961).
9. E. L. Jordan, J. Electrochem. Soc., **108**, 478 (1961).
10. G. Smolinsky and R.E. Dean, Mater. Lett., **4**, 256 (1986).
11. R. A. Levy, P. K. Gallagher, and F. Schrey, J. Electrochem. Soc., **134**, 1744 (1987).
12. B. Gelernt, Semicond. Int., **13**, 83 (1990).
13. A. K. Hochberg and D. L. O'Meara, J. Electrochem. Soc., **136**, 1843, (1989).
14. D. T. C. Huo, M.F.Yan, and P.D. Foo, J. Vac. Sci. Technol., **A9** (5), 2602 (1991).
15. D. V. Tsu, G. Lucovsky, and B. N. Davidson, Phy. Rev., **B40**, 1795 (1989).

IV. CHEMICAL PROPERTIES OF Si SURFACES RELATED TO OXIDATION AND OXIDE DEPOSITION

INTRODUCTION

G. Lucovsky
Deptartment of Physics
North Carolina State University, Raleigh, NC 27695

M. Morita
Department of Electronics
Tohoku University, Sendai 980, Japan

This chapter consists of six papers that deal with the chemical properties of Si surfaces that are relevant to surface cleaning and passivation prior to oxidation and oxide deposition. These issues are of importance in device technology, because the pre-oxidation and pre-deposition surface preparation can have a significant impact on the electrical properties of the Si-SiO_2 interface region: for example, the density of interfacial defects, D_{it}, the breakdown field, and the lateral transport properties of carriers in the Si region that is in contact with the oxide.

The paper by Hirose et al, uses FT-IR-ATR and XPS to study the chemical bonding properties of ultra thin oxides, < 1.3 nm thick, formed in pure water after HF and BHF surface treatments. It identifies pre-oxidation conditions that minimize dihydride bonding and F-atom attachment on at Si(100)-SiO_2 interfaces. The paper by Higashi defines experimental conditions for generating atomically perfect surfaces on Si(111) wafers. Techniques used to confirm surface perfect include scanning tunneling microscopy and IR spectroscopy. The study emphasizes that there is a narrow range of pH for HF based solutions that will remove native oxides, and not promote detectable surface damage on Si(111) wafers.

The paper by Morita and Ohmi stresses the importance of a low temperature, 300°C, oxidation prior to the final high temperature oxidation, such as 900°C, in promoting high quality electrical performance, and high reliability in Si–SiO_2 device structures. This process is similar in concept to

were cleaned in organic solution, boiled in $HCl:H_2O_2:H_2O=3:11:86$ for 10min and dipped in $NH_4OH:H_2O_2:H_2O=1:1:4$ (n-type) or 3:3:7 (p-type) for 2min. Samples were dipped in a 4.5% HF solution or in pH modified BHF for removing native oxide (~9Å). Finally, in order to eliminate the influence of surface Si-F bonds on oxidation kinetics, the surface was rinsed in pure water for 1min. The wafer was oxidized in different environments such as pure water (17MΩ, DOC 9ppm), clean room air (23°C, 1.2% water vapor concentration) or chemical solutions (H_2SO_4/H_2O_2 or hot HNO_3). The oxidized surface was characterized by FT-IR-ATR and XPS.

FT-IR-ATR and XPS Analysis

Chemical bonding states of ultra-thin native oxide and its interface were analyzed by the ATR spectra. A Ge crystal prism in contact with a Si wafer surface was used for the multiple internal reflection medium. For the p-polarized ATR measurements, the infrared absorption due to molecular vibration normal to the surface is enhanced by a factor of $(\varepsilon^{Si})^2/\varepsilon^2$, where ε^{Si} and ε are the dielectric constant for Si and the surface oxide layer, respectively[7].

At each step of wafer storage time the chemical bonding features of the silicon surface were studied by MgKα (1253.6eV) x-ray excited Si_{2P} and O_{1S} photoelectron spectra. The escape angle θ between the photoelectron detector axis and the direction normal to the Si surface was kept at 75° for the surface sensitive measurements. The oxide thickness d_{ox} is given by the integrated photoelectron intensity of chemically shifted Si_{2P} spectrum N_{OX} and that of metallic Si signal N_{Si} as follows:

$$d_{ox}=\lambda_{ox}\cos\theta \ln\left(k\frac{N_{ox}}{N_{Si}}+1\right). \qquad (1)$$

Here, λ_{ox} is the escape depth of Si_{2P} photoelectron from SiO_2 (25Å)[8]. The value of k=2.086 is estimated by the atomic densities of Si and SiO_2 and their photoelectron escape depths. Note that in eq. (1) the SiO_2/Si interface is assumed to be atomically abrupt. This assumption appears to be basically reasonable as discussed later. The deconvolution of chemically shifted Si_{2P} spectrum was carried out as follows: The Si_{2P} spectrum of a hydrogen-terminated Si(100) surface treated in a 4.5% HF solution was used as a reference of the bulk Si_{2P} signal. As is discussed later, an HF treated Si surface without water rinse is terminated mainly by hydrogen bonds and partly (about 12%) by fluorine bonds. The Si_{2P} signal peak related to a Si-H bond is thought to appear at a binding energy of 0.35eV higher than the $Si_{2P3/2}$ peak as revealed by high resolution XPS measurements[9]. Since the HF treated Si(100) surface is covered mainly with SiH_2 and SiH_3 bonds as shown later, a hydrogen related Si_{2P} peak might appear at a binding energy of about 1eV higher than the bulk Si_{2P} peak. However, the resolution of the present XPS system (FWHM=1.15eV at $Ag_{3d5/2}$) is not enough to distinguish the Si^{1+} suboxide peak from a hydrogen related Si_{2P} signal. Also, the oxidation of the silicon backbonds of a SiH_x unit causes a further chemical shift of the signal arising from O_ySiH_x (x+y=4). Hence, we tried to extract the hydrogen related signal by subtracting the bulk Si_{2P} signal from a measured spectrum and decomposing the residual spectrum to Si^{1+}, Si^{2+} and Si^{3+} components whose chemical shifts are 0.95, 1.75 and 2.48eV, respectively[10]. If the Si_{2P} component is still remaining, it could be associated with the hydrogen related signal.

RESULTS AND DISCUSSION

Hydrogen and Fluorine Bonds on Chemically Cleaned Si Surface

Fluorine atoms are existing on HF or BHF treated Si surfaces. It is likely that the surface fluorine forms Si-F bond because of its high reactivity and the F_{1S} binding energy at 686.2eV which corresponds to Si-F[11]. Also, fluorine appears to remain selectivity at reactive sites such as atomic steps on the surface because the fluorine coverage is almost constant (~12%) for the HF concentration range from 0.5 to 20%[2]. Figure 1(a) represents the XPS signal intensity ratio of F_{1S} to Si_{2P} which reflects the surface fluorine coverage. The coverage is reduced to nearly zero at pH=5.3 by adding NH_4OH to a NH_4F:HF=7:1 solution. Further increase of the NH_4OH concentration results in the increase of fluorine bonds, and also N_{1S} signal becomes observable, indicating the existence of surface reaction products such as $(NH_4)_2SiF_6$. In the figure, the fluorine coverage for a pH-modified BHF-treated Si(111) wafer is about 1/3 or less with respect to that of diluted HF treated Si with ~12% fluorine coverage. If it is assumed that fluorine bonds remain at reactive sites on the surface, the BHF(pH=5.3)-treated silicon surface whose fluorine coverage is nearly zero could be atomically flat without any high density atomic steps which can be the starting points

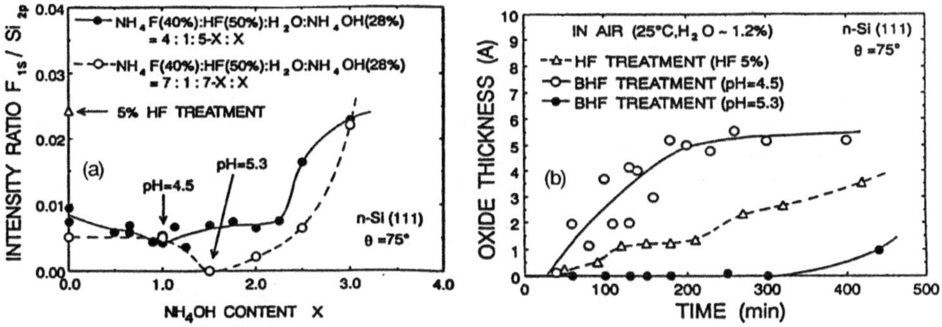

Figure 1. Surface chemical bonding features of HF or BHF treated Si(111) surfaces. (a) Integrated intensity ratio of F_{1S} to Si_{2P} versus NH_4OH content for wafers treated in pH modified BHF. (b) Native oxide layer thickness versus air expose time for Si (111) surfaces cleaned by 5% HF or BHF (pH=4.5, 5.3).

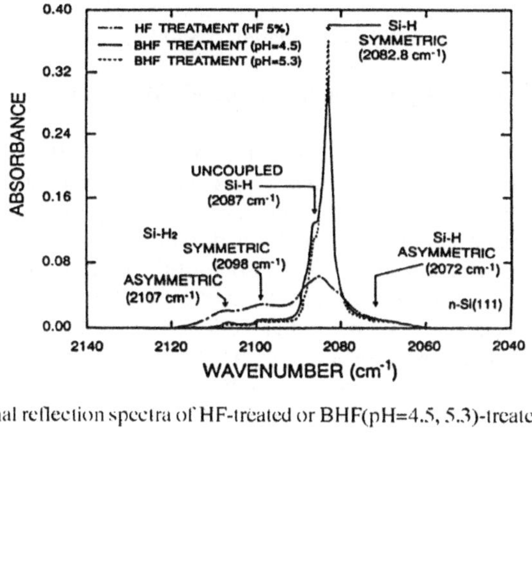

Figure 2. Internal reflection spectra of HF-treated or BHF(pH=4.5, 5.3)-treated Si(111) surfaces.

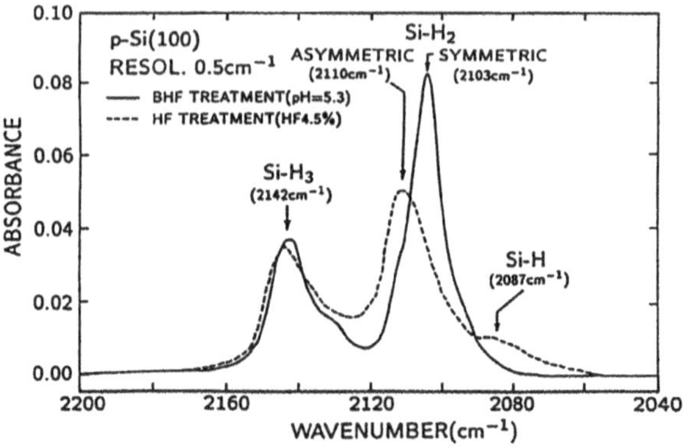

Figure 3. Internal reflection spectra of HF-treated or BHF(pH=5.3)-treated Si(100) surface.

of the oxidation. In fact, the BHF (pH=5.3)-treated Si(111) surface is hardly oxidized for 300min in clean room air as shown in Fig. 1(b). Correspondingly, the p-polarized ATR spectrum of the BHF (pH=5.3) treated surface in Fig.2 is dominated by a very sharp vibrational line at 2082.8cm^{-1} which refers to the monohydride bond and the s-polarized spectrum is extremely weak, indicating that the surface Si-H bond configuration is normal to the Si(111) surface to form the 1x1 structure. Therefore, we may conclude that the Si(111) surface cleaned by BHF with pH=5.3 is nearly step-free or atomically flat. The main peak height for the surface treated by BHF with pH=4.5 for a NH_4F:HF=4:1 solution is a little lower than the case of pH=5.3 for NH_4F:HF=7:1, while it is much sharper than that of 5% HF treated Si. Nevertheless, the oxidation rate for the HF treated Si surface, which is significantly rough, is slower than the BHF (pH=4.5) treated Si probably because the higher fluorine coverage for the HF treated Si results in the effective passivation of reactive sites such as atomic steps or microfacets by forming Si-F bonds. It is likely that the fluorine bond induces the polar nature at the neighboring silicon bonds and stabilizes the silicon network[2].

ATR spectra of HF or BHF treated Si(100) surface are compared in Fig. 3. An ideal Si(100) surface should exhibit the SiH_2 absorption band only. However, rather strong SiH_3 absorption is observed because the (100) surface is rather rough compared to (111). The SiH_2 absorption for the BHF treated surface is stronger than the HF treated surface, showing that the BHF cleaned surface is flatter than the HF treated case. Si(100) wafers are widely used for fabricating ULSIs. Hence, we will mainly investigate here the native oxide growth on atomically rough Si(100) surfaces as compared to Si(111).

Oxidation Mechanism of Hydrogen-Terminated Si Surface

Silicon oxidation which slowly proceeds in pure water was first analyzed. The native oxide thickness as a function of storage time in pure water is shown in Fig.4. The oxidation curve exhibits the plateaus where the oxide thickness apparently saturates at about 2, 4, 6 and 8Å. These saturated oxide thickness coincides well with a structural model of the SiO_2/Si interface proposed by Herman et al.[12]. It is also confirmed from the ATR spectra that the hydrogen-terminated Si surface is oxidized from the backbond of surface Si-H bond as already reported[13]. This implies that the initial oxidation of the hydrogen terminated surface predominantly proceeds parallel to the surface. Therefore, it is reasonable to speculate that the oxidation rate along the surface is faster than the perpendicular direction. The plateaus in Fig. 4 could be explained by a model that the oxidation is initiated at chemically reactive step edges and propagates parallel to the surface, being promoted by the layer-by-layer reaction[14].

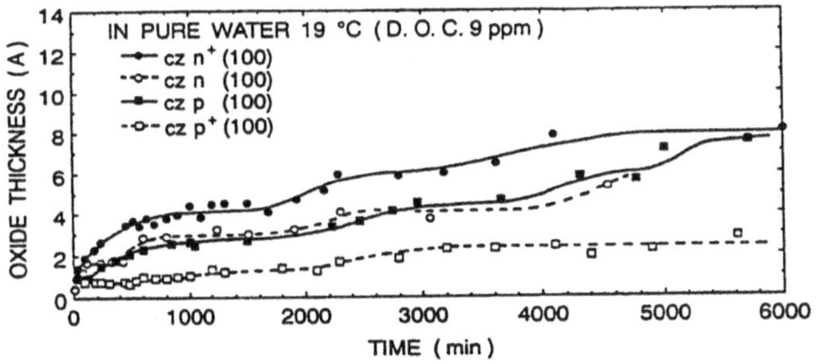

Figure 4. Native oxide layer thickness on HF treated Si(100) as a function of oxidation time in pure water.

The plateau region in the figure indicates the transition time during which oxygen penetrates into the next Si layer to be oxidized. Similar oxidation kinetics is also observed for the oxidation in clean room air.

The oxidation rate for n^+-Si is significantly faster than that of p^+, while those of n and p-type Si are in between. The Si oxidation is known to be enhanced by forming O_2^- ions through free electron tunneling from the Si conduction band to adsorbed O_2 molecule. In the case that the empty electron state of adsorbed O_2 is located near the Fermi level of p-type Si substrate (E_v + 0.2eV), the electron transfer from easily occurs for n^+-, n- and p-type Si in this order in agreement with the observed result. The formation of O_2^- ions also induces the surface electric field which assists the oxidation. A calculated Debye length for a donor concentration of 5.0 x $10^{18}cm^{-3}$ is 18Å which provides the surface electric field as high as ~10^5V/cm for the band bending of a few kT. This value is enough to significantly enhance the oxidation rate for n^+-Si[15].

Infrared Spectra of Native Oxide Grown in Pure Water

The p-polarized ATR and transmission spectra for 150Å thick thermal oxide are compared in Fig. 5. The LO phonon band at ~1200cm^{-1} is predominant in the ATR spectrum, while the TO phonon band at ~1080cm^{-1} is intense in the transmission spectrum. It should be noted that infrared spectra of oxide taken by transmission or p-polarized internal

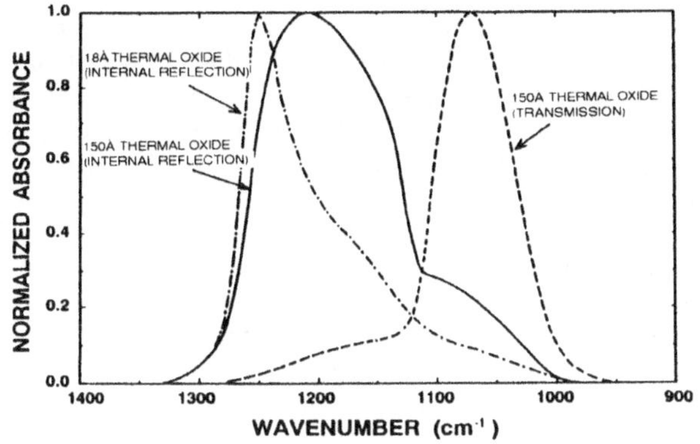

Figure 5. P-polarized ATR and transmission spectra for 150Å thick thermal oxide together with 18Å thermal oxide.

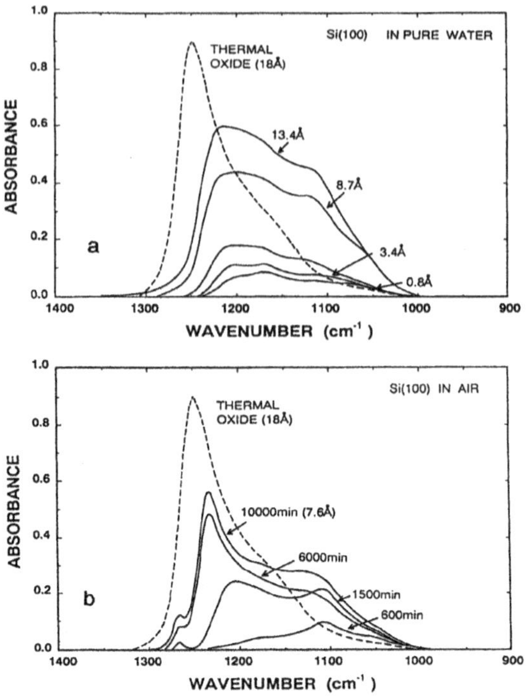

Figure 6. The Si-O-Si stretching mode spectra of native oxide grown in pure water(a) and in clean room air(b).

reflection technique do not necessarily coincide with each other because the LO phonon band at ~1200cm^{-1} is strongly enhanced with the p-polarized internal reflection measurement, while no enhancement occurs for the transmission measurement because it is equivalent to s-polarized. A p-polarized ATR spectrum for 18Å thick thermal oxide is also shown in the figure. The LO phonon band for the 150Å thick oxide shifts toward the lower wavenumber side with respect to the 18Å thick oxide. This implies that the SiO_2/Si interface structure is relaxed in the thicker oxide.

The Si-O-Si stretching absorption spectra obtained by ATR for native oxide formed in pure water are shown in Fig. 6(a). The absorption peak shifts to higher wavenumber side as the oxide thickness increases. The LO phonon peak for the 13.4Å thick oxide remains at 1210cm^{-1} which is still lower than that of the 18Å thick thermal oxide peak at 1250cm^{-1}, and the TO phonon band is rather broad. This indicates that the native oxide/Si interface is structurally relaxed despite room temperature oxidation. The ATR spectra for native oxide grown in clean room air also exhibits the LO phonon band shifting to higher wavenumber side (1230cm^{-1}) with the increase of oxide thickness (Fig. 6(b)) and the TO phonon band is rather sharp compared to Fig. 6(a), suggesting the difference of the interface structures for the both types of oxides. This difference could be related to the different SiH_x-bond configurations in the interface as revealed by the hydrogen stretching vibration mode at 2100~2300cm^{-1} in Fig. 7. The absorption peaks are assigned by assuming that the electronegativity of atom group connected to SiH_x bond is proportional to the wavenumber shift[16]. The absorptions by SiH_2 at 2110cm^{-1} and SiH_3 at 2142cm^{-1} become weak with increasing the oxide thickness and the absorbance at the higher wavenumber side is enhanced because the

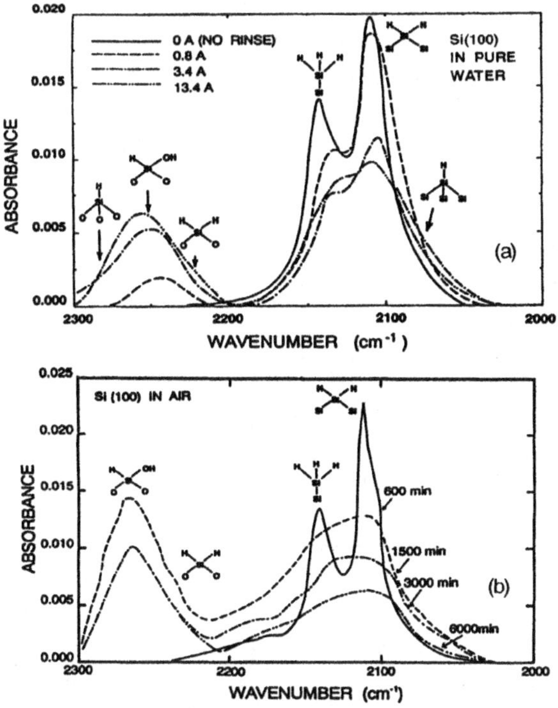

Figure 7. SiH_x stretching mode spectra of native oxide grown in pure water (a) and in clean room air (b).

oxidation starts from the backbond of SiH_x bond. The SiH_2 and SiH_3 signal intensities for the oxide grown in pure water remain at a saturated value, while they are progressively reduced for the oxidation in air. This is because the hydrogen penetration into the oxide is significant in the case of pure water oxidation. The oxidation of the backbond of the surface Si-H bond is caused through the oxygen insertion to Si-H to form SiOH on the top surface[13], so that the saturated SiH_x bond density in Fig. 7(a) indicates the existence of hydrogen bonds in the native oxide/Si interface to relax the strained bonds. On the other hand, the air oxidation can not supply enough hydrogen from the ambient, showing the smooth decrease of SiH_x absorption band during the air oxidation (Fig. 7(b)).

Si_{2P} Spectrum Associated with Interface SiH_x Bond

The existence of SiH_x bonds in the interface is examined by XPS. The Si_{2P} spectrum taken for the oxide formed in a solution of H_2SO_4:H_2O_2=4:1 is compared with that for the oxide grown in pure water. The Si_{2P} peak of the water grown oxide shifts to 103.5eV even for a very thin oxide (5.7Å), and the signal due to the lower level suboxides such as Si^{1+}or Si^{2+} is very weak (Fig. 8). In contrast to this, the H_2SO_4+H_2O_2 treated oxide, in which hydrogen bonds are hardly incorporated[17], shows a significant amount of Si^{1+} and Si^{2+} suboxide signals, and the main peak shift is a little small compared to the water oxide. When the water oxide is annealed at 500°C for 10min in UHV($6x10^{-9}$ Torr), the hydrogen effusion is confirmed by the ATR analysis in which SiH_2 and SiH_3 absorptions are dramatically reduced by annealing. Correspondingly, the Si_{2P} signal at ~101.4eV is enhanced with annealing. The enhanced signal component at about 2 eV higher than the bulk Si_{2P} signal

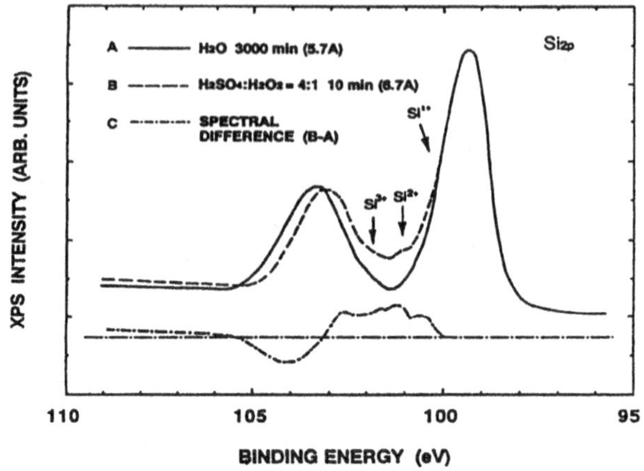

Figure 8. Comparison of Si_{2p} photoelectron spectrum of native oxide formed in a solution of $H_2SO_4+H_2O_2$ with that for the oxide grown in pure water.

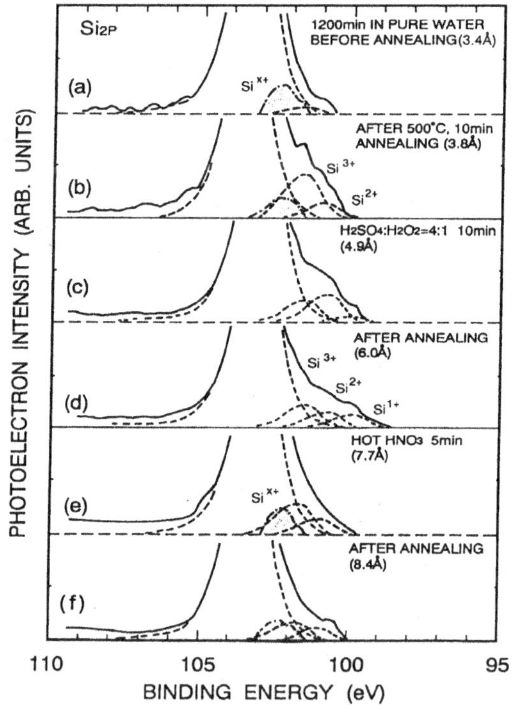

Figure 9. Si_{2p} photoelectron spectra and their deconvoluted spectra for native oxide grown in pure water (a), $H_2SO_4+H_2O_2$ (c) and hot HNO_3 (e). The spectra for the samples annealed at 500°C are shown in (b), (d) and (f). The bulk Si signal is removed.

shows the increase of Si^{2+} in the SiO_2/Si interface region as confirmed by the angle resolved analysis. Therefore, the Si^{2+} signal which must be predominantly detected in the hydrogen free SiO_2/Si(100) interface is eliminated in the water oxide probably due to the existence of

SiH_x in the interface region. And the annealing reduces the SiH_x bonds to enhance the Si^{2+} suboxide signal.

The deconvoluted spectra of the Si_{2P} signal for native oxides grown in pure water, $H_2SO_4+H_2O_2$ and hot HNO_3 are compared in Fig. 9. In the water oxide, the Si^{2+} and Si^{3+} signal is significantly enhanced by UHV annealing at 500°C. After the vacuum annealing, the Si^{4+} signal of the chemical oxides shifts to 0.2~0.3eV higher energy side. SiH_x bonds in the chemical oxide grown in hot HNO_3 is not significantly effused by 500°C annealing, and hence the Si_{2P} spectra are not very much changed. In the figure, the extra spectral yield denoted by Si^{x+} appears at about 3eV higher energy side with respect to the bulk Si peak for both the water oxide and hot HNO_3 oxide. The Si^{x+} signal is not observed for the $H_2SO_4+H_2O_2$ oxide. This Si^{x+} spectrum has been assigned as O_3-SiH bond because the signal is missing for the oxide grown in a $H_2SO_4+H_2O_2$ solution[17]. In the water oxidation and hot HNO_3 oxide enough H^+ ions are supplied together with oxidant, and hence a part of Si-Si bonds in the interface are cleaved to form $SiH_x(x=1\sim3)$ so as to relax the strained bonds. This is consistent with the result of Fig. 6, where the LO phonon peak position of the water grown oxide is located at the lower wavenumber side than the thermal oxide.

CONCLUSIONS

The BHF treated Si(111) surface is atomically flat and chemically stable. The HF treated Si(111) and Si(100) surfaces are not flat, but rather stable due to fluorine passivation of reactive sites such as atomic steps and microfacets. The rough surface with a low level of fluorine coverage is easily oxidized.

It is shown that the layer-by-layer oxidation proceeds in pure water for the hydrogen terminated Si(100) surface and a significant amount of SiH_x bonds are incorporated in the native oxide/Si interface. It is likely that the strained bonds in the water grown native oxide/Si interface region are relaxed by incorporated SiH_x bonds.

ACKNOWLEDGEMENTS

The authors express their thanks to Mr. K. Sawara, Sumitomo Chemical Co. Ltd., for his help in the experimental work. A part of this work has been done in Research Center for Integrated Systems, Hiroshima University. One of the authors (M. Hirose) wants to thank Dr. Y. J. Chabal, AT&T Bell Laboratories, for his valuable discussion on the ATR measurements.

REFERENCES

1. T. Takahagi, I. Nagai, A. Ishitani, H. Kuroda and Y. Nagasawa, J. Appl. Phys. 64: 3516 (1988).
2. T. Sunada, T. Yasaka, M. Takakura, T. Sugiyama, S. Miyazaki, and M. Hirose, Jpn. J. Appl. Phys. 29:L2408 (1990).
3. G. S. Higashi, Y. J. Chabal, G. W. Trucks and K. Raghavachari, Appl. Phys. Lett. 56:656 (1990).
4. S. Watanabe, M. Shigeno, N. Nakayama and T. Ito Extended Abstracts of 1991 Intern. Conf. on Solid State Devices and Materials (1991, Yokohama) p. 502.
5. T. Yasaka, K. Kanda, K. Sawara, S. Miyazaki and M. Hirose, Jpn. J. Appl. Phys. 30: 3567 (1991).
6. K. Sawara, T. Yasaka, S. Miyazaki and M. Hirose, Tech. Digest of Intern. Workshop on Science and Technology for Surface Reaction Process (Tokyo, 1992) p. 93.
7. Y. J. Chabal, Surf. Sci. Reports 8:211 (1988).
8. S. I. Raider and R. Flitsch, IBM J. Res. Develop. 22:294 (1978).
9. M. Niwano, H. Katakuwa, Y. Takeda, Y. Takakuwa, N. Miyamoto, A. Hirakawa and K. Yagi, J. Vac. Sci. Technol. A9:195 (1991).
10. F. J. Himpsel, F. R. McFeely, A. Taleb-Irrahimi, A. Yarmoff and G. Hollinger, Phys. Rev. B 38:6084 (1988).

11. T. Takahagi, A. Ishitani, H. Kuroda and Y. Nagasawa, J. Appl. Phys. 69:803 (1991).
12. F. Herman, J. P.Batra and V. Kasowski: "The Physics of SiO_2 and Its Interface," S. Pantelides, ed., Pergamon Press, NY (1978) p.333.
13. T. Takahagi, A. Ishitani, H. Kuroda and Y. Nagasawa, J. Appl. Phys. 68:2187 (1990).
14. T. Yasaka, M. Takakura, S. Miyazaki and M. Hirose, Proc. of Mat. Res. Soc. Symp. vol. 222 (Anaheim, 1991) p.225.
15. P. J. Jorgensen, J. Chem. Phys, 37:874 (1962)
16. G. Lucovsky, Solid State Commun. 29:571 (1979).
17. K. Sugiyama, T. Igarashi, K. Moriki, Y. Nagasawa, T. Aoyama, R. Sugino, T. Ito and T. Hattori, Jpn. J. Appl. Phys. 29:L2401 (1990).

UNDERSTANDING THE SURFACE CHEMICAL AND STRUCTURAL IMPLICATIONS OF HF SOLUTION CLEANING OF SILICON

G. S. Higashi

AT&T Bell Laboratories
Murray Hill, New Jersey 07974

ABSTRACT

Dissolution of Si surface oxides in HF acid lead to Si surfaces which are atomically clean and terminated/passivated with hydrogen. Hydrogen terminates the Si dangling bonds rather than fluorine because the polar nature of the Si-F bond makes it susceptible to subsequent HF attack. The subsequent attack then frees SiF_x species into the solution while simultaneously forming the hydrogen termination. The surface structure of the resulting H-termination varies with crystal orientation (Si(100) vs. Si(111)) as well as with solution pH and is studied using infrared absorption spectroscopy and scanning tunneling microscopy. An optimum pH of 7.8 results in atomically well ordered H-passivated Si(111) surfaces. The role of OH^- in the formation of these well ordered surfaces is elucidated in water rinsing experiments.

INTRODUCTION

Increasingly more stringent requirements in all areas of semiconductor processing are being imposed with each new generation of integrated circuit. Lithography, plasma etching, oxidation, and chemical and physical vapor deposition all come to mind with many technical problems to be overcome. We must always remember, however, that nearly every processing step is either preceded by or followed by a wafer cleaning step.[1] In fact, it is predicted that nearly 40% of all processing steps will be related to wafer cleaning or interface preparation in future generations of integrated circuits. Metal, hydrocarbon, surface oxide, and particle contaminants must all be controlled in a successful fabrication sequence.

There are two fundamental types of surface cleans. There are cleans which remove contaminants and simultaneously grow passivating oxides and there are cleans which remove surface oxides. The most widely used cleaning solutions which grow passivating oxides use mixtures of hydrogen peroxide with acids or bases. The most widely used of these is the RCA clean,[1,2] which is a sequential etch in a basic peroxide ($NH_4OH:H_2O_2$)

followed by an acidic peroxide solution ($HCl:H_2O_2$). Another frequently used acidic peroxide is the Piranha etch ($H_2SO_4:H_2O_2$). These cleans effectively remove both metallic contaminants and hydrocarbon residues and leave Si surfaces covered with a passivating oxide layer which renders these surfaces hydrophillic. In instances where surface oxides need to be removed, solutions containing HF acid are used. The HF chemistry is unique in that it readily dissolves these surface oxides but does not attack the underlying Si. The most commonly used HF solutions are either dilute aqueous solutions (10-100:1 H_2O:HF) or dilute buffered solutions where NH_4F is used as the buffer. The surfaces surfaces produced using these solutions are hydrophobic and remarkably resistant to oxidation.[3,4]

The chemically grown hydrophillic oxide surfaces are relatively well understood. They are comprised of a thin layer of SiO_2 (12-15 Å thick) which is somewhat like the SiO_2 grown in conventional oxidation furnaces except that its stoichiometry is more highly variable and it is permeated with water and/or OH units. SiO_2 is a highly polar compound and thus is naturally hydrophillic. The hydrophobic surface produced by HF acid etching has only recently begun to be understood.[3-8] A common misconception in the industry which was propagated for decades was that these surfaces were terminated with fluorine. We now know that fluorine is a minority species on these surfaces[5,6] and that it is hydrogen termination which explains why these surfaces are hydrophobic and nonreactive to exposure to atmosphere.

The object of this paper is to bring you up to date with our present understanding of HF acid etching. This is accomplished by first discussing the chemical mechanism by which the surface becomes hydrogen terminated. This is followed by studies of surfaces prepared by etching in dilute HF using infrared absorption spectroscopy to identify the various SiH species encountered on Si(111) and Si(100) surfaces. A comparison is then made to the spectra obtained using concentrated HF and 7:1 buffered HF on these surfaces. This leads to a study of the pH dependence of the surface morphologies obtained where atomically perfect hydrogen terminated Si(111) are produced. The mechanism by which these surfaces are produced are finally discussed in the context of some recent results involving water rinsing.

MECHANISM OF HYDROGEN TERMINATION

Part of the confusion concerning fluorine termination of the Si following HF acid etching arises because the accepted mechanism of the dissolution of SiO_2 leads automatically to fluorine terminated Si. Fig. 1a illustrates the manner in which HF attacks Si-O bonds leaving Si-F behind on the surface. First, remember that HF is for the most part not dissociated as ions in solution but is left as an intact molecule. This molecule easily (low activation barrier) can insert itself between the Si and O of the Si-O bond. This insertion is greatly facilitated by the highly polar nature of the Si-O bond, which the highly polar HF molecule can take advantage of during its attack. Thus, very naturally, the positively charged H-atom gets associated with the negatively charged O-atom and the negatively charged F-atom get associated with the positively charged Si-atom of the Si-O bond. This liberates H_2O into the solution and leaves Si-F in its place on the surface (Fig. 1b). The Si-F bond is the strongest single bond known in chemistry with a bond energy of ~6 eV. The bond strength of the Si-H is only ~3.5 eV and leads one to the false conclusion that this must mean that the F-terminated surface must be more stable than the H-terminated surface.

Ubara, Imura and Hiraki[9] were the first to propose a mechanism to get around this dilemma. They recognized that the Si-F bond must be highly polar because of the large electronegativity difference between these atoms. They then proposed that the Si-F bond caused bond polarization of the Si-Si back bond allowing HF attack of the back bond as illustrated in Fig. 1c. In this case, SiF_x species would be released into the solution leaving

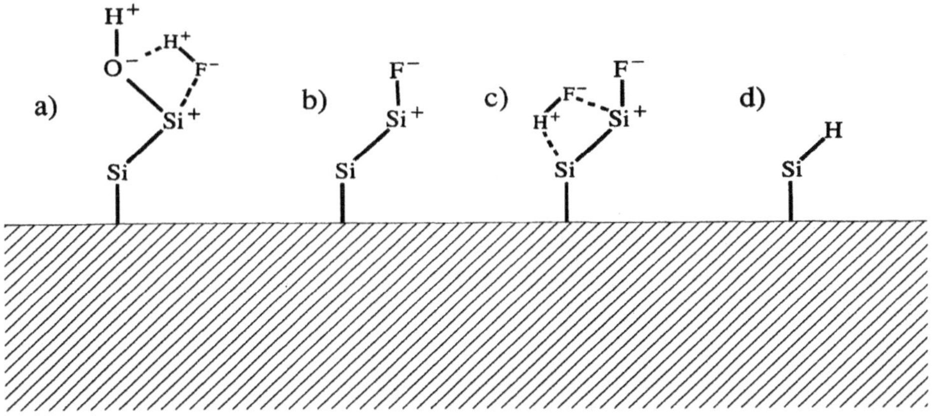

Fig. 1. Schematic representation of the HF surface reaction sequence which ultimately leads to H-terminated Si surfaces.

Si-H behind on the surface as shown in Fig. 1d. These basic concepts were confirmed using state-of-the-art *ab-initio* molecular orbital calculations of the activation energies of these types of reactions by Trucks *et. al.*[10] In those calculations, we found an activation energy of ~1.0 eV for reactions of the type shown in Fig. 1c. Low activation energies such as these were explained precisely because of the charge transfer proposed by Ubara *et. al.* In the absence of that charge transfer, as is the case for the non-polar Si-H bonds, the activation energy of the Si-Si back bond attack was found to be 1.6 eV, 0.6 eV higher in energy than for fluorinated Si species. The lesson is very simple. HF attacks polar species very effectively but is much less effective against non-polar species. In this sense, not only do we understand why it is hydrogen and not fluorine which terminates the Si dangling bonds after HF acid etching but we now also understand why HF dissolves oxide so readily but leaves the Si untouched.

DILUTE HF

We begin our discussion of the H-terminations obtained using dilute aqueous HF solutions. These experiments are described in detail in references 8 and 11. In essence, a Piranha etch is used to clean the samples and grow a passivating oxide which is then removed in a dilute HF solution. There is, of course, an intervening rinse in deionized water between these two steps, but it is important to point out that wafers are not rinsed subsequent to the HF treatments. This is possible since the wafers come out of the dilute HF solution completely dry due to the hydrophobic nature of these surfaces. The resulting hydrogen termination is then studied using multiple-internal-reflection Fourier-transform infrared spectroscopy to obtain high sensitivity to the Si-H stretching vibrations.

The infrared spectrum, in the region of the Si-H stretch, for a Si(111) surface treated in this manner is shown in Fig. 2. The absorption spectrum observed in Fig. 2b is quite complex, exhibiting many well defined modes ~$10 cm^{-1}$ in width. Interpreting such a complex mode structure is extremely difficult without additional information. Definitive mode assignments come from isotopic substitution experiments combined with *ab-initio* force constant/normal mode analyses on model compounds.[8] The isotopic substitution experiments allow one to decouple modes which are coupled to one another, simplifying the observed spectra. Examples of these spectra are shown in Figs. 1a and 1a′, where DF solutions were prepared containing trace amounts of HF. In this manner, H-atoms on the

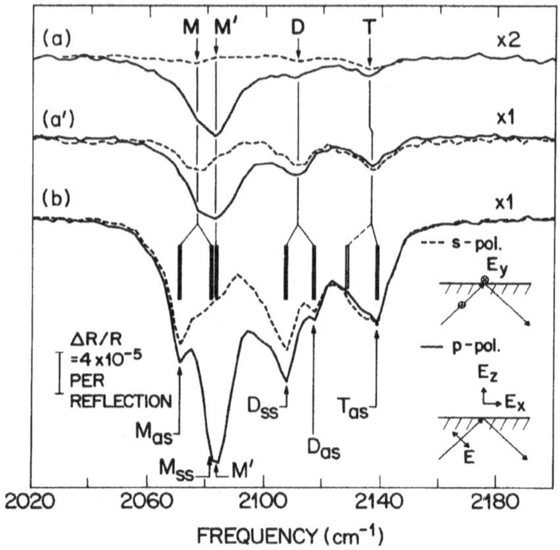

Fig. 2. Polarized IR spectra of Si(111) silicon-hydrogen stretching vibrations for 3 isotopic concentrations; (a) 10% H : 90% D, (a') 25% H : 75% D, (b) 100% H. The vertical bars in bold represent the calculated coupled mode splittings from the observed isolated frequencies (**M** = 2077 cm^{-1}, **D** = 2111 cm^{-1}, **T** = 2137 cm^{-1}).

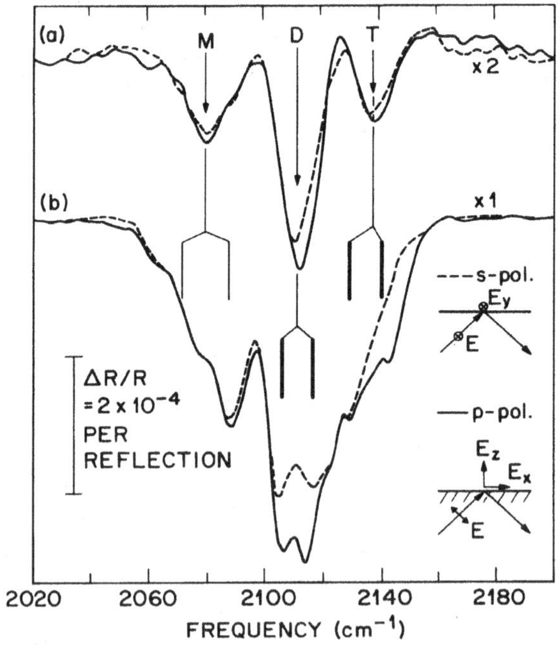

Fig. 3. Polarized IR spectra of silicon-hydrogen stretching vibrations at 2 isotopic concentrations for Si(100); (a) 20% H : 80 % D and (b) 100% H.

surface never find themselves adjacent to other H-atoms but are surrounded by D-atoms. The H-atom mass is sufficiently different from the D-atom masses that near neighbor interactions are turned off and mode splitting is suppressed. These "isolated" spectra show that there are three fundamental modes on this surface, one near 2077 cm^{-1}, another at 2111 cm^{-1}, and one at 2137 cm^{-1}. The *ab-initio* calculations show that these modes are associated with Si-monohydrides (M and M′), Si-dihydrides (D), and Si-trihydrides (T), respectively. Support for these mode assignments come not only from the relative positions of M and M′, D, and T, but from the correspondence to the calculated and measured individual mode splittings.

With this understanding of the constituents of the modes observed, conclusions can now be drawn about the Si(111) surface structure after HF treatment. The mode M′ which appears only in p-polarization is oriented perpendicular to the surface and must be associated with a Si-monohydride located on a terrace type of position. In contrast, modes M and D, associated with coupled Si-monohydrides and Si-dihydrides, respectively, are polarized in the plane of the surface and must be associated with a step edges. The trihydride modes, also polarized in the plane of the surface, could also be associated with terrace locations but were argued to be located near step edges due to an in plane anisotropy observed for intentionally miscut Si(111) samples in reference 11. Thus, of the modes observed less than 50% are associated with terraces, implying that these surfaces must be atomically rough.

The spectra obtained for Si(100) samples treated in dilute HF look quite similar (Fig. 3), except that dihydride modes predominate rather than monohydride modes. One observes, once again, all three type of modes, Si-monohydride (M), Si-dihydride (D), and Si-trihydride (T). In this case, however, terrace monohydride (M′) cannot exist and only coupled monohydride (M) is observed. This surface is approximately 50% dihydride and 25% each, mono- and tri-hydride, again implying a high degree of surface roughness.

CONCENTRATED HF AND 7:1 BUFFERED HF

Surfaces prepared in concentrated HF (49%) and 7:1 buffered HF are surveyed next. Concentrated HF is not often used in industry but is an interesting candidate for study since it has a negligible Si etch rate ($\lesssim$0.3 Å/min).[12)] HF solutions buffered in NH_4F are commonly used in industry to obtain highly controlled etch rates at a solution pH (typically pH=5.0, 7:1 buffered HF) for which photoresist is relatively stable. 7:1 buffered HF is made up of 7 parts 40% NH_4F in water to 1 part 49% HF in water. In these experiments, after the 1000Å sacrificial oxide is removed, a chemical oxidation in an $HCl{:}H_2O_2$ solution (5:1:1 $H_2O{:}HCl{:}H_2O_2$ at 80°C) is used just prior to the HF treatments. It should also be noted that $H_2SO_4{:}H_2O_2$ oxidations give similar results and that spectra are obtained directly from the solutions without rinsing in deionized water.

The infrared spectrum for a Si(100) surface treated in concentrated HF is shown in Fig. 4a. The general features of the spectrum are identical to those in Fig. 3b except that the sharp modes have been washed out, leading one to postulate that this surface is somewhat more highly disordered. 7:1 buffered HF immersion for 2-4 minutes, on the other hand, results in the dramatic changes shown in Fig. 4b. Rather than being dominated by dihydride modes, most of the spectral weight has moved into the monohydride region of the infrared spectrum. It is known that buffered HF etches Si more rapidly than concentrated HF (~0.45Å/min.).[13)] This leads one to the conclusion that the dihydride like features were etched away to give the monohydride features and that the monohydrides are simply more stable than the dihydrides. This also leads one to the idea that these Si(100) surfaces must, in a certain sense, be faceted along Si(111) crystal planes. One must be cautious with this interpretation since the Si(111) terrace mode identified earlier (indicated by the arrow in Fig. 4b) is only a minority species on this surface. The faceting, if one can call it that, must only be a few atoms deep. Prolonged immersions in 7:1, however, does

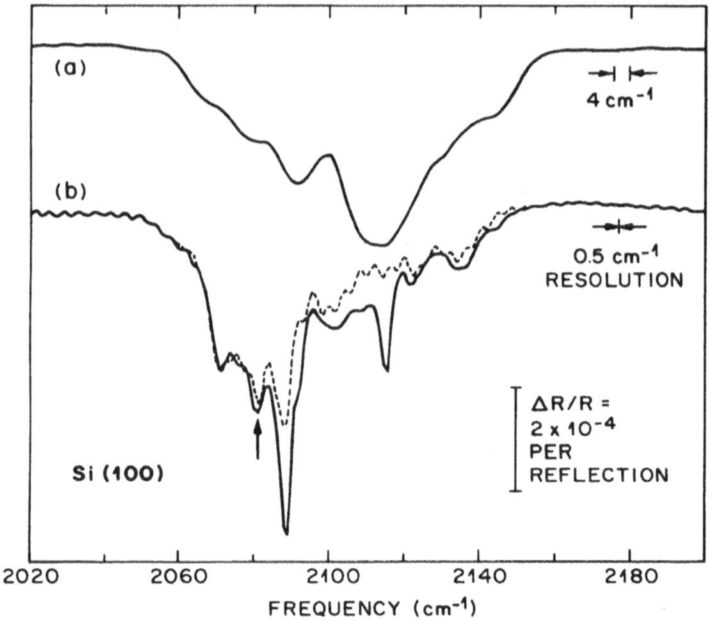

Fig. 4 (a) Internal reflection spectrum of Si(100) upon oxide removal with concentrated HF; (b) p-polarization (solid) and s-polarization (dashed) spectra of the surface subsequently treated in buffered HF (pH = 5.0).

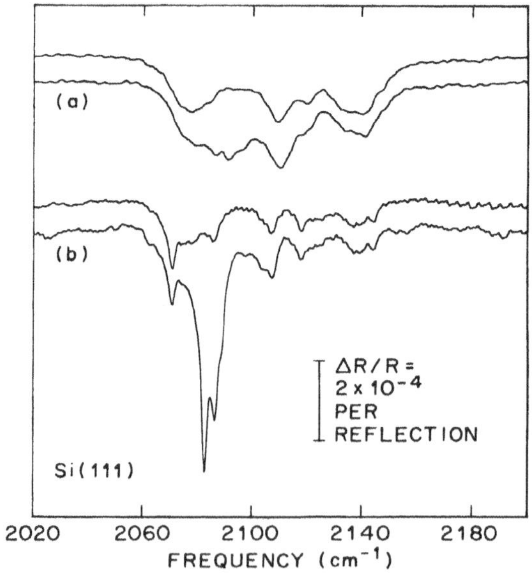

Fig. 5 (a) Internal reflection spectrum of Si(111) upon oxide removal with concentrated HF in p-polarization (lower) and s-polarization (upper); (b) p-polarization (lower) and s-polarization (upper) spectra of the surface subsequently treated in buffered HF (pH = 5.0).

cause the Si(111) terrace mode to grow, implying real facet formation.

The spectrum of a Si(111) surface treated in concentrated HF in shown in Fig. 5a. This spectrum is quite a bit different than that observed for dilute HF in Fig. 2b. The perpendicular monohydride terrace mode appears to be broadened and shifted to the blue of its original 2083 cm^{-1} position. It also appears to be somewhat decreased in intensity, leading to nearly equal concentrations of mono-, di-, and tri-hydride species. In contrast, the Si(111) terrace monohydride mode appears to be greatly enhanced after immersion in 7:1 buffered HF (Fig. 5b) with all of the other modes being suppressed in intensity. It is also clear from this spectrum that the species we have been calling the terrace monohydride mode is actually made up of a number of modes. As with the buffered HF experiments on the Si(100) surface, it is clear that this solution favors the formation of (111) crystal planes.

IDEAL Si(111) MONOHYDRIDE TERMINATION

At this point, it is apparent that different HF solutions behave quite differently. The question is, "Why?" One might imagine that the HF concentration might have something to do with what is going on, but the dilute HF and the 7:1 buffered HF have similar concentrations but very different spectra. Another possible answer is the solution pH. Concentrated HF has a pH~1. Highly diluted HF (1%) has a pH~2. But the pH of 7:1 buffered HF is substantially higher (pH~5). With this kind of thinking in mind, a study of the effect of the HF solution pH was undertaken.[14)] 7:1 buffered HF was used as the starting solution. The pH was lowered by adding HCl to the solution and raised by adding NH_4OH.

At a solution pH of between 8-9, just slightly basic, the spectra shown in Fig. 6 were obtained. The solid line of Fig. 6a shows the p-polarized data taken using this solution.

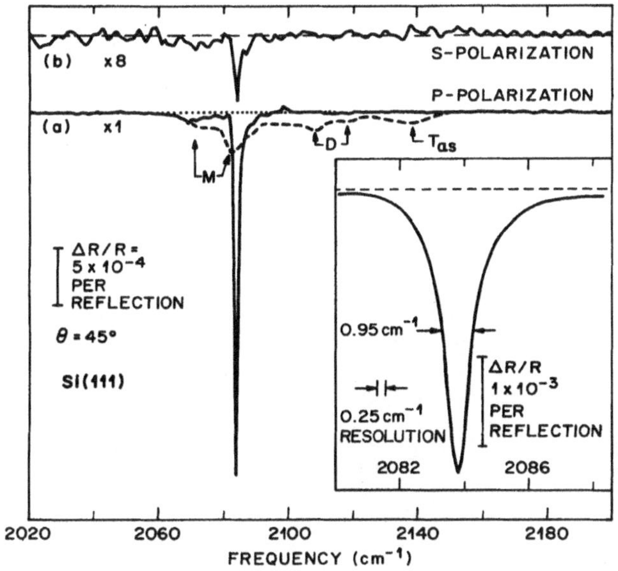

Fig. 6. Internal reflection spectra of HF treated Si(111) surfaces: (a) Surface treated with pH modified buffered HF (pH ~ 9) (solid curve: resolution = 0.5 cm^{-1}) and with dilute HF (100:1 H_2O:HF) (dashed curve: resolution = 1 cm^{-1}); (b) s-polarization for surface treatment with pH modified buffered HF (pH ~ 9) (resolution = 2 cm^{-1}).

The broken line is the data of Fig. 2b for dilute HF replotted for reference purposes. Basically, all modes except a highly homogeneous terrace monohydride have disappeared. The terrace monohydride mode at $2083.7 cm^{-1}$ is shown in the inset to have full width at half maximum (FWHM) of $0.95 cm^{-1}$, making it one of the most well ordered surfaces ever produced. The perfection of this surface is also seen quite dramatically in a magnified view of the data taken in s-polarization (Fig. 6b), where the absence of modes implies the absence of step edges. The only feature observed in s-polarization coincides with the peak in p-polarization and is due to the finite rejection-ratio of our wire grid polarizer. The last item of note is that the integrated strength of this mode is comparable to or slightly larger than the integrated strength of the entire dilute HF spectrum. The mode comprises the entire oscillator strength of the surface.

Further experimentation has shown that the most reproducible manner in which to make these ideally terminated Si(111) surfaces is to use NH_4F directly out of the bottle (40% in water).[15,16] NH_4F has a pH = 7.8 and has been found to give the best surface order, as measured by infrared absorption, to date (FWHM = $0.04 cm^{-1}$ at 40K).[17] This linewidth was estimated to correlate to ideal monohydride domains hundreds of angstroms in extent. These experiments also showed that rinsing in high quality deionized water subsequent to the NH_4F treatment was absolutely essential to obtaining good results. The mechanism by which these ideally terminated Si(111) terraces are formed has been shown to involve preferential attack of step edges.[14,18] This step edge attack is best explained as an attack by OH^- ions in the solution, giving rise to the dramatic pH dependence of the etching/passivation. This idea is not surprising given the highly anisotropic etch rates of basic solutions, which favor the formation of Si(111) crystal planes.

Of course, infrared absorption is only one of a host of surface science probes which have been applied to this surface. The consensus is that this surface is remarkably flat and well ordered.[15-21] Perhaps the most graphic illustration of the surface order achieved come from measurements using scanning tunneling microscopy (STM).[16,19] A constant-current image, taken at a bias voltage of ~ 2 Volts tunneling from the tip to the empty states of the sample, is shown in Fig. 7. The hexagonal array of the Si(111) surface is clearly observed in this image with a Si atom spacing of 3.8Å, in good agreement with the

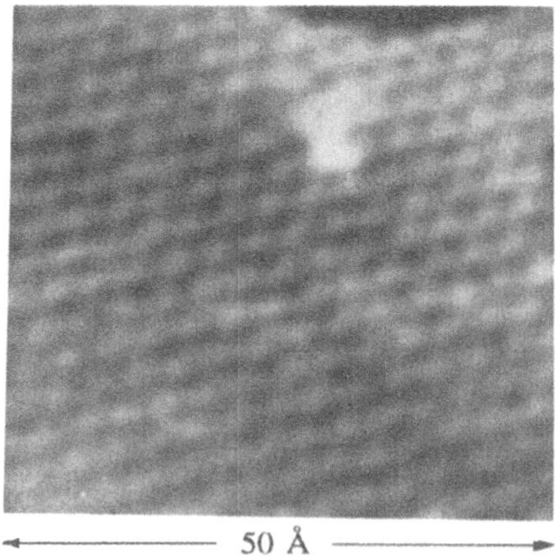

Fig. 7. STM image of Si(111) treated in NH_4F. Transition from light to dark in the grey scale image represents a height change of 0.5Å.

expected Si lattice constant. The single surface defect seen in the upper portion of the image is most likely a physisorbed molecule and indicates that the contamination level of this surface is somewhat less that 1 atomic percent. The apparent surface corrugation (peak to trough) is only 0.07Å according to the STM measurement. This is due to the absence of mid-gap surface states, which give the high contrast seen on Si surfaces without H passivation.

EFFECTS OF WATER RINSING

Changing the pH of a solution, of course, changes much more than the OH^- concentration. The concentrations of HF, HF_2^-, H^+ and F^- are also drastically altered.[22] In our original study of HF etching as a function of solution pH,[14] we argued that OH^- was attacking step edges using data from a water rinsing experiment. In that paper, clear evidence was shown for attack of dihydride species as a result of the room temperature water treatment. More recently, Watanabe *et al.* have shown that boiling water treatments have a Si(111) surface smoothing effect similar to that produced by treatments in NH_4F.[23] To illustrate how dramatic the effects of these water treatments can be data from the dissolution of a thermally grown oxide is shown in Fig. 8. In that experiment, a 1000 Å dry oxide was grown at 1000°C on a Si(111) surface and given a 30 min inert gas anneal at the growth temperature. The spectrum obtained after dissolving that oxide in concentrated HF (with no water rinse) is shown in Fig. 8a. Stretches in the dihydride region of the spectrum dominate the IR absorption. Subsequent to the room temperature water rinse, however, only the terrace monohydride mode survives with any significant intensity (Fig. 8b). The dramatic modification of the infrared spectrum is a clear indication that water is playing a much more active role in the chemistry than previously thought. The effects in this particular example are influenced by the high degree of perfection of the original Si/SiO_2 interface. On ordinary polished or chemically oxidized Si(111) wafers, the defect modes (dihydrides, trihydrides and coupled monohydrides) cannot be completely removed by room temperature water rinsing.

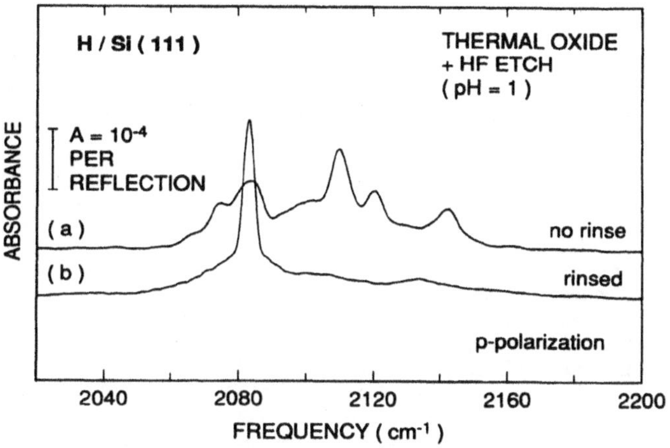

Fig. 8. P-polarized infrared absorption spectra of thermally oxidized Si(111) after (a) etching in concentrated HF, and (b) subsequent rinsing in deionized water.

It is apparent that water is playing a much more active role than previously thought. A surface reaction with the OH^- probably best explains the water rinsing results observed. An important fact pointed out by Watanabe *et al.*,[23] is that these surfaces are still completely terminated with hydrogen and exhibit very low levels of oxygen and carbon. The inescapable conclusion of this finding is that OH^- must also lead to hydrogen termination of the Si surface. In hindsight, this result may not be so surprising since there are certain chemical similarities between HF and OH^-. Both F and O make strong bonds to Si. Both HF and OH^- are extremely polar diatomics and would both experience the Coulombic barrier lowering in the attack of polar Si-Si bonds.

SUMMARY

Extensive studies using infrared spectroscopy to probe HF treated Si surfaces have been performed. Detailed mode assignments tell us that there are predominantly three basic types of species on the surfaces when etched in either concentrated or dilute HF, monohydrides, dihydrides, and trihydrides. Both the Si(100) and the Si(111) are atomically rough with comparable concentrations of mono-, di-, and tri-hydrides. Si(111) surfaces show two types of monohydrides, monohydrides on step edges and monohydrides on terraces. By appropriate choice of the HF solution pH, ideally monohydride terminated Si(111) surfaces can be formed. The high degree of surface perfection achieved rival any in surface science even though these surfaces are prepared in liquid solution and transfered at atmospheric pressure to the vacuum apparatus. Thus, one should think of HF etching as a hydrogen termination/passivation process rather than an etching process alone. We are only just beginning to understand HF etching/passivation, but it is quite clear that the polar nature of Si-F bonds makes these species unstable on these surfaces due to back bond attack and ultimately lead to these highly stable H-terminated surfaces. The mechanism by which these atomically perfect Si(111) surfaces are produced is also only just starting to be understood. Step edge attack by OH^- is the most likely explanation. This attack can lead to ideally H-terminated Si(111) surfaces under certain circumstances even in water alone.

Acknowledgments

It is a pleasure to acknowledge my long time collaborators in this work, R. S. Becker, Y. J. Chabal, and K. Raghavachari. I would also like to thank V. A. Burrows, P. Dumas, P. Jakob, and G. W. Trucks, for stimulating discussions and their contributions to my present understanding of this problem. P. Jakob deserves special thanks for providing the spectra used in Fig. 5 and Fig. 8. S. B. Christman, E. E. Chaban, A. J. Becker, and R. D. Yadvish are also gratefully acknowledged for their technical support.

REFERENCES

1. W. Kern: J. Electrochem. Soc. **137** (1990) 1887.
2. W. Kern and D. Puotinen: RCA Rev. **31** (1970) 187.
3. M. Grundner and H. Jacob: Appl. Phys. A **39** (1986) 73.
4. P. J. Grunthaner, F. J. Grunthaner, R. W. Fathauer, T. L. Lin, M. H. Hecht, L. D. Bell, W. J. Kaiser, F. D. Schowengerdt, and J. H. Mazur: Thin Solid Films **183** (1989) 197.
5. D. B. Fenner, D. K. Biegelsen, and R. D. Bringans: J. Appl. Phys. **66** (1989) 419.
6. T. Takahagi, I. Nagai, A. Ishitani, H. Kuroda, and Y. Nagasawa: J. Appl. Phys. **64** (1988) 3516.
7. E. Yablonovitch, D. L. Allara, C. C. Chang, T. Gmitter, and T. B. Bright: Phys. Rev. Lett. **57** (1986) 249.
8. V. A. Burrows, Y. J. Chabal, G. S. Higashi, K. Raghavachari, and S. B. Christman: Appl. Phys. Lett. **53** (1988) 998.
9. H. Ubara, T. Imura, and A. Hiraki: Solid State Comm. **50** (1984) 673.

10. G. W. Trucks, K. Raghavachari, G. S. Higashi, and Y. J. Chabal: **65** (1990) 504.
11. Y. J. Chabal, G. S. Higashi, K. Raghavachari, and V. A. Burrows: J. Vac. Sci. Technol. A **7** (1989) 2104.
12. S. M. Hu and D. R. Kerr: J. Electrochem. Soc. **114** (1967) 414.
13. W. Hoffmeister: Int. J. Appl. Radiat. Isot. **2** (1969) 139.
14. G. S. Higashi, Y. J. Chabal, G. W. Trucks, and K. Raghavachari: Appl. Phys. Lett. **56** (1990) 656.
15. P. Dumas, Y. J. Chabal, and G. S. Higashi: Phys. Rev. Lett. **65** (1990) 1124.
16. G. S. Higashi, R. S. Becker, Y. J. Chabal, and A. J. Becker: Appl. Phys. Lett. **58** (1991) 1656.
17. P. Jakob, Y. J. Chabal, K. Raghavachari, P. Dumas: submitted to Appl. Phys. Lett.
18. P. Jakob and Y. J. Chabal: J. Chem. Phys. **95**, 2897 (1991).
19. R. S. Becker, G. S. Higashi, Y. J. Chabal, and A. J. Becker: Phys. Rev. Lett. **65** (1990) 1917.
20. R. B. Doak, Y. J. Chabal, G. S. Higashi, and P. Dumas: J. Electron Spectroscopy and Related Phenomena **54/55** (1990) 291.
21. P. Dumas and Y. J. Chabal: Chem. Phys. Lett. **181** (1991) 537.
22. J.S. Judge, J. Electrochem. Soc. **118**, 1772 (1971).
23. S. Watanabe, N. Nakayama and T. Ito, Appl. Phys. Lett. **59**, 1458 (1991).

PRE-GATE OXIDE Si SURFACE CONTROL

Mizuho Morita and Tadahiro Ohmi

Department of Electronic Engineering
Faculty of Engineering, Tohoku University
Sendai 980, Japan

INTRODUCTION

Understanding and controlling of the factors which determine the integrity of very thin gate oxide films is becoming increasingly important with the scaling down of metal-oxide-semiconductor (MOS) ultra large scale integration (ULSI) devices. Existence of microcontamination such as particles, organic impurities and metallic impurities on a Si surface is well known to degrade the electrical insulating performance and the reliability of the gate oxide. Additionally, as the gate oxide becomes thinner, the influence of the Si surface microroughness and the native oxide on the integrity of the gate oxide can no longer be neglected.

Some correlations between electrical properties of the oxide film and the Si surface microroughness right before oxidation or the Si–SiO_2 interface microroughness have been reported so far. Electrical properties of very thin oxides depend on the microroughness of the Si wafer surface prepared by a chemical-mechanical polishing technique.[1] The Si surface microroughness is increased by an $NH_4OH/H_2O_2/H_2O$ cleaning process during wet chemical cleaning to remove microcontamination. However, the Si surface smoothness can be maintained by using NH_4OH–content–reduced solution with excellent cleaning capability.[2-4] Anneal of Si wafers in sufficiently low concentration oxygen causes etching and consequent roughening of the Si surface.[5] Thermal oxidation generally increases the microroughness of the Si–SiO_2 interface.[6,7] The microroughness of the Si surface before oxidation or the Si–SiO_2 interface is reported to affect to the dielectric breakdown statistics of oxide films,[1-3,5,7] Hall mobility in MOS–inversion layer[6,7] or channel electron mobility of metal–oxide–semiconductor field effect transistors (MOSFET).[4] Thus, it is important to prepare the Si wafer with a microscopic flat surface and to establish accompanying device fabrication processes to suppress roughening of the Si surface and the Si–SiO_2 interface.

Furthermore, the presence of native oxide and/or unintentionally grown preoxide cannot be neglected when evaluating the electrical insulating performance and the reliability of very thin gate oxides, because the thickness ratio of preoxide to net thermal oxide relatively increases as the gate oxide becomes thinner. We have investigated the

The Physics and Chemistry of SiO_2 and the Si-SiO_2 Interface 2
Edited by C.R. Helms and B.E. Deal, Plenum Press, New York, 1993

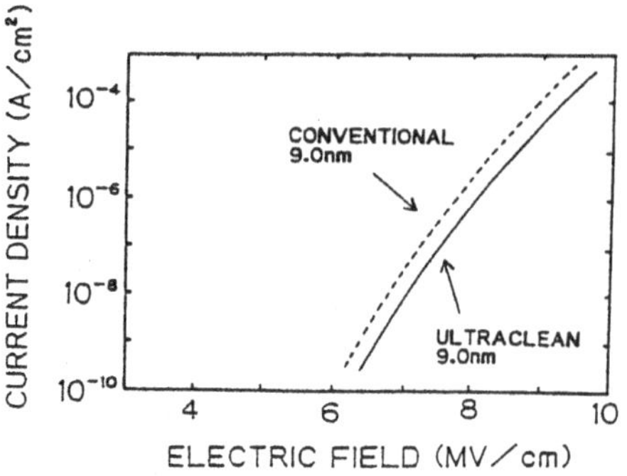

Fig. 4 Current density–average electric field characteristics of n^+–polycrystalline Si/SiO_2 /p–Si(100) diodes under negatively biased metal electrodes for 9.0 nm–thick oxides.

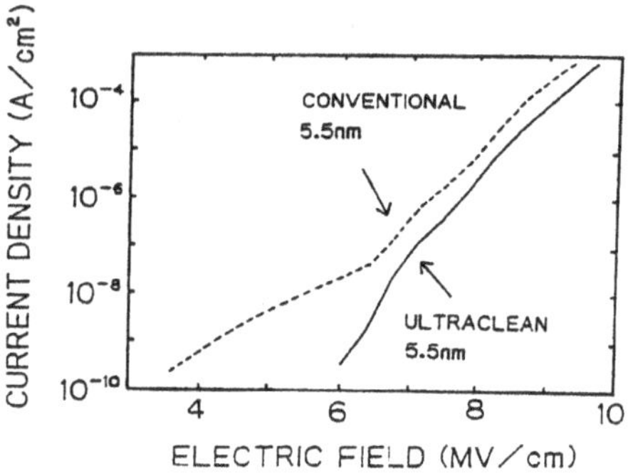

Fig. 5 Current density–average electric field characteristics of n^+–polycrystalline Si/SiO_2 /p– Si(100) diodes under negatively biased metal electrodes for 5.5 nm–thick oxides.

after the formation of intentional preoxide with the thickness of approximately 0.4 nm at 300 °C for passivation.[4] The oxide growth at 900 °C obeys a simple parabolic law. The kinetics can be explained by the Deal–Grove model even for oxide thicknesses thinner than 30 nm,[20–22] where the parabolic rate constant B (diffusion controlled) determined from plots of Fig. 2 is 2.3×10^{-4} $\mu m^2/h$ at 900 °C. This result suggests that the preoxide is located on the outer surface of thermal oxide. Figure 3 shows the inverse logarithmic plots of the data shown in Fig. 2. The oxidation at temperatures from 300 to 700 °C may be controlled by the inverse logarithmic growth law.[23] The kinetics of the oxidation at 200 °C is complicated, probably because hydrogen atoms terminating the Si surface remain at 200 °C.[18]

Figures 4 and 5 show the current density–average electric field characteristics of MOS diodes with n^+–polycrystalline Si/SiO_2/p–Si(100) structure under the negatively biased metal electrodes for 9.0 nm and 5.5 nm oxides, respectively. The current level through the ultraclean oxide with 0.4 nm preoxide is lower than that through the conventional dry oxide with 1.4 nm preoxide for 9.0 and 5.5 nm–thick oxides. The current of the ultraclean oxide over the range and of the conventional dry oxide at the average electric field higher than 6.4 MV/cm is considered to be dominated by the Fowler–Nordheim tunneling of electrons from n^+–polycrystalline Si to SiO_2, because the barrier height for electrons tunneling at n^+–polycrystalline Si/SiO_2 interface is lower than that for holes tunneling at SiO_2/p–Si. This result exhibits that the barrier height of electrons tunneling at n^+–polycrystalline Si/SiO_2 interface for the ultraclean oxide is higher than that for the conventional dry oxide. At the average electric field lower than 6.4 MV/cm, the leakage current is easily observed in the conventional dry oxide of 5.5 nm. The current through conventional dry oxide is increased in the low field region as the thickness is reduced. However, the current level through ultraclean oxide is not significantly increased even for 5.5 nm–thick oxide. In fact, for ultraclean oxide, the current density–average electric field characteristics for 5.5 nm nearly overlaps that for 9.0 nm. The leakage current is also speculated to result in electron conduction from the barrier height for electrons at n^+–polycrystalline Si/SiO_2 interface being low compared with that for holes at SiO_2/p–Si. Therefore, these current level differences can be concluded to be caused by the presence of preoxide located at the outer surface of thermal oxide.

Figure 6 shows the effect of the preoxide thickness in ultraclean oxide on the

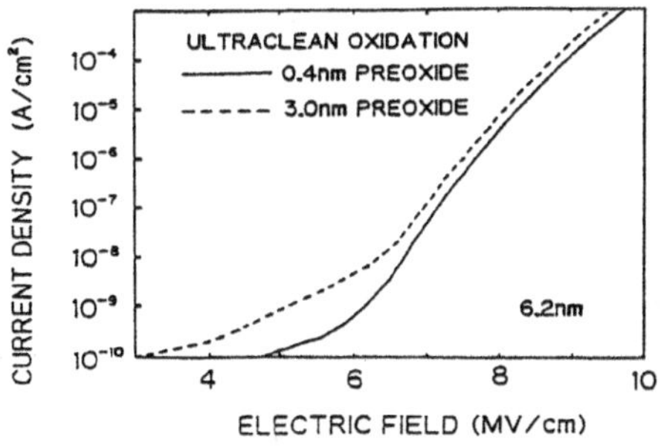

Fig. 6 Preoxide thickness dependance of current density–average electric field characteristics of n^+–polycrystalline $Si/SiO_2/p$–Si diodes under negatively biased metal electrodes for 6.2 nm ultraclean oxides.

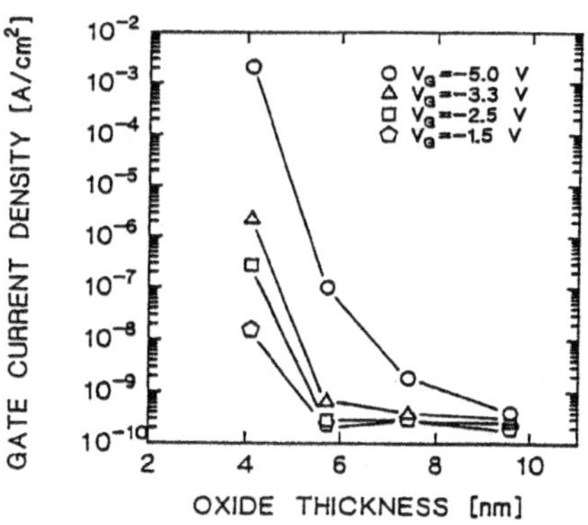

Fig. 7 Thickness dependance of current density through ultraclean oxides for various gate voltages(V_G).

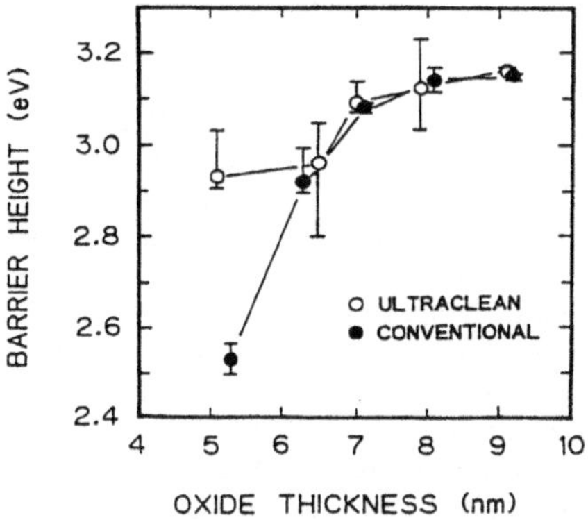

Fig. 8 Energy barrier height at Si–SiO_2 interface for electrons emission from Si to SiO_2 as a function of oxide thickness.

current density–average electric field characteristics of n^+–polycrystalline $Si/SiO_2/p$–Si(100) MOS diodes. The current level of the ultraclean oxide with 0.4 nm preoxide is lower than that of the ultraclean oxide with 3.0 nm preoxide over the range. The current level through the ultraclean oxide with 0.4 nm preoxide is not significantly increased, while the current through the ultraclean oxide with 3.0 nm preoxide is enhanced at low field region(≤6.5 MV/cm). This result proves that the presence of preoxide in very thin gate oxide films induces the degradation of the electrical insulating performance. Figure 7 shows the oxide thickness dependence of current density through the ultraclean oxides with 0.4 nm preoxide for various gate voltages. The current density is kept to be lower than 1×10^{-9} A/cm^2 even for 5.5 nm oxide at the gate voltage of −3.3 V or lower, while the gate current increases with decreasing oxide thickness at −5 V.

Figure 8 shows the energy barrier height at Si/SiO_2 interface for electrons emission

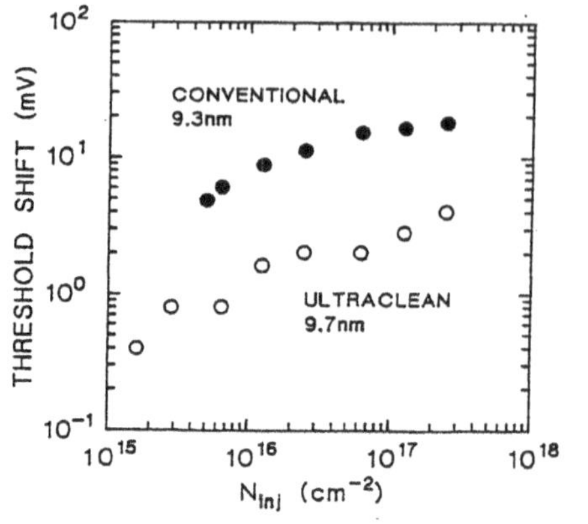

Fig. 9 Threshold voltage shift of n–MOSFET as a function of the number of injected electrons.

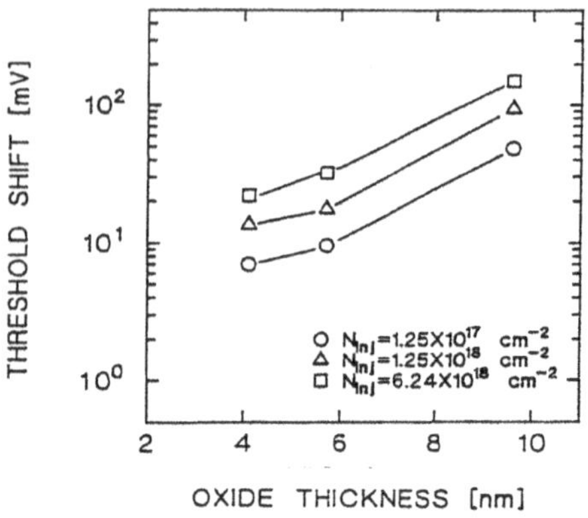

Fig.10 Thickness dependance of threshold voltage shift of n–MOSFET with ultraclean oxides for some numbers of injected electrons.

from Si to SiO_2 as a function of oxide thickness, which was derived from the current dominated by Fowler–Nordheim tunneling in the current density–average electric field characteristics of Al/SiO_2/n–Si(100) MOS diodes under positively biased metal electrodes. Although the barrier height for the conventional dry oxide with 1.4 nm preoxide is drastically decreased as the thickness is thinner, the height for the ultraclean oxide with 0.4 nm preoxide is only slightly decreased and is higher than 2.9 eV for 5.1 nm oxides. The current level through the ultraclean oxide with the thickness of 5.1 nm is actually lower than that through conventional dry oxide. This result probably implies that the barrier height for the conventional dry oxide is apparently low because the current is due to Frenkel–Poole emission along with Fowler–Nordheim tunneling, or that the oxidation mechanism is influenced by the presence of preoxide located at the outer surface of thermal oxide which consequently gives a different oxide structure near the SiO_2/Si interface from that of the ultraclean oxide. Therefore, it is clear that the electrical insulating capability of the preoxide–controlled ultraclean oxide can be preserved even for very thin oxide.

The threshold voltage shift of n–MOSFET with 10 nm oxides as a function of the number of injected electrons is shown in Fig. 9, where the channel length is 48.4 μm, and the channel width 100 μm. Hot electrons were injected from Si substrate into the SiO_2 under the substrate voltage of –10 V and the gate oxide field of 5 MV/cm.[24,25] The shift for the ultraclean oxide(9.7nm) with 0.4 nm preoxide is smaller than that for conventional dry oxide(9.3nm) with 1.4 nm preoxide, indicating that the ultraclean oxide has higher reliability. Figure 10 shows the oxide thickness dependence of the threshold voltage shift of n–MOSFET with preoxide–controlled ultraclean oxides for various amounts of injected electrons. The threshold shift decreases with decreasing the oxide thickness. This manifests that the reliability of oxides is enhanced as the thickness gets thinner.

CONCLUSION

We have demonstrated that the preoxide grown during the wafer temperature ramp–up before thermal oxidation affects the integrity of very thin oxide films. Very thin ultraclean oxide formed by controlling the preoxide growth has high electrical insulating performance and high reliability. This work signifies that pre-gate oxide Si surface control has strong potential in the formation of very thin gate oxide with higher electrical insulating performance and higher reliability.

ACKNOWLEDGEMENTS

The authors are grateful to A. Teramoto , K. Makihara , K. Nakamura and K. S. Tseng for help in this work. The majority of this work was carried out at the Super Clean Room of the Laboratory for Microelectronics, Research Institute of Electrical Communication, Tohoku University.

REFERENCES

1. M. Morita, A. Teramoto, K. Makihara, T. Ohmi, Y. Nakazato, A. Uchiyama, and T. Abe, in ULSI Science and Technology/1991, edited by J. M. Andrews and G. K. Celler, The Electrochemical Society, Pennington, NJ, 1991, p.400.
2. T. Ohmi, M. Miyashita, M. Itano, T. Imaoka, and I. Kawanabe, IEEE Trans. Electron Devices 39, 537(1992).
3. M. Miyashita, K. Makihara, T. Tsuga, and T. Ohmi, to be submitted to J. Electrochem. Soc.
4. T. Ohmi, K. Kotani, A. Teramoto, and M. Miyashita, IEEE Electron Device Lett. 12, 652(1991).
5. M. Offenberg, M. Liehr, G. W. Rubloff, and K. Holloway, Appl. Phis. Lett. 57, 1254(1990).
6. P. O. Harn and M. Henzler, J. Vac. Sci. Technol. A2, 574(1984).
7. P. O. Harn, M. Grandner, A. Schnegg, and H. Jacob, in The Physics and Chemistry of SiO_2 and the Si-SiO_2 Interface, edited by C. R. Helms and B. E. Deal, Plenum, New York, 1988, p.401.
8. T. Ohmi, M. Morita, E. Hasegawa, M. Kawakami, and K. Suma, in ULSI Science and Technology/1989, edited by C. M. Osburn and J. M. Andrews, The Electrochemical Society, Pennington, NJ, 1989, P.327.
9. M.Morita, T Ohmi, E. Hasegawa, M. Kawakami, and K. Suma, Appl. Phys. Lett. 55, 562(1989).
10. M. Morita, T. Ohmi, E. Hasegawa, M. Kawakami, and M. Ohwada, J. Appl. Phys. 68, 1272(1990).
11. M.Morita, T. Ohmi, E. Hasegawa, and A. Teramoto, Jpn. J. Appl. Phys. 29, L2392(1990).
12. M. Morita, T. Ohmi, E. Hasegawa, M. Kawakami, and K. Suma, in Digest, Technical Papers, 1989 VLSI Symposium, Kyoto, 1989, p.75.
13. T. Ohmi, M. Morita, A. Teramoto, K. Makihara, and T. S. Tseng, Appl. Phys. Lett. 60, 2126 (1992).
14. T. Ohmi, M. Morita, and T. Hattori, in The Physics and Chemistry of SiO_2 and the Si-SiO_2 Interface, edited by C.R.Helms and B.E.Deal, Plenum, New York, 1988. p.413.
15. M. Morita, T. Ohmi, and E. Hasegawa, Solid-State Electron. 33, 143(1990).
16. M.Morita, T. Ohmi, E. Hasegawa, A. Teramoto, S. Kawajiri, in Ultra Clean Technology, Submicron ULSI Process Technology III, edited by T. Ohmi and T. Nitta, Realize, Tokyo, 1990, p.125(in Japanese).
17. I. Itoh, K. Mukaiyama, T. Ohmi, M. Morita, and K. Makihara, in Extended Abstracts of the 180th Electrochemical Society Meeting, Phoenix, 1991, p.635.

18. N. Yabumoto, K. Saito, M. Morita, and T. Ohmi, Jpn. J. Appl. Phys. 30, L419(1991).
19. N. Yabumoto, K. Minegishi, K. Sato, and H. Harada, Ultra Clean Technology 1, 13(1990).
20. B. E. Deal and A. S. Grove, J. Appl. Phys. 36, 3770(1965).
21. B. E. Deal, in The Physics and Chemistry of SiO_2 and the Si–SiO_2 Interface, edited by C. R. Helms and B. E. Deal, Plenum, New York, 1988. p.5.
22. C. J. Han and C. R. Helms, J. Electrochem. Soc. 134, 1297(1987).
23. N. Cabrera and N. F. Mott, Rept. Progr. Phys. 12, 163(1948).
24. C. A. Bosselear, Solid–State Electron. 16, 648(1973).
25. J. F. Verwey, J. Appl. Phys. 44, 2681(1973).

CHEMICAL STRUCTURES OF NATIVE OXIDES FORMED DURING WET CHEMICAL TREATMENTS ON NH_4F TREATED Si(111) SURFACES

Takeo Hattori and Hiroki Ogawa

Department of Electrical and Electronic Engineering
Musashi Institute of Technology
1-28-1 Tamazutsumi, Setagaya-ku, Tokyo 158, Japan

ABSTRACT

Chemical structures of native oxides formed during wet chemical treatments on NH_4F treated Si(111) surface were investigated using X-ray Photoelectron Spectroscopy (XPS) and Fourier Transformed Infrared Attenuated Total Reflection(FT-IR-ATR). It was found that the amounts of Si-H bonds in native oxids and those at native oxide/silicon substrate interface are negligibly small in the case of native oxides formed in H_2SO_4-H_2O_2-H_2O solution. Based on this discovery, it was confirmed that native oxides can be characterized by the amount of Si-H bonds in the native oxides.

INTRODUCTION

In order to remove organic contaminants, heavy metal impurities and particles on silicon wafer surfaces, so called RCA standard clean[1] and its modified RCA standard clean[2] has been used extensively. The key idea in RCA standard clean is to passivate the chemically cleaned reactive silicon surface to impure chemical solution by including H_2O_2 in the cleaning solution. In earlier studies it was found that the chemical structures of native oxides formed during wet chemical treatments can be characterized by the amount of Si-H bonds in native oxides.[3,4] However, the deconvolution of Si2p photoelectron spectra was not satisfactory because the deconvoluted spectra were not resolved completely with each other. Also, the amount of Si-H bonds on the DF treated silicon surface, which was used as a reference sample, was not negligibly small for the evaluation of Si-H bonds at native oxide/silicon interface. It is therefore the purpose of the present study to confirm the existence of Si-H bonds in native oxides by measuring difference photoelectron spectrum and difference infrared absorption spectrum.

The Physics and Chemistry of SiO_2 and the Si-SiO_2 Interface 2
Edited by C.R. Helms and B.E. Deal, Plenum Press, New York, 1993

EXPERIMENTAL DETAILS

The native oxides studied were prepared as follows using four kinds of treatments listed in Table 1. Firstly, Si(111) wafers were treated in NH_4OH. Native oxides formed by this treatment were removed by buffered hydrofluoric acid solution(10%HF:40%NH_4F=1:7). Secondly, native oxides were formed in HCl. Native oxides formed by this treatment were removed by the treatment in 40%NH_4F for 6.5 minutes to obtain atomically flat H-terminated Si(111) surfaces.[5] Then, four kinds of native oxides were formed by the four kinds of treatments listed in Table 1. The cleaning in deionized water was performed after all the chemical treatments.

Table 1 Treatments studied

Treatment	Composition	Temperature	Time
HCl	1 HCl	37-65	10
	1 H_2O_2		
	1 H_2O		
H_2SO_4	4 H_2SO_4	85-90	10
	1 H_2O_2		
HNO_3	HNO_3	46-60	5
NH_4OH	0.05 NH_4OH	63-80	10
	1 H_2O_2		
	5 H_2O		

Infrared absorption spectra were measured using JIR-5500 manufactured by JEOL. The native oxides in this case were formed on a 0.5 mm x 55 mm x 52.5 mm 2-6 Ω cm n-type vicinal Si(111) silicon wafer with 45 degrees bevels on each of the long sides. This vicinal Si(111) surface was prepared by cutting 4 degrees off the (111) plane to the $< 11\bar{2} >$ direction. The infrared radiation from the interferometer is focused at normal incidence onto the input bevel, and is internally reflected nearly 100 times, and exits the output bevel to be collected and refocused onto the photodetector. Highly resolved X-ray photoelctron spectra were measured using ESCA-300 manufactured by Scienta Instrument AB furnished with monochromatized Al K α radiation at acceptance angle of 3.3 degrees, pass energy of 150 eV and slit width of 0.8 mm. The native oxides used for the XPS studies were formed on 0.5 mm x 12 mm x 40 mm 2-6 Ω cm n-type Si(111) surfaces. Si2p, 01s and C1s spectra were measured. In order to clarify the structural change in depth direction, angle resolved photoelectron spectra were measured with photoelectron take off angle of 2, 15 and 90 degrees.

EXPERIMENTAL RESULTS AND DISCUSSIONS

Figure 1 shows infrared absorption spectra measured with resolution of 0.5 cm^{-1} for 40% NH_4F treated vicinal Si(111) surface with < 112 > parallel to propagating direction of infrared. Here, infrared absorption spectrum obtained for native oxide formed in H_2SO_4 was used as a

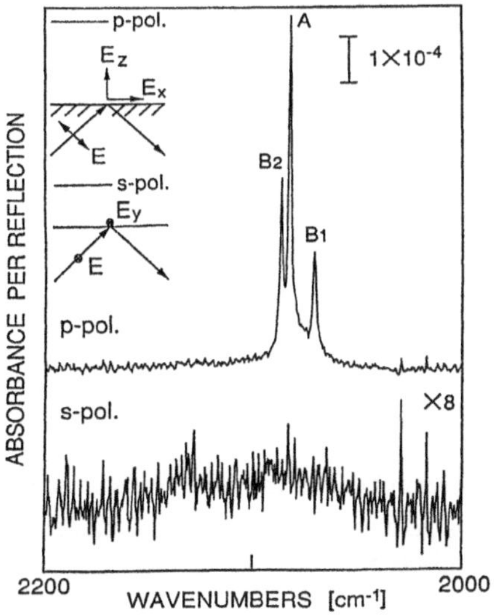

Fig. 1 p- and s- polarized infrared absorption spectra of 40% NH_4F-treated vicinal Si(111) surfaces. Here, infrared absorption spectrum obtained for native oxide formed in H_2SO_4 is used as a reference spectrum.

reference spectrum. This figure can be explained based on the experimental studies[6] such that silicon monohydrides exists on the terrace in addition to small number of silicon monohydrides formed at step edges. Namely, the p-polarized infrared absorption spectrum, which appears at 2071.0 cm^{-1}, 2083.3 cm^{-1} and 2087.1 cm^{-1}, corresponds to band B_1, A and B_2 assigned by Jakob et al.,[6] respectively. On the other hand s-polarized infrared absorption spectrum does not appear. This implies no Si-H bonds in the (110) plane. The line width of p-polarized infrared absorption at 2083.7 cm^{-1} is 1.2 cm^{-1} and is quite close to the minimum value of 0.7 cm^{-1} obtained so far.[7] Therefore, H-terminated Si(111) surface with atomically flat terrace must be produced by the present treatment and was used for the study of the chemical structures of native oxides formed during wet chemical treatments.

Figure 2 shows infrared absorption spectra obtained for three kinds of native oxides formed in HNO_3, HCl an NH_4OH. Here, infrared absorption spectrum obtained for native oxide formed in H_2SO_4 was used as a reference spectrum. According to this figure infrared absorption arising from Si-H bonds in native oxide films[3] appears at 2260 cm^{-1} in the case of native oxides formed in HNO_3 and HCl. The amount of Si-H bonds in native oxide formed in HNO_3 is larger than that in native oxide formed in HCl, while those in native oxide formed in NH_4OH are negligibly small. Furthermore, infrared absorption, which appears in the range from 2000 to 2200 cm^{-1}, is deduced to arise mostly from monohydrides,[6] dihydrides,[6] trihydrides[6] and those including Si-O-Si bonds localized at the interface[8] and is described in the following as infrared absorption arising from Si-H bonds at the interface. The infrared absorption arising from Si-H bonds at the interface is largest in the case of native oxide formed in HNO_3. It should be noted that the infrared absorption arising from Si-H bonds at the interface is not negligibly small, but appreciable in the case of native oxide formed in basic solution of NH_4OH.

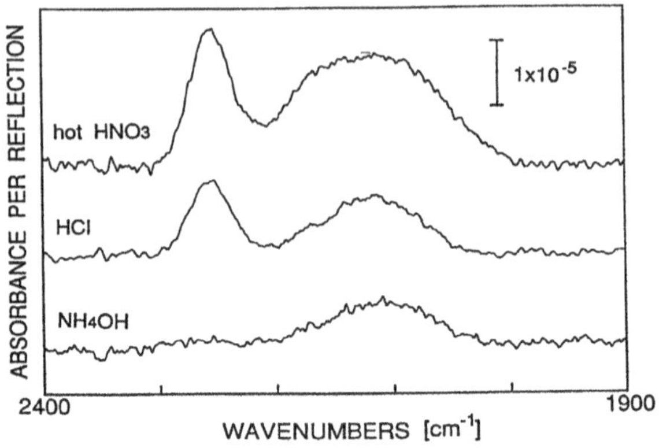

Fig. 2 Infrared absorption spectra obtained for three kinds of native oxides formed in HNO_3, HCl and NH_4OH. Here, infrared absorption spectrum obtained for native oxide formed in H_2SO_4 is used as a reference spectrum.

In Figs. 1 and 2 the infrared absorption spectrum obtained for native oxide formed in H_2SO_4 was used as reference spectrum, because the amount of Si-H bonds at and near the native oxide silicon interface is negligibly small as explained in the following. Figure 3 shows infrared absorption spectrum for native oxide formed in H_2SO_4. The reference spectrum in this case was obtained for the silicon surface terminated with deutrium, which was prepared by dipping silicon wafer in 1% DF solution. This DF solution was prepared by mixing small amount of HF with heavy water having purity of 99.7%. Therefore, infrared absorption at 2260 cm^{-1} arising from Si-H bonds in native oxide formed in H_2SO_4 is negligibly small. The p-polarized infrared absorption arising from Si-H bonds at native oxide/silicon substrate interface formed in H_2SO_4 is negligibly small because almost no infrared absorption is observed in Fig. 1 except the sharp infrared absorption. Therefore, the amounts of Si-H bonds in native oxides

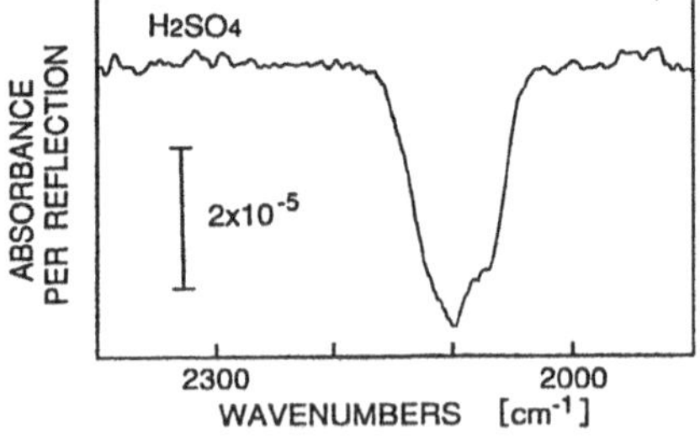

Fig. 3 Infrared absorption spectrum for native oxide formed in H_2SO_4. Here, infrared absorption spectrum obtained for silicon surface terminated with deutrium is used as a reference spectrum.

and those at the interface estimated from Fig. 2 are quite close to true amounts.

Figure 4(a) shows $Si2p_{3/2}$ photoelectron spectrum for native oxide formed in H_2SO_4. Here, spin orbit partner $Si2p_{1/2}$ is removed.[9] Figure 4(b) is obtained by subtracting spectrum in Fig. 4(a) from $Si2p_{3/2}$ photoelectron spectrum for native oxide formed in HNO_3. Here, the spectral intensities for silicon in silicon dioxide are adjusted to be equal to each other. Si^{x+} spectrum shown in Fig. 4(b) has the chemical shift, which is close to that determined by Sugiyama et al.[3], and the value of Full Width Hall Maximum(FWHM) listed in Table 2 and is considered to arise from silicon atom bonded to one hydrogen and three oxygen atoms.

Figure 5 shows $Si2p_{3/2}$ photoelectron spectra measured for four kinds of native oxides. Figure 5(a) was measured at photoelectron take off angle of 2 degrees, while Figure 5(b) was measured at photoelectron take off angle of 15 degrees. Here, the spectral intensities for silicon in silicon dioxide are adjusted to be equal to each other. The chemical shifts and the values of FWHM of Si^{1+}, Si^{y+}, Si^{2+} and Si^{3+} spectra were determined in the case of native oxide formed in H_2SO_4 assuming Gaussian line shape for these spectra and skew Gaussian line shape for silicon substrate spectrum[10] and are listed in Table 2. The chemical shifts of Si^{1+}, Si^{2+} and Si^{3+} are close to those determined by Himpsel et al.[11] The chemical shift of Si^{y+} is close to those determined by Terada et al.[12] and is considered to arise from silicon bonded to one hydrogen, one oxygen and two silicon atoms. Using the chemical shifts and the values of FWHM listed in Table 2, the deconvoluted spectra shown in Fig. 5 were obtained. According to this figure Si^{x+} spectra appear in the $Si2p_{3/2}$ photoelectron spectra measured for native oxide formed in HNO_3 and that formed in HCl. Thickness of native oxide formed in

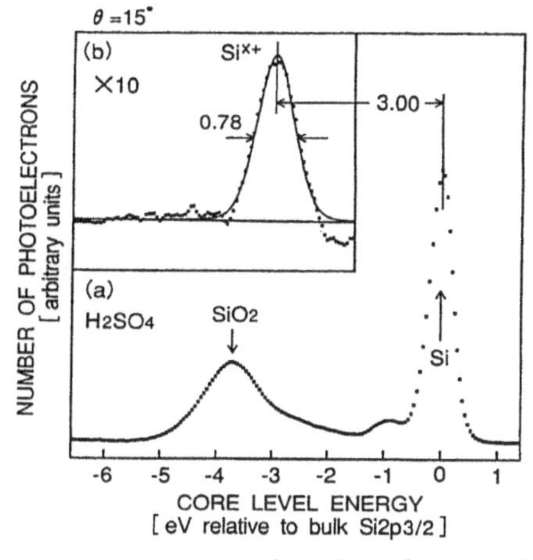

Fig. 4 $Si2p_{3/2}$ photoelectron spectra measured at photoelectron take off angle of 15 degrees: (a) reference spectrum meaured for native oxide formed in H_2SO_4; b) difference spectrum obtained by subtracting reference spectrum from the spectrum measured for native oxide formed in HNO_3.

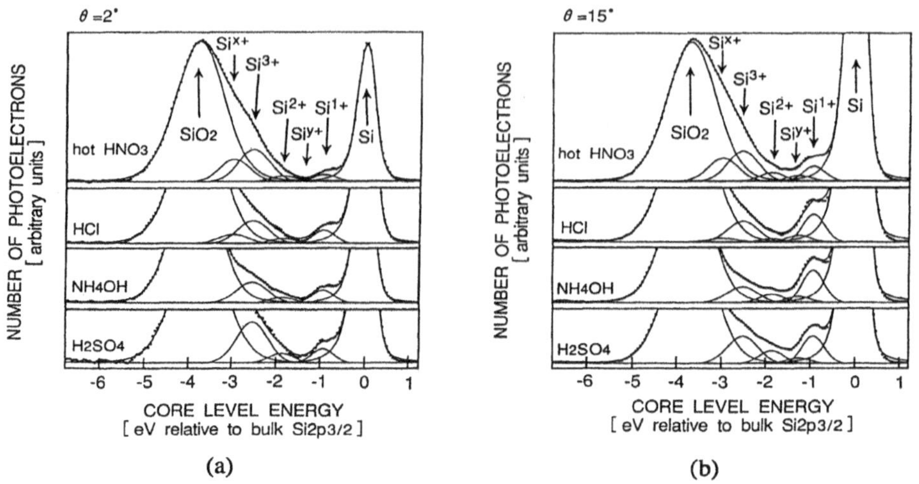

Fig. 5 $Si2p_{3/2}$ photoelectron spectra for four kinds of native oxides measured at photoelectron take off angle of (a) 2 degrees, and (b) 15 degrees.

HNO_3, HCl, NH_4OH and H_2SO_4 evaluated with the assumption that the electron escape depth of native oxides is equal to that of thermal oxide is 0.58, 0.46, 0.53 and 0.43 nm, respectively. These thicknesses were determined from the spectral intensity ratio of (NO/NS). Here, NO is the $Si2p_{3/2}$ spectral intensity of silicon in silicon oxide and NS is the $Si2p_{3/2}$ spectral intensity of silicon in silicon substrate. Here, electron escape depth in silicon substrate and that in thermally grown silicon dioxide is 2.7 nm and 3.4 nm, respectively. Therefore, the thickness of these native oxides is approximately equal to each other. By performing the same kind of analysis as that described previously[13] the amount of Si^{x+} in native oxides can be deduced. Si^{x+} spectral intensity obtained for native oxide formed in HNO_3 is larger than that obtained for native oxide formed in HCl. These results agree with those obtained from infrared absorption spectra shown in Fig. 2. In the case of native oxide formed in HNO3, the distribution of Si-H bonds in the oxide film is uniform, because Si^{x+} spectral intensity obtained

Table 2 Chemical shifts and the values of FWHM

intermediate states	Chemical Shift [eV]	FWHM[eV]
Si^{1+}	0.95	0.58
Si^{y+}	1.30	0.65
Si^{2+}	1.89	0.65
Si^{3+}	2.56	0.80
Si^{x+}	3.00	0.78

for take off angle of 2 degrees is approximately equal to that obtained for take off angle of 15 degrees. On the other hand in the case of native oxide formed in HCl, Si-H bonds mostly localize near the surface of the oxide film, because Si^{x+} spectral intensity for take off angle of 2 degrees is larger than that for take off angle of 15 degrees. Si^{x+} spectra does not appear for native oxides formed in H_2SO_4 and NH_4OH. The amounts of Si-H bonds in the native oxides determined from $Si2p_{3/2}$ spectrum in Fig. 5 by assuming uniform distribution of Si-H bonds in native oxides are shown in Table 3.

Table 3 Amounts of Si-H bonds in native oxides [$x10^{-14}cm^{-2}$]

treatment	IR	XPS
HNO_3	0.9	1.3
HCl	0.4	0.2
NH_4OH	0.0	0.0
H_2SO_4	0.0	0.0

CONCLUSION

The chemical structures of native oxides formed during various wet chemical treatments were investigated by XPS and FT-IR-ATR. It was found that the amounts of Si-H bonds in native oxide and at native oxide/silicon substrate interface are negligibly small in the case of native oxides formed in H_2SO_4-H_2O_2 solution. Based on this discovery it was found that the chemical structures can be characterized by the amounts of Si-H bonds in the native oxide. Namely, the amount of Si-H bonds in native oxide is largest in the case of treatment in HNO_3, the second largest is the case of treatment in HCl, the third largest is the case of treatment in NH_4OH and the smallest is the case of treatment in H_2SO_4.

The XPS studies were performed on native oxides formed on Si(111) surface, while the infrared absorption studies were performed on native oxides formed on vicinal Si(111) surface. Qualitative discussions will be not affected by the slight difference in orientation. However, the effect of atomic steps on the amount of Si-H bonds at and near the native oxide/silicon substrate interface should be clarified in the future.

ACKNOWLEDGEMENTS

The one of the authors (T. H.) expresses his hearty thanks to Dr. Kenichi Katayama for supplying silicon wafers used in the preset study. Part of this work was supported by a 1990-1991 Grant-in-Aid for Specially Promoted Research (No. 01065003) from the Ministry of Education, Science and Culture of Japan.

REFERENCES

1. W. Kern : J. Electrochem. Soc. 137 (1990) 1887. and references therein.
2. T. Ohmi, M. Miyashita, M. Itano, T. Imaoka and I. Kawanabe: to be published in IEEE Trans. Electron Device 39(1992) No.3.
3. K. Sugiyama, T. Igarashi, K. Moriki, Y. Nagasawa, T. Aoyama, R. Sugino, T. Ito and T. Hattori: Jpn. J. Appl. Phys. 29 (1990) L2401.
4. H. Ogawa, N. Terada, K. Sugiyama, K. Moriki, N. Miyata, T. Aoyama, R. Sugino,T. Ito and T. Hattori: to be published in Appl. Surf. Sci.(1991)
5. G. S. Higashi, R. S. Becker, Y. J. Chabal and A. J. Becker: Appl. Phys. Lett. 58 (1991) 1656.
6. See Fig. 4(a) in the paper by P. Jakob and Y. J. Chabal appeared on J. Chem. Phys. 95 (1991) 2897.
7. Y. J. Chabal : private communication.
8. Y. Nagasawa, H. Ishida, T. Takahagi, A. Ishitani and H. Kuroda: Solid State Electronics 33, Suppl. (1990) 129.
9. G. Hollinger and F. J. Himpsel: Appl. Phys. Lett. 44 (1984) 93.
10. K. Siegbahn: J. Electron Spectrosc. Relat. Phenom. 36 (1985) 113.
11. F. J. Himpsel, F. R. MacFeely, A. Taleb-Ibrahimi, J. A. Yarmoff and G. Hollinger: Phys. Rev. B38 (1988) 6084.
12. N. Terada, H. Ogawa, K. Moriki, A. Teramoto, K. Makihara, M. Morita, T. Ohmi and T. Hattori: Jpn. J. Appl. Phys.
13. T. Hattori, K. Takase, H. Yamagishi, R. Sugino, Y. Nara and T. Ito: Jpn. J. Appl. Phys. 28 (1989) L296.

SILICON SURFACE ANALYSIS AND VERY THIN SILICON OXIDE CHARACTERIZATION AFTER HF/ETHANOL PREOXIDATION SILICON CLEANING

B. Garrido[+], F. Gessinn*, J.L. Prom*, J.R. Morante[+],
J. Samitier[+] and G. Sarrabayrouse*

[+]LCMM, Facultat de Física, Universitat de Barcelona, Diagonal 645-647,
08028 Barcelona, Spain
*LAAS-CNRS, 7, Avenue du Colonel Roche, 31077 Tolouse Cx, France

ABSTRACT

A comparison between the silicon HF/Ethanol cleaning and the HF/H_2O one has been performed. Ellipsometric measurements, show the presence of an overlayer 0.3 nm thick versus the 1.2-2.0 nm in the HF/H_2O case. Atomic Force Microscopy shows a drastic reduction of the surface roughness. The subsequently thermally grown SiO_2 layers (<100 Å) analyzed by IR absorption show a different bond angle distribution.

I - INTRODUCTION

Silicon surface cleaning is recognized as a key issue to the development of future integrated circuits because it affects their reliability in a number of ways. For example thin oxide layers[1,2] or metal contacts[3] show cleaning dependent electrical properties. The case of thin oxide layers is crucial. Indeed their thickness will decrease down to the 5-7 nm range as a result of the transistor dimension shrinkage. For such thin layers the quality of the oxide-silicon interface is an important parameter and breakdown has been shown influenced not only by the contamination level[4] but also by the interface morphology[5].

For a better control of their thickness thin oxide growth is performed at low temperature ($\leq$ 900 °C) during a short period. Consequently relaxation does not occur and the characteristics of the silicon surface prior to oxidation are of an increasing importance.

Because of the silicon ability to grow an uncontroled native overlayer in the ambient atmosphere it is necesary either to perform a preoxidation gas phase in-situ cleaning, well adapted to single wafer cluster tools, or to produce a non reactive silicon surface as a result

The Physics and Chemistry of SiO_2 and the Si-SiO_2 Interface 2
Edited by C.R. Helms and B.E. Deal, Plenum Press, New York, 1993

of the cleaning operation. It has been shown recently that a HF/Ethanol clean generates a hydrogen passivated silicon surface[6] with a very low contamination level which improve the electrical properties of the subsequently grown oxide[7]. In this paper we give aditional information upon the HF/Ethanol preoxidation cleaning compared to the more common HF/H_2O step.

II - EXPERIMENT

P-type <100> oriented silicon wafers have been degreased in a H_2SO_4/H_2O mixture and etched in HF/H_2O prior to oxidation at 1150 °C in dry oxigen to grow a 350 nm thick sacrificial silica layer[8] subsequently etched in buffered HF. Then a RCA[9] sequence has been applied and the wafers have been separated into two batches. Wafers of the first batch have been inmersed in a HF/H_2O, 1:10 mixture during 30" at 20 °C and subsequently rinsed in DI water. Wafers of the second batch have been cleaned similarly but in a HF/Ethanol, 1:10 mixture and no rinse.

The surface of the wafers has been characterized by ellipsometry at 405 nm (RUDOLPH AUTO ELII equipment) and Atomic Force Microscopy (Nanoscope II microscope). For some wafers of each batch the thick oxide has been locally etched and a thin oxide layer has been grown after cleaning, at 900 °C in dry O_2. Some wafers have been annealed at 1050 °C in N_2 during 30". Chromium deposition and annealing in forming gas have been performed in order to define Metal-Oxide-Silicon capacitors which have been characterized by capacitance voltage measurements using a HP 4284 Impedance analyzer, and current-voltage and constant current injection measurements using a HP 4145 analyzer. The microscopical structure of the thin oxide layer has been analyzed by Fourier Transform Infrared Spectroscopy (FTIR) with a BOMEM DA3 spectrometer and a MB-120 Spectrometer with an Infrared Microscope, being possible in both cases to measure in transmission an reflexion mode. We have choosen the reflection-absorption technique to get a high signal/noise ratio, because for SiO_2 layers thinner than 10 nm, the sensitivity is lower in a conventional transmission geometry.

III- RESULTS AND DISCUSSION

III-1 Characterization of the silicon surface

The ellipsometric results have been interpreted in terms of a non absorbing matrix silica layer with refractive index equal to 1.47 on top of the silicon substrate. The native overlayer if any has been found thinner (0.3-0.4 nm) in the case of the HF/Ethanol clean than in the HF/H_2O one (1.2-2 nm)., this last result being consistent with those obtained by E.A. Irene and coworkers[10]. In the former case the overlayer is of the order of or less than one monolayer (equivalent silica) in consistency with previous results concerning Auger analysis of similar samples[7] which have shown a decrease down to the noise level of the oxigen peak after a HF/Ethanol clean. The morphology of the silicon surface as viewed by Atomic Force Microscopy is shown in figure 1. A drastic reduction of the roughness can be seen as a result of the HF/Ethanol clean. The analyzed area in Figure 1 is about 1 μm x 1 μm. The same analysis has been done at differents locations on a wafer and on diferent wafers. The result shown on Figure 1 is typical of all our observations. The microroughness of the silicon surface has already been shown to be dependent upon mixing ratio in the $NH_4OH/H_2O_2/H_2O$ cleaning[11]. This shows that in order to control the surface roughness, the whole cleaning sequence has to be optimized.

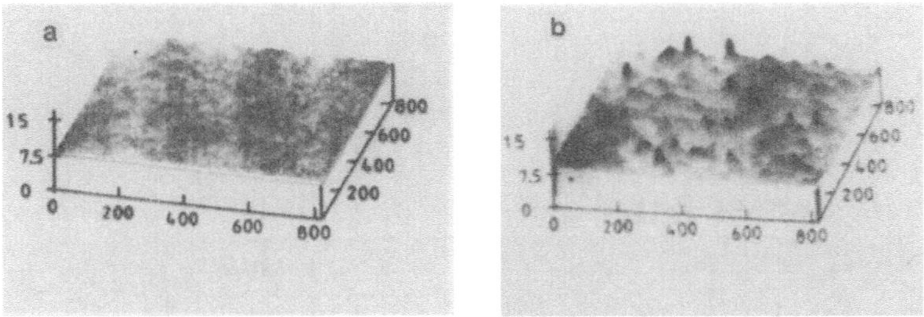

Figure 1. Silicon surface topography by Atomic Force Microscopy for HF/Ethanol(a) and HF/H_2O(b) samples. Full scale : X:850 nm, Y:850 nm, Z:15 nm.

III-2 Characterization of the grown oxide layer

Current voltage (I-V) curves of capacitors with an annealed oxide 3.5-10 nm thick (as measured by capacitance-voltage experiment[12]) have been recorded for a number of samples on each wafer. One important effect of the HF/Ethanol treatment is to drastically reduce the dispersion of the curves mainly for the thinnest layers as illustrated on figure 2. A low dispersion of the I-V curves is significant either of a good uniformity for the parameters which control the conduction (area equal to 40 μm x 40 μm). However it has to be noticed that in the case where Fowler-Nordheim conduction applies -i.e. for layers thicker than 5 nm- the current voltage curves of HF/Ethanol cleaned samples can be quasi-idealy fitted on the basis of the thickness obtained by capacitance-voltage measurements[13]. Because the capacitance and the current do not show the same dependency versus thickness, this is in favour of a uniform layer thickness across the device area. Such an effect on the current-voltage curves of the HF/Ethanol clean compared to the HF/HNO_3/H_2O one has already been mentioned[14].

The results by Kassmi[13] et al. show that the improvement certainly comes from a reduction of parasitic currents al low field, which could be associated with localiced conduction paths with a very low metal-semiconductor barrier height. Indeed current-voltage curves for the HF/Ethanol samples are the same as the curves associated with the less conducting HF/H_2O type samples. Furthermore at high field and in both cases the Fowler-Nordheim law with a barrier height equal to about 3 eV applies. At lower currents and in the case of HF/H_2O samples the total current appears to be dominated by excess currents. The improvement of the current reproducibility is certainly to be regarded as a consecuence of the reduction of the roughness of the silicon surface.

Another effect of the HF/Ethanol treatment is to improve the breakdown strengh of the silica layers. Figure 3 shows the charge injected before breakdown Q_{BD} as a function of the injected current. A clear increase of Q_{BD} at low field can be interpreted by a decrease of the contribution of local defects in consistency with the results mentioned above.

The structural analysis of silicon oxide layers with FTIR spectroscopy has been performed in the mid-infrared spectral region, where the spectra of SiO_2 show three characteristic bands at frequencies around 450, 800 and 1075 cm^{-1}, corresponding to transverse optical (TO) modes arising from the rocking, bending and asymmetric stretching motions of the Si-O-Si unit. Due to sensitivity reasons, we have limited our analysis in the region between 850-1500 cm^{-1} where de asymmetric stretching mode (TO_3) exhibits a pronounced TO-LO splitting[15], which is observed in oblique incidence as a result of the normal component of the light electric field.

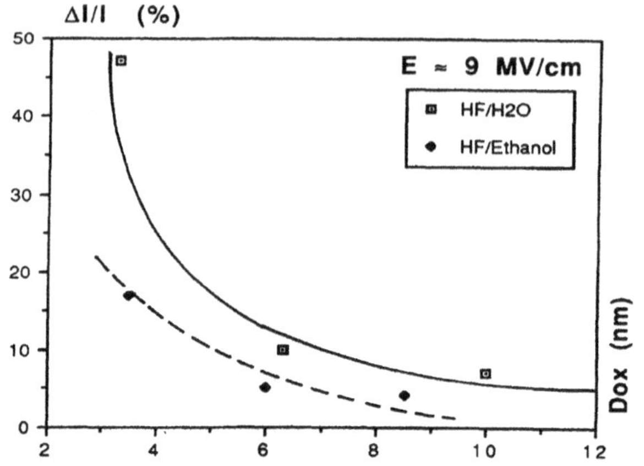

Figure 2. Dispersion of the current-voltage measurements for an electric field E equal to about 900 V/cm, versus the insulator thickness. Preoxidation cleaning type as a parameter.

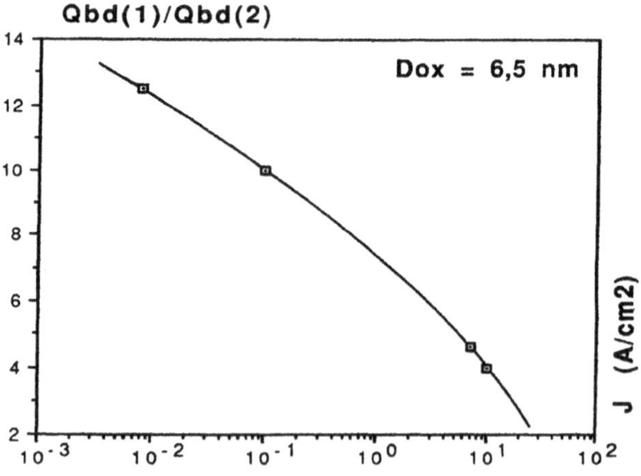

Figure 3. Ratio of the charge injected at breakdown in HF/Ethanol type capacitors Q_{bd} (1) to that in HF/H_2O type capacitors Q_{bd} (2) versus the injection current.

In order to analyze the IR spectra obtained and also to compare the spectra with the vibrational frequencies previously reported by other authors -usually obtained at 0° angle in transmission mode-, we have developped a computer multilayer simulation program, which takes into account that peak positions and shape of reflexion and transmission absorption bands depend on geometrical features such as angle, polarization and layer thickness[16]. From the TO_3 peak position and according with the central force model we have determined the averaged intrinsic strain in both as-grown and annealed oxide layers. This model link the TO_3 vibrational frequency with the mean Si-O-Si bond angle[17](2θ) by

$$\nu = \nu_0 \sin\theta \qquad (1)$$

where $\nu_0 = 1134$ cm^{-1}. The intrinsic stress modifies the Si-O-Si bond angle distribution and, as a result, the Si-Si distance by

$$d_{Si-Si} = 2\, r_0 \sin\theta \qquad (2)$$

where $r_0 = 0.16$ nm is the Si-O bond length. Therefore, d_{Si-Si} is directly related with the oxide strain ϵ by[18]

$$\epsilon = \frac{d_{Si-Si}}{(d_{Si-Si})_0} - 1 \qquad (3)$$

where $(d_{Si-Si})_0$ correspond to the fully relaxed SiO_2 value ($2\theta = 144°$). The averaged stress can be calculated using the thermal SiO_2 Young's modulus and Poisson's ratio.

In table I we report the peak position, FWHM (full width at half maximum) and the calculated stress values for the as-grown and annealed samples. The oxide layers submitted to the different cleaning do not show significative differences in the TO_3 peak position an FWHM. However, an increase in the TO_3 peak position and in consequence a diminution of the mean intrinsic stress is obtained when the layer thickness increases and also after the annealing treatment. The value reached after the annealing process is also found to be thickness dependent. This behaviour is in agreement with previous IR analysis of thin thermal silicon oxide[20] and can be understood in the framework of the viscoelastic relaxation model[21].

In spite of the fact that the FWHM do not show significant variation between both type of cleanings, the TO_3 absorption band for as-grown samples shows little changes in its shape. To analyze these we have attempted firstly a deconvolution with two Gaussians, a method already used[22,23], although the significance of each component is not well known. All the spectra can be consistently deconvoluted by two Gaussian components with 0% Lorentzian shape. Figure 4 show the deconvolution behaviour for 105 Å thick as-grown samples after HF/Ethanol or HF/H_2O cleanings. We have observed that the distance between the maximum of both Gaussian components is almost constant, but their area ratios change with the cleaning process (Figure 5). We have reported previously a similar behaviour for thin oxide layers grown after an RCA and HF/H_2O cleanings[24]; in all cases, the IR spectra obtained after HF cleanings showed a shape more Gaussian -small high component- than the others. A decrease of layer thickness leads to an enhancement of these differences which are related with modifications in the SiO_2 amorphous network.

After the annealing treatment both HF/Ethanol and HF/H_2O absorption peaks are found to have the same shape and position (Table I); only remains the thickness dependence. Accordingly we can conclude that the bond network structure relaxes in a way independent of the initial state.

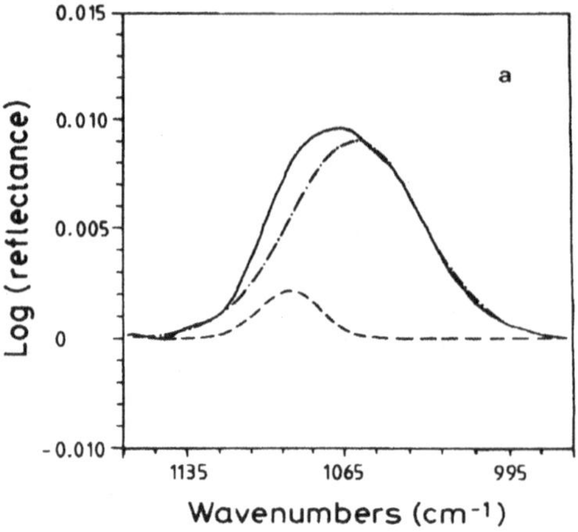

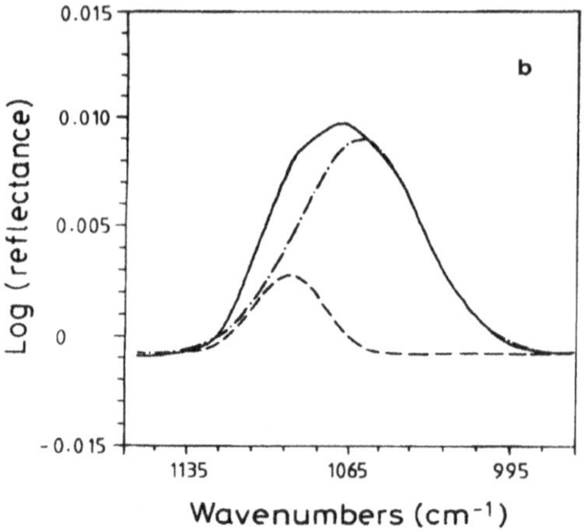

Figure 4. Deconvolution behaviour for 105 Å thick as-grown samples after (a) HF/Ethanol and (b) HF/H_2O cleanings.

Table I. Peak position, FWHM and stress values for unannealed (NA) and annealed (A) and for the 55 Å and 105 Å thick samples.

Sample	TO_3 ($cm^{-1}\pm0.5$)	FWHM ($cm^{-1}\pm1$)	σ (10^9 dyn/cm^2)
H_2O NA 55	1065	74	-12.5
Eth NA 55	1066	73	-11.6
H_2O A 55	1070.5	71	-7.4
Eth A 55	1070.5	71	-7.4
H_2O NA 105	1068	75	-9.7
Eth NA 105	1068	75	-9.7
H_2O A 105	1074.5	70	-3.7
Eth A 105	1074.5	70	-3.7

To investigate the significance of the Gaussian component area ratios, we have also analyzed the shape by calculating the momentum of the distribution. The most relevant conclusions are : a) the as-grown samples have a more symmetric distribution than the annealed samples and b) the as-grown HF/Ethanol samples have a flatter top band distribution around the mean value. These results are in good agreement with the deconvolution behaviour : more symmetrical distributions are related with the prevalence of one gaussian component while a flat distribution with a more equality of the Gaussian areas.

The XPS analysis of the Si-SiO_2 interface layer shows (table 2) a smaller suboxide concentration for the samples cleaned with HF/Ethanol, even after the annealing treatment.

This result is in good agreement with the above reported for the silicon surface characterization. Moreover, taking into account that the grown temperature, 900°C, is too low to allow a significative relaxation process in the silicon oxide during the growth, this result could explain the differences in the silicon oxide structure observed by FTIR in the as-grown samples. In this sense, the different silicon surface roughness would determine the tetrahedra bonding arrangement.

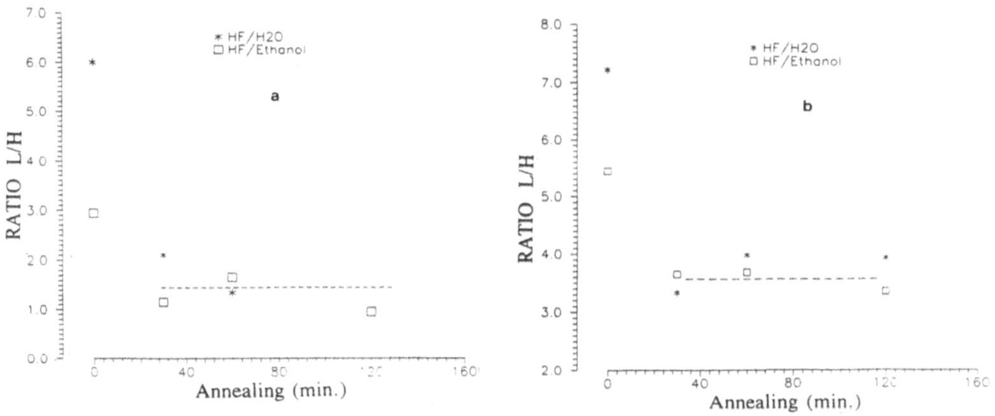

Figure 5: Area ratio low/high component after the deconvolution process. (a) 55 Å thick samples and (b) 105 Å thick samples.

Table II. Results of the XPS analysis of the Si-SiO$_2$ interface. NA : non annealed, A : annealed.

-	HF/H$_2$O NA	HF/H$_2$O A	HF/Et NA	HF/Et A
Si^{+1}:Si^{+2}:Si^{+3}	1:1:3	1.2:1:3.2	2.3:1:2.8	0.9:1:1.9
Monolayers	2.7	2.9	2.0	1.7

In conclusion, although the electrical characteristics for the annealed HF/Ethanol samples are better than the HF/H$_2$O ones, the chemical structure observed by FTIR is very similar. However, we can expect significative variations of the point defect concentration as a function of the structure evolution with the annealing treatment depending on the initial cleaning procedure.

ACKNOWLEDGMENTS

The authors acknowledge the skillfull asistance of F.Rossel from "Service Team" at LAAS-CNRS during the fabrication of the samples, and J.Portillo and J.Gual for XPS results. This work has been partially supported by "Dirección General de Investigación Científica y Técnica del Ministerio de Educación y Ciencia, Spain" and "Direction de la Cooperation Scientifique et Technique du Ministere de Relations Exterieures" and Groupement GCIS du CNRS (France).

REFERENCES

1. J. Ruzyllo, Preoxidation surface treatments in thermal oxidation of silicon, in: The Physics and Chemistry of SiO_2 Interface", C.R.Helms and B.E.Deal, ed., Plenum ess, New York, USA (1988).
2. M. Heyns, C. Hassenak, R. de Keersmaecker and R. Falster, The impact of Si-surface characteristics on MOS device field, in:"Semiconductor cleaning technology,1989", J. Ruzillo and R.E. Novak, ed., The Electrochem. Soc., Pennington NJ, USA, (1990).
3. M. Miyawaki, S. Yoshitake and T. Ohmi, Improvement of Aluminium-Si contacts performance in native-oxyde free processing, IEEE Trans. on Electron Dev. Lett, EDL 11, 10:448 (1990).
4. I.G. Mc Gillivray, J.M. Robertson and A.J. Walton, in: "Semiconductor Silicon 86", H.R.Huff, T. Abe and B. Kolbesen, ed., The Electrochem. Soc., Pennington, NJ,USA (1986).
5. J.P. Gambino, J.N. Nguyen, B. Cunningham and J.F. Shepard, Dielectric breakdown of SiO_2 grown on rough Si surfaces, in: "The Physics and Chemistry of SiO_2 and the Si-SiO_2 Interfaces", C.R. Helms and B.E. Deal, ed., Plenum Press, New York, USA (1988).
6. P.J. Grunthaner, F.J. Grunthaner, R.W. Fathaner, T.L. Lin, M.H. Hecht, L.D. Bell and W.K. Kaiser, Hydrogen-terminated silicon substrates for low-temperature molecular beam epitaxy, Thin Solid Films, **183**:197 (1989).
7. J.L. Prom, J. Castagne, G. Sarrabayrouse and A. Muñoz-Yague, Influence of the preoxidation cleaning on the electrical properties of thin SiO_2 layers, IEE Proc., Part I, **135**:20 (1988).
8. M. Itsumi and F. Kiyosumi, Origin and elimination of defects in SiO_2 thermally grown on Czochralski silicon substrates, Appl. Phys. Lett., **40**:496 (1982).
9. W. Kern and D.A. Puotinen, Cleaning solutions based on hydrogen peroxide for use in silicon semiconductor technology, RCA Review, **31**:187 (1970).
10. G. Gould and E.A. Irene, An in-situ study of aqueous HF treatment of silicon by contact angle measurement and ellipsometry, J. Electrochem. Soc., **135**:1535 (1988).
11. T. Ohmi, K. Kotani, A. Teramoto and M. Miyasticta, Dependence of electron channel mobility on Si-SiO_2 interface microroughness. IEEE Electron Device Lett., **12**:652 (1991).
12. S. Duzelier, G. Sarrabayrouse, J.L. Prom and G. Hollinger, Thickness measurement of thin oxide layers in MOS capacitors, Electronics Lett., **27**:1399 (1991).
13. K. Kassmi, J.L. Prom and G. Sarrabayrouse, Electrical conduction in MOS capacitors with an ultra-thin oxide layer, Sol. St. Electron., **34**: 509 (1991).
14. J.L. Prom, P. Morfouli, K. Kassmi, G. Pananatzakis and G. Sarrabayrouse, Improvement of the electrical properties of Metal-SiO_2-Silicon capacities by a preoxidation HF/ethanol clean of the substrates, IEE Proc., Part G, **138**:321 (1991).
15. P. Lange, Evidence for disorder-induced vibrational mode coupling in thin amorphous SiO_2 films. J. Appl. Phys. **66**:201 (1989).
16. B. Garrido, J.A. Moreno, J. Samitier and J.R. Morante, Accurate infrared spectroscopy analysis in back-side damaged wafers, sent to the E-MRS Spring Meeting, Strasbourg (1992).
17. A. Lehmann, L. Schumann, K. Hubner, Optical phonons in amorphous silicon oxides, Phys. stat. sol. b, **117**:689 (1983).
18. J.T. Fitch, G. Lucovsky, E. Kobeda, E.A. Irene, Effects of thermal history on stress-related properties of very thin films of thermally grown silicon dioxide, J. Vac. Sci. Technol. **B 7**:153 (1989)
19. C.H. Bjorkman, J.T. Fitch, G. Lucovsky, Correlation between midgap interface state density and thickness-averaged oxide stress and strain at Si/SiO_2 interfaces formed by thermal oxidation of Si, Appl. Phys. Lett, **56**:1983 (1990).
20. J.T. Fitch, C.H. Bjorkman, G. Lucovsky, F.H. Pollak, X. Yin, Intrinsic stress and stress gradients at the SiO_2/Si interface in structures prepared by thermal oxidation of Si and subjected to rapid thermal annealing, J. Vac. Sci. Technol. **B 7**:775 (1989).
21. E.A. Irene, E. Tierney, J. Angilello, A viscous flow model to explain the appearance of high density thermal SiO_2 at low oxidation temperatures, J. Electrochem. Soc., **129**:2594 (1982).
22. W. Boyd, Deconvolution of the infrared absorption peak of the vibrational stretching mode of silicon dioxide: Evidence for structural order?, Appl. Phys. Lett., **51**:418 (1987).
23. S. Hayoshi, S. Tanimoto, K. Yamamoto, Analysis of surface oxides of gas-evaporated Si small particles with infrared spectroscopy, high-resolution electron microscopy, and X-ray photoemision spectroscopy", J. Appl. Phys, **68**:5300 (1990).
24. B. Garrido, J. Samitier, J.R. Morante, L. Fonseca, F. Campabadal, Influence of the silicon wafer cleaning treatment on the Si/SiO_2 interfaces analyzed by infrared spectroscopy, to be published in Appl. Surf. Science, (1992).

EFFECTS OF METALLIC IMPURITIES UPON THIN GATE OXIDE INTEGRITY AND RELATED BULK PROPERTIES IN CZ Si

K.-C. Cho[1], J.-G. Park[1], Y.-S. Kwak[1], D.-J. Lee[1], D.-S. Lim[1], C.-K. Shin[1], S. Hahn[2], and W. L. Smith[3]

1. Samsung Electronics
   Quality Control 1st Section
   Kihung Plant, Kihung
   Kyungki-do, Korea

2. Department of Materials Science and Engineering
   Stanford University
   Stanford, California, 94305 USA

3. Therma-Wave, Inc.
   47320 Mission Falls court
   Fremont, California 94539 USA

ABSTRACT

In this study, we investigated effects of various well-known metallic impurities such as Fe, Cu, and Al (contaminated by a spin coating method with concentration levels ranged from 1×10^{10} to 1×10^{12} atoms/cm^2) upon: 1) thin oxide integrity; 2) total oxide charge; and 3) various bulk electrical and structural parameters. The oxide integrity was determined from the breakdown field strength and the time dependent dielectric breakdown characteristics of MOS capacitors with the thermally-grown 23 nm thick SiO_2 films. In addition, in order to correlate bulk structural and electrical parameters with the oxide integrity, two different types of lifetime measurements were carried out for these wafers. Some of these samples were further characterized for metal-induced surface and bulk microdefects by transmission electron microscopy and thermal wave modulated optical reflectance (using both mapping and imaging modes). Based upon our oxide integrity and bulk electrical and structural characterization data, we conclude that Fe is the most harmful metallic impurity among the three metallic elements investigated in this study.

The Physics and Chemistry of SiO_2 and the Si-SiO_2 Interface 2
Edited by C.R. Helms and B.E. Deal, Plenum Press, New York, 1993

INTRODUCTION

The reduction in device dimension and the increase in complexity of VLSI and especially ULSI technology require strictest control of unintended contamination. Particularly dangerous are heavy metal contaminants in terms of yield and reliability of finished microelectronic devices. The primary electrical effect of metallic impurities is the introduction of energy levels close to the middle of silicon bandgap. Since these levels act as recombination centers, metallic impurities cause a decrease in minority carrier lifetime and and an increase in the leakage currents of p-n junctions [1]. When the impurities exceed their solubility limits in silicon, metallic clusters (mostly silicides) generate and represent a dielectric constant discontinuity and reduce the breakdown field strength of the thin oxide film. In addition, metallic impurities generate some other types of defects such as stacking faults and other precipitates and make dislocations and stacking faults electrically active [2].

Even though the effects of metallic impurities upon microelectronic devices are qualitatively well-known, the question of what levels of specific metallic impurities are acceptable in VLSI/ULSI MOS devices has not been answered satisfactorily. In addition, our understanding on potential correlations between various surface and bulk properties due to metallic impurity contamination is still lacking. In this investigation, we studied effects of various well-known metallic impurities such as Fe, Cu, and Al (contaminated by a spin coating method [3] with concentration levels ranged from 1×10^{10} to 1×10^{12} atoms/cm^2) upon 1) thin oxide integrity, 2) total oxide charge, and 3) various bulk electrical and structural parameters. The oxide integrity was determined from the breakdown field strength measurements. In addition, in order to correlate bulk structural and electrical parameters with the oxide integrity, various measurements such as 1) total oxide charge, 2) MOS C-t generation lifetime, 3) bulk minority carrier recombination lifetime and/or diffusion length using: 1) laser microwave photoconductivity decay method ("μ-PCD"); 2) surface photovoltage ("SPV"); and 3) photo-induced current scanning method ("ELYMAT"). In addition, some of these samples were further characterized for metal-induced surface and bulk microdefects by transmission electron microscopy (TEM) and thermal wave modulated optical reflectance method (using both mapping and imaging modes).

EXPERIMENTAL

The samples used in this study were p-type, res. ~ 5 ohm-cm, (100)-oriented, 150 mm diameter Czochralski (CZ) Si wafers. These wafers were frontside-contaminated with well-known metallic impurities such as Al, Cu, and Fe with concentration levels from 1×10^{10} to 1×10^{12} atoms/cm^2 by a spin coating method. After the contamination process, surface metallic impurity concentration was measured with vapor phase decomposition atomic absorption spectrophotometry (VPDAAS) and/or total reflection x-ray fluorescence (TXRF), and found to be generally in good agreement with the process recipe (± 10%). These contaminated wafers were separated into two groups (Group A and Group B). Group A wafers were subjected to gate oxidation cycle (dry O_2, 950°C oxide film thickness ~ 23 nm). Group B wafers were processed through simulated Mb DRAM thermal cycles up to field oxidation. After the field oxidation cycle, these wafers were oxide film stripped by HF treatment, RCA cleaned and subjected to the same gate oxidation cycle as Group A wafers. During the course of this investigation, we also used p-type res ~ 5 ohm-cm (100)-oriented, 150 mm diameter Cz Si wafers as Reference wafers.

Thin gate oxide integrity, total oxide charges and various bulk electrical and structural defect properties of both Group A and Group B Wafers were evaluated. The oxide integrity was determined from the measurement of average breakdown field strength and time-dependent-dielectric breakdown characteristics while the total oxide charges of thin gate oxide were measured by Semi Test Inc.'s SCA System. In order to correlate bulk structural and electrical parameters with the oxide integrity, MOS C-t generation lifetime, and bulk minority carrier recombination lifetime and/or diffusion length measurements were carried out. In addition, some of these samples were further characterized for metal-induced surface and bulk microdefects by transmission electron microscopy (TEM) and thermal wave modulated optical reflectance (using both mapping and imaging modes).

RESULTS AND DISCUSSION

Figs 1 and 2 show effects of Fe, Cu, and Al impurities (Concentration ranged from 1×10^{10} to 1×10^{12} atoms/ cm^2) upon average breakdown field strength after gate oxidation only and after Mb DRAM simulated cycle + gate oxidation, respectively. The data show that:

1) for the case of gate oxidation only, regardless of impurity types and levels in this study, average breakdown field strength is minimally affected; and
2) in the case of Fe contamination, after the Mb DRAM simulated thermal cycle + gate oxidation, substantial degradation of average breakdown field strength (ΔE_{bd} ~ 4 MV/cm for the case of contamination level of 10^{12} atoms/cm^2 compared to that of Reference wafer) is seen while no recognizable change in breakdown field strength is observed for both Cu and Al contaminated samples.

Effects of these same metallic impurities upon time-dependent-dielectric breakdown characteristics were evaluated. Our data (not shown) basically show the same trends as those of average breakdown field strength with respect to the level of metallic contamination.

Variations of bulk minority carrier recombination lifetime with respect to concentration levels of all three metallic impurities after gate oxidation only are displayed in Figs. 3 - 5. The data in these figures show that, for both Fe and Cu contaminated samples, bulk recombination lifetime monotonically and rather gradually decreases with an increase in contamination level whereas no degradation is observed for Al contaminated samples. Figs. 6 - 8 show variations of bulk lifetime with respect to contamination level of the same metallic impurities after the Mb DRAM simulated thermal anneal + gate oxidation cycle. In this case, unlike the data shown in Figs. 3 and 4, we observe substantial degradation in bulk lifetime with an increase in contamination level for Fe contaminated samples while a rather small decrease in lifetime is seen for Cu contaminated wafers (Figs. 6 and 7). For Al contaminated samples, no recognizable changes in lifetime with respect to the contamination level are seen (Fig. 8).

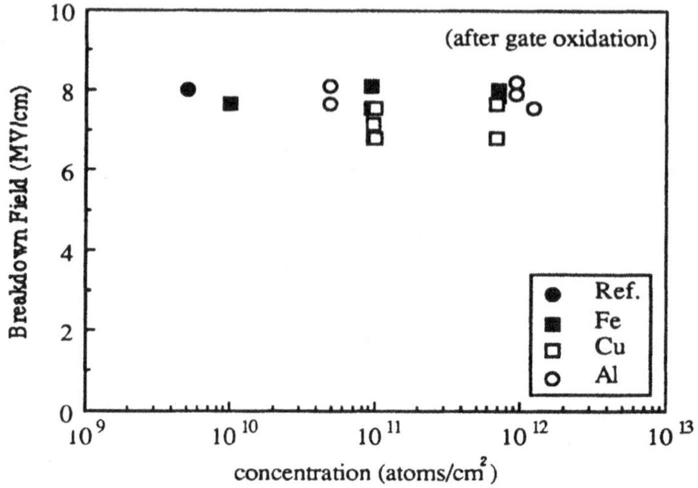

Fig. 1. Variation of average breakdown field strength with respect to contamination level after gate oxidation only.

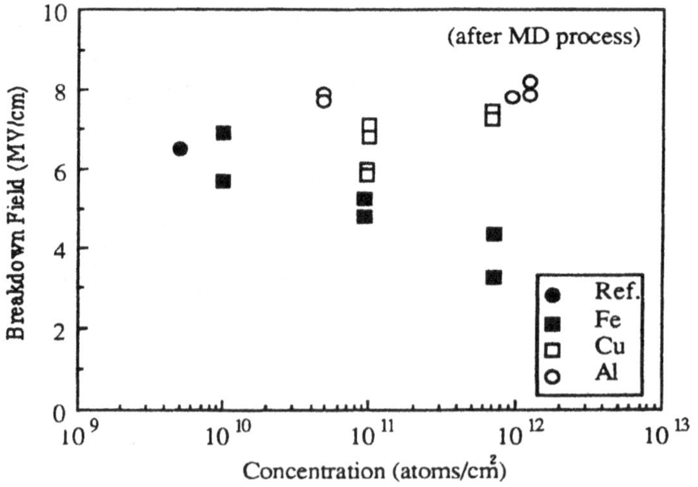

Fig. 2. Variation of average breakdown field strength with respect to contamination level after MD process.

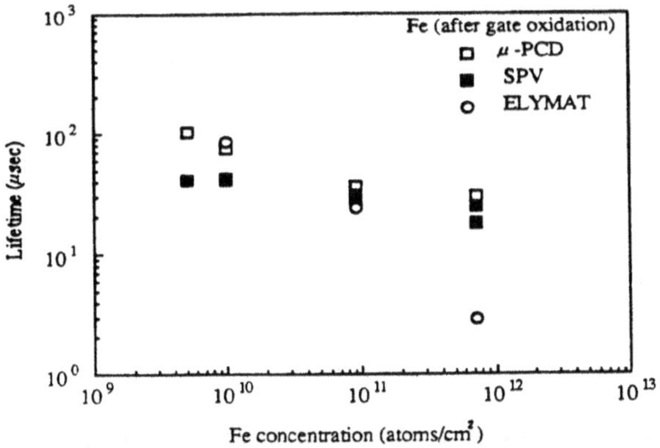

Fig. 3. Variation of bulk recombination lifetime with respect to Fe contamination level after gate oxidation only.

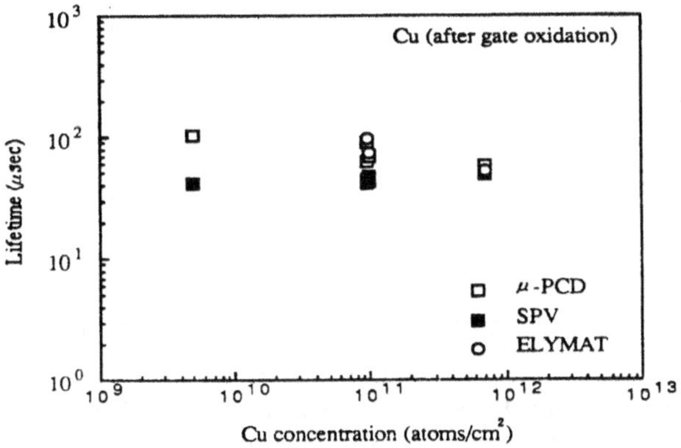

Fig. 4. Variation of bulk recombination lifetime with respect to Cu contamination level after gate oxidation only.

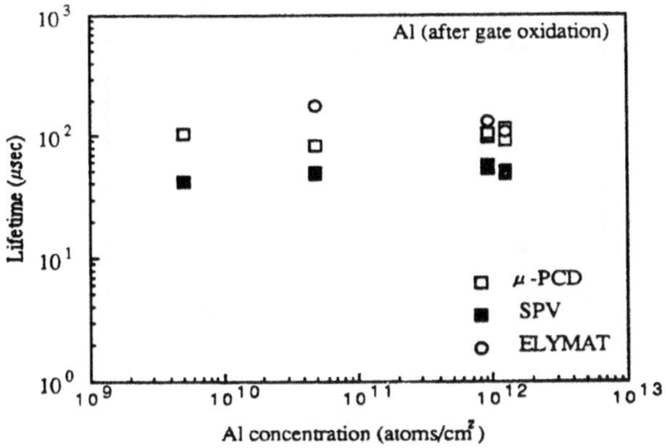

Fig. 5. Variation of bulk recombination lifetime with respect to Al contamination level after gate oxidation only.

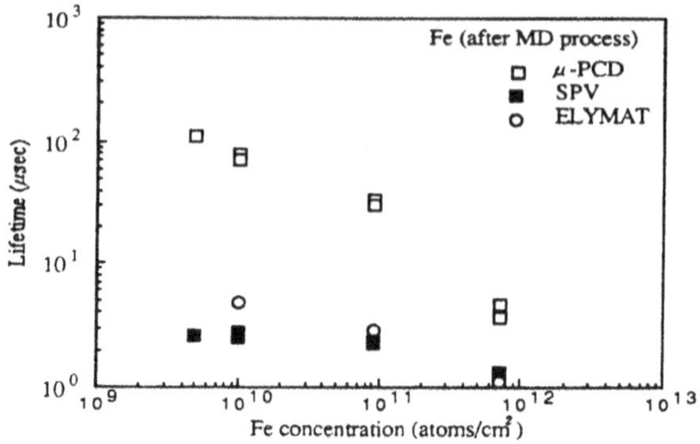

Fig. 6. Variation of bulk recombination lifetime with respect to Fe contamination level after MD process.

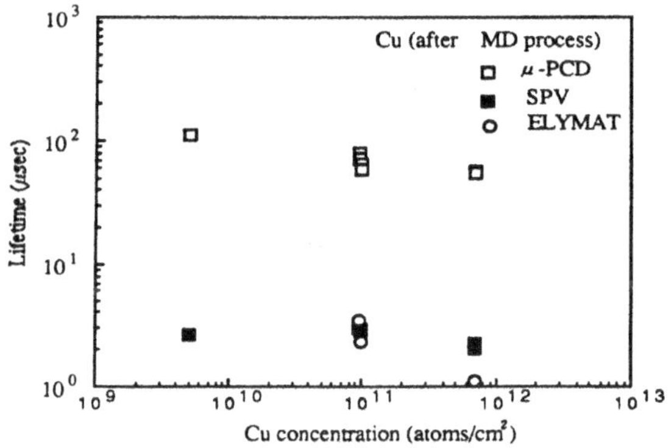

Fig. 7. Variation of bulk recombination lifetime with respect to Cu contamination level after MD process.

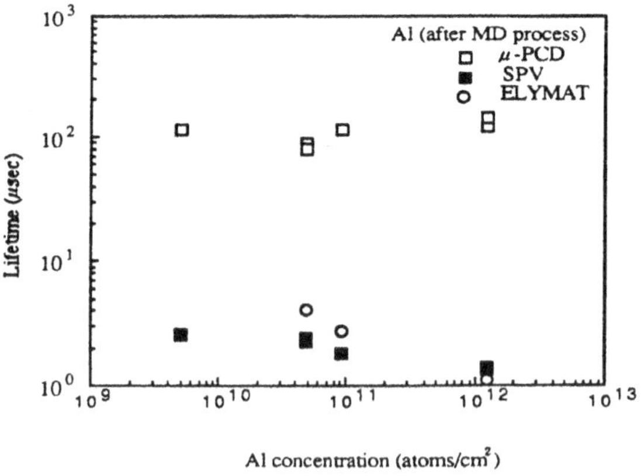

Fig. 8. Variation of bulk recombination lifetime with respect to Al contamination level after MD process.

Effects of metal contamination upon MOS C-t generation lifetime were monitored. In the case of gate oxidation only, regardless of types of metallic impurities, no recognizable changes in generation lifetime with respect to contamination level (data not shown) were observed. However, for the Fe contaminated samples after the Mb DRAM simulated thermal anneal + gate oxidation, MOS C-t generation lifetime quickly degraded with an increase in contamination level whereas small decrease in generation lifetime was observed from Al contaminated samples (Figs. 9 and 10). In the case of Cu contamination up to 10^{12} atoms/cm^2, MOS C-t generation lifetime was pretty constant (Fig. 11).

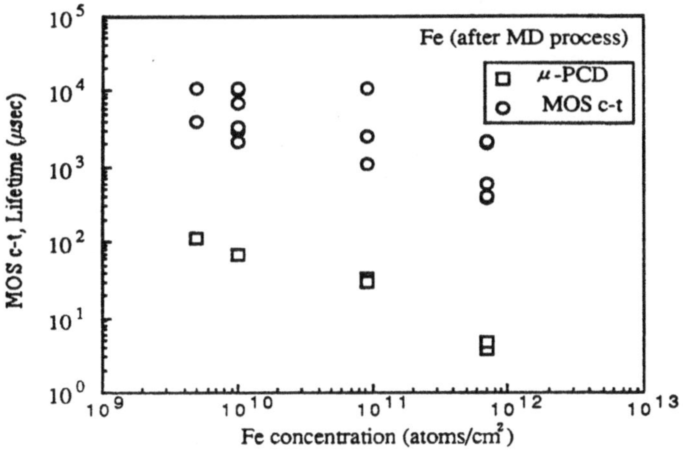

Fig. 9.Variation of MOS C-t generation lifetime with respect to Fe contamination level after MD process. Bulk recombination lifetime variation is included as a comparison.

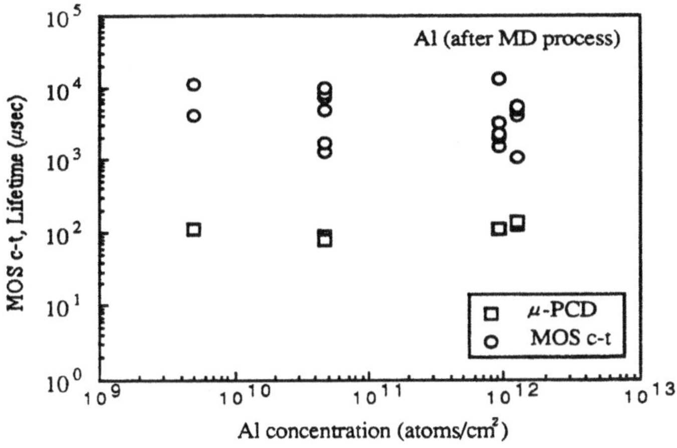

Fig. 10.Variation of MOS C-t generation lifetime with respect to Al contamination level after MD process. Bulk recombination lifetime variation is included as a comparison.

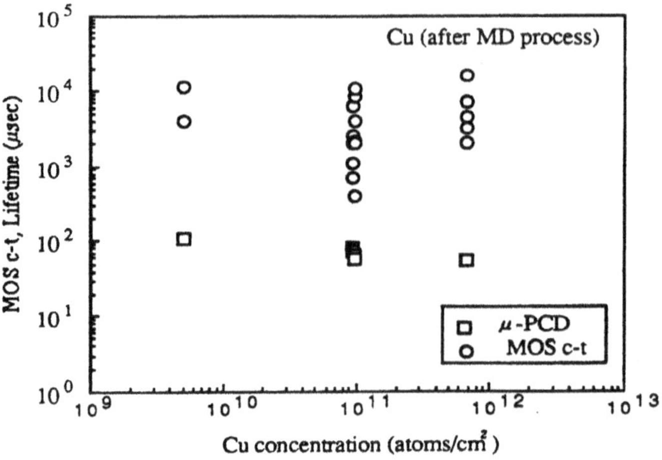

Fig. 11.Variation of MOS C-t generation lifetime with respect to Cu contamination level after MD process. Bulk recombination lifetime variation is included as a comparison.

Figs. 12 and 13 show variations of total oxide charge of Si/23 nm thick gate oxide system with respect to contamination level of three metallic impurities after two types of thermal processing. The data show that total oxide charge for Fe and Cu contaminated samples is negative and around 10^{11}q/ cm^2. This value stayed constant regardless of contamination level and thermal anneal cycles used. However, for Al contaminated samples, the total oxide charge is also negative in sign and increases with an increase in contamination level for both types of thermal processing.

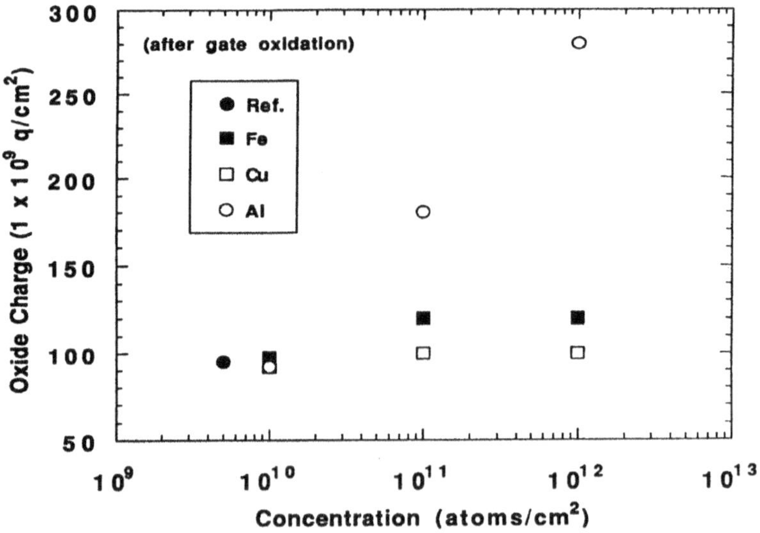

Fig. 12.Variation of total oxide charge with respect to contamination level after gate oxidation cycle only.

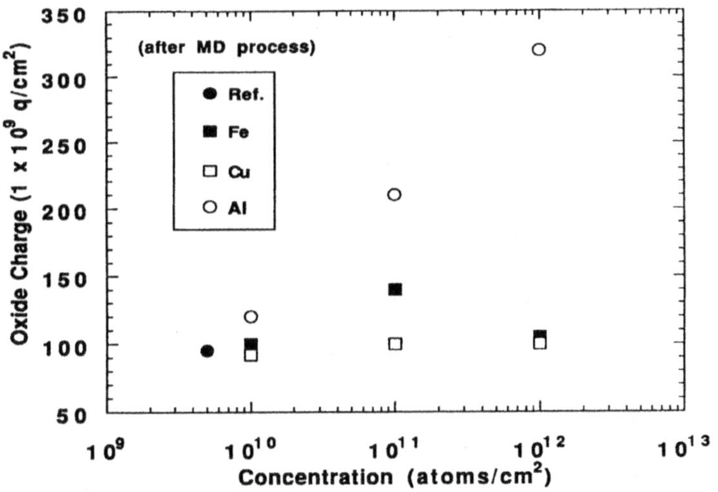

Fig. 13.Variation of total oxide charge with respect to contamination level after MD process.

Finally, metal-induced surface and bulk microdefects formed during two types of thermal treatments were investigated with thermal wave modulated optical reflectance and transmission electron microscopy. During the course of this investigation, we find out that most of the samples contained almost no defects mainly due to rather low concentration level used. Only Fe contaminated samples (with a level of 10^{12} atoms/cm^2) exhibited both metal-induced surface and bulk microdefects after the Mb DRAM simulated thermal anneal + gate oxidation cycle. Figs. 14 and 15 show both TW defect images due to Fe-induced surface and bulk microdefects and typical TEM micrographs corresponding to the TW defect images. Since TW is only sensitive to well-developed metal silicides and associated crystallographic defects [11], we conclude that TEM defect images in Fig. 15 are $FeSi_2$ and/or metal-crystallographic defect complexes.

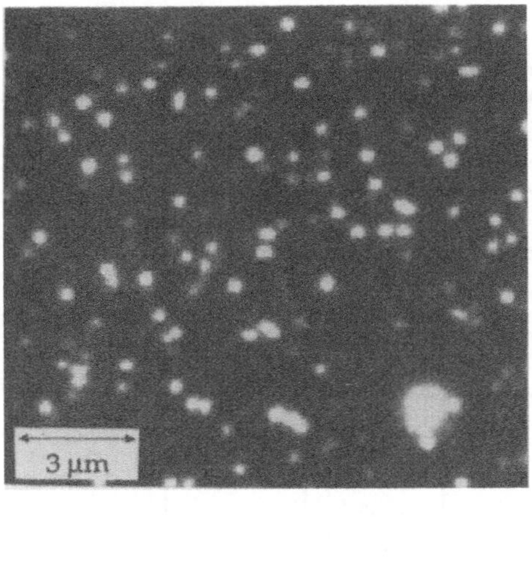

Fig. 14.Typical TW defect image due to Fe contamination level of 10^{12} atoms/cm^2 after MD process.

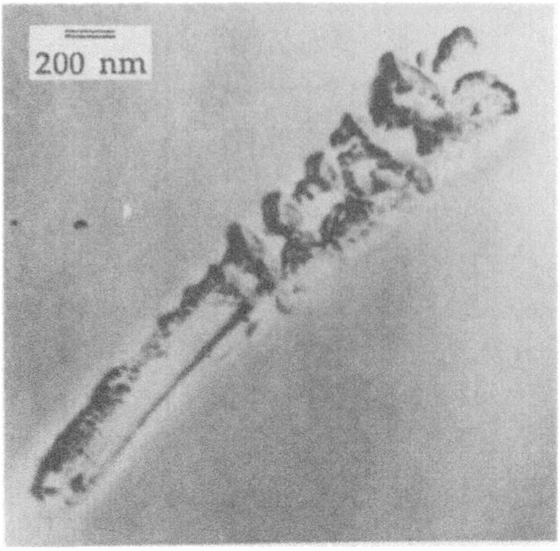

Fig. 15. Typical TEM defect morphology for Fe contaminated samples (with level of 10^{12} atoms/cm^{2}) after MD process. Besides this morphology, we also observed bulk stacking faults in this kind of sample.

SUMMARY

In this study, we investigated effects of various well-known metallic impurities such as Fe, Cu, and Al (contaminated by a spin coating method with concentration levels ranged from 1×10^{10} to 1×10^{12} atoms/cm^{2}) upon 1) thin oxide integrity, 2) total oxide charge, and 3) various bulk electrical and structural defect parameters. Table I lists a summary of our results. (In this table, we also add the data generated from our recent Au contamination studies). Based upon our oxide integrity and bulk electrical and structural characterization data, we conclude that Fe is the most harmful metallic impurity among the four metallic elements investigated in this study.

Table I. A summary of experimental results.

(-- strong negative response, - medium negative response 0 no change, + medium positive response).

Impurity Type	Fe	Au	Cu	Al
BVOX	- -	-	0	0
TDDB	- -	-	-	0
OXIDE CHARGE	0	0	0	+
TW	+	0	+	0
MOS C-t	- -	- -	0	-
SPV Diffusion Length	- -	-	-	0
ELYMAT Diffusion Length	- -	-	-	0
μ-PCD Lifetime	- -	-	-	0
TEM	$FeSi_2$ ppt Stacking Faults	No Defects	Small CuSi ppt (only contamination level $\geq 10^{15}$ atoms/cm^{2})	No Defects

References

1. G. H. Schwttke, K. Brack, and E. W. Hearn, *Microelectron. Reliab.* 10, 467 (1971).
2. H. Strack, K. R. Mayer and B. O. Kolbesen, *Solid-State Electron.* 22. 135 (1979).
3. S. Hahn, W. L. Smith, H. Suga, R. Meinecke, R. R. Kola and G. A. Rozgonyi, *J. Crystal Growth* 103, 206 (1990).

V. CHEMICAL, STRUCTURAL, AND MICROROUGHNESS EFFECTS AT THE Si-SiO$_2$ INTERFACE

INTRODUCTION

T. Hattori
Dept. of Electrical Engineering
Musashi Inst. of Technology, Tokyo, Japan

C. R. Helms
Dept. of Electrical Engineering
Stanford University, Stanford, CA 94305

The papers included in this chapter all deal with the properties of the Si/SiO$_2$ interface at atomic level dimensions. In particular the understanding and control of the quality of the bulk wafer, polishing, cleaning, native oxide, and thermal oxidation process, which is critical, especially for the scaling of MOS, ULSI devices, is discussed. Specifically, studies of the effects of roughness, oxidation temperature and crystallographic orientation on the distribution of intermediate oxidation states in ultrathin oxide, the critical investigation of high-resolution transmission electron microscope image of the SiO$_2$/Si interface, the dependence of surface microroughness on type of silicon substrates, the oxidation induced changes in the Si surface microroughness, the effect of surface microroughness on gate oxide leakage currents, the effect of solidification induced defects in CZ-silicon on thin gate oxide integrity are described.

In the first paper, recent work using electron spectroscopies to investigate the oxidation states of the interface Si is described in the invited paper of Himpsel from IBM. The recent controversy concerning the crystalline nature of SiO$_2$ near the interface was discussed in the invited paper by Akatsu et al, from IBM, Japan. They showed that microroughness could lead to effects in TEM images that could be interpreted as due to crystalline SiO$_2$, even if none was present. The effect of microroughness on device properties, including dielectric breakdown, is discussed in the next invited paper from Tohoku University, by Ohmi et al. They showed that the effect of the roughness on device propeties depends on the type of wafer used (CZ, FZ, epi). In the paper by Chonko et al from Motorola, the effect of roughness on both the wafer/SiO_2 as well as poly/SiO_2 interfaces is reported. The benefits of sacrificial oxidations in obtaining a smooth interface is discussed in the paper by Tseung and Tobin, also from Motorola. The last two papers in this chapter deal with the effect of "bulk" Si defects on gate oxide integrity. In both papers, one by Suga et al from Mitsubishi, and one from a group at Samsung, J.-S. Park et al, the detrimental effect of bulk defects on oxide properties is quantatively discussed.

LOCAL BONDING AT SiO_2/Si INTERFACES

F. J. Himpsel, D. A. Lapiano-Smith, and J.F. Morar

IBM Research Division
Thomas J. Watson Research Center
P.O. Box 218
Yorktown Heights, NY 10598

J. Bevk

AT&T Bell Laboratories
Murray Hill, NJ 07974

The distribution of oxidation states at various SiO_2/Si interfaces is determined using surface-sensitive, high-resolution Si2p core level spectroscopy with synchrotron radiation. It provides a critical test of current structural models for the interface. The dependences on crystallographic orientation, substrate roughness, and growth temperature are investigated. We also examine recent reports of layer-by-layer growth of SiO_2.

INTRODUCTION AND OVERVIEW

The structure of the SiO_2/Si interface remains an unresolved issue. Compared to other interfaces, a major obstacle is the amorphous structure of the SiO_2 overlayer, which prevents many standard structural techniques from being applicable. Although there have been reports of a crystalline oxide structure close to the interface,[1-7] a variety of structural models[1-14] are still competing with each other. Even the question of a unique bonding configuration at the interface remains open. Substrate roughness and the preparation method may affect the interface structure. While we have studied in previous work[12-14] the stable interface structure that is formed at temperatures and pressures close to the thermal equilibrium line[15] between oxide-free and oxide-covered Si, we focus here on possible metastable structures that are obtained far away from equilibrium. A good example is oxidation at room temperature and high pressure of oxygen.

We are addressing these questions using a local probe, which is sensitive to the nearest neighbor bonding, and does not require long range order. By analyzing

The Physics and Chemistry of SiO_2 and the Si-SiO_2 Interface 2
Edited by C.R. Helms and B.E. Deal, Plenum Press, New York, 1993

the core level shifts and intensities of the Si2p level[16] induced by oxygen neighbors one is able to count the number of oxygen ligands and determine the abundance of Si atoms with different bond configurations. In the following, we will first give an overview of the technique, and then proceed to recent results on open questions relating to the uniqueness of the interface structure. Oxides are prepared on surfaces with varying roughness and at different temperatures. Essentially, roughness is found to have an effect similar to elevated temperature. Both convert lower oxidation states of Si to higher ones and increase the oxygen uptake by facilitating insertion of oxygen into Si backbonds. A unique interface structure is obtained after annealing, independent of the starting conditions. It only depends on the crystallographic orientation of the substrate. In addition, we examine recent reports of layer-by-layer growth of the oxide, in order to explore new avenues for precisely-controlled growth of ultrathin oxide layers.

Si2p CORE LEVEL SPECTROSCOPY

Using chemically-shifted Si2p components for surface and interface analysis has been reviewed in Ref. 16, and with particular emphasis on Si oxidation in Ref. 13. Here we briefly recall the salient points, using the data in Fig. 1 as an example. The Si2p photoelectron spectrum of oxidized Si can be decomposed into 5 well-resolved components, corresponding to the oxidation states 0 to 4^+ of Si, or in other words, to the number of oxygen atoms bonding to a Si atom. The 0, 2^+, 3^+, and 4^+ states are already visible in the raw data (top curve), while the 1^+ oxidation state is masked by the $j=1/2$ spin-orbit partner of the main, $j=3/2$ Si^0 line of the Si substrate. For a quantitative analysis we subtract an extrinsic, secondary electron background (full line) plus an intrinsic, energy loss background (dashed line). The former is measured by reducing the photon energy by 10eV, but keeping the same kinetic energy window. The latter is taken proportional to the integral curve of the Si2p spectrum, corresponding to a step-like loss function. After background subtraction the spectrum is decomposed into the $j=1/2$ and $j=3/2$ components. This is a straight-forward mathematical procedure (see Ref. 16) once the spin-orbital splitting and the intensity ratio are known. These are determined from reference surfaces,[13,17] such as clean Si and H-terminated Si. Throughout this paper we use a splitting of 0.59eV and an intensity ratio of 0.52 with a photon energy of 130eV. The latter is chosen to optimize the surface and interface sensitivity. The $Si2p_{3/2}$ spectrum clearly exhibits all oxidation states, spaced by about 0.9eV. Estimates of the number of Si atoms in various oxidation states can be obtained from the peak areas, if an enhancement of the cross §[13] is taken into account for the higher oxidation states (Table 1).

DISTRIBUTION OF OXIDATION STATES VERSUS ROUGHNESS, TEMPERATURE, AND CRYSTALLOGRAPHIC ORIENTATION

The effect of roughness on the interface oxide structure is demonstrated for Si(100) in Fig. 3 (compare also Ref. 18 for the effect of roughness). The clean Si(100)2×1 substrates in Fig. 3 are obtained from identical wafers, but prepared in two different ways. A rough surface with a broad Si2p spectrum (probably due to inhomogeneous Fermi level pinning) is obtained at temperatures above

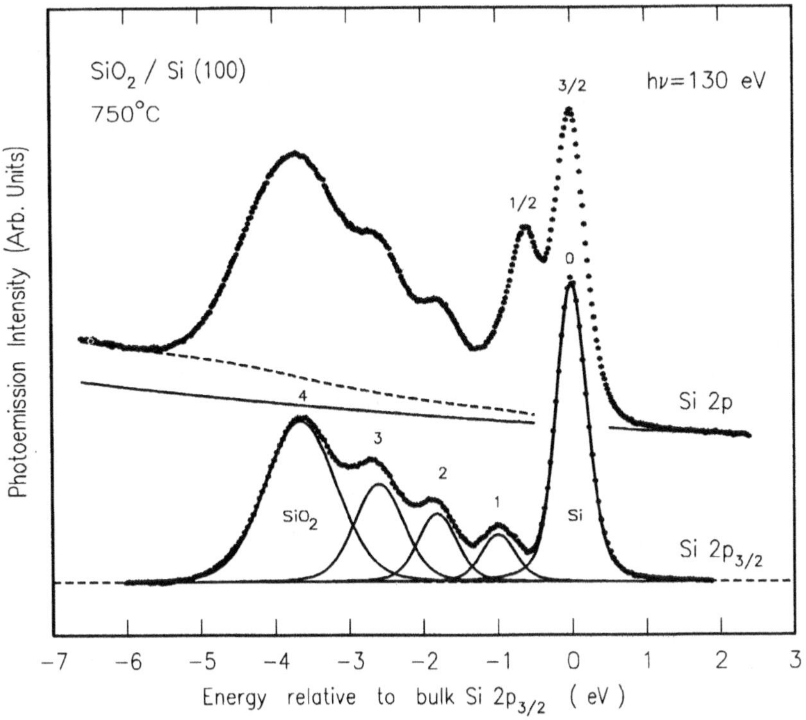

Fig. 1 High resolution Si2p photoelectron spectrum, taken at the escape depth minimum in order to resolve the intermediate oxidation states 1 to 3 at the SiO_2/Si interface. The raw data (top) are converted to the Si2p$_{3/2}$ core level spectrum (bottom) by background subtraction and spin-orbit decomposition. The SiO_2 film was formed on Si(100) by exposure to 2×10^{-5} Torr dry O_2 at 750°C for 20 sec. This spectrum represents the high temperature equilibrium interface structure.

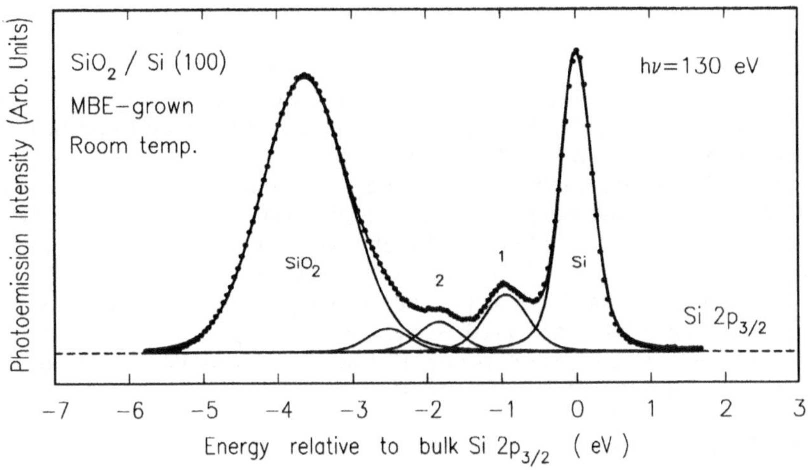

Fig. 2 Similar to Fig. 1, except for using an ultrasmooth, MBE-grown substrate, and growing the oxide at room temperature (1atm O_2 for 5 hours). This spectrum represents a metastable, low temperature interface structure.

Table 1. Abundance of various oxidation states of Si, obtained from Si2p areas at a photon energy of 130eV, after normalizing to the bulk Si line and dividing by the relative cross sections σ from Ref. 13. Differences in the depth distribution are not taken into account.

	Si^{1+}	Si^{2+}	Si^{3+}	Si^{4+}
σ	1.0	1.1	1.7	2.2
1atm O_2 5h room temp. (Fig. 2):				
Si(100) smooth MBE	0.26	0.12	0.06	1.03
50Torr O_2 2.5min room temp. (Fig. 3):				
Si(111) smooth	0.28	0.18	0.10	0.13
Si(100) smooth	0.28	0.28	0.14	0.18
Si(100) rough	0.16	0.24	0.30	0.34
Anneal 500°C 2 min:				
Si(111) smooth	0.21	0.14	0.13	0.20
Si(100) smooth	0.21	0.19	0.18	0.26
Si(100) rough	0.12	0.16	0.27	0.48
Anneal 700°C 1 min:				
Si(111) smooth	0.18	0.12	0.25	0.18
Si(100) smooth	0.20	0.22	0.29	0.20
Si(100) rough	0.10	0.16	0.30	0.46
2×10^{-5}Torr O_2 20sec 750°C (Ref. 13 and Fig. 1):				
Si(111)	0.21	0.12	0.29	0.46
Si(100)	0.18	0.26	0.30	0.53

1050°C, which are needed to remove the native oxide. Such high temperatures can be avoided by H-passivation via a 10% HF dip, which completely removes SiO_2 and leaves only a small fraction of a monolayer of suboxides,[17] which can be heated off at 850°C. Such surfaces exhibit sharp Si2p lines and a well-pronounced surface core level emission.[16] Even smoother surfaces are obtained by molecular beam epitaxy[2,3,5-7] (MBE). Corresponding data are shown in Fig. 2. A smooth Si(111)7×7 surface was obtained by dipping in 40% NH_4F and heat-cleaning in vacuum at 1050°C. All oxidations were performed in pure O_2, with H_2O frozen out by a liquid nitrogen trap.

For oxidation at room temperature the general trend is a population of higher oxidation states with increasing substrate roughness (Figs. 2,3). These differences largely disappear after annealing, and higher oxidation states become populated for all substrates (Fig. 3). It appears that roughness anticipates the effect of annealing by facilitating the incorporation of oxygen into Si backbonds. By the same token, roughness is observed to increase the oxygen uptake (compare the two Si(100) room temperature spectra in Fig. 3). The same effect is again seen when oxidizing at elevated temperature (Fig. 1).

Apart from the distribution of oxidation states we may consider the width of the interface, represented by the number of oxygen atoms in intermediate oxidation states (1^+, 2^+, 3^+). It is the same for the rough and smooth Si(100) substrates in Fig. 3, but decreases by about 40% for the ultrasmooth,

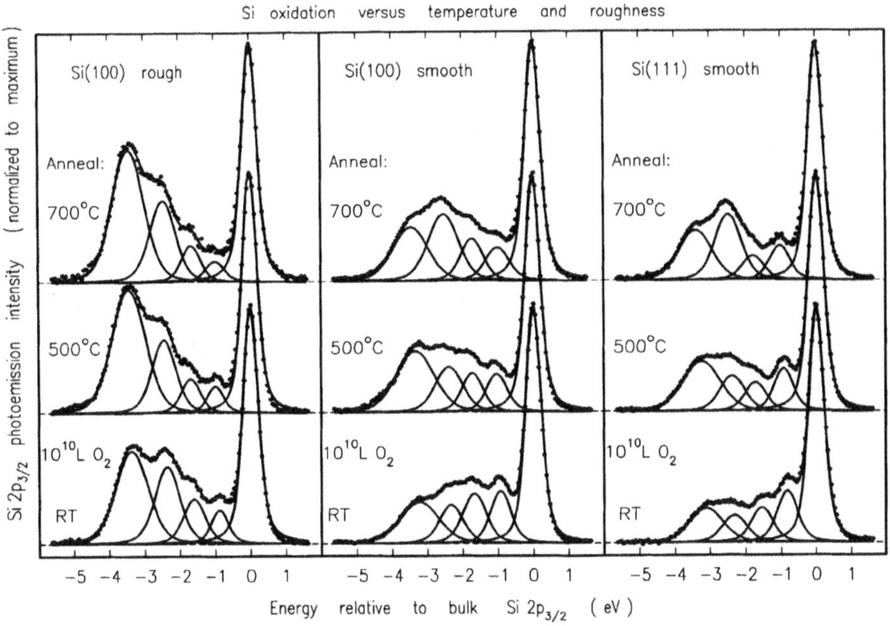

Fig. 3 Distribution of oxidation states versus roughness, temperature, and crystallographic orientation. For room temperature exposure (50 Torr dry O_2 for 2.5 min) the lower oxidation states dominate, as long as the surface is smooth. Higher oxidation states become populated at rough surfaces, or by annealing (2min at 500°C and 1min at 700°C). Si(111) has less Si^{2+} than Si(100).

MBE-grown Si(100) substrate in Fig. 2. In order to obtain an absolute number one may use a discrete layer model with an escape depth of 3.3Å at $h\nu = 130$eV that has been used previously in the analysis of surface core levels on Si (see Ref. 13). The intensity of the outermost Si(100) layer is 34% of the total intensity (i.e., surface plus bulk), while the second layer contains 22% of the total intensity. The analogous numbers for the Si(111) surface are 38% and 24%, taking the average between the two types of (111) planes. The same argument may be applied to the SiO_2/Si interface, although one should be aware that differences in depth distribution between the oxidation states will distort the result, and that the additional oxygen content at the interface changes the escape depth (for a more thorough analysis see Ref. 13). The simple model gives 0.9 monolayer for the interface width on MBE-grown Si(100) at room temperature (Fig. 2), suggesting an atomically-abrupt interface. On the Si(111) surface the number of Si atoms in intermediate oxidation states is found to be about 20% less than on Si(100), as far as the thin oxides in Fig. 3 are concerned. However, it increases to a value 40% larger than on Si(100) for thicker oxides (see Refs. 13, 14), while the width of the (100) surface remains stable. The changing intensity of intermediate oxides on the Si(111) interface does not necessarily reflect a change in interface width. It could also be caused by incomplete oxide coverage at low thickness, which would imply that the SiO_2/Si(111) interface is wider than the SiO_2/Si(100) interface, even in the low coverage regime.

The effect of temperature is to convert lower oxidation states to higher ones (see the annealing results in Fig. 3). An analogous shift towards higher oxidation states has been found previously in the chemisorption regime.[19] The room temperature oxides convert after annealing to structures similar to those obtained near thermal equilibrium (compare Table 1 and Fig. 1). For thin films the conversion begins below 500°C and is complete at 700°C. Thicker films require higher temperatures. Summing over all oxidation states in Table 1 we find that the total number of oxidized Si atoms remains roughly constant with annealing, until desorption begins at about 750°C. The interface width, i.e. the sum over 1^+, 2^+, and 3^+, also remains constant. The differences between rough and smooth surfaces in Fig. 3 largely disappear after annealing, e.g. the distribution of oxidation states becomes comparable. Thus we reach a unique equilibrium structure from the point of view of local bonding, which is independent of the starting conditions. It would be interesting to see if that is also true for the ultrasmooth, MBE-grown substrate in Fig. 2, where an annealing experiment was not performed. For such substrates there are differences in electrical properties associated with the surface roughness before oxidation.

It almost goes without saying that the oxygen uptake increases at elevated temperature. For example, a dose of only 400L O_2 is needed at 750°C in Fig. 1 to grow a film of the same thickness as for a dose of 10^{10}L at room temperature exposure in Fig. 3.

The crystallographic dependence in the distribution of oxidation states shows up as a lack of Si^{2+} for Si(111) compared to Si(100) (compare similar results in Refs. 11, 13, 14). The oxidation states reflect the number of broken bonds at these surfaces (one or three for Si(111), two for Si(100)). At room temperature the difference has almost disappeared, showing a need for thermal activation in order to reach the stable equilibrium structure.

STRUCTURAL MODELS

A variety of structural models have been proposed for the SiO_2/Si interface, particularly for the (100) orientation. Many of them can be ruled out, because they are inconsistent with the observed distribution of oxidation states. In the following we will only consider structural models that are able to explain our core level results. Thereby we distinguish between the metastable, room temperature structure (represented by the MBE-grown Si(100) sample in Fig. 2), and the equilibrium, high-temperature structure (represented by Si(100) oxidized at 750°C in Fig. 1).

For the room temperature structure we have an abrupt interface with most of the Si atoms bonding to a single oxygen. Such a configuration can be obtained by having Si-Si dimers at the interface, similar to those at the clean Si(100)2×1 surface. Dimerization reduces the interface bond density by a factor of two, which helps to match the two times lower bond density of SiO_2 to that of Si. Such a dimer structure may be connected to various crystalline or amorphous phases of SiO_2. There is also evidence for a dimerization from half order spots in x-ray diffraction.[5,7] Note that the truncated bulk structure would give at least half a layer Si^{2+} when connected to various oxide structures, due to the factor of two lower areal bond density of SiO_2. This is incompatible with the data.

The high temperature equilibrium phase is characterized by the appearance of higher oxidation states, particularly Si^{3+}. At the same time the width of the interface goes up. The half order x-ray diffraction spots go away at high temperature,[5,7] indicating a qualitative structural change. Several models[10,13] are able to explain this behavior by an interface extended by Si^{3+} protrusions. It appears that the stable interface structure requires a significant proportion of all oxidation states. Such a distribution may be helpful in minimizing the stresses that are associated with the two-fold reduction in the density of silicon atoms going from Si to SiO_2 (see Ref. 10).

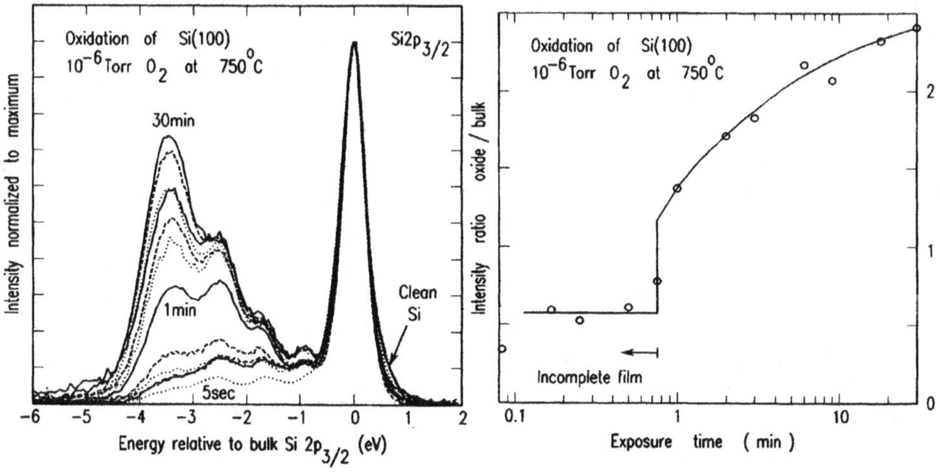

Fig. 4 Uptake of oxygen on Si(100) under conditions similar to a previous report[21] of layer by layer growth. The area of all oxidation states (1 to 4) relative to the bulk line is plotted versus exposure time to 10^{-6} Torr dry O_2 at 750°C. The step at 1 min is due to the completion of a continuous oxide film at about 3 layers coverage.

LAYER-BY-LAYER GROWTH OF SiO_2

There have been a number of reports of stepwise growth of SiO_2, both for native oxide growth at room temperature,[20], and for high temperature growth[21,22] A recent Si2p XPS experiment[21] finds a stepwise increase in the intensity of oxidized Si atoms during oxidation in 10^{-6}Torr of O_2 at 862°C. We have attempted to reproduce this experiment with our better energy and depth resolution (see Fig. 4). However, under the temperature-pressure conditions quoted in Ref. 21 we find that the surface remains free of oxide. This is consistent with previous results on oxygen uptake,[15] where the parameters in Ref. 21 put us clearly on the oxide-free side of the pressure-temperature equilibrium curve. Therefore, we have reduced the temperature to 750°C, where we are on the oxide side of the equilibrium curve. Also, the samples were cooled under oxygen after each exposure, in order to prevent partial desorption of the oxide film. Even so, we do not find clear evidence for step-like uptake (see Fig. 4). Three steps have been reported in Ref. 21, each corresponding to one layer of SiO_2. We only find a single, large step at about 1min exposure, which might correspond to the third step in Ref. 21. For lower exposures the film becomes dicontinuous. The thickness of the SiO_2 layer at 1min exposure is obtained analogous to our previous discussion on Si2p areas. It corresponds to 1.5 layers of oxidized Si atoms, which translates into about 3 oxide layers when taking the lower density of Si atoms in SiO_2 into account. The discontinuous nature of the oxide layer below 1min exposure becomes evident from the surface core level emission[16] of clean Si(100)2x1 that appears at the high energy tail of the Si2p spectrum. With such an incomplete oxide layer we find significant desorption at our operating temperature of 750°C. Essentially, Si suboxides (e.g. SiO) desorb at lower temperature, and SiO_2 converts to SiO at the edge of an open area (compare the desorption mechanism found in Ref. 23). Thus, the morphology of the growing surface needs to be taken into account when looking for layer-by-layer growth at elevated temperature.

REFERENCES

1. F. Rochet, S. Rigo, M. Froment, C. D'Anterroches, C. Maillot, H. Roulet, and G. Dufour, Adv. Phys. **35**, 237 (1986); F. Rochet, M. Froment, C. D'Anterroches, H. Roulet and G. Dufour, Philos. Mag. B **59**, 339 (1989).
2. A. Ourmazd, D. W. Taylor, J. A. Rentschler and J. Bevk, Phys. Rev. Lett. **59**, 213 (1987); A. Ourmazd, J.A. Rentschler, and J. Bevk, Appl. Phys. Lett. **53**, 743 (1988); For an alternative interpretation of the doubled periodicity at the interface in TEM see: H. Akatsu and I. Ohdomari, Appl. Surf. Sci. **41/42**, 357 (1989).
3. A. Ourmazd, P. H. Fuoss, J. Bevk and J. F. Morar, Appl. Surf. Sci. **41/42**, 365 (1989).
4. P. H. Fuoss, L. J. Norton, S. Brennan and A. Fischer-Colbrie, Phys. Rev. Lett. **60**, 600 (1988).
5. I. Hirosawa, K. Akimoto, T. Tatsumi, J. Mizuki and J. Matsui, J. of Cryst. Growth **103**, 150 (1990).
6. G. Renaud, P. H. Fuoss, A. Ourmazd, J. Bevk, B. S. Freer and P. O. Hahn, Appl. Phys. Lett. **58**, 1044 (1991).
7. T. A. Rabedeau, I. M. Tidswell, P. S. Pershan, J. Bevk and B. S. Freer, Appl. Phys. Lett. **59**, 706 (1991).
8. F. Herman and R. V. Kasowski, J. Vac. Sci. Technol. **19**, 395 (1981).
9. M. Hane, Y. Miyamoto and A. Oshiyama, Phys. Rev. B **41**, 12637 (1990).
10. I. Ohdomari, H. Akatsu, Y. Yamakoshi and K. Kishimoto, J. Appl. Phys. **62**, 3751 (1987).
11. F. J. Grunthaner and P. J. Grunthaner, Mat. Sci. Reports, **1**, 65 (1986).
12. G. Hollinger and F. J. Himpsel, Appl. Phys. Lett. **44**, 93 (1984).
13. F. J. Himpsel, F. R. McFeely, A. Taleb-Ibrahimi, J. A. Yarmoff and G. Hollinger, Phys. Rev. B **38**, 6084 (1988).
14. F. J. Himpsel, F. R. McFeely, A. Taleb-Ibrahimi, J. A. Yarmoff and G. Hollinger, *The Physics and Chemistry of SiO_2 and the Si-SiO_2 Interface*, ed. C. R. Helms and B. E. Deal (Plenum, 1988), p. 219.
15. F. W. Smith and G. Ghidini, J. Electrochem. Soc. **129**, 1301 (1982).

16. F. J. Himpsel, B. S. Meyerson, F. R. McFeely, J. F. Morar, A. Taleb-Ibrahimi and J. A. Yarmoff, Proceedings of the Enrico Fermi School on *Photoemission and Absorption Spectroscopy of Solids and Interfaces with Synchrotron Radiation,* ed. M. Campagna and R. Rosei, North Holland (Amsterdam 1990), p. 203.

17. B. S. Meyerson, F. J. Himpsel, and K. J. Uram, Appl. Phys. Lett. **57**, 1034 (1990).

18. G. Hollinger, R. Saoudi, P. Ferret and M. Pitaval, *The Physics and Chemistry of* SiO_2 *and the* $Si\text{-}SiO_2$ *Interface*, ed. C. R. Helms and B. E. Deal (Plenum, 1988), p. 211.

19. M. Tabe, T. T. Chiang, I. Lindau and W. E. Spicer, Phys. Rev. B **34**, 2706 (1986); G. Hollinger and F. J. Himpsel, J. Vac. Sci. Technol. A **1**, 640 (1983); W. Braun and H. Kuhlenbeck, Surf. Sci. **180**, 279 (1987); M. Nakazawa and H. Sekiyama, Appl. Phys. Lett. **56**, 2108 (1990); M. Niwano, H. Katakura, Y. Takeda, Y. Takakuwa and N. Miyamoto, J. Vac. Sci. Technol. A **9**, 195 (1991).

20. E. A. Taft, J. Electrochem. Soc. **135**, 1023 (1988); M. Morita, T. Ohmi, E. Hasegawa, M. Kawakami and M. Ohwada, J. Appl. Phys. **68**, 1272 (1990); T. Yasaka, M. Takakura, S. Miyazaki and M. Hirose, Mat. Res. Soc. Symp. Proc. **222**, 225 (1991).

21. V. D. Borman, E. P. Gusev, Y. Y. Lebedinskii and V. I. Troyan, Phys. Rev. Lett. **67**, 2387 (1991).

22. T. Hattori, H. Nohira, Y. Tamura and H. Ogawa, to be published in Jpn. J. Appl. Phys.

23. R. Tromp, G. W. Rubloff, P. Balk and F. K. LeGoues, Phys. Rev. Lett. **55**, 2332 (1985).

HIGH-RESOLUTION TRANSMISSION ELECTRON MICROSCOPE IMAGE OF THE SiO_2/(001)Si INTERFACE

Hiroyuki Akatsu,[1] Yasuyuki Sumi,[2] and Iwao Ohdomari[2]

[1]IBM Research, Tokyo Research Laboratory
5-19 Sanbancho, Chiyoda-ku, Tokyo 102 Japan
[2]School of Science and Engineering, Waseda University
Ohkubo 3-4-1, Shinjuku-ku, Tokyo 169 Japan

INTRODUCTION

As gate oxide becomes ever thinner, it is increasingly important to elucidate the morphology of the SiO_2/(001)Si interface, because slight interface roughness affects the dielectric properties of the oxide film, such as the leakage current and breakdown. It is therefore necessary, for the sake of future device technology, to evaluate the interface morphology and clarify the mechanism by which the interface roughness is formed.

High-resolution transmission electron microscopy (HRTEM) is one method of observing the interface structure without removing the oxide. A number of papers have been published on the SiO_2/Si interface morphology observed by HRTEM and its relation to the electrical properties of the thin SiO_2 film.[1,2,3,4,5] As a result of these efforts, it is generally understood that the interface morphology observed by HRTEM depends on the crystallographic orientation of the c-Si substrate; that is, the SiO_2/(111)Si interface is observed to be quite flat,[2] whereas the SiO_2/(001)Si interface is roughened.[1,3,4,5]

In spite of these efforts, however, the actual morphology of the SiO_2/Si interface is still unclear. This is mainly due to the difficulty of analyzing an HRTEM image. HRTEM image is not a simple projected image of the atomic positions inside the specimen, but is formed by (a) electrons multiply scattered inside the specimen, and (b) further modulation of the electron wave function caused by imperfections in the HRTEM optical system, such as, defocusing, current fluctuation, and lens aberrations. In addition, since the scattering conditions near the SiO_2/Si interface are different from those in the crystalline bulk, because of the abrupt termination of the periodic atom arrangement in c-Si, it is probable that the c-Si lattice image near the interface is different from that inside the bulk. Therefore, it is very

The Physics and Chemistry of SiO_2 and the Si-SiO_2 Interface 2
Edited by C.R. Helms and B.E. Deal, Plenum Press, New York, 1993

difficult to deduce the actual morphology of the SiO_2/Si interface directly from the HRTEM images.

In this work, we will try to evaluate the SiO_2/(001)Si interface morphology by comparing HRTEM images and optical diffraction patterns (ODPs) obtained experimentally from several regions of various thickness in a specimen with simulated ones obtained by using some interface structural models. The interface morphology of a model is determined by calculating the interface distortion energy in such a way as to lower the excess distortion energy due to the formation of the interface.[6] A model of an SiO_2/(001)Si interface with thin epitaxial tridymite is also used for the simulation, to allow examination of the atomic arrangement at the interface exhibiting the 110 period image.

EXPERIMENT

Boron-doped (001)Si wafers were used throughout the experiments. RCA cleaning followed by etching with a $HF+NH_4F$ solution was performed for preoxidation cleaning of the wafers. The oxidation was done in a dry-oxygen-flow furnace heated to 950°C. After the oxidation, the wafers were gradually cooled down to room temperature. The flow gas was changed from oxygen to Ar at the onset of the cooling-down period. It took about 1 hour to lower the wafer temperature to room temperature.

In order to protect the interface layer from damage suffered during preparation of the TEM specimen, polycrystalline Si was deposited on top of the oxide layer by chemical vapor deposition.

A TEM specimen for cross-sectional observation was made by slicing a pair of wafers, which were bonded together with epoxy glue, in a direction parallel to the [110] direction of the c-Si substrate. The sliced specimen was thinned by mechanical polishing and further Ar ion milling.

HRTEM observation was made by using a JEM2000EX device operated at 200 keV in the [110] direction of the c-Si substrate.

An ODP was taken by using a small laser spot, which was beamed only onto the lattice image near the SiO_2/(001)Si interface in the negative film of an HRTEM image.

VARIATION OF AN HRTEM IMAGE NEAR AN SiO_2/(001)Si INTERFACE ACCORDING TO THE SPECIMEN THICKNESS

Figure 1 shows HRTEM images of the SiO_2/(001)Si interface and corresponding ODPs. HRTEM images are taken in two different regions of a specimen, with different specimen thicknesses. The specimen thickness of the "thin" region is thinner than that of the region exhibiting the first thickness fringe, while that of the "thick" region is thicker.

The thickness of each region in a specimen is estimated from the thickness fringes and the ODP of the lattice image of the c-Si substrate. We can identify a region that is thinner than the region exhibiting the first thickness fringe by using the 002 spots of c-Si in ODPs. The 002 spots are not observed in the ODP for a very thin region of a specimen because the intensity of the 002 spots in the electron

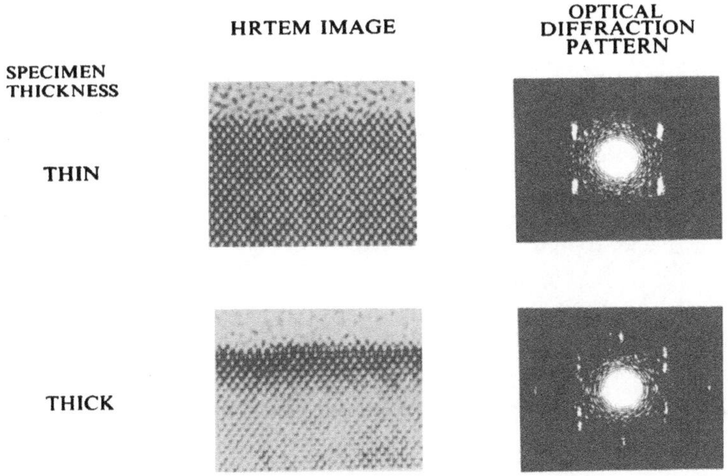

Figure 1. HRTEM images near the SiO_2/(001)Si interface and corresponding optical-diffraction patterns for thin and thick specimen.[7]

diffraction pattern of c-Si is very weak and does not contribute to the image formation for the thickness.[7]

As shown in the figure, HRTEM images near the interface and their ODPs differ according to the specimen thickness. For the thin region, a lattice image is characteristic of the c-Si substrate below the interface, and changes abruptly to a granular image representative of amorphous SiO_2. Interface roughness is clearly observed. The ODP of this region shows a streak in the direction perpendicular to the interface. This is understood to be the termination effect of the periodic lattice image of c-Si at the interface. Conversely, for the thick region, a periodic image different from that of the c-Si substrate is observed between the granular image of SiO_2 and the lattice image of the c-Si substrate. The specific layer is about 4 monolayers (ML) thick, with strong 110 periodicity, and is observed along the roughened interface. This 110 period image is observed for various defocusing conditions. It is also noticeable that the ODP taken from the 110 period image exhibits anomalous 111 spot splitting in the direction perpendicular to the interface. This splitting cannot be explained by the simple termination effect of the atomic arrangement at the SiO_2/(001)Si interface.

SIMULATING AN HRTEM IMAGE NEAR AN SiO_2/Si INTERFACE

HRTEM images and ODPs are simulated in two models by the following methods.

The multiple electron diffraction in the specimen is simulated by the multi-slice method.[8] The cell size is 3.07 nm × 4.34 nm, which corresponds to 8 units in the [110] direction and 8 units in the <001> direction, respectively. a-SiO_2 region is approximated by a vacuum area in this simulation. The slice thickness for the multi-slice calculation is 0.192 nm.

The effect of lens aberrations and the defocusing condition of the HRTEM optical system are simulated by the method proposed by Horiuchi.[9] The electron

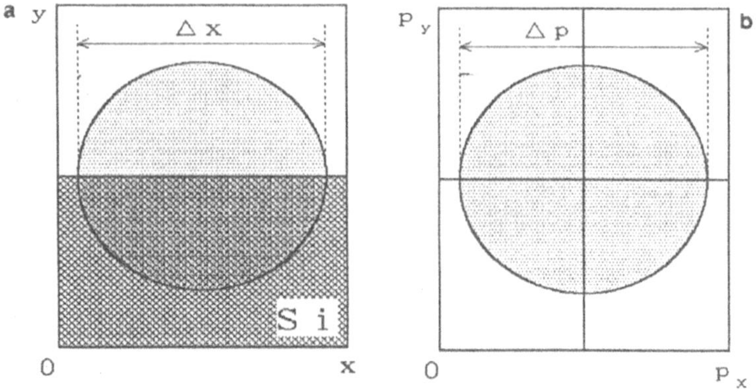

Figure 2. Distributions of the electron wave function (a) in the real space and (b) in the momentum space.

irradiation angle effect is, however, introduced as a broadening in the spacial frequency of the incident electron wave function. The probability density in the spacial frequency in the direction parallel to the SiO_2/(001)Si interface plane is approximated with a Gaussian function.

In the conventional artificial superlattice method, a planar electron wave is used for the incident electron beam of the multislice calculation, and the irradiation angle effect is simulated after the multislice calculation. Therefore, the simulated result reflects two interfaces formed at the top and bottom of the model, because of the discrete Fourier transformation property.

As shown in Figure 2, since the probability density of the incident electron beam in the real space is constrained within a certain area according to the Gaussian

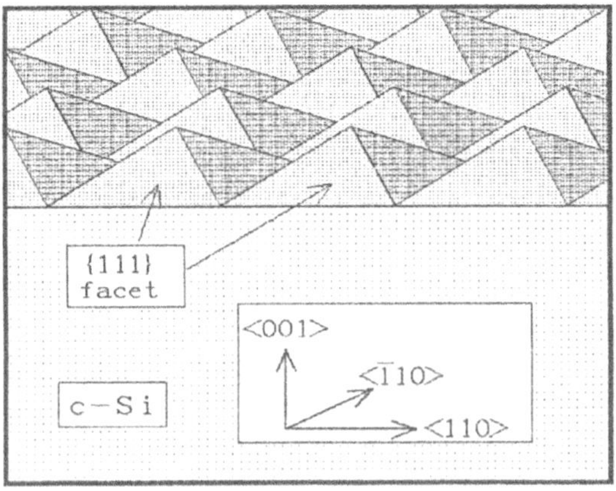

Figure 3. Schematic drawing of the interface morphology of the {111}-facet model.[7]

distribution of its spacial frequency in the momentum space, we can specifically simulate the diffraction of electrons which pass near the target SiO_2/Si interface in the present simulation method. The real HRTEM image corresponds to the summation of the intensity formed by electrons which have a limited areal distribution and irradiate various positions on a specimen surface. Here, the irradiation angle is assumed to be 0.5 mrad.

The ODP of the HRTEM image is simulated as a Fourier transformed image of the simulated HRTEM image.

HRTEM IMAGE OF A ROUGHENED SiO_2/(001)Si INTERFACE

Figure 3 shows a schematic drawing of the SiO_2/(001)Si interface morphology of the {111}-facet model.[7] Interface distortion energy calculation using the ball-and-stick models showed that distortion energy caused by the formation of the interface is lower in a {111}-facet model than in an atomically flat SiO_2/(001)Si interface.[6] In order to simplify the modeling, the heights of the Si protrusions are set to be the same while they are displaced randomly in a direction perpendicular to the observed direction, in order to avoid artificial alignment of the roughness. Three {111}-facet models with Si protrusions 4,6, and 8 ML high have been used.

The simulated HRTEM images of the {111}-facet models are shown in Figure 4.[7] The top row shows the images for the "thin" region while the bottom row shows those for the "thick" region. The far left column shows the corresponding experimental results.

For the thin models, there is no drastic change in the HRTEM image, even near the interface. The c-Si lattice image continues up to the interface. The image contrast becomes vague for regions with roughness. These features are very consistent with the experimental results. The apparent roughness height in the HRTEM images is smaller than the actual roughness height of the model.

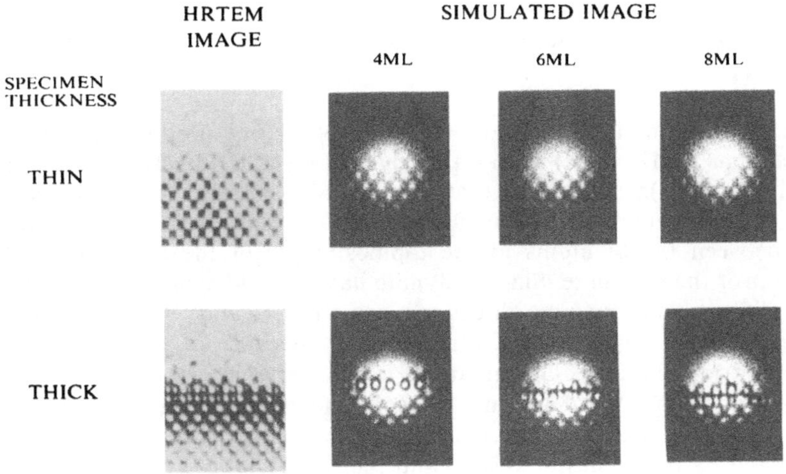

Figure 4. Results of HRTEM image simulation using {111}-facet models wit different roughness heights.[7]

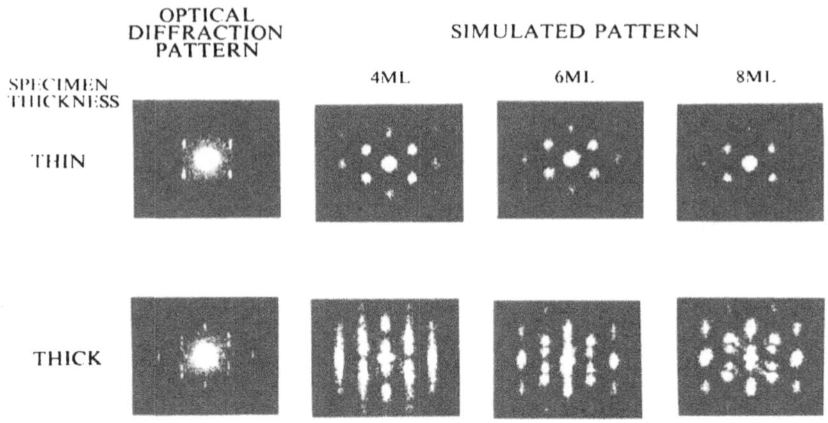

Figure 5. Result of optical-diffraction pattern simulation using the {111}-facet models with different roughness heights.[7]

On the other hand, for the thick models, an HRTEM image exhibiting strong 110 periodicity is reproduced in the roughness region. The best correspondence of the experimental images with simulated ones is obtained when the roughness height is more than 6 ML. For a roughness height of 8 ML, a normal c-Si lattice image appears above the 110 image. This is also agree well with experimental observation.

The simulated ODPs for the {111}-facet models are shown in Figure 5, along with the corresponding experimental results.[7] Although only the streak of each spot in the direction perpendicular to the interface is observed for the thin models, 111 spot splitting is not reproduced. For the thick models, 111 spot splitting is clearly reproduced. This is also evidence of good agreement between simulation and experiment.

HRTEM IMAGE OF AN SiO_2/(001)Si INTERFACE WITH THIN EPITAXIAL TRIDYMITE

The interface model is basically the same as the one proposed by Ourmazd et al.[10] The thin (2 ML thick) epitaxial tridymite layer is directly connected on an atomically flat (001)Si surface so that the $(1\bar{1}0)$Si is parallel to the (100)tridymite, and the (110)Si is parallel to the (002)tridymite, respectively. Oxygen atoms are inserted between the Si atoms in the topmost layer of the c-Si and in the first silicon layer of the tridymite. Since tridymite has two fold symmetry, we will denote the two different observation directions as $<110>$ and $<1\bar{1}0>$, respectively, following the notation used by Ourmazd et al. Because of the lattice mismatch between the tridymite and c-Si, the atoms in the tridymite are displaced from their equilibrium positions. We determine the coordinates of the atoms in the tridymite layer so as to minimize the interface distortion energy. The force constants are approximated by the values of a-SiO_2, and the equilibrium oxygen bond angle is set at 180°.

Figure 6 shows the planar defects in the tridymite layer. Since the compressed lattice constant of the epitaxial tridymite layer in the $<110>$ direction of the Si

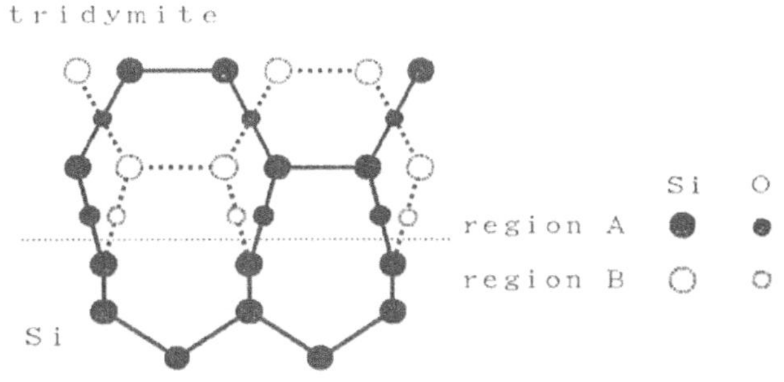

Figure 6. Two ways of the tridymite epitaxy on top of the (001)Si surface.

substrate is double that in the direction of the c- Si, there are two ways to connect the tridymite layer on top of the (001)c-Si surface by the shifting the tridymite layer in the <110> direction of the c-Si by the (110)Si interplanar distance, as shown in the figure. Here we denote the two regions with different type of connection as "region A" and "region B", respectively. The boundaries between the two regions are the planar defects in the tridymite layer.

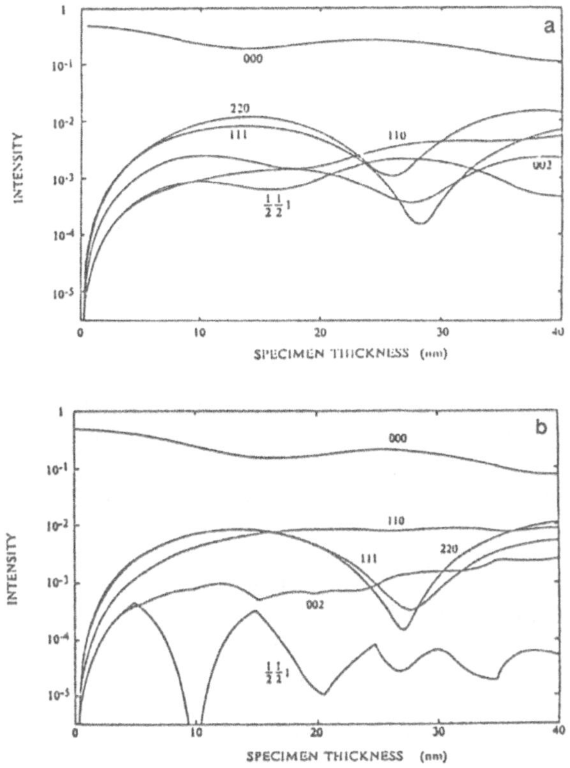

Figure 7. The variations in the diffraction spot intensity as a function of thickness (a) of the model with no planar defects and (b) of the model with planar defects of every 5 nm.

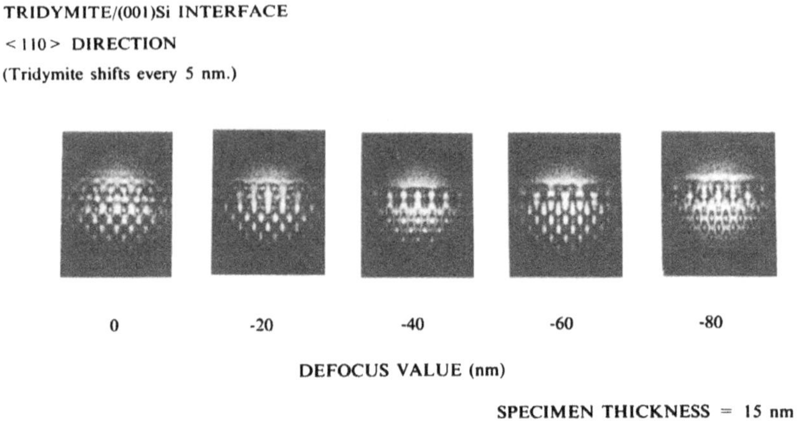

Figure 8. Simulated HRTEM images of a 15-nm-thick tridymite/(001)Si model with planar defects for various defocusing conditions.

Ourmazd et al. reported that they succeeded in reproducing the 110 period image by introducing planar defects in the tridymite layer.[10] However, they did not discuss the defect density in their paper but merely stated that it was "extensively observed in oxides." In this section, we will discuss the planar defect density needed to reproduce the 110 period image for various specimen thickness, and the dangling bond density near the SiO_2/(001)Si interface.

We made two models with epitaxial tridymite on the (001)Si surface. One has no planar defects, and the other has the planar defects of every 5 nm in the <110> direction.

Figure 7 shows the variation in the diffraction spot intensity as a function of the model thickness. The electron acceleration voltage is 400 keV for this simulation. For the model with no planar defects, the intensity of $\frac{1}{2}\frac{1}{2}1$ reflection is very strong. Thus the HRTEM image exhibits a period double that of the 110 period image. On the other hand, the intensity of the $\frac{1}{2}\frac{1}{2}1$ reflection is diminished when the thicknesses of "region A" and "region B" are almost the same for the model with planar defects. This is because the electron waves reflected from "region A" and "region B" interfere destructively with each other. When the total thicknesses of regions A and B are almost the same, the resulting HRTEM image exhibits a 110 period image because of the low $\frac{1}{2}\frac{1}{2}1$ reflection intensity. However, when the total thicknesses of regions A and B are different - for example, when the sample thickness is around 15 nm - the intensity of the $\frac{1}{2}\frac{1}{2}1$ reflection is still strong. Figure 8 shows the simulated HRTEM images of a 15-nm-thick model with planar defects for various defocusing conditions. The tridymite layer clearly exhibits the double period image except for the particular defocusing condition of -40 nm. These results show that planar defects every 5 nm in the <110> direction are not enough to reproduce the 110 period image at a specimen thickness of around 15 nm, where the 110 period image is observed experimentally.[10]

The simulation also reveals that a tridymite image with double period is observed for a very thin specimen. Therefore, if there is epitaxial tridymite at the

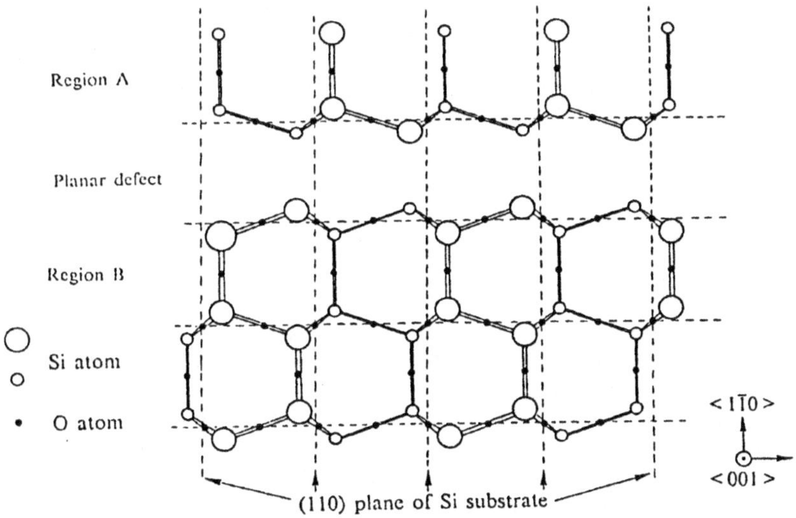

Figure 9. Atomic arrangement at the planar defect plane in the epitaxial tridymite grown over the (001)Si.

interface, the double period image of tridymite is frequently observed instead of the 110 period image. In order to reproduce the 110 period image for various specimen thicknesses, we have to assume very frequent planar defects in the tridymite layer.

As regards bonds in the tridymite, the planar defect is a source of a great number of dangling bonds because the bonds sticking out from "region A" and "region B" cannot be connected in the defect plane, as shown in Figure 9. If we assume that the planar defect density is $2x10^6/cm^2$ and that not all the bonds protruding from Si atoms into the defect planes are terminated, the dangling bond density is about $10^{14}/cm^2$. Therefore, the frequent planar defects which is required to exhibit the 110 period image increase the dangling bond density. This makes the interface unstable.

CONCLUSIONS

The specimen thickness dependence of the HRTEM images and their ODPs was reproduced by simulation, using a {111}-facet model. Particularly for the thick specimens, the 110 period image observed between the c-Si lattice image and the SiO_2 granular image was reproduced without assuming the existence of crystalline-phase SiO_2 between them. The height of the roughness of our sample is more than 6 ML.

In the case of the tridymite model, the planar defect density must be very high to exhibit the 110 period image for various specimen thickness. However, the dangling bond density increases with the planar defect density, because the defect plane is a source of dangling bonds.

In view of these results, we conclude that the SiO_2/(001)Si interface is roughened even though the initial Si surface is atomically flat. The driving force to form the roughness is the reduction of the interface distortion energy.

ACKNOWLEDGEMENTS

The authors would like to express their gratitude to S. Horiuchi of the National Institute for Research in Inorganic Materials for fruitful discussion on the HRTEM image simulation method.

REFERENCES

1. O.L. Krivanek, D.C. Tsui, T.T. Sheng, and A. Kamgar, A High Resolution Electron Microscopy Study of the Si-SiO_2 Interface, in:"Phisics of SiO_2 and Its Interface, " S.T. Pantelides, ed., Pergamon, New York (1978).
2. J.H. Mazur, R. Gronsky, and J. Washburn, Oxidation of Singular and Vicinal Surfaces of Silicon : The Structure of Si-SiO_2 Interface, Proc.SPIE 463:88 (1984).
3. A.H. Carim and A. Bhattacharyya,Si/SiO_2 Interface Roughness:Structural Observations and Electrical Consequences, Appl.Phys.Lett. 46:872 (1985).
4. S.M. Goodnick, D.K. Ferry, C.W. Wilmsen, Z. Liliental, D. Fathy, and O.L. Krivanek, Surface Roughness at the Si(100)-SiO_2 Interface, Phys.Rev. B32:8171 (1985).
5. N.M. Ravindra, D. Fathy, J. Marayan, J.K. Srivastava, and E.A. Irene, Two-step Oxidation Process in Silicon, Material Letters 4:337 (1986).
6. I. Ohdomari, H. Akatsu, Y. Yamakoshi, and K. Kishimoto, The Structural Models of the Si/SiO_2 Interface, J.Non-Cryst.Solids 89:239 (1987).
7. H. Akatsu, Y. Sumi, and I. Ohdomari, Evaluation of SiO2/(001)Si Interface Roughness using High-resolution Transmission Electron Microscopy and Simulation, Phys.Rev. B44:1616 (1991).
8. J.M. Cowley and A.F. Moodie, The Scattering of Electrons by Atoms and Crystals, Acta Crystallogro. 10:609 (1957).
9. S. Horiuchi. "High-Resolution Transmission Electron Microscopy," Kyoritsu, Tokyo (1988).
10. A. Ourmazd, D.W. Taylor, J.A. Rentschler, and J. Bevk, Si→SiO_2 Transformation : Interfacial Structure and Mechanism, Phys.Rev.Lett. 59:213 (1987).

DEPENDENCE OF SURFACE MICROROUGHNESS ON TYPES OF SILICON SUBSTRATES

T. Ohmi, T. Tsuga and J. Takano

Department of Electronic Engineering
Faculty of Engineering
Tohoku University
Sendai 980, Japan

Abstract

The dielectric breakdown field intensity is affected by the surface microroughness of silicon substrate. It has been found that the increase of surface microroughness due to the APM cleaning varies among the wafer types such as Cz, FZ and epitaxial(EPI). The surface microroughness is caused by the point defect such as vacancy cluster. In the case of the n type Cz wafer, the surface microroughness is increased even in the dilute HF cleaning, however this problem is perfectly resolved by inject H_2O_2 into the dilute HF.

Introduction

Cleanliness of silicon wafer surface is an essential requirement for deep submicron ULSI fabrication. In order to establish high performance future process technology, perfect control of surface and interface become very important. The ultraclean wafer surface must be free from particles, organic impurities, metallic impurities, native oxide, adsorbed impurity molecules and surface microroughness[1],[2]. So far, silicon substrate surface cleaning procedures have generally been carried out based on the concept of RCA cleaning[3], which has confirmed to eliminate the silicon wafer surface contaminations such as particles, organic and metallic impurities. It has already been reported that the surface microroughness of substrates is mainly caused by an alkaline solution treatment such as NH_4OH in $NH_4OH/H_2O_2/H_2O$ cleaning (APM cleaning) process and that the increase of surface microroughness is suppressed with $NH_4OH/H_2O_2/H_2O$ solution in which the NH_4OH content is reduced[4],[5]. It has been found that the increase of surface microroughness strongly affects on the electrical characteristics such as the dielectric breakdown field intensity and charge to breakdown of MOS diode having thin oxide film. The APM cleaning reduced NH_4OH mixing ratio has effectively removed the particles and metallic impurities on silicon surface[6]-[8]. Thus, although it has been cleared that the increase of surface microroughness caused by the alkaline solution treatment, the reason has been

The Physics and Chemistry of SiO_2 and the Si-SiO_2 Interface 2
Edited by C.R. Helms and B.E. Deal, Plenum Press, New York, 1993

unknown. In this paper, it is cleared that the increase of surface microroughness due to the APM cleaning varies among the wafer types such as Cz, FZ and epitaxial (EPI) wafers, and that the increase of surface microroughness is caused by the point defect such as vacancy cluster. Further, it is found that the increase of surface microroughness with the dilute HF, in particular on n type Cz wafer, is completely suppressed by injecting H_2O_2 into dilute HF solution, and that metallic impurity such as Copper is removed.

Surface microroughness in Wet Chemical Cleaning

Silicon wafer surface have generally been cleaned with wet chemical solutions based on the concept of RCA cleaning. The wet chemical cleaning processes consist of H_2SO_4/H_2O_2, $NH_4OH/H_2O_2/H_2O$ and $HCl/H_2O_2/H_2O$ solutions. Theses solutions have removed the silicon wafer surface contaminations such as particles, organic and metallic impurities. However, this procedure increases the surface microroughness. Figure 1 shows the typical STM images in 1x1μm^2 area, where Fig.1 (a) and (b) are n type Cz wafer surfaces before and after the RCA cleaning respectively. In this experiment, the RCA cleaning was employed with the conventional composition as follows : H_2SO_4(98%):H_2O_2 (30%)=4:1, NH_4OH(28%):H_2O_2(30%):H_2O=1:1:5, HCl(36%):H_2O_2(30%):H_2O=1:1:6, where these cleaning were carried out for 5 minutes at about 130°C, for 10 minutes at 85°C and for 10 minutes at 85°C, respectively. The n type Cz (100) wafer doped with phosphorus (P) was used. Its resistivity was 8–12 Ω/cm and average surface microroughness was approximately 0.1~0.2 nm. The surface microroughness was evaluated with Scanning Tunnelling Microscope (STM : SAM 3100, SEIKO Electric Inc.) before and after the cleaning process. As shown in Fig.1, there is a clear difference of the surface

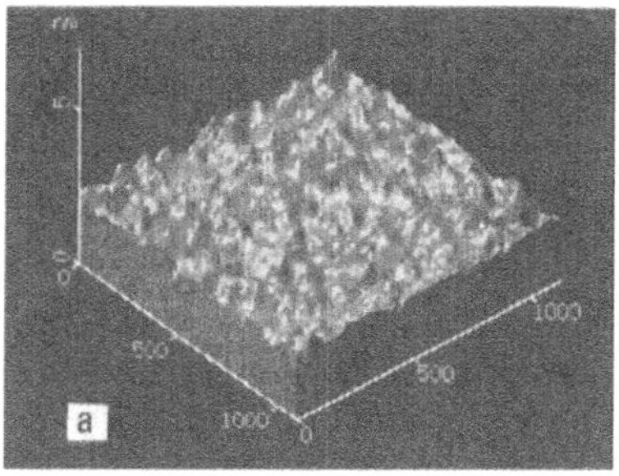

BEFORE CLEANING

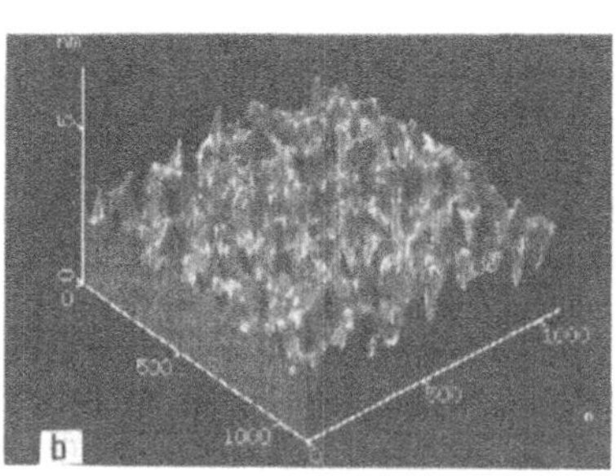

AFTER RCA CLEANING

Fig.1. Typical STM images of wafer surfaces before(a) and after(b) treated in an entire RCA cleaning having a mixing ratio of 1:1:5 in APM solution.

microroughness between the wafer before the cleaning (a) and the wafer after the cleaning (b). This result indicates that the increase of surface microroughness is caused by the wet chemical cleaning process.

It was investigated which process of the RCA cleaning affects the surface microroughness. Figure 2 shows the surface microroughness of the n type Cz wafer when it was treated with the H_2SO_4/H_2O_2 solution (SPM cleaning) having the mixing ratio of 4:1, the $HCl/H_2O_2/H_2O$ solution (HPM cleaning) having the mixing ratio of 1:1:6 and the $NH_4OH/H_2O_2/H_2O$ solution (APM cleaning) having the mixing ratio of 1:1:5. The SPM cleaning was repeated 4 times. Figure 2 indicates that neither the SPM cleaning nor the HPM cleaning increases the surface microroughness, and that the APM cleaning increases the surface microroughness. A slight increase of the surface microroughness shown in the case of the SPM cleaning is mainly caused by the 4 cycle repetition of the SPM cleaning and the advanced buffered hydrogen fluoride (BHF) cleaning, where the advanced BHF is buffered hydrogen fluoride having the NH_4F concentration of 17%, the HF concentration of 0.17% and the surfactant concentration of 400 ppm[9][10]. Figure 3 shows the surface microroughness of n type Cz wafer which was treated with the APM solutions for 10 minutes with the various mixing ratio from 1:1:5 to 0.05:1:5. The temperature during the APM cleaning was maintained at 80~90°C. As shown in Fig.3, the surface microroughness on the Cz wafer gradually increases as the NH_4OH mixing ratio in the APM cleaning solution increases. The increase and the fluctuation of the surface microroughness on the Cz wafer are found to be suppressed in the same control level when the NH_4OH mixing ratio in the APM cleaning solution is reduced to 0.05. It has already been confirmed that the APM cleaning can remove particles and metallic impurities in a more effective manner if the NH_4OH concentration in the APM solution is reduced to 5% of the conventional level. However, the surface microroughness increased after going through the entire RCA cleaning process even if the NH_4OH mixing ratio in the APM solution was reduced to the

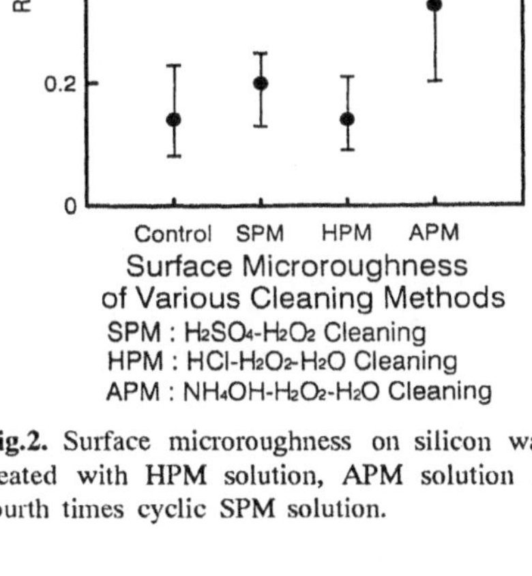

Fig.2. Surface microroughness on silicon wafer treated with HPM solution, APM solution and fourth times cyclic SPM solution.

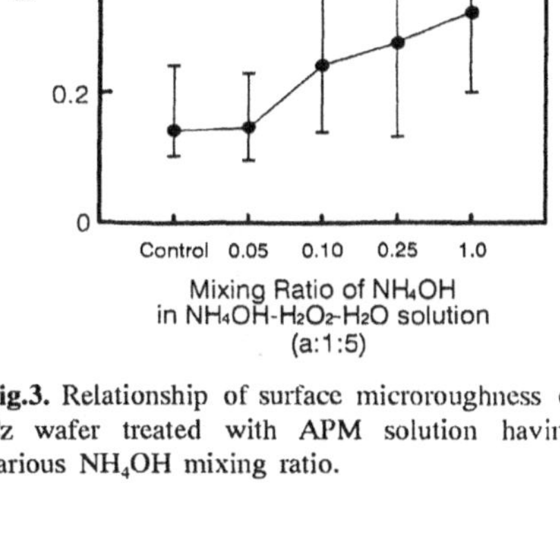

Fig.3. Relationship of surface microroughness of Cz wafer treated with APM solution having various NH_4OH mixing ratio.

level of 0.05:1:5. The surface microroughness increased in the hot ultrapure water rinsing of the conventional RCA cleaning procedure. When a room temperature ultrapure water rinsing was introduced right after the APM cleaning, the surface microroughness did not increase at all. Therefore, in order to maintain the surface smoothness, the room temperature ultrapure water rinsing is definitely required right after the APM cleaning, followed by the hot ultrapure water rinsing[6]-[8]. Further it has been revealed that this combination of room temperature ultrapure water rinsing and the hot ultrapure water rinsing makes it easy to remove organic materials[11]-[13].

Next it will discuss the native oxide removal process which is the final step of the wet cleaning process. At present, the native oxide is mainly removed with the diluted HF cleaning. However, it has been pointed out that in order to add the capability to remove Cu adsorbed onto the surface, the HF–H_2O_2 cleaning is more effective[14][15]. Figure 4 shows the relationship between the H_2O_2 concentration in the HF–H_2O_2 cleaning and the surface microroughness of n type Cz wafer. In this experiment, the HF concentration was fixed at 0.5% and the H_2O_2 concentration was varied from 0% to 10%. The immersion time was set at 10 minutes. It is demonstrated that the surface microroughness increases when the wafer is treated with diluted HF solution without containing H_2O_2. When the H_2O_2 concentration in the diluted HF exceeded 0.1%, the surface microroughness is suppressed

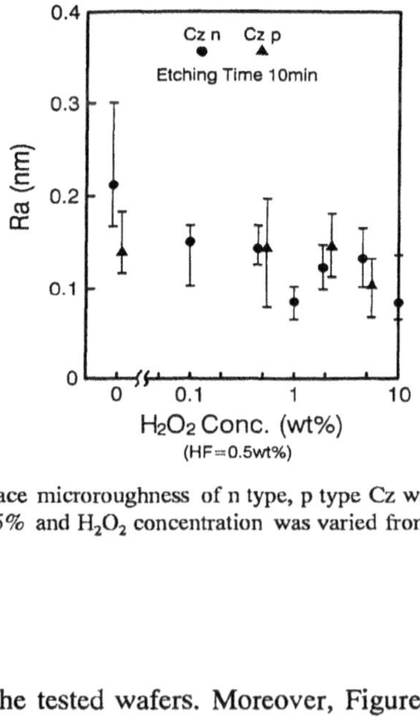

Fig.4. Relationship between surface microroughness of n type, p type Cz wafers and HF–H_2O_2 cleaning. The HF concentration was fixed at 0.5% and H_2O_2 concentration was varied from 0 to 10%. The etching time was set at 10 minutes.

lower than 0.13 nm in all the tested wafers. Moreover, Figure 5 shows that the HF–H_2O_2 cleaning can remove Cu, which has not been removed with the diluted HF cleaning. In this experiment, the Si surface was contaminated with Cu and it was treated with two different cleaning processes as follows : the diluted HF cleaning and HF–H_2O_2 cleaning. Then the Cu concentration on the silicon surface was measured with Total Reflection X-Ray Fluorescence Spectrometry (TRXRF : TREX 610, Technos Inc.). Figure 5 indicates that the diluted HF cleaning can hardly remove Cu but that the HF–H_2O_2 cleaning can effectively remove Cu as the H_2O_2 concentration increases. Besides higher Cu removal efficiency can be obtained when the HF–H_2O_2 cleaning time is increased from 1 minute to 60 minutes. It has been found that employing the HF–H_2O_2 cleaning to remove the native oxide not only enhances the Cu removal efficiency but also maintains the surface microroughness more effectively.

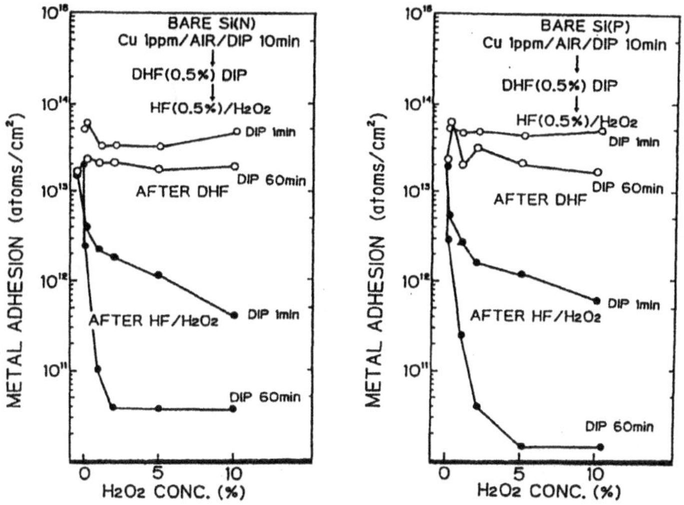

Fig.5. Metallic impurity (Cu) removal efficiency by HF–H_2O_2 cleaning and diluted HF cleaning. HF concentration was 0.5%, H_2O_2 concentration varied from 0 to 10%, and dipping time were set at 1 minutes and 60 minutes.

Influence of Surface Microroughness on Electrical Characteristics of Thin Oxide Films

As the semiconductor device dimensions are getting smaller, the thickness of oxide films such as gate oxide and storage capacitor oxide is continuously going to decrease. The thin oxide film quality must be improved to guarantee device performance and reliability. In this experiment, it is revealed that the increase of surface microroughness degrades the electrical characteristics of MOS diodes having the thin oxide film. Figure 6 shows the dependence of dielectric breakdown field intensity (E_{BD})of MOS diodes on the average surface microroughness (Ra) of Cz, FZ and epitaxial (EPI) wafers. The wafers having various levels of surface microroughness were oxidized at 900° C under the ultraclean purity oxygen ambience. The MOS diodes having an area of $1.6x10^{-4}cm^2$ have been fabricated by evaporating aluminum films on oxide films having the thickness of 10 nm for substrates having various surface microroughness. The judgment electric current was $1x10^{-4}$ A. As shown in Fig.6, the electric breakdown field intensity (E_{BD}) gets higher as the surface microroughness decreases. Moreover, no low field breakdown phenomena such as A and B mode breakdown[2] are observed on the wafer with small Ra. It is seen from Fig.6 that the reliability of the gate oxide primarily depends on the surface microroughness. It has been revealed that the Si surface must be as smooth as possible in order to improved the reliability of thin oxide. Besides it has been found the surface microroughness on the $Si-SiO_2$ interface affects not only the breakdown voltage of oxide but also the charge to breakdown(Q_{BD}) and the electron channel mobility of MOSFET[6]-[8],[16].

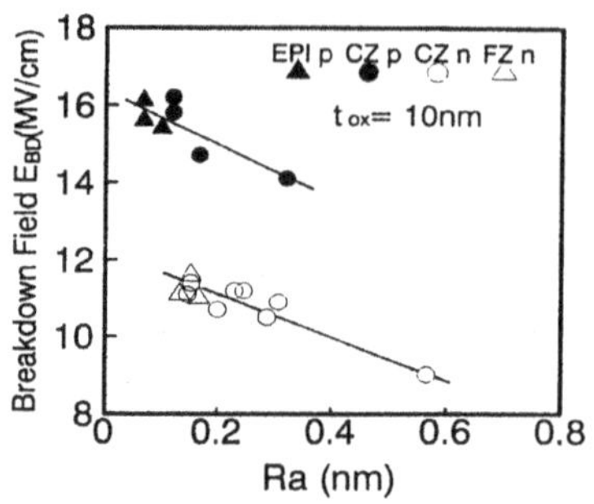

Fig.6. Dielectric breakdown field intensity(E_{BD}) as a function of average surface microroughness having oxide thickness of 10 nm formed on n type Cz, n type FZ, p type Cz and p type EPI wafers.

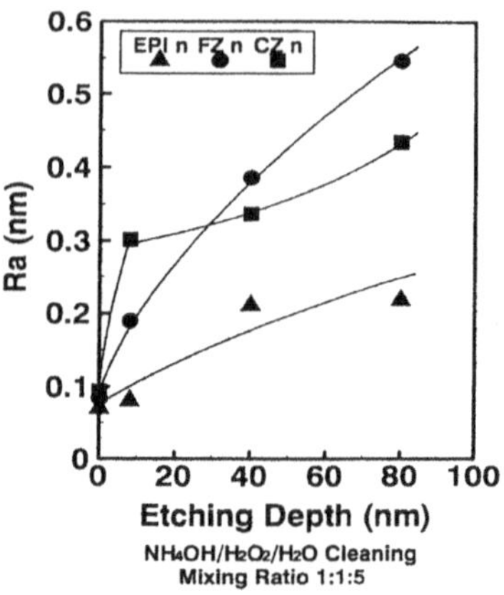

Fig.7. Surface microroughness on n type Cz, FZ and EPI wafers treated with cyclic APM cleaning ($NH_4OH:H_2O_2:H_2O$=1:1:5).

Dependence of Vacancy Cluster into Silicon Wafer of Different Types on Surface Microroughness

The previous sections pointed out the important conditions for the cleaning process to suppress the increase of surface microroughness and the dependence of electrical characteristics of thin oxide films on surface microroughness. In this section, it will be revealed that the difference of silicon wafer types such as Cz, FZ and epitaxial (EPI) wafers affects on the increase of surface microroughness in the cleaning process. And it will be explained the reason why the surface microroughness increases on the silicon wafers of different types. Figure 7 shows the surface microroughness as a function of silicon etching depth by the repetition of the APM cleaning for 10 minutes at 85°C, and Fig.8 shows the STM images of Cz, FZ and EPI wafers after 10 times repetition of APM cleaning. The mixing ratio of APM cleaning solution was $NH_4OH:H_2O_2:H_2O$=1:1:5. The increase of surface microroughness in EPI wafer is very much limited, while the surface microroughness of FZ and Cz wafers gradually increases as the wafer is etched.

As we have seen from Figs.7 and 8, the increase of surface microroughness depends on difference of silicon wafer types. This is supposed that the increase of surface microroughness is caused by the difference of wafer quality such as silicon vacancy cluster concentration. We consider that the silicon vacancy is not point defect, because the surface microroughness does not immediately increase by the repetition of APM cleaning. In addition to this, we consider that the increase of surface microroughness depends on the silicon vacancy cluster concentration more than the oxygen concentration strongly. The oxygen concentration in Cz and FZ wafers is $10^{18}cm^{-3}$ and $10^{16}cm^{-3}$, respectively. Despite the concentration difference between them is $10^{2}cm^{-3}$, the behaviors of surface microroughness changing of them to the repetition number of APM cleaning are not so different. In order to satisfy the effect of silicon vacancy cluster concentration, phosphorus and antimony atoms were diffused into Cz, FZ and EPI wafers. Figure 9 shows the amount of phosphorus atoms diffused into Cz, FZ and EPI wafers at 900°C for 30 min in N_2 gas[17]. The phosphorus atoms were diffused into these wafers from PSG films having 500 nm thickness, where the PSG films were deposited with normal pressure chemical vapor deposition (NP–CVD) under atmospheric pressure at 300°C (5%PH_3/N_2:5%SiH_4/N_2=1:5 volume). The amount of diffused phosphorus atoms was obtained by repeating the anodic oxidation and four–point probe measurements. The amount of phosphorus atoms in EPI wafer is less than that of Cz or FZ wafers. It has been supposed that the amount of phosphorus atoms diffused into silicon wafer depends on the amount of silicon vacancy cluster. In order to confirm this fact, antimony atoms were diffused into these wafers. Because the diffusion of antimony atoms into wafer strongly depends on the amount of vacancy into silicon wafer. Figure 10 shows the amount of diffused antimony atoms in these wafers at 1000°C for 24 hours in N_2 gas. The antimony atoms were diffused from OCD film (Sb–20220–SG, Tokyo Ohka Kogyo Co., Ltd.) having 370nm thickness into these wafers. The amount of diffused antimony atoms was also obtained by repeating the anodic oxidation and four–point probe measurements. The amount of antimony atoms in EPI wafer is also less than that of Cz or FZ wafers. These indicate that silicon vacancy concentration in EPI wafer is considered to be lower compared with Cz or FZ wafers, because, in particular, the diffusion of antimony is considered to be mainly promoted through vacancy model. Therefore, it is concluded that the surface microroughness strongly depends on the silicon vacancy concentration.

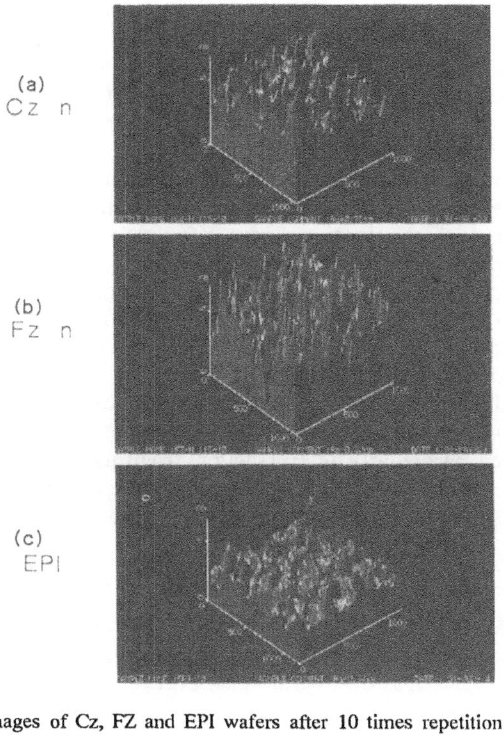

Fig.8. The STM images of Cz, FZ and EPI wafers after 10 times repetition of APM cleaning.

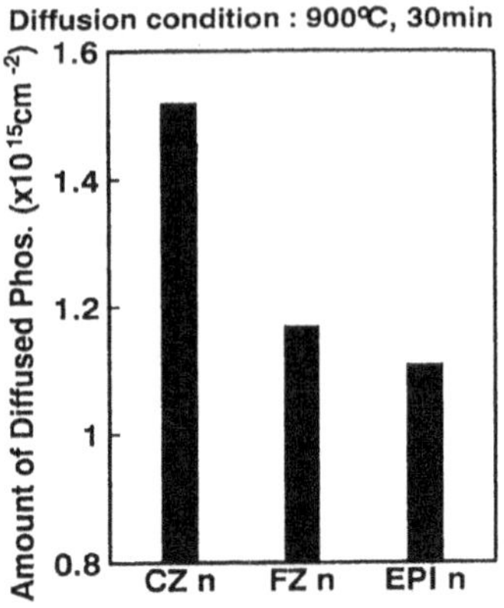

Fig.9. The amount of phosphorus atom diffused into Cz, FZ and EPI wafers at 900°C for 30 minutes.

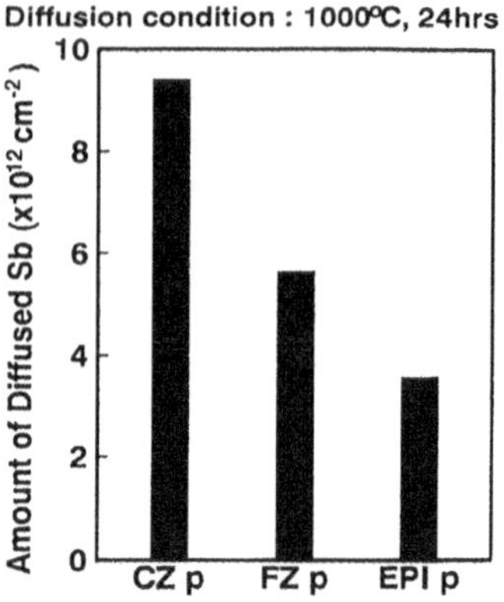

Fig.10. The amount of antimony atom diffused into Cz, FZ and EPI wafers at 1000°C for 24 hours.

Conclusion

It has been found that the quality of thin oxide is strongly affected by the microroughness of the substrate surface: the breakdown field intensity(E_{BD}) gets higher as the substrate surface gets smoother. This means that the improvement of the surface smoothness is the first step to improved the reliability of the ULSI devices. It has been revealed that the smoothness of the substrate surface is degraded in the wet chemical cleaning process, in particular in the APM cleaning ($NH_4OH:H_2O_2:H_2O$ = 1:1:5). It has been found that the degradation of the surface microroughness can be completely suppressed by changing the mixing ratio of APM cleaning to 0.05:1:5 and by introducing the room temperature ultrapure water rinsing between the APM cleaning and the hot ultrapure rinsing which is effective to remove NH_4OH adhering on wafers and wafer carriers. In the case of the n type Cz wafer, the surface microroughness is found to be degraded even in the diluted HF cleaning. However this problem is perfectly resolved by injecting over 0.1% H_2O_2 into diluted HF. It has been found that the diluted HF cleaning should be replaced with the HF-H_2O_2 cleaning in order to remove metallic impurities such as Cu. The electron negativity of these metallic impurities is higher than that of silicon and they can not be removed in the diluted HF cleaning presently employed at the final stage of wet process. We have demonstrated that the increase of the surface microroughness strongly depends on the silicon vacancy concentration in wafer. The increase of the surface microroughness is confirmed to be limited particularly in EPI wafer, resulting in highly reliable thin oxide formation in EPI wafer. These results been recommend the epitaxial substrate as the wafer for the fabrication of future sub-half-micron ULSI devices.

Acknowledgment

This work was carried out at the Super Clean Room of the Laboratory for Microelectronics, Research Institute of Electrical Communication, Tohoku University

References

[1]T.Ohmi,Technical Digest,1989,International Electron Devices Meeting, Washington, D.C., pp.49-52, December 1989.

[2]T.Ohmi, Microelectronic Engineering, vol.10, pp.163–176, 1991.
[3]W.Kern and D.A.Pauotien, RCA Review 31, pp.187–205, June 1970.
[4] T.Ohmi, Foresightedness in RCA Cleaning Concept and Their Present, 8th Workshop on ULSI ultra clean technology Advanced Wet Chemical Processing 1 proceeding, pp.5–15, Dec.,1990.
[5]H.Mishima, T.Yasui, T.Mizuniwa, M.Abe and T.Ohmi, IEEE Transaction on Semiconductor Manufacturing, 2, [3] pp.65–75, August 1989.
[6]M.Miyashita, M.Itano, T.Imaoka, I.Kawanabe and T.Ohmi, Extended Abstracts of 179th ECS meeting, Washington, DC, May 1991.
[7]M.Miyashita, M.Itano, T.Imaoka, I.Kawanabe and T.Ohmi, Digest of Technical Papers, 1991 Symposium on VLSI Technology, Oiso, pp.45–46, May 1991.
[8]T.Ohmi, M.Miyashita, M.Itano, T.Imaoka and I.Kawanabe, "Dependence of Thin-Oxide Films Quality on Surface Microroughness", IEEE Trans. on Electron Devices, vol.39, No.3, pp.537–545, March 1992.
[9]H.Kikuyama, N.Miki, K.Saka, J.Takano, I.Kawanabe, M.Miyashita and T.Ohmi, IEEE Transactions on Semiconductor Manufacturing, vol.3, No.3, pp.99–108, August 1990.
[10]H.Kikuyama, N.Miki, K.Saka, J.Takano, I.Kawanabe, M.Miyashita and T.Ohmi, IEEE Transactions on Semiconductor Manufacturing, vol.4, No.1, pp.26–35, February 1991.
[11]R.Rosenberg, M.Itano, I.Kawanabe, F.W.Kern.Jr, M.Miyashita and T.Ohmi, Proceedings of Microcontamination 90, pp.273–291, October 1990.
[12]M.Itano, F.W.Kern,Jr, M.Miyashita, I.Kawanabe, R.W.Rosenberg and T.Ohmi, submitted to IEEE Trans. on Semiconductor Manufacturing.
[13]M.Itano and T.Ohmi, submitted to IEEE Trans. on Semiconductor Manufacturing.
[14]T.Shimono, M.Morita, Y.Muramatsu, M.Tsuji, 8th Workshop on ULSI Clean Technology, Advanced Wet Chemical Processing I, pp.59–68, December 1990.
[15]Y.Shimanuki, J.Ryuta, E.Morita, H.Furuya and T.Tanaka Proceedings of STEP/Microroughness '91, Tokyo, April 1991.
[16]T.Ohmi, K.Kotani, A.Teramoto and M.Miyashita, "Dependence of Electron Channel Mobility on Si–SiO_2 Interface Microroughness", IEEE Electron Device Letters, vol.12, No.12, December 1991.
[17] M.Yoshida, Oyo Buturi vol.50, No.1, pp.2–17, 1981.

THE EFFECT OF SURFACE ROUGHNESS ON GATE OXIDE LEAKAGE CURRENTS

Mark Chonko[1] and Vidya Kaushik[2]

[1] Advanced Product Research and Development Laboratory
Motorola, Inc. 3501 Ed Bluestein Blvd., Austin TX 78721

[2] Technology Assessment Laboratory, Motorola, Inc.
3501 Ed Bluestein Blvd., Austin TX 78721

ABSTRACT

We have studied the effects of silicon surface roughness on 150 Å gate oxide characteristics. Varying degrees of roughness were induced, before oxidation, by timed wet silicon etching. For short etch times, the initial roughness is reproduced at the top oxide surface while the substrate interface is smoothed. Roughening results in increased electron injection from the poly/oxide interface. Injection from this interface increases rapidly with increasing roughness (etch time). At longer times, the substrate interface begins to retain the etched roughness and its injection level also begins to increase. The non-symmetric evolution of IV characteristics can be used to qualitatively evaluate process induced surface roughness. This technique is used to compare the potential of two common oxide etching solutions to roughen a silicon surface.

INTRODUCTION

Roughness at the SiO_2/Si interface and its consequences have been studied extensively. Although most of the work has concentrated on oxidation induced roughness,[1-3] recent results have shown the importance of the impact of pre-oxidation cleaning chemistries.[4,5] We have studied the effect of pre-oxidation surface roughness on the characteristics of a 150 Å gate oxide. Surface roughness was induced by timed wet silicon etches. The electron injection characteristics for the poly interface and substrate interface evolve differently as the roughness (etch time) increases. For short times, the poly interface retains the roughness of the initial surface, resulting in increased electron injection currents. The substrate interface is smoothed by

The Physics and Chemistry of SiO_2 and the Si-SiO_2 Interface 2
Edited by C.R. Helms and B.E. Deal, Plenum Press, New York, 1993

the oxidation and its electron injection levels remain constant. At longer etch times, the substrate interface retains the etched surface roughness causing the IV characteristics to degrade as well. This asymmetric development of IV characteristics can be used effectively to qualitatively evaluate process induced surface roughness. It has been used to demonstrate the superiority of buffered HF solutions over H_2O:HF solutions with respect to surface roughening.

EXPERIMENTAL

MOS capacitors were fabricated on 125 mm, <100>, n-type silicon. A Framed Mask Poly Buffered LOCOS isolation process was used.[6] After removing a sacrificial oxide in 10:1 buffered oxide etch, the wafers were etched for various times in a 750:1 Nitric:HF solution to roughen the substrate. Gate oxides of 150 Å thickness were grown at 900°C in 3% HCl. Prior to oxidation a modified RCA cleaning process was used. Phosphorus diffused (PH_3) polysilicon was deposited, patterned and RIE etched to form the gate electrodes.

Current-voltage measurements were performed with an HP4145 parametric analyzer under illumination. TEM micrographs were obtained using a JEOL 2000 FX II microscope.

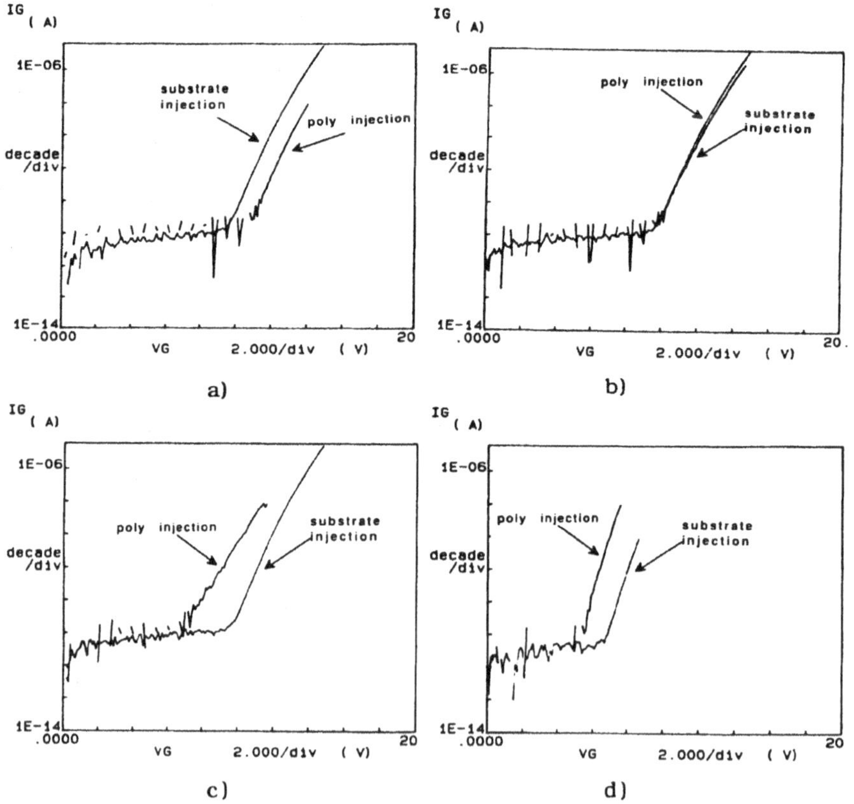

Figure 1. IV curves for gate poly injection and substrate injection for a) 0 seconds b) 30 seconds c) 60 seconds and d) 120 seconds of 750:1 nitric:HF etching prior to oxidation

RESULTS

Figure 1 shows the current-voltage characteristics for samples etched up to 120 seconds in the 750:1 solution. The polysilicon interface rapidly becomes leaky with etch time, however, the substrate interface shows increased electron injection only for the longest etch time. The relationship between the two interfaces is emphasized in Figure 2. Here the electric field required to maintain a fixed current density of 1.5×10^{-5} A/cm^2 is shown for both polarities as a function of etch time.

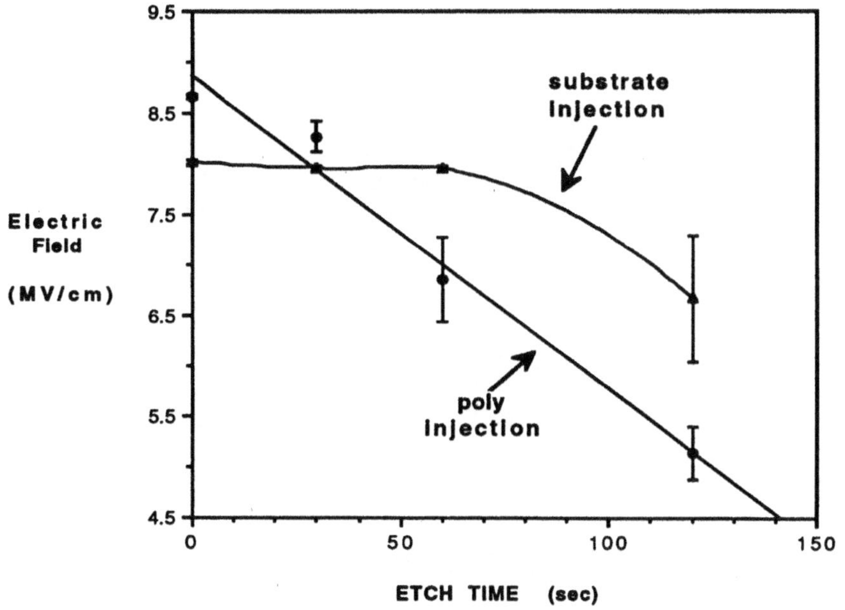

Figure 2. The electric field required for a current density of 1.5×10^{-2} A cm^{-2}, for both injection from both interfaces as a function of etch time.

DISCUSSION

The asymmetry of the IV characteristics with polarity is similar to that of oxides grown on polysilicon where the current enhancement is due to high electric fields at asperities.[7] The results presented above can also be explained in terms of interface roughness. The 750:1 nitric:HF etch leaves the unoxidized substrate rough. This roughness is maintained at the free surface of the growing oxide. Here any sharp features will result in increased electron current when the gate electrode is biased negatively. The substrate smoothing is caused by the rounding of sharp features by the gate oxidation process.[8] This process continues as the amount of roughness increases so that the gate injection is enhanced while the substrate injection remains at the level characteristic of unetched silicon. Only when the roughness becomes severe is the substrate injection enhanced. For this case, the resulting IVs are very similar to those of oxides grown on plasma-etched surfaces.[9,10] Figure 3 is a TEM cross-section of a capacitor which received a 60 second 750:1 etch. The poly/oxide interface shows a much greater level of short range roughness than does the substrate interface, as reflected in the electrical data.

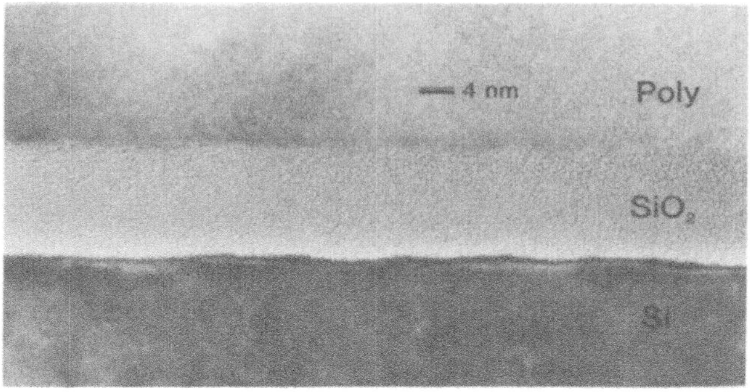

a

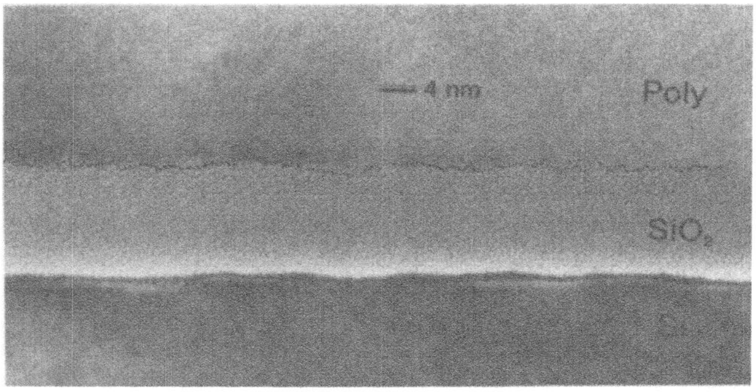

b

Figure 3. a)TEM cross-sections showing both interfaces of a 60 second etch sample. b) shows the poly/oxide interface highlighted.

This asymmetric response of the IV curves can be used to qualitatively evaluate process induced surface roughness. The ability of two common oxide etchants to roughen a silicon substrate was investigated using this technique. A 150 Å sacrificial oxide was grown and removed using a 10:1 Buffered Oxide Etch or a 10:1 H_2O:HF solution. Wafers were subjected to various amounts of overetch and 150 Å gate oxide capacitors were then fabricated. The injection characteristics are shown in Figure 4. It is clear that the 10:1 H_2O:HF roughens the silicon substrate more than the BOE. This is supported by recent atomic force microscopy and scanning tunneling microscopy results.[11,12] The use of H_2O:HF solutions may have an adverse impact on the quality of very thin gate oxides.

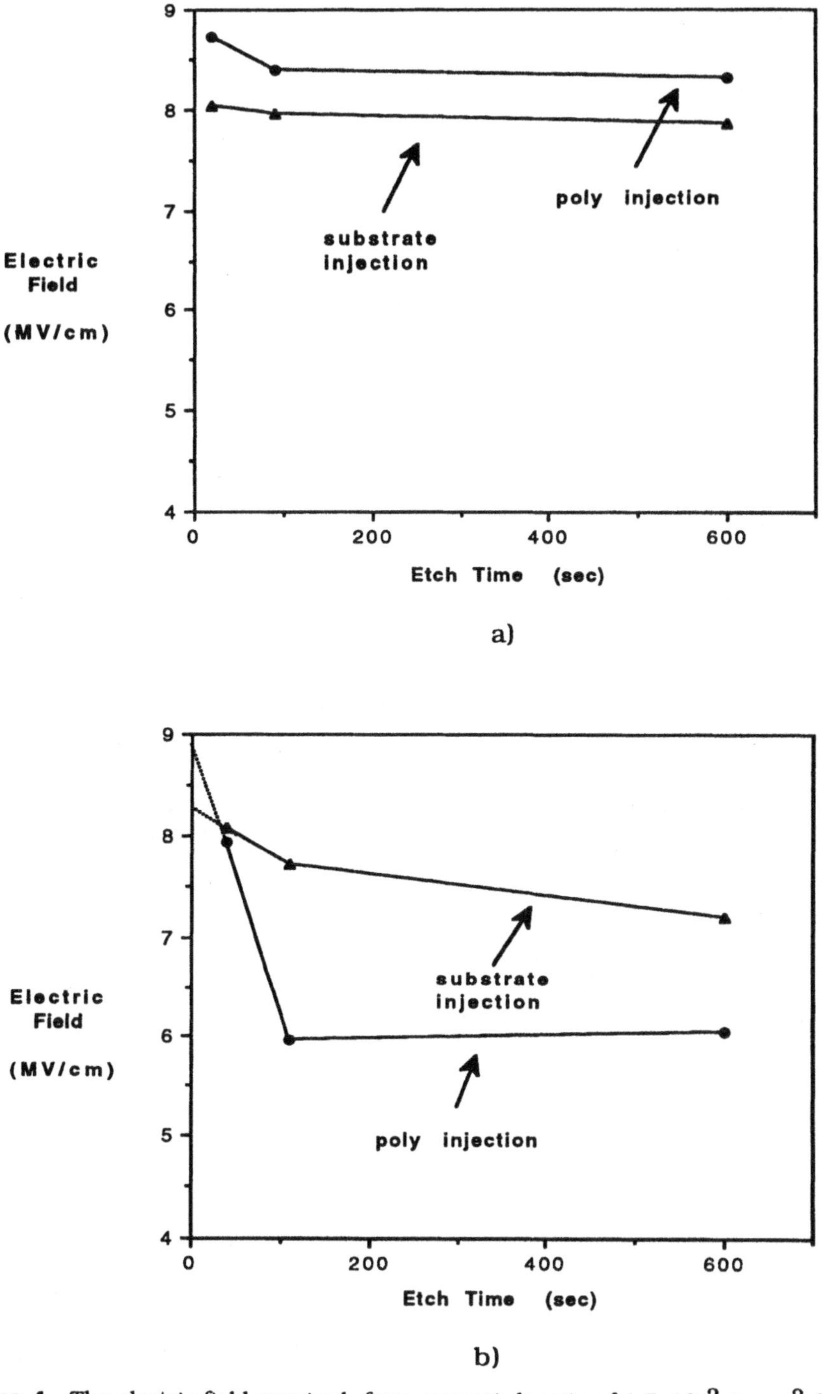

Figure 4. The electric field required for a current density of $1.5x10^{-2}$ A cm^{-2} for two different oxide etchants. a) 10:1 BOE and b) 10:1 H2O:HF

SUMMARY

We have investigated the effect of surface roughness on the IV characteristics at both the substrate and gate electrode interfaces. For relatively small degrees of roughness, the free surface of the growing oxide film retains the character of the unoxidized substrate while the silicon interface is smoothed. This results in increased leakage currents when electron injection occurs from the gate electrode as compared to substrate injection. The resulting asymmetry in the IV characteristics can be used to qualitatively evaluate process induced surface roughness and it was demonstrated that buffered oxide etches are less damaging to silicon substrates than H_2O:HF solutions.

ACKNOWLEDGEMENTS

Valuable discussions with Phil Tobin are gratefully acknowledged.

REFERENCES

1. A.H Carim and R Sinclair, *J. Electrochem. Soc.*, **134**, 741 (1987).
2. P.O. Hahn et al, *The Physics and Chemistry of SiO2 and the SiO2 Interface*, ed. by C.R. Helms and B.E. Deal, 401, (Plenum, New York, 1988).
3. G. Hollinger and R. Saoudi, *ibid*, 211.
4. M. Miyashita et al, *1991 Symposium on VLSI Technology*, **6-2**, 45 (1991).
5. M. Meuris et al, *ECS Extended Abstracts*, **91-2**, 775 (1991).
6. B.Y. Nguyen et al, *ECS Extended Abstracts*, **90-1**, 614 (1990).
7. R.M. Anderson and D.R. Kerr, *J. of App. Physics*, **48**, 4834 (1977).
8. K. Yamabe and K. Imai, *IEEE Trans. on Electron Dev.*, **ED-34**, 1681 (1987).
9. J.P. Gambino et al, *The Physics and Chemistry of SiO2 and the SiO2 Interface*, ed. by C.R. Helms and B.E. Deal, 445, (Plenum, New York, 1988).
10. Y. Fong et al, *IEEE Trans. on Electron Dev.*. **37**, (1990).
11. Y. Kim and C.M. Lieber, *J. Am. Chem. Soc.*, **113**, 2333 (1991).
12. G.S. Higashi et al, *Appl. Phys. Lett.*, **58**, 1656 (1991).

A DOUBLE SACRIFICIAL OXIDE PROCESS FOR SMOOTHER 150 Å SiO_2 GATE OXIDE INTERFACES

Hsing-Huang Tseng and Philip J. Tobin

Motorola Inc.
Advanced Products Research and Development Laboratory
MD: K-10, 3501 Ed Bluestein Blvd., Austin, Texas 78721

ABSTRACT

We have found that a double sacrificial oxide process provides a smoother active surface than the single sacrificial oxide process. The onset of Fowler-Nordheim tunneling (injected from polysilicon/gate oxide interface) for gate oxide grown using a double sacrificial oxide process occurs at a higher field than that for the single sacrificial oxide case. This implies that the double sacrificial oxide process results in a smoother polysilicon/gate oxide interface. Further, the interface state density of the gate oxide/substrate interface is reduced for the double sacrificial oxide process.

INTRODUCTION

As the fabrication process utilized in submicron technology becomes more complicated, process-induced gate oxide damage may result in an increased incidence of low electrical field breakdown, trap generation in the bulk and increased interface state density. The gate oxide damage may in turn result in device degradation problems such as threshold voltage scatter. In this paper, a double sacrificial oxide process [1] is shown to reduce the gate oxide area defect density and the interface state density dramatically as compared with a single sacrificial oxide process.

DEVICE FABRICATION

After etching an 800 Å first sacrificial oxide grown at 1000 °C in dry O2 and HCl ambient, a 150 Å second sacrificial oxide is grown in dry O2 and HCl ambient at 900 °C and etched. This double sacrificial oxide process is compared to a single sacrificial oxide process consisting of the first oxide growth and etch only. Capacitors are built using a 0.5 μm technology [2] to study the oxide characteristics. The effect of the second sacrificial oxide on the roughness of two interfaces (polysilicon/gate oxide and gate oxide/silicon substrate) and defect density reduction are investigated.

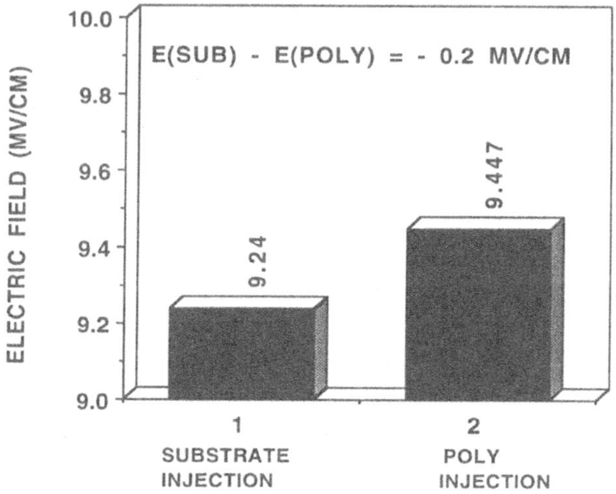

Figure 1. Electric field for F-N tunneling at J= 1.6 E-3 A/sq cm. Double sacrificial oxide process

RESULTS AND DISCUSSION

Figures 1 and 2 show the electric field for Fowler-Nordheim tunneling at a current density of 1.6×10^{-3} A/cm^2 for single and double sacrificial oxide process, respectively. Here the current is injected for both polarities. The electric fields for substrate injection are close for both single and double sacrificial oxide processes. However, the electric fields for current injected from the polysilicon/SiO2 interface for the single sacrificial oxide process is lower than that for the double sacrificial oxide process.

Figure 3 shows typical current-voltage curves with current injected from the polysilicon/SiO2 interface for samples with single and double sacrificial oxide process. For a fixed current density, the single sacrificial oxide process results in a smaller voltage than the double sacrificial oxide case. This implies that the polysilicon/SiO2 interface is rougher for the single sacrificial oxide process than for the double process. It has been reported that the degree of roughness of the polysilicon/SiO2 interface reflects the active surface roughness level before growing gate oxide [3]. Therefore the double sacrificial oxide process results in a smoother active surface prior to gate oxide growth than the case of a single sacrificial oxide process.

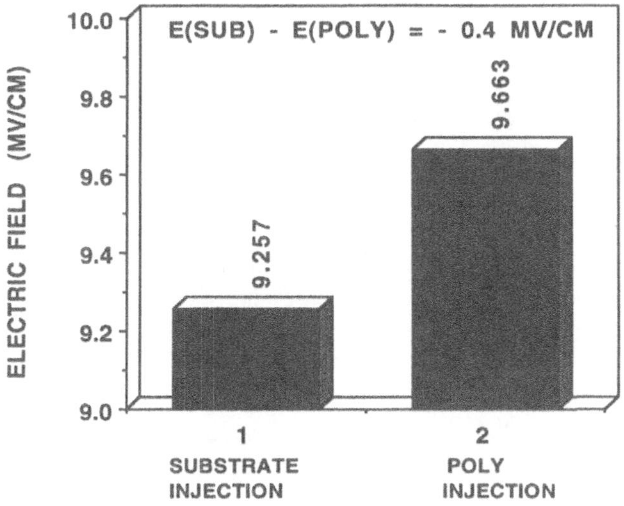

Figure 2. Electric field for F-N tunneling at J= 1.6 E-3 A/sq cm. Single sacrificial oxide process

A reduction in roughness also occurs at the atomic scale at the SiO2/substrate interface, since the mid-gap interface state density (Dit) for the double sacrificial oxide process is 2.5X smaller [4] and also has a 4X smaller standard deviation than that for the single sacrificial oxide process (Figure 4).

The breakdown histograms of a relatively large area capacitor (0.018 cm^2) are shown in Figure 5. There is a 2.5X lower incidence of low voltage failures for gate oxide grown after the double sacrificial oxide process demonstrating the beneficial effect. The smooth SiO2/Si interface results in fewer defects for the double sacrificial oxide process.

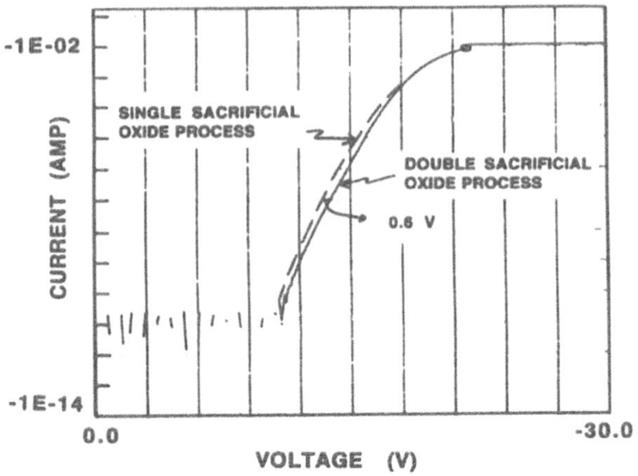

Figure 3. Current-voltage curves with current injected from the polysilicon/SiO2 interface

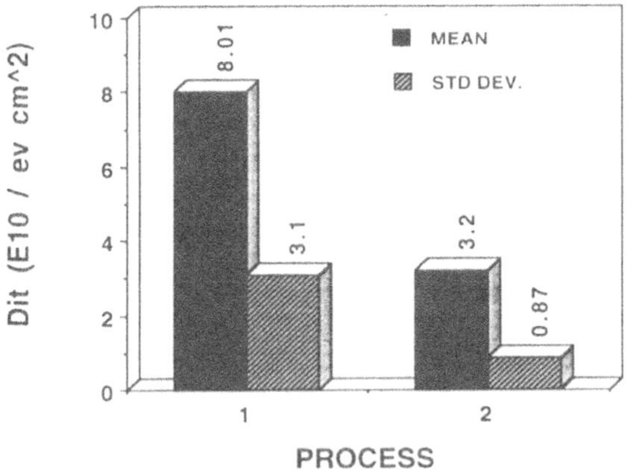

Figure 4. Interface state density comparison. Process 1: single sacrificial oxide process. Process 2: double sacrificial oxide process

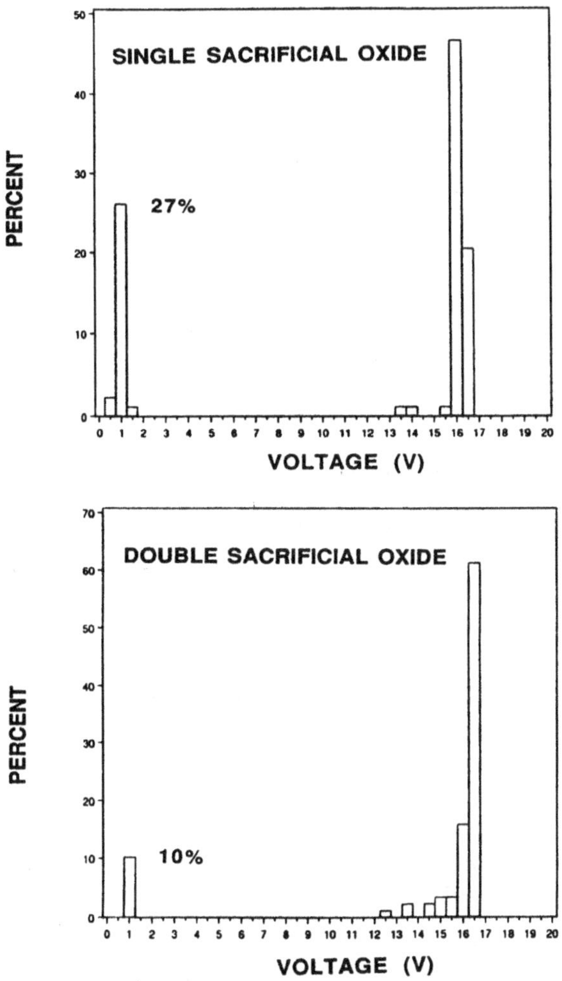

Figure 5. Time zero breakdown histograms for large area capacitors

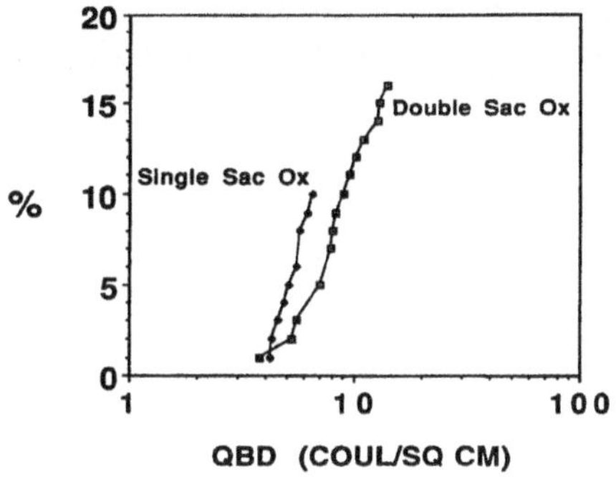

Figure 6. Accumulative plot of the charge to breakdown for field edge intensive capacitors

Figure 6 shows the accumulative plot of the charge to breakdown (Qbd) for field edge intensive capacitors for both single and double sacrificial oxide processes. The constant current density under stressing is - 0.1 A/cm^2. The double sacrificial oxide process results in about 2X longer time to breakdown than that of single sacrificial oxide process. The field oxide edge thinning problem is probably reduced for double sacrificial oxide process [1].

SUMMARY

Several attractive gate oxide characteristics achieved by implementing a double sacrificial oxide process have been observed and discussed. The double sacrificial oxide process reduces the roughness of the active surface on which gate oxide is grown. This results in a higher field onset of Fowler-Nordheim tunneling with electron injection from the polysilicon, a lower interface state density and a lower gate oxide defect density. The double sacrificial oxide process also reduces the field oxide edge thinning problem and a longer time to breakdown for field edge intensive capacitors is achieved.

Acknowledgement

Helpful discussions with Carlos Mazuré in the early part of this study are acknowledged.

References

1. I. J. Voors, K. Osinski, F. H. A. Vollebregt and C. A. Seams, " Gate Oxide Reliability in a Sealed Interface Local Oxidation Scheme", ESSDERC, 19th European Solid State Device Research Conference, Berlin, P.361 (1989).
2. J. Hayden, et. al., " A High-performance Sub-half Micron CMOS Technology for Fast SRAMs", IEDM, P.417 (1989).
3. M A. Chonko and V. Kaushik, " The Effect of Surface Roughness on Gate Oxide Leakage Currents", this symposium.
4. P. O. Hahn and M. Henzler, " The Si-SiO2 interface: correlation of atomic structure and electrical properties", J. of Vac. Sci. Technol. A 2 (2), P. 574 (1984).

EFFECT OF SOLIDIFICATION INDUCED DEFECTS IN CZ - SILICON UPON THIN GATE OXIDE INTEGRITY

Hisaaki Suga , Hidenobu Abe , Hiroshi Koya , Toshihiro Yoshimi , Isamu Suzuki , Hideo Yoshioka and Norio Kagawa [1]

Evaluation Technology Department
[1] Process Technology Department
Technology Division
Mitsubishi Materials Silicon Corporation
314 Nishisangao , Noda - shi , Chiba - ken , Japan 278

ABSTRACT

In this study , dielectric breakdown strength of silicon dioxide film of 5 to 25nm in thickness is revealed to be determined only by the amount of tiny solidification induced defects in magnetic field applied and conventional CZ silicon single crystals. Nucleus of oxygen induced stacking fault , heavy metal impurities on the surface and surface microroughness are less sensitive factors for it. The deteriorated integrity of the thin films is recovered only by the high temperature annealing at 1250°C under oxygen ambient.

INTRODUCTION

In highly integrated LSI device fabrication process , a particulate contamination , process margin failure , reliability of thin gate oxide film , and capacitor leakage caused by defect or metallic contamination are important factors for device production yield. Following two factors are considered to be originated from the silicon substrate , a lot of research works have been done extensively [1)~9)]. Reliability of the thin oxide film is examined by the application of high electric field and the deterioration of the breakdown strength is considered to be caused by the film quality change , the inhomogeneities in the film , microroughness of Si / SiO_2

The Physics and Chemistry of SiO_2 and the Si-SiO_2 Interface 2
Edited by C.R. Helms and B.E. Deal, Plenum Press, New York, 1993

interface and so on. However surface roughness and surface contamination are not now serious problem in commercial silicon wafers for the 4M bit DRAM device fabrication yield but growth condition of single crystals itself currently controls the quality of the gate oxide.

MSIL / MMC group already showed the effect of such substrate factors as COP nucleus [8], resistivity [4], film thickness [6], surface pit [7] and nucleus of oxygen precipitate [9] in substrate on the breakdown voltage of the gate oxide. As a result, the breakdown voltage is clarified to be strongly dependent on the growth condition and the superiority is essentially kept in device fabrication heat history. Present paper describes the effect of on the gate oxide strength on the basic of consistent data obtainded from the crystals of which the information about the crystallization and thermal history before a device process are well known.

EXPERIMENTAL

Samples of Czochralski silicon wafers with <100> orientation, p type, 10Ω · cm and 6 inches diameter are employed. Growth condition of 40 kg charge in a 16 inches diameter silica crucible, pull rates ranged from 0.35~2.10mm / min and targeted oxygen contents in silicon ingot of 1.5×10^{18}atoms / cm^3 (old ASTM) for CZ and 0.7×10^{18}atoms/cm^3 for MCZ.

Pulled crystals receive slicing, lapping, donor annihilation heat treatment, mirror polishing and modified RCA cleaning. Microroughness obtained from 1μm×1μm area and 500μm×500μm area is controled around 0.2nm and 0.4nm, respectively.

Wafer annealing at high temperatures is performed in dry O_2 or N_2 ambient and then surface layers of 15μm in depth are polished off in order to eliminate the roughened surface by annealing. In the oxide film breakdown strength test, oxide films of 5~60nm in thickness are prepared by wet or dry oxidation at 850°C or 900°C. The breakdown is judged from the leakage of 100μA / cm^2 between substrate and Al or polysilicon electrode of 5mmϕ. 196 points for each wafer are examined.

Recombination lifetime is obtained from photoconductive decay curve of the substrate with oxide film.

Detection method of solidification induced defects are small pit counting by 0.1μm class laser particle counter (Hitachi LS - 6000) after repeated SC1 cleaning[10,11] and flow pattern[12] counting caused by inserting wafer into Secco's etchant[13] for 30min. SC1 cleaning is done by using a $NH_4OH / H_2O_2 / H_2O$ (1 : 1 : 5) solution at 85°C for 20min.

RESULTS

1. Breakdown field and pull rate

Fig.1 shows the degeneration of breakdown field of thin gate oxide film by raising pull rate of substrate silicon single crystals with oxygen content of about 1.5×10^{18}atoms / cm^3. This degeneration nature is not affected by

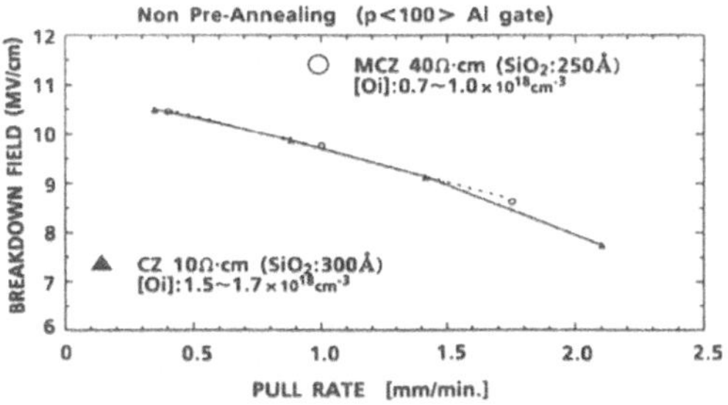

Figure 1 The breakdown field degeneration by raising up of pull rate.

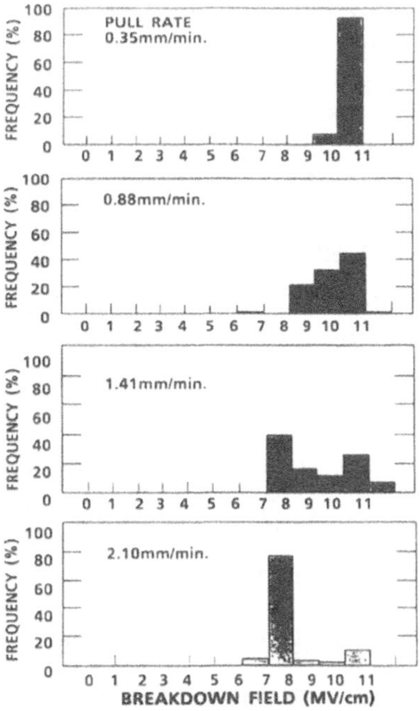

Figure 2 Histograms of the breakdown frequency showing the degeneration by pull rate.

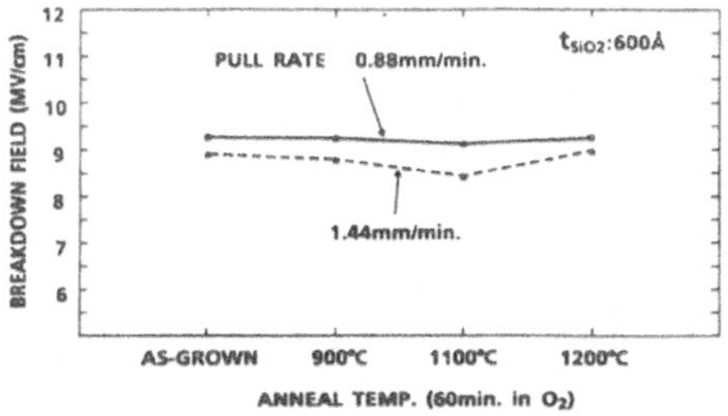

Figure 3 No recovery of the breakdown field of degenerated wafer by annealing below 1200°C.

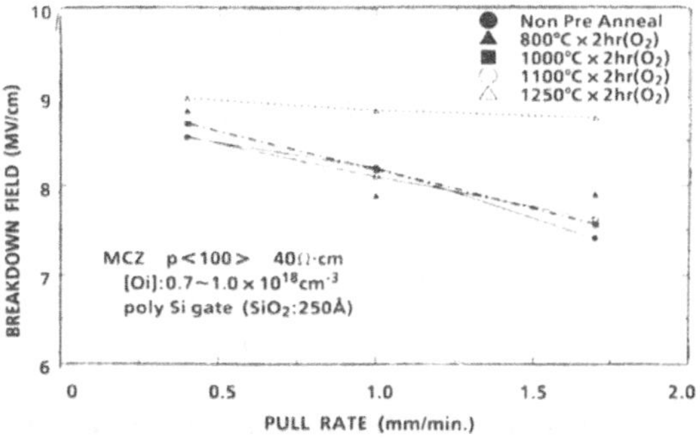

Figure 4 Recovery of the breakdown field by high temperature annealing.

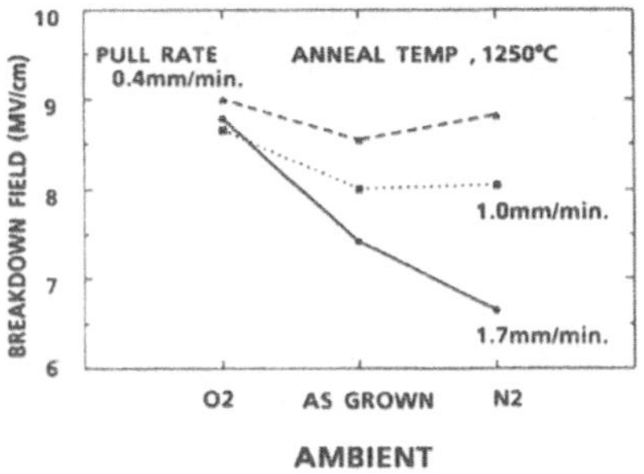

Figure 5 Annealing effect of the breakdown field in different ambient.

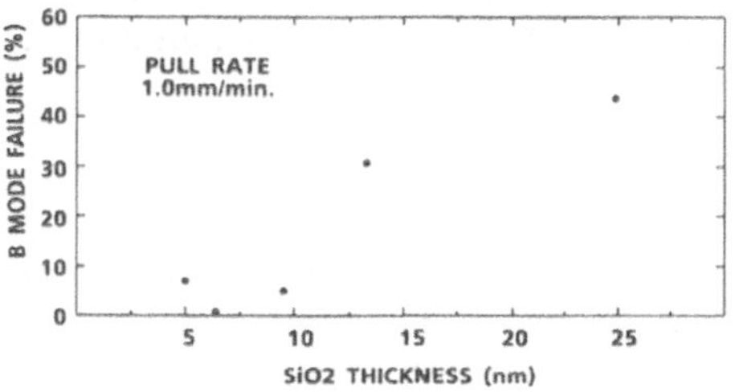

Figure 6 Thin oxide film thickness dependence on the breakdown.

the oxygen content and the application of magnetic field to silicon melt on the crystallization.

Fig.2 shows the histograms of breakdown field of thin oxide films which are produced on polished CZ - silicon wafers of different pull rate. Raising - up of pull rate increases the occurrence of B - mode failure in the breakdown.

2. Recovery of the breakdown field by annealing

Fig.3 and 4 shows the effectiveness of annealing on the breakdown field of degenerated crystals. As shown in Fig.3 annealings below 1200°C can not affect the significant recovery of breakdown field even in O_2 ambient. Fig.4 shows the remarkable recovery of breakdown field only after annealing at 1250°C.

Fig.5 shows the different recovery rate with the annealing atmosphere of dry O_2 or N_2. Annealing under oxygen atmosphere explicitly recovers the degenerated breakdown field , but under nitrogen atmosphere the recover rate is slow or inversely the degeneration occurs.

3. Film thickness dependence of the breakdown field

Fig.6 shows the film thickness dependence on the breakdown field of a degenerated substrate. B - mode failure can not be confirmed in the thin oxide cases below 10nm. Murakami et al [6] showed the same density of B - mode defects included in oxide film of thicker than 15nm as ours. According to their data , the expected number of the B - mode defect in thin film of 10nm in thickness under a 5mmϕ gate is only about 1.5 pieces.

4. Breakdown field and lifetime

Fig.7 shows the independence of breakdown field on the recombination lifetime. Silicon wafers of same ingot but with a different lifetime were examined , but this kind of grown - in defect which causes the accelaration of the electron - hole recombination and OSF generation does not affect the breakdown behavior of thin film.

5. Breakdown field and surface cleanliness

Fig.8 shows the relation between the breakdown field and surface cleanliness. In our cleaning procedure before oxidation , the cleanliness does not affect the breakdown behavior of thin film measured.

6. Grown - in defect

(1) COP

Fig.9 shows the COP nuclei in substrates tend to increase with pull rate. This is consistent with previous data [10]. Substrate of the slowest pull rate includes the COP nuclei of 10^3 pieces / cm^3 in density. Laser particle counter can detect a latex ball which is larger than 0.1μm. It is already confirmed

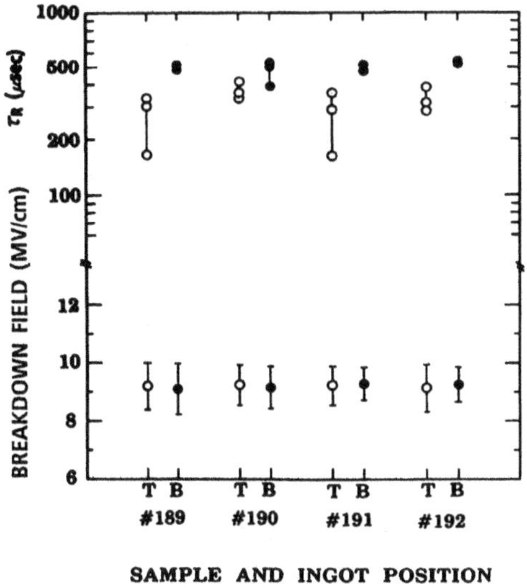

Figure 7 Lifetime dependence on the breakdown field.

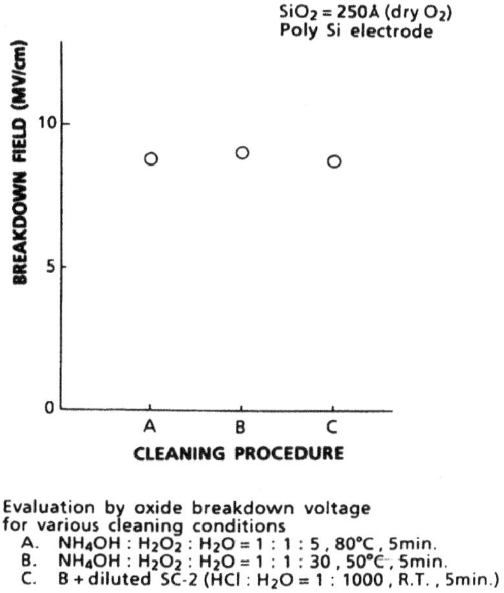

Figure 8 Surface cleanliness dependence on the breakdown field.

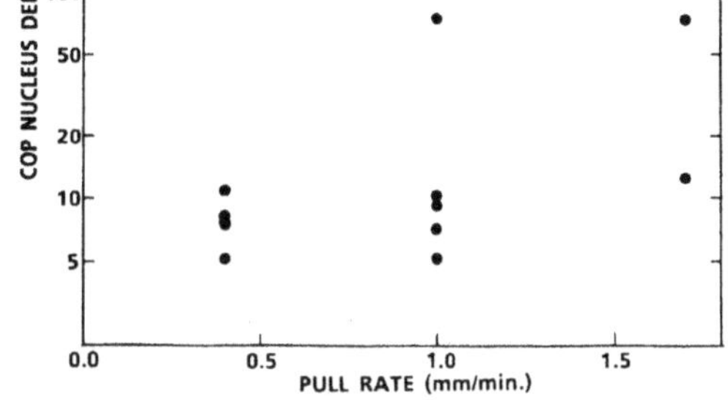

Figure 9 Relationship between COP nucleus density and pull rate.

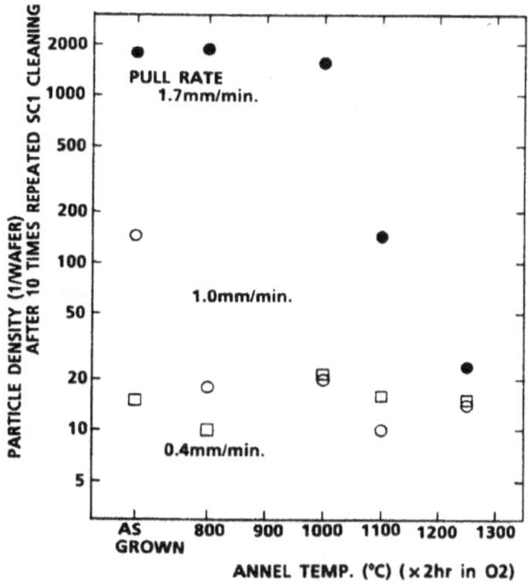

Figure 10 Annealing effect of particle on wafers with different pull rate.

that in the cleaning process of our wafer , particles which the counter detects is mostly a surface pit [11]. Wafers which is cleaned in a simple step of SC1 cleaning does not always show a high surface pit density. Shiota et al negated the dependence of this kind of pit (COP) emarged on the wafer surface on breakdown field experimentally [7].

Fig.10 shows the decrease of COP density begins on the wafer annealing at 1100°C.

(2) Flow pattern

Fig.11 shows the increase of flow pattern density with pull rate. Flow patterns also imply the existence of chemical inhomogeneities but in thicker layer of ~30μm in thickness [12] than COP case. Akiyama shows that the flow pattern shape is the wedgy terrace of height of 0.1μm along with ditch[14]. Apex of wedge is only a common point in the flow pattern produced by step - wise wafer rotation etching. Higher terrace than 0.1μm has rounded the wedge apex , and finally disappear even in 30min Secco's etching.

(3) B - mode failure defect estimated by the breakdown voltage

Fig.12 shows the relation between the density of defects which cause B - mode failure and pull rate. The defect density estimated[15] ranges between $0.1{\sim}2\times10^6$ pieces / cm^3.

7. Breakdown failure and solidification induced defect

Fig.13 shows the weak correlation between B - mode failure defect and COP nuclei. The change in two orders of magnitude in COP nucleus density does not reflect linearly on the B - mode defect density , and the absolute value of both defects is out of accordance.

Fig.14 shows the relation between breakdown field of as - grown and annealed wafers and particle number on wafer after 10 times repeated SC1 cleanings. Wafers with low particle number do not always show the high breakdown field. Points in a broken circle indicate the wafers annealed at 1250°C for 2hr. Degenerated wafers steadily recover the breakdown field in this case together with the reduction of particles.

DISCUSSION

By the increase of pull rate above 0.5mm / min , defect incorporation on solidification increases and their fast aggregation occurs above 1250°C. In this temperature range the defect aggregation can be thermally decomposed. Once beyond this range , a defects aggregation becomes stable in the lower temperature heat history. This picture is depicted from the breakdown failure feature in pull rate change and annealing recovery experiment. But the identification of defects which dominate the breakdown is incomplete. Although nuclei of oxygen precipitates were explicitly negated as a candidate for a breakdown defect [9], a nucleus for COP or flow pattern is

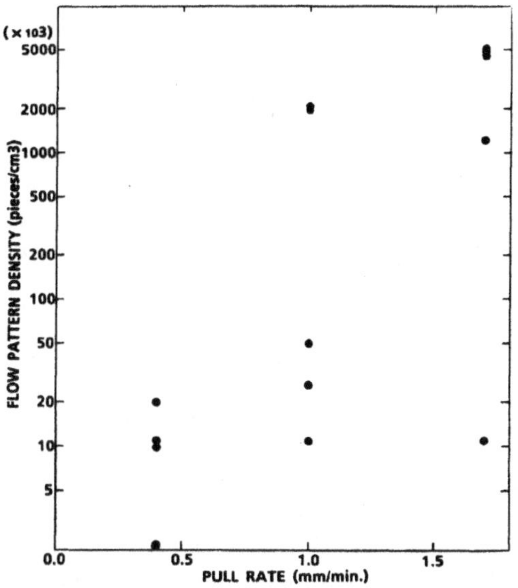

Figure 11 Change of flow pattern density of wafers with different pull rate.

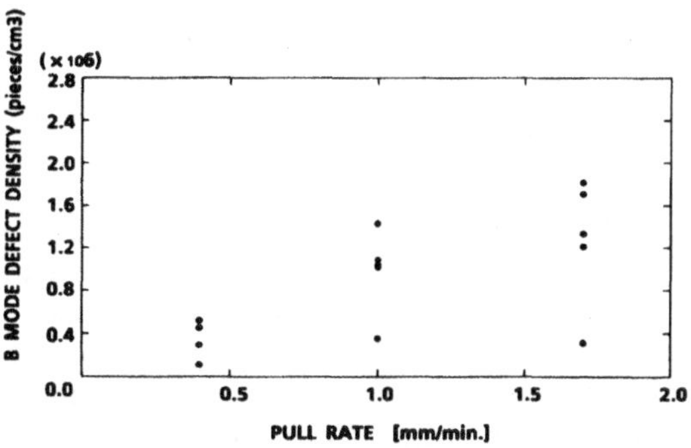

Figure 12 Defect density calculated by B-mode failure dependent on pull rate.

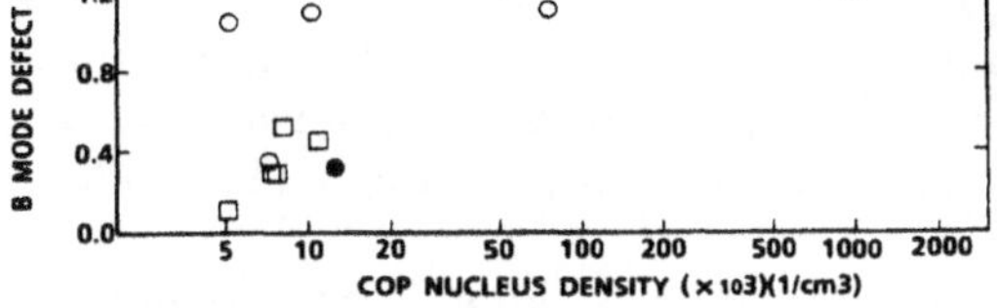

Figure 13 Relationship between B mode failure defect density and COP nucleus density.

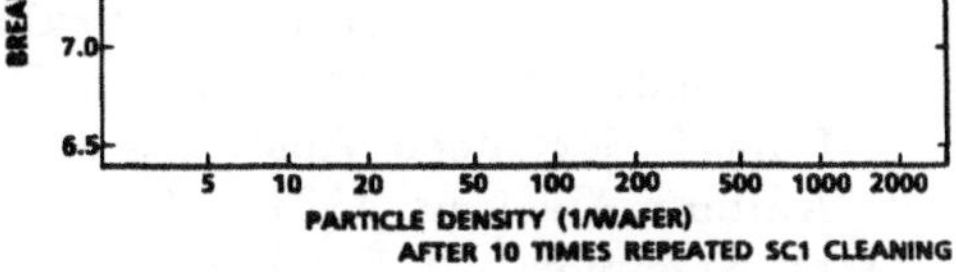

Figure 14 Relationship between the breakdown field and particle density after repeated SC1 cleanings.

still probable for B - mode failure defects. Crystal Originated Particle and flow pattern associated with solidification may be a kind of traces of such solidification induced defects.

The nature of B - mode failure defect seems to be like large vacant - agglomeration, because only massive interstitial silicon injection at 1250°C steadily recover the breakdown failure of the degenerated wafer. Matsui by position annihilation study clarifies no difference of S parameter values which means the vacancy concentration among our wafers with different pull speed [16]. Vacant atom agglomerates may exist independently on the equilibrium vacancy in the silicon matrix. Microroughness revealed by atomic force microscopy remains constant among our wafers of different pull speed. Experimental detection of microvoids of $10^5 \sim 10^6$ pieces / cm^3 in density and of which size is around 5 to 10nm would be essential to prove the substrate dependent degeneration of thin oxide film.

CONCLUSION

Deterioration of the gate oxide by fast pulling of silicon ingot and improvement by the high temperature annealing of wafers under oxygen ambient is expected to be caused by the incorporation of solidification induced defects, their agglomeration and annihilation at high temperatures. Although the evidence of the defect for deteriorating thin oxide film is not cleared by the COP or flow pattern behavior, they are still suspected defects.

REFERENCES

1) H. Suga and K. Murai, "Effect of Bulk Defects in Silicon on SiO_2 Film Break - down," Emerging Semiconductor Technology, ASTM STP 960, D. C. Gupta and P. H. Langer, Eds. American Society for Testing and Materials, 1986, P336 - 352
2) M Morita, A. Teramoto, K. Makihara, T. Ohmi, Y. Nakazato, A. Uchiyama and T. Abe, Abstract 320, The Electrochemical Society Extended Abstracts, Vol. 91 - 1, Washington, D.C., May 5 - 10, 1991
3) H. Hasegawa, K. Arai, Y. Murakami, K. Fusegawa and T. Shingyouji, "Influence of Gate Materials on Dielectric Break - down of Thermal Oxide Films", Extended Abstracts of Japan Soc. of Applied Phys. (The 52nd Autumn Meeting, 1991), 10p - E - 3
4) M. Nagamori, T. Tsuchiya, Y. Murakami, H. Abe and T. Shingyouji, " Influence of Conductivity Type and Dopant Density on Dielectric Break - down of Thermal Oxide Film", Extended Abstracts of Japan Soc. of Applied Phys. (The 52nd Autumn Meeting, 1991), 10p - E - 4
5) T. Shiota, H. Shibasaki, Y. Murakami and T. Shingyouji, " Gate Area Dependence of Dielectric Break - down Characteristics of Thermal Oxide Films", Extended Abstracts of Japan Soc. of Applied Phys. (The 52nd Autumn Meeting, 1991), 10p - E - 5

6) Y. Murakami , N. Yokoo , T. Shiota and T. Shingyouji , " Thickness Dependence of Dielectric Break - down Characteristics of Thermal Oxide Films " , Extended Abstracts of Japan Soc. of Applied Phys. (The 52nd Autumn Meeting , 1991) , 10p - E - 6
7) T. Shiota , E. Morita , H. Shibasaki , T. Shingyouji and Y. Shimanuki , " The effect of Quantity of COP on the Break - down Characteristics of Oxide " , Extended Abstracts of Japan Soc. of Applied Phys. and Related Soc., (The 39th Spring Meeting , 1992) , 29p - ZD - 10
8) T. Shiota , E. Morita , T. Shingyouji and Y. Shimanuki , " The Influence of Nucleus of COP on the Break - down Characteristics of Oxide " Extended Abstracts of Japan Soc. of Applied Phys. and Related Soc., (The 39th Spring Meeting , 1992) , 29p - ZD - 11
9) Y. Murakami , K. Arai , H. Furuya , T. Shingyouji and Y. Shimanuki , " Influence of Oxygen Precipitate Nuclei Density on Dielectric Break - down of Thermal Oxide Films " Extended Abstracts of Japan Soc. of Applied Phys. and Related Soc., (The 39th Spring Meeting , 1992) , 29p - ZD - 12
10)J. Ryuta , E. Morita , T. Tanaka and Y. Shimanuki , " Effect of Crystal Pulling Rate on Formation of Crystal - Originated Particles on Si Wafers " , Japan. J. Appl. Phys. , 31 (1992) , P293 - L295
11)J. Ryuta , E. Morita , T. Tanaka and Y. Shimanuki , " Crystal - Originated Singularities on Si Wafer Surface after SC1 Cleaning " , Japan. J. Appl. Phys. , 29 (1990) , No.11 , L1947 - 1949
12)H. Yamagishi , I. Fusegawa , N. Fujimaki and M. Katayama , " Recognition of D defects in silicon single crystals by preferential etching and effect on gate oxide integrity " , Semicond. Sci. Technol. , 7 (1992) A135 - A140
13)Secco D'Aragona F. J. Electrochem. Soc., 119 (1972) , 948
14)K. Akiyama (Mitsushi Materials Co.) , unpublished
15)C. M. Osburn and D. W. Ormond : J. Electrochem. Soc., 119 (1972) , 591
16)M. Matsui (Nagoya University) , unpublished

EFFECTS OF D-DEFECTS IN CZ SILICON UPON THIN GATE OXIDE INTEGRITY

J.-G. Park,[1] S.-P. Choi,[1] G.-S. Lee,[1] Y.-J. Jeong,[1]
Y.-S. Kwak,[1] C.-K. Shin,[1] S. Hahn,[2] W. L. Smith,[3] and P. Mascher[4]

[1] Quality Control Division
Samsung Electronics
San #24 Nongseo-Lee
Kihung-Eup, Yongin-Gun
Kyungki-Do, Korea

[2] Department of Materials Science and Engineering
Stanford University
Stanford, California 94305 U.S.A.

[3] Therma-Wave, Inc.
47320 Mission Falls Court
Fremont, California 94539 U.S.A.

[4] Department of Engineering Physics
McMaster University
Hamilton, Ontario
Canada L8S 4M1

ABSTRACT

In this study, using oxide breakdown voltage and time-dependent-dielectric breakdown measurements, thermal wave modulated reflectance (both mapping and imaging modes), positron annihilation spectroscopy and chemical etching/optical microscopy, we investigated:

- effects of D-defects upon oxide integrity,
- possible oxide breakdown mechanism due to D-defects, and
- nature of D-defects.

Our data show that: (1) D-defects in the Si substrate affect both oxide breakdown voltage and time-dependent-dielectric breakdown characteristics of thin gate oxide; (2) D-defects can induce rough Si/SiO_2 interfaces during various DRAM processes – an increase in the interface roughness is the main cause for poor oxide integrity; (3) the variation of monovacancy-type defects with respect to the high temperature thermal annealing closely matches that of D-defects; and (4) D-defects are not of an interstitial nature and are possibly vacancy-related defects.

INTRODUCTION

Since particles on Si wafer degrade VLSI device performance, many efforts have been made to reduce the number of particles. The minimum size of the particles to be eliminated has been related to the linewidth of the VLSI devices. It has been supposed that the particles counted by widely used laser particle counters are dust on the wafers. Therefore, the cleaning process of wafers has been studied mainly to find the ways to remove the particles. The SC1 cleaning [1] (cleaning by $NH_4OH/H_2O_2/H_2O$ chemical mixture solution) is known to be effective in reducing the amount of dust. However, J. Ryuta and coworkers have recently claimed that most of the small particles left after the SC1 cleaning cycles are not dust on the wafers but pits formed during the SC1 cleaning [2]. It is suggested that such pits originate from some grown-in defects in the Si crystals. These grown-in defects (called "COP," meaning crystal-originated particles) have been shown to affect the thin gate oxide integrity [2]. Since these defects are induced during solidification process, various researchers believe they are point defect clusters. Point defects in silicon single crystals have been studied very extensively; two types of crystal defects are well known, namely, A- and D-defects. A-defects were reported as silicon interstitial dislocation loops by transmission electron microscopy (TEM) observation [3]. In contrast, it is difficult to confirm the composition of D-defects. Abe recently investigated such defects in detail by a technique involving inner diffusions of vacancies or of interstitial silicon atoms after annealing in various kinds of ambient [4]. He reported that D-defects contained a vacancy agglomerate. However, these point defects cannot be recognized directly. A copper decoration method is usually used, but the content of the point defects can change during annealing around 1273°K; it is very difficult to count their densities. Yamagishi and coworkers reported developing the preferential etching method to delineate D-defects without involving a heat treatment at high temperature [5]. They claimed that:

- D-defects appeared as "flow patterns" after the preferential etching treatment
- D-defects affect oxide integrity; higher defect density leads to lower average breakdown voltage and poorer time-dependent-dielectric breakdown characteristics

- D-defect density depends upon crystal pull rate; the higher the crystal pull rate, the higher the defect density
- D-defect density and COP density are linearly correlated.

Even though both COPs and D-defects appear to share many common features, their direct structural correlations have not yet been made. In addition, it is not clear how both defects lead to oxide failures. In this study, using oxide breakdown voltage and time-dependent-dielectric breakdown measurements, thermal wave modulated optical reflectance (mapping and imaging modes), positron annihilation spectroscopy and chemical etching/optical microscopy, we investigated:

- effects of D-defects upon oxide integrity
- possible oxide failure mechanism due to D-defects
- nature of D-defects.

EXPERIMENTAL

The samples used in this investigation were p-type (res., ~5 ohm-cm), (100)-oriented, 150 mm diameter Cz Si wafers with interstitial oxygen concentration, O_i, of (12.75±0.5) ppma (ASTM F121-81). Carbon concentration, C_S, of these wafers was typically below instrument sensitivity limit (≤ 0.05 ppma). In order to investigate effects of D-defects upon thin gate oxide integrity, two different types of experiments were carried out in this study, namely, Test I and Test II.

In Test I, we used only the wafers specially prepared by one specific commercial material supplier. These wafers were chosen from four ingots grown with four different average pull speeds (0.4, 0.5, 0.55 and 1 mm/min, respectively). After modified RCA cleaning, as-received wafers were submitted into Mb DRAM processes. Subsequently, the wafers were further processed with gate oxidation (oxide thickness ~23 nm), polysilicon deposition, $POCl_3$ diffusion and patterning to fabricate test device structures for gate oxide integrity test. Both oxide breakdown voltage and time-dependent-dielectric breakdown characteristics were measured with HP 4142 B Semiconductor Device Analyzer After the oxide integrity test, these device structures were further characterized by using emission microscope for multilayer inspection (EMMI) and a focus ion beam system (FIB). Finally, these samples were preferentially etched to observe flow pattern defects with Secco etchant for 30 min.

In Test II, the wafers from four commercial wafer vendors (Vendors A, B, C, and D) were used. First, some of the as-received wafers were tested with Secco etching for 30 min. to determine the flow pattern defect density, and with PAS lifetime method to determine the various vacancy-related defect lifetime signatures inside the bulk of the samples. The remainder of as-received wafers was annealed at 1200°C for 2 h in O_2 diluted with N_2 ambients. The annealed wafers were characterized by both Secco etching and PAS lifetime methods in order to get potential correlations between flow pattern defects and vacancy-related defects.

The positron lifetime spectra were obtained using state of the art spectrometers with time resolution FWHM of between 205 and 230 ps. Each spectrum contained at least 6×10^6 counts. The positron sources were rather weak, typically 3.5 μC_i of $^{22}NaCl$, deposited on thin (1-3 μm) Al-foil. Numerical analysis of the lifetime data was performed using the programs PATFIT 88 [10] or PFPOSFIT [11]; agreement of results on test spectra was always found to be within statistical uncertainty. After correction for annihilations in the Al-foil and the NaCl itself, all of the spectra contained three lifetime components, the longest of which, however, was an essentially constant term of about 1.5 ns with no more than 0.3% intensity. This term could easily be separated from the actual spectra, leaving two significant terms from the samples themselves.

RESULTS AND DISCUSSION

1. Effects of Crystal Pull Rate Upon Oxide Integrity

Figs. 1 and 2 show effects of crystal growth rate upon average oxide breakdown field strength, oxide breakdown yield and time-dependent-dielectric breakdown (TDDB) characteristics. In Fig. 1, the breakdown yield is defined as the percentage of MOS devices whose breakdown field strength is larger than 6 MV/cm. Both figures clearly demonstrate that wafers from the ingot grown with fast pull speed growth program exhibit poorer oxide integrity compared to those from the crystal grown by slow pull speed growth program.

2. Oxide Breakdown Mechanism Due To D-Defects

Miyashita and coworkers [6] recently reported that there are two types of surface microdefect clusters, namely, undecorated and metal-decorated clusters, and that only metal-decorated surface microdefect clusters can act as weak spots in oxide, thus contributing to low oxide breakdown distribution. Our experience during this investigation shows that a big drop in average oxide breakdown field strength ($\geq$ 3-4 MV/cm) is generally observed whenever there are line problems due to metallic impurities, whereas rather small changes in the breakdown field strength (at most 1-1½ MV/cm) occur due to D-defects. Therefore, we believe the degradation in oxide integrity due to D-defects is not likely caused by metal decoration.

In this study, using both EMMI and FIB, we further characterized defect morphologies which cause the B-mode breakdown failure. Fig. 3a shows the

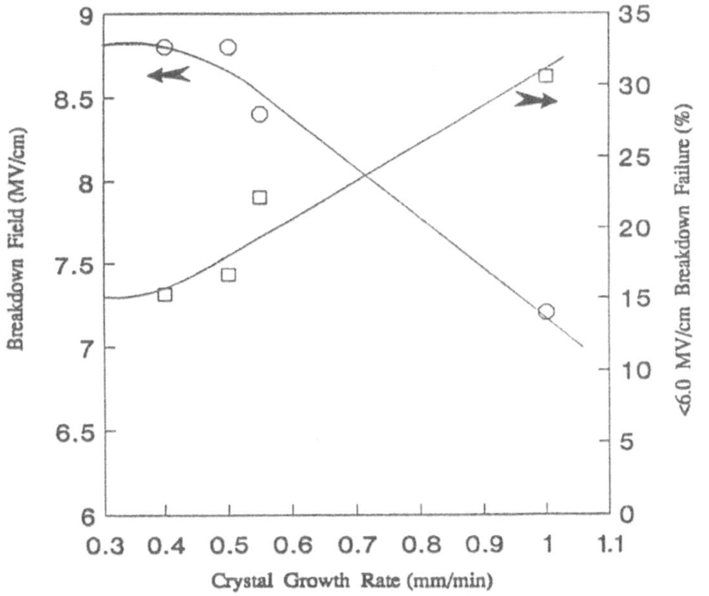

Fig. 1 **Effects of crystal pull rate upon average oxide breakdown field strength and MOS oxide breakdown yield.**

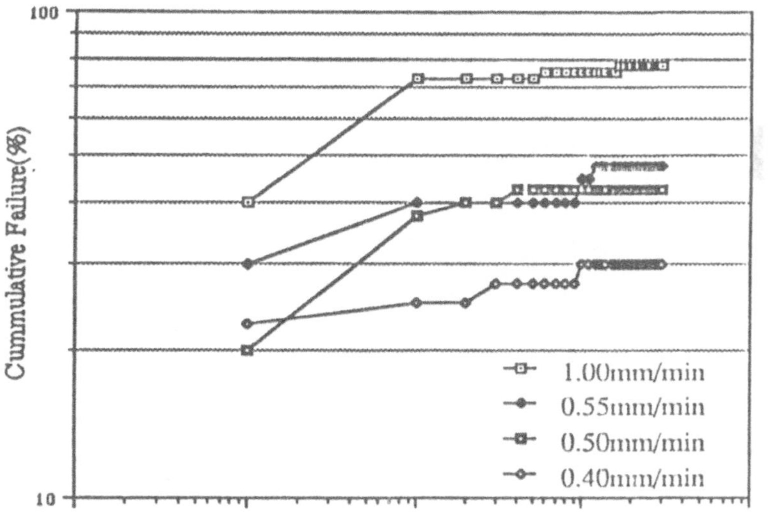

Fig. 2 **Effects of crystal pull rate upon time-dependent-dielectric breakdown.**

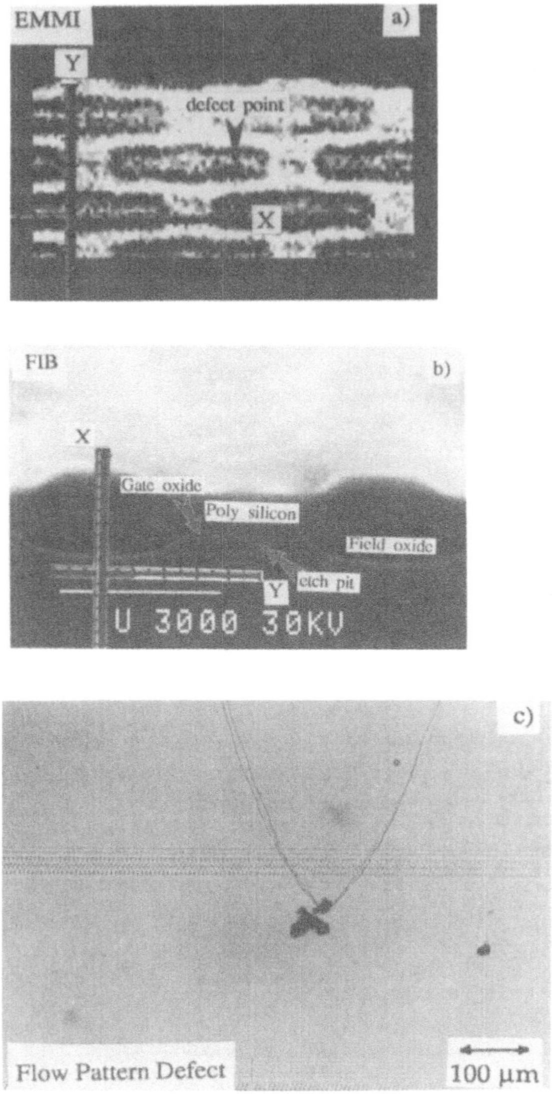

Fig. 3 **a) Typical EMMI micrograph which contains the device with failed oxide structure due to D-defects, (b) cross-section SEM micrograph with FIB for the device with failed oxide structure due to D-defects, and (c) flow pattern defect beneath the device structure with poor oxide layer.**

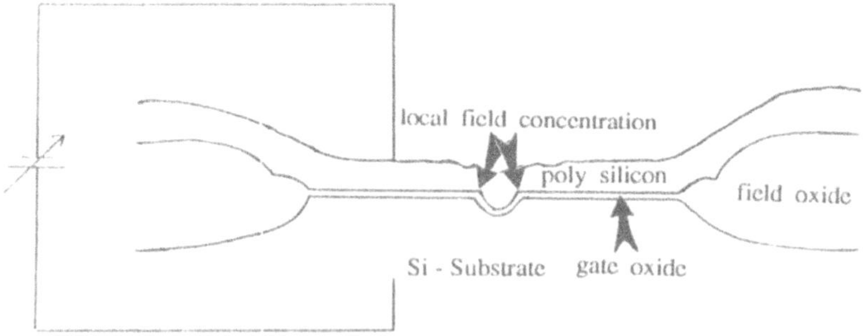

Fig. 4 **Oxide breakdown mechanism due to D-defects.**

typical EMMI micrograph which contains the device with failed oxide structure due to D-defects. Cross section SEM micrograph with FIB displays that almost all the D-defect induced oxide failures are originated from etch pits (shown in Fig. 3b). Fig. 3c also shows the flow pattern defect (beneath the device structure with poor oxide layer) observed after stripping off polysilicon layer and underneath oxide layer by Secco etching for 30 min. Based upon these experimental data, the D-defect induced oxide breakdown mechanism can be considered as follows. D-defects create shallow etch pits on wafer surface during DRAM processes that include various etching and oxidation steps. These etch pits degrade interface roughness between Si and SiO_2. The rough surface causes accelerated local field concentration at the edge of etch pits under the biased voltage, which in turn leads to B-mode type oxide failures (Fig. 4).

3. Nature of D-Defects

The exact nature of D-defects is one of the most hotly debated scientific issues in Si material industry. Up to now, D-defects were generally believed to contain vacancy agglomerates [4]. However, based upon their TEM evaluation, Takeno and coworkers [7] have recently claimed that D-defects in Cz Si are dislocation clusters and dislocation loops. In addition, they reported these dislocation loops are of interstitial types. We believe this controversy mainly comes from the fact that there is a very limited number of characterization techniques sensitive enough to identify vacancy-type defect structures. In Test II, we attempted to further investigate the nature of D-defects by thermal wave imager [which is very sensitive to a disorder in crystallinity near the surface (up to 3-5 μm)] and positron annihilation spectroscopy [sensitive to vacancy-related defect clusters (up to 200 μm)].

As described in the Experimental Section above, both as-received and high temperature (at 1200°C for 2h in O_2 diluted with N_2 ambients) annealed wafers from four different commercial material supplies were further analyzed by both thermal wave imager and PAS lifetime method. After both thermal wave imager and PAS lifetime measurements, we etched off 60 μm of wafer surface layer by the mixed etchant, and the flow pattern defect density was measured via the preferential etching procedure.

In PAS lifetime measurement, two types of positron lifetimes have been observed for all the wafers, both as-received and annealed (besides 218 ± 2 ps characteristic lifetime from perfect Si lattice).

One type is shorter lifetime ($\tau_1 = 113 \pm 15$ ps), which has been reported to be associated with positrons trapped at interstitial-type clusters (some researchers call these defects "oxygen-related defects" [8]). The intensity of this lifetime component is rather small ($I_1 = 6.13\%$) and does not change significantly, even after the anneal at 1200°C.

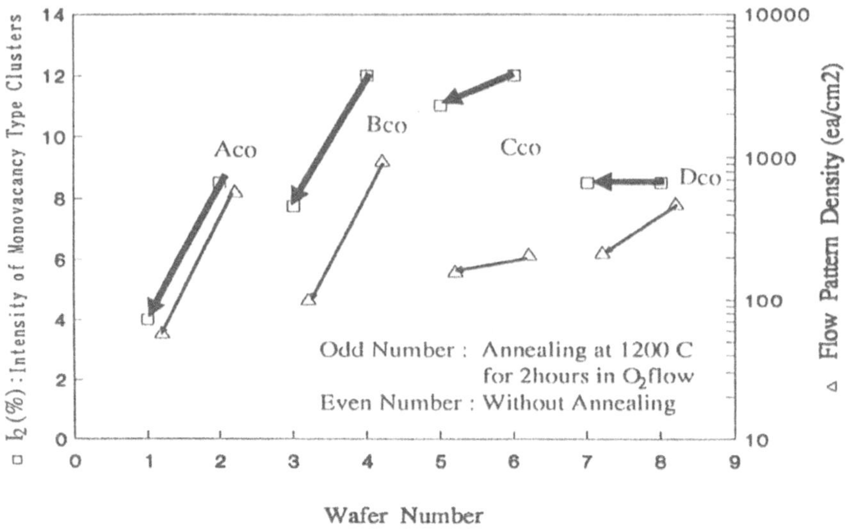

Fig. 5 **Intensity variation of monovacancy-type defect component by PAS lifetime method due to high temperature anneal at 1200°C for the wafers from the four commercial Si material vendors A, B, C, and D. In addition, the variation of flow pattern defect density due to the same high temperature thermal anneal is shown.**

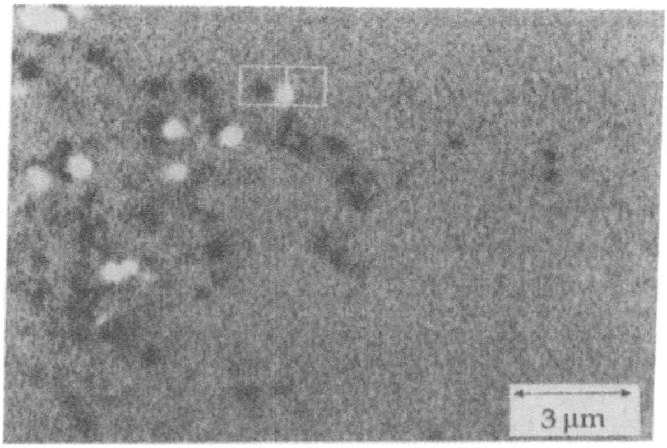

Fig. 6 **Typical TW defect data for the wafers which contain high density of D-defects (density > 1000/cm^2).**

Another type is longer lifetime (τ_2 = 273 ps), which is known to be associated with positrons trapped at monovacancy-type defect clusters. The intensity variation of this lifetime component due to the high temperature anneal at 1200°C for the wafers from the four commercial Si material vendors A, B, C, and D is shown in Fig. 5. In addition, we also include the variation of flow pattern defect density due to the same high temperature anneal as a comparison. The data in Fig. 7 show that high temperature anneal affects monovacancy-type defect (τ_2) component intensity. A substantial decrease in τ_2-component intensity was observed for the wafers from Vendors A and B, whereas a small decrease or no recognizable changes in τ_2-component intensity was observed for the wafers from Vendors C and D. In addition, interesting correlations between τ_2-component intensity and D-defect density have been observed. A big drop in τ_2-component intensity corresponds to a large reduction in D-defect density, whereas a small change in τ_2-component intensity leads to a small change in D-defect density. The real causes for this interesting difference in defect behavior among four commercial vendors have not been completely understood. One possibility is a difference in the crystal growth procedure (and subsequent thermal anneal procedures) used. Subsequent TW imager measurements for these samples show that TW image data for the wafers with high D-defect density generally contain numerous dark TW defects, indicating that the recombination time [of the electron-hole (e-h) pairs injected by the laser in the TW apparatus] is greater at the dark defect sites than it is in the surrounding regular Si lattice. A typical TW image of a defective region of Si is shown in Fig. 6. We note that these dark defect images are distributed inhomogeneously (Fig. 6). We also note that some defects are bright instead of dark. It is believed from previous other studies [6] that the higher-than-background TW signal results from strong e-h recombination at D-defects that are metal-contaminated. Additional studies are under way to confirm this effect.

The PAS observation of longer lifetime in the presence of D-defects is important to explain the characteristic dark appearance of the D-defects in the TW image. No other Si defect has ever been observed to exhibit lower-than-background TW signal, consistent with the unique increase in e-h lifetime at the D-defect site. All other known Si defects cause decreased e-h lifetime and correspondingly higher-than-background TW signal.

Therefore, based upon both PAS lifetime and preliminary TW dark defect image data, we infer that D-defects are not interstitial defect clusters but vacancy-type defect clusters.

SUMMARY

In this study, using oxide breakdown voltage and time-dependent-dielectric breakdown measurements, thermal wave imaging, positron annihilation spectroscopy and chemical etching/optical microscopy, we investigated:

- effects of D-defects upon oxide integrity
- possible oxide breakdown mechanism due to D-defects
- nature of D-defects.

Our experimental data show that

1. the crystal pull rate affects both oxide breakdown field strength (related to the B-mode type failures) and time-dependent-dielectric breakdown characteristics of thin gate oxide. The degradation of these device characteristics is caused by D-defects in the Si substrate;
2. D-defects can induce rough Si/SiO_2 interfaces during various DRAM processes. An increase in the interface roughness accelerates the local field stress concentration and eventually leads to lower breakdown field strength and poor time-dependent-dielectric breakdown characteristics;
3. thermal annealing at 1200°C for 2h in O_2 diluted by N_2 atmosphere decreases the D-defect density and the concentration of monovacancy-type defects (by positron annihilation spectroscopy lifetime technique). The variation of monovacancy-type defects with respect to the high temperature thermal annealing closely matches the D-defects; and
4. both thermal wave imaging and positron annihilation spectroscopy data indicate that D-defects are not of interstitial nature, and allow us to infer that the D-defects are vacancy-related defects.

REFERENCES

1. W. Kern and D. A. Puotinen, *RCA Rev.* 31:207 (1970).
2. J. Ryuta, E. Morita, T. Tanaka, and Y. Shimanuki, *Jpn. J. Appl. Phys.* 29:L1947 (1990).
3. H. Föll, U. Gösele, and B. O. Kolbesen, *in*: "Semiconductor Silicon 1977," H. R. Huff and E. Sirtl, eds., Electrochemical Society, Princeton (1977), p. 565.
4. T. Abe, *Ohyo Butsuri* 59:272 (1990) [in Japanese].
5. H. Yamagishi. I. Fusegawa, N. Fujimaki, and M. Katayama, *Semicond. Sci. Technol.* 7:A135 (1992).
6. M. Miyashita, H. Hiratsuka, and Y. Matsushita, *in*: "Defects in Silicon-II," ECS PV 91-9, W. M. Bullis, F. Shimura, and U. Gösele, eds., Pennington, NJ (1991), p. 407.
7. H. Takeno, S. Ushio, and T. Takenaka, presented at Jpn. Appl. Phys. Soc. Meeting, 1991 October.
8. S. Dannefaer and D. Kerr, *J. Appl. Phys.* 60:1313 (1986).

OXIDATION INDUCED CHANGES IN THE SI SURFACE MICROROUGHNESS

V. Nayar, R. Jackson, A. J. Pidduck and C. Pickering

Defence Research Agency (Malvern)
St. Andrews Road, Malvern, Worcs., WR14 3PS, UK

INTRODUCTION

Microroughness and surface defects play an increasingly important part in the performance of metal-oxide-semiconductor microelectronic devices. This is the case, in particular for small geometry devices where the properties of the oxide/Si interface have more effect on the operation of the device. The importance of microroughness has been recognised by a number of authors[1-4]. However, the cause and pattern of its evolution is still unclear. Prior to gate oxidation, in the device fabrication schedule, the active surface region is likely to undergo a large number of processes. A common step is a stress relief or sacrificial oxidation. In this study the results from Scanning Optical Microscopy in differential phase contrast mode (SOM-dpc), Spectroscopic Ellipsometry (SE) and Atomic Force Microscopy (AFM) are used to study the appearance of roughness resulting from thermal oxidation. These three techniques are sensitive to roughness at different lateral length scales, $<0.15\mu m$ approximately in the case of SE and from $\approx 0.5\mu m$ to tens of μm for the SOM. The AFM bridges the gap between the SE and SOM. All of them are sensitive to vertical scales of just a few Angstroms. It will be shown that oxidation results in the appearance of random roughness on different length scales. However, oxidation can also smooth the Si {100} surface under certain conditions and over certain length scales.

EXPERIMENTAL

N-Type ($\approx 10\Omega cm$, CZ) 100mm diameter Si wafers were cleaned in a automated spray cleaning system (FSI Mercury) using a modified RCA clean[5]. A number of wafers were then thermally oxidised under a variety of process conditions listed below.

Process 1) In a steam ambient to $\approx$1000nm at 1150°C, (here after sacrificial oxidation).

Process 2) In a dry oxygen ambient to $\approx$25nm at 900°C. Also some wafers from 1) were stripped of oxide using HF solution, cleaned and re-oxidised along with the fresh wafers which only received a clean.

Process 3) In a dry oxygen ambient to $\approx$100nm at 1050°C.

Process 4) In a steam ambient to $\approx$100nm at 1050°C.

SE was carried out at an incident angle of 75.1° using a commercial ellipsometer from SOPRA. A detailed description of the data collection can be found in the paper by Nayar et al.[6]. The ellipsometric data were analysed according to the method described by Aspnes and Theeten[7]. The interfacial layer thickness ($T_{int.}$) was modelled by using a 50%SiO_2/50%Si Bruggeman Effective Medium Approximation to calculate its optical behaviour. The oxide thickness (T_{oxide}), its relative density (D, in comparison to fully relaxed vitreous silica) and the interlayer thickness were calculated. SOM-dpc images of the samples

The Physics and Chemistry of SiO_2 and the Si-SiO_2 Interface 2
Edited by C.R. Helms and B.E. Deal, Plenum Press, New York, 1993

with the oxide intact and after removal by etching in dilute HF solution were also collected. SOM-dpc is a reflection mode microscope and is sensitive to surface slope. All the SOM-dpc images were collected at the same instrument settings. We have previously shown[8] that the SOM-dpc image collected from a Si surface with a native oxide is due to the oxide/Si interface. However, for thicker oxides interference effects due to the oxide affect the reflectivity of the samples. The effect of the oxide therefore needs to be considered carefully. Contact mode AFM imaging in air was carried out with a Digital Instruments Nanoscope. The grown oxide/Si interface was studied by etching the oxide with diluted HF (48%HF,1:7,H_2O) solution. The surfaces were over-etched for 60 seconds followed by a 60 second rinse in deionised H_2O. Typically, the force exerted by the tip on the etched surface was 30-60nN. The image quality was largely unaffected by the force level. All AFM images were collected under the same scanning conditions and at the same elapsed time after HF etching. The surfaces of the 1050°C wet and dry oxides were measured without any treatment. For the oxide surfaces the force was considerably higher on the order of 300-400nN. This was due to the strong interaction of the Si_3N_4 AFM tip and the thermal oxide. Indeed imaging the oxide surface often led to both surface and tip damage. Further work is underway to optimise scanning conditions for oxide surfaces. The AFM data has been analysed to determine the root mean square roughness (σ nm) as a function of the analysed area. Graphs of σ (averaged over several positions within the scanned region) versus area are used below to illustrate the roughness present at different length scales.

RESULTS

Process 1) Figure 1a shows a typical SOM-dpc image from the surface of an as-received wafer. Notable surface features are observed. These have been previously related to the polishing process applied by the wafer supplier. By way of comparison the oxide/Si interface resulting from the extended wet oxidation at 1150°C is shown in figure 1b. A significant smoothing of the oxide/Si interface is evident. The only distinct features observed were occasional very shallow pits. This observation of smoothing on all length scales was confirmed by AFM imaging, Figures 2a and b, and is clearly demonstrated by the σ data in Figure 3 (error bars indicate the typical spread in σ for that area). Curves of σ versus area are applicable only for surfaces displaying "uniform" roughness. Surfaces with inhomogeneous roughness (i.e. dominated by localised defects) lead to a much larger statistical spread in σ for the lower end of the length scales. Generally, at the upper end of the σ data there is a rise. This is a scanning artefact and presents an upper limitation on the length scales that can be sampled with the desired height resolution. Additionally, the very much smoother sacrificial Si surface is a test of the height resolution and sensitivity of the AFM.

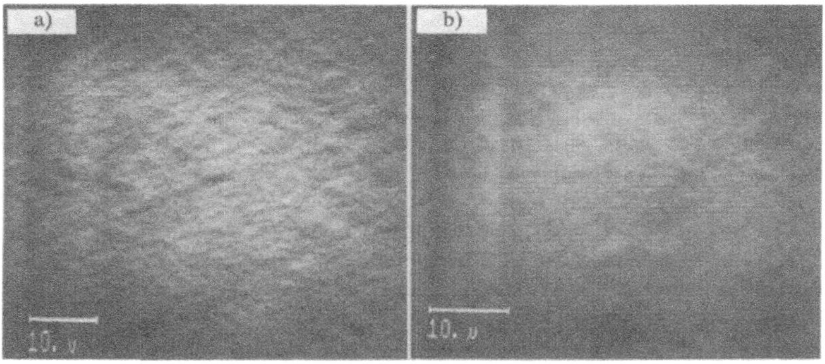

Figure 1. Typical SOM-dpc images of a) unprocessed wafer surface b) sacrificial oxidised wafer after HF strip.

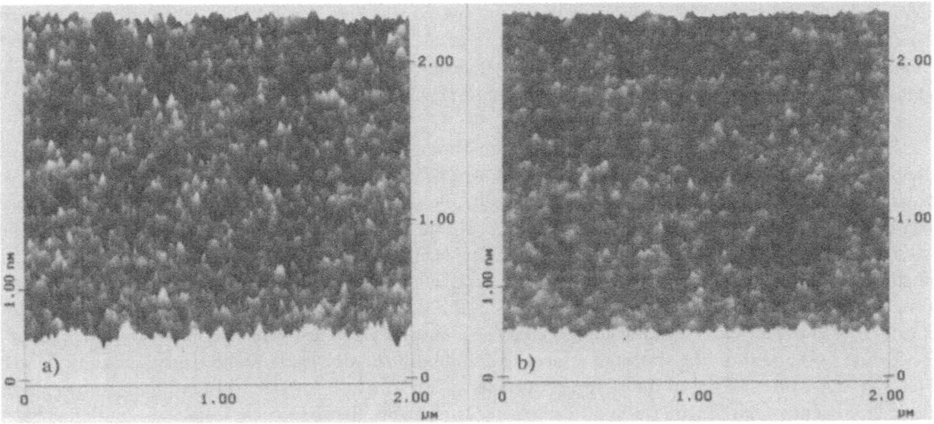

Figure 2. Typical AFM images of a) unprocessed wafer surface b) sacrificial oxidised wafer after HF strip.

Table 1 Results from spectroscopic ellipsometric analysis.

Growth Process	T_{oxide} (Å)	D	$T_{Int.}$ (Å)
2) Fresh	259.7±1.7	0.993±0.009	3.0±1.5
2) Sacrificial	263.1±1.7	0.990±0.013	2.9±1.8
3) Dry	1049.3±1.2	1.009±0.001	7.8±2.0
3) Wet	1125.0±1.8	1.015±0.001	6.7±2.9

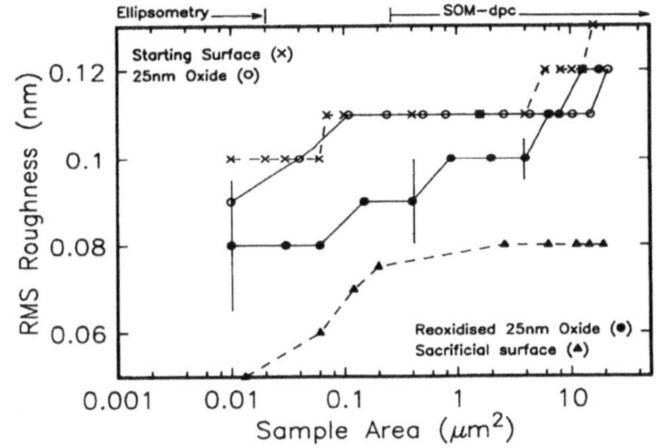

Figure 3. σ versus area for Si surfaces prior to and after 25nm oxidation.

Process 2) Oxidation to 25nm did result in roughening at the oxide/Si interface, although only subtle changes were observed. In figure 3 σ curves for the cleaned and oxidised surface overlaps that of the as-received surface. However, a much clearer picture of the impact of oxidation to 25nm at 900°C is obtained from the reoxidised wafer surface where an increase in roughness is obvious. The two curves for the 25nm oxides suggest that the fine scale roughness (i.e. below about 0.15µm) is similar. This is further emphasised when the ellipsometric data are considered, Table 1. The calculated interlayer thickness was about 0.3nm for both oxide layers. The AFM σ value cannot be directly equated with the ellipsometric interlayer thickness, for reasons which are discussed below. The approximate ranges of sensitivity for the different techniques are also shown in Figure 3.

Process 3,4) Figures 4a and b show SOM-dpc images of the dry and wet (respectively) oxidised samples, with the 100nm oxides intact. Firstly, it was found that the surface topography after either wet or dry oxidation to $\approx$100nm was similar. Secondly, it was evident that these 100nm oxide surfaces are very different to the typical starting wafer surface and that the roughness is dominated by a random sub-micron texture. Wider scale surface structure seen in figure 1 is not visible. To gain some insight on whether the image collected is due to the oxide surface or the oxide/Si interface oxide stripped samples were also studied, Figures 5a and b. It is clear that the random roughness is also present at the oxide/Si interface. However, some wider scale features (over several μms) are now visible. These are similar to the features observed on the starting surface. Additionally, the dry oxide/Si interface is slightly rougher, from the SOM-dpc images of the stripped surfaces. Confirmation for this can be found in figure 6 and Table 2. σ versus area for the wet and dry interfaces illustrate that fine scale roughness is similar but on wider scales ($>0.20\mu m$ or $0.04\mu m^2$) significantly higher for the dry oxide.

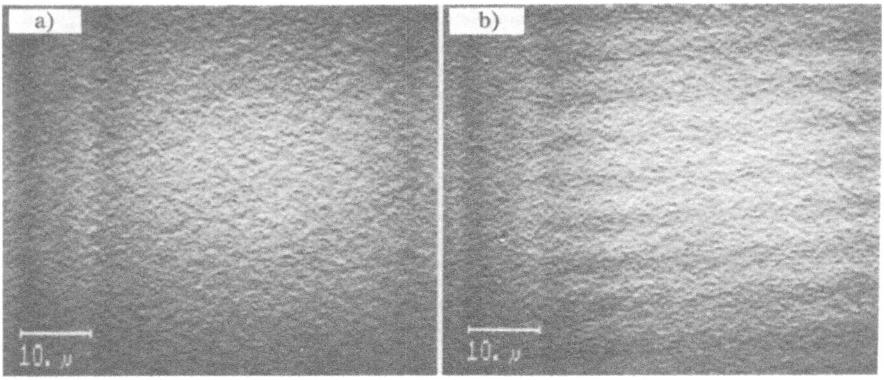

Figure 4. SOM-dpc image after oxidation to $\approx$100nm at 1050°C, a) dry growth and b) wet growth.

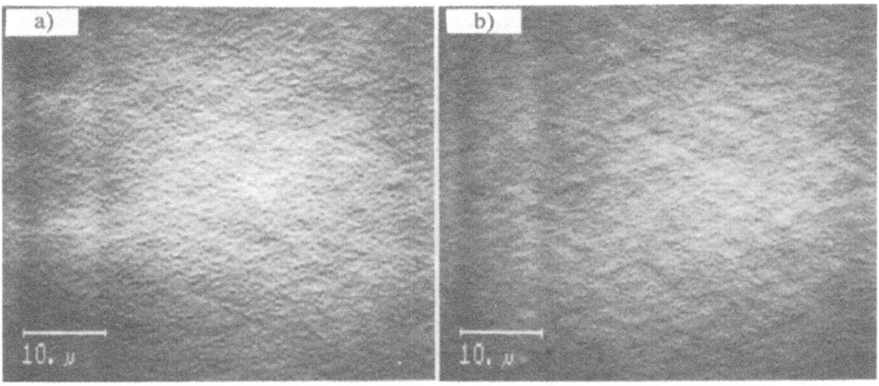

Figure 5. SOM-dpc image after etch removal of 100nm oxides, a) dry growth and b) wet growth.

Table 2 AFM roughness parameters for the wet and dry 1050°C oxides.

Growth Process	R_a(nm)	R_z(nm)
3) dry oxide surface	0.18	0.95
4) wet oxide surface	0.18	0.86
3) dry oxide/Si interface	0.10	0.53
4) wet oxide/Si interface	0.08	0.42

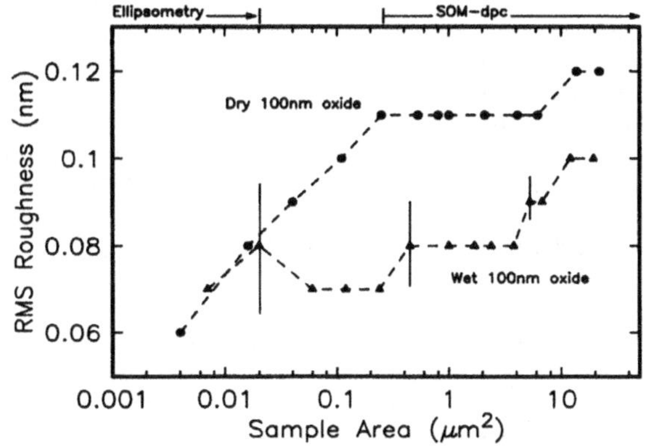

Figure 6. σ versus area for Si surfaces after etch removal of 100nm oxides.

The larger SOM-dpc signal observed with the oxide intact in comparison to that after oxide removal may be due to local lateral fluctuations in the oxide thickness (and hence surface roughness) or density. Preliminary AFM measurements of the oxide surfaces revealed a higher level of roughness compared to the oxide/Si interface, see Table 2 (R_a is the average height and R_z is the average difference between the five highest peaks and five lowest valleys) and also the starting wafer surface. AFM confirms that the roughness levels and topography were similar for the dry and wet oxide surfaces. The area imaged by AFM overlaps the lateral length scales to which the SOM-dpc technique is sensitive. The SE data fitting again showed that the interfacial layer thicknesses and the oxide densities of the wet and dry oxides were similar. The wet oxide was calculated to be slightly denser and have a slightly thinner interlayer than the dry oxide. Interlayer thicknesses for these oxides are higher than those for the 25nm oxides. A similar result for the interlayer thickness of dry oxides was also found by Jellison[9]. As noted above relating the calculation of interlayer thickness by ellipsometry to the AFM data is difficult. It should be noted that SE is particularly sensitive below lateral length scales $\approx 0.15\mu m$ and the fall off of sensitivity is gradual with increasing lateral distances. Therefore, a surface with large undulations but over a distance of $0.5\mu m$ would only result in an interlayer thickness which was related to the average height changes within lateral distances $< 0.15\mu m$. Finally, it should also be noted that SE would also be sensitive to any sub-oxide region at the interface because of the optical equivalence of sub-stoichiometric silicon oxides to the very fine scale layers modelled using the Effective Medium Approximation.

DISCUSSION AND CONCLUSION

Broadly, the data collected suggest that the oxide/Si interface can be both roughened and smoothed by thermal oxidation procedures. Oxidation to 25nm at 900°C leads to an increase in roughness which can easily be masked by that present on the starting surface. Residual polishing marks over a few μms are largely unaffected but a weak additional random roughness does appear.

Wet and dry oxidation to ≈100nm at 1050°C produces oxides with surprisingly similar physical properties given the difference in thermal history. The wet oxidation took ≈10 minutes as compared to ≈80 minutes for the dry oxidation and the oxide growth rate is very different at the end of these periods. Also, the proposed mechanisms for wet and dry oxidation are very different[10]. Thus the similarities must originate from some other over-riding factor. There is however, significant difference in the interfacial roughness (see Figures 5 and 6).

The origin of the roughening is likely to be linked to the oxide formation mechanism or alternatively the intrinsic nature of the wafer surface and near surface region. Some possibilities are listed here. In competition with roughening must also be smoothing mechanisms such as annealing.

i) Roughening may simply be a statistical consequence of the random arrival of diffusing oxidant species at the interface. Diffusion of oxidant species relates to the oxide thickness and its structure. It was noted above that roughening over μm length scales was observed even for thin 25nm oxidation. Therefore roughening is unlikely to be related to random arrival of oxidant simply because it is very improbable that oxidising species diffuse laterally over such large distances within the oxide.

ii) The wafer surface and near surface region may have grown in structural defects or impurities which result in local variations in the oxidation rate and hence roughening. Hence as a test it would be expected that roughness evolution would depend on the origin and preprocessing of the wafer. However, similar forms of roughening are observed for 25nm oxide with very different starting surfaces once the differences are accounted for.

iii) An increase in the roughness of the oxide/Si interface may mean an inhomogeneity in the local interface reaction rate exists. It is well know that thermal oxides are compressively stressed. This stress originates from the volume expansion due to oxide formation. The average stress is inevitably a consequence of microscopic events at the oxide/Si interface. Reaction at any particular interfacial site requires an accommodation of the increased volume by the surrounding oxide and Si (largely the oxide). If the effect of this accommodation is to change the balance between the rate of reaction along and normal to the interface then the roughness consequently alters. In a previous study[8] we determined that the density of thin dry oxides decreased rapidly with oxide thickness from 5nm to 20nm at both 900°C and 1050°C. The origin of this change is proposed to be strain induced plastic deformation as described by Rafferty et al.[11]. Evidence for significant changes in the physical properties of thin oxides has been reported in many studies[6,9,12]. Further work is necessary to deduce the origin of roughening of the oxide/Si interface and whether it is linked to changes in the physical properties of oxide or underlying Si. Certainly many of the oxidation/annealing studies should be reconsidered in the light of the effect interfacial layers and oxide density variations upon ellipsometric data analysis.

In conclusion, the observations of changes in the surface topography are similar to those reported in previous studies[1,2]. However, these studies did not give a detailed topographic representation of the surface and as such the interpretation of roughness statistics was difficult. Even simple methods such as σ versus area plots inadequately convey fully representative information. It was shown above that process induced roughness exists on different length scales and requires different analytical tools. Combining optical, non-destructive methods such as the spectroscopic ellipsometer and scanning optical microscope with the atomic force microscope (which is a contact mode technique) should lead to an improved understanding of both processing induced surface effects and the oxidation of Si.

ACKNOWLEDGEMENT

The material used in this study was produced in DRA Electronics Division's Silicon Processing and Evaluation Laboratory (SPEL). This work was performed under the DTI IED2/1/1769 project.

REFERENCES

1. P O Hahn, M Grundner, A Schnegg and A Jacob, "Correlation of surface morphology and chemical state of Si surfaces to electrical properties", 436, in "Proceedings of INFOS'91", F Koch and A Spitzer eds., North-Holland, Amsterdam (1989).

2. G J Pietsch and M Henzler, "Continuous roughness characterisation from atomic to micron distances: angle resolved photon and electron scattering", 457, in-see ref. 1.

3. M Heyns, C Hasenack, R de Keersmacker and R Falster, "Impact of silicon surface characteristics on MOS device yield for ULSI", in Microelectronic Engineering 10, 235 (1991).
4. M Morita, A Teramoto, K Makihara, T Ohmi, Y Nakazato, A Uchiyama and T Abe, "Effects of Si wafer surface micro-roughness on electrical properties of very-thin gate oxide films", in "Proceedings 2nd Symposium on ULSI Science and Technology", Electrochemical Society Proc, 91-11, (1991).
5. W Kern and D A Poutinen, "Cleaning solutions based on hydrogen peroxide for luse in silicon semiconductor technology", in RCA Review V31, 187 (1970).
6. V Nayar, C. Pickering and A M Hodge, "Determination of the properties of thin thermal SiO_2 on silicon by spectroscopic ellipsometry", in Thin Solid Films 195, 185 (1991).
7. D E Aspnes and J B Theeten, "Spectroscopic analysis of the interface between Si and its thermally grown oxide", J Electrochem. Soc. V127, No. 6, 1359 (1980).
8. A J Pidduck and V Nayar, "Optical imaging of microroughness on polished Si wafers", Applied Physics A53, 557 (1991).
9. G E Jellison Jr., "Examination of thin SiO_2 films on Si using spectroscopic polarisation modulation ellipsometry", J Applied Physics, 69 (11), 7627, (1991).
10. E A Irene, "Models for the oxidation of Si", in CRC Critical Revs. in Sol. St. and Mat. Sci. V14 No.2, 175 (1988).
11. C S Rafferty, L Borucki and R W Dutton, "Plastic flow during thermal oxidation of silicon", in Appl. Phys. Lett. 54(16), 1516 (1989).
12. J T Fitch, C H Bjorkman, J J Sumakeris and G Lucovsky, "The effects of intrinsic in-plane stress on the local atomic structure of thermally grown SiO_2", in Mat. Res. Soc. Symp. Proc. 130, 289 (1989).
13. L M Landsberger and W A Tiller, "Refractive index, relaxation times and the viscoelastic model in dry-grown SiO_2 films on Si", in Appl. Phys. Lett. 51(18), 1416 (1987).

VI. NOVEL STRUCTURES, PROCESSES, AND PHENOMENA

INTRODUCTION

A. Goodman
Code 1114, Office of Naval Research
800 N. Quincy Street, Arlington, VA 22217

W. T. Lynch
Semiconductor Research Corporation
79 Alexander Drive, Bldg. 4401, Suite 300
P. O. Box 12053, Research Triangle Park, NC 27709

The major theme of this chapter of the symposium is new materials, device structures, fabrication and synthesis methods, and their properties, all linked by the common thread of SiO_2 and the Si-SiO_2 interface. Here, new is intended to convey the meaning not in current high-volume production.

In the lead-off paper, S. Cristoveanu provides a comprehensive review of the properties of SIMOX (separation by implantation of oxygen) and related materials systems and device structures. The current status of SIMOX and other SOI (silicon-on-insulator) technologies is placed in perspective; this includes a discussion of defects, reliability, radiation effects, transient effects, and other issues that are currently being addressed, and must be resolved, in order to make SIMOX a mainstream technology. Next, G. J. Dunn describes the beneficial effects of nitridation and reoxidation on SiO_2 intended for gate-insulator use. The reduction of hot carrier damage and the consequent improvement in reliability for submicron n-channel devices are discussed in some detail as well as the potential for p-channel device improvement. In the third paper, L. K. Wang, C. C. Hsu, and W. Chang consider the use of PECVD (plasma enhanced chemical vapor deposition) for fabricating the gate oxide for MOS applications where conventional thermal oxidation is not an option; e.g., when the semiconductor is either Si-Ge or a III-V material. The effect on device reliability of hot carrier damage is compared for thin PECVD and thermal oxides for both p-channel and n-channel transistors. The observation of enhanced interface trap generation in PECVD oxide n-channel transistors and reduced electron trapping in PECVD oxide p-channel transistors is attributed to the larger density of hydrogenated dangling bonds in PECVD oxide. R. B. Klein and N. S. Saks consider the trapping of both electrons and holes in ONO (oxide-nitride-oxide) gate-dielectric layers. The authors find large densities of both electron and hole traps with very large (Coulombic-attractive) capture cross-sections and no significant temperature dependence of the trap parameters. Significant differences between the characteristics of ONO and reoxidized nitrided oxide are explained by the very different fabrication procedures.

The effect of fluorine incorporation in SiO_2 gate insulators is next considered by D. Kouvatsos, R. J. Jaccodine, and F. A. Stevie, and provides detailed SIMS profiles of fluorine incorporation and motion in NF_3-enhanced oxides. Pulsed switching of the NF_3 during oxidation shows that the F is essentially immobilized within the oxide structure being formed. This also confirms the F retention at the interface for pre-oxidation HF cleans. Small quantities of F at the interface significantly reduce D_{it}, especially for surface potentials near the band edge. R. Tsu discusses extreme quantum confinement and superlattice structures and their possible applications to Si/SiO_2 structures. The physics of extreme quantum confinement require new considerations beyond those used in the present day treatment of superlattices and quantum wells. For example, bulk band structure and effective mass approximations may not apply without some modification, the effective dielectric constant may be reduced, scattering may be reduced, and the binding energy of hydrogenic impurities may be increased. This paper discusses some of these new effects, and the reasons for their appearance.

In the last paper, M. Chonko, D. Vandenberg, and D. Keitz consider the integrity of very thin silicon films deposited on SiO_2, and the ranges of deposition pressure and temperature for LPCVD silicon such that the deposited thin layers protect underlying oxides from HF etch. The authors find that amorphously deposited (low temperature; i.e. $T<560°C$) films are continuous and can provide protection in the 500 Å thickness range for a continuous range of deposition pressures. Such films can be used in enhanced Si device structures. They also find that films deposited between 590°C and 625°C do not ensure protection against an oxide etch.

PROPERTIES OF SIMOX AND RELATED SYSTEMS

Sorin Cristoloveanu and Thierry Ouisse*

Laboratoire de Physique des Composants à Semiconducteurs (UA-CNRS)
Institut National Polytechnique de Grenoble
ENSERG, BP 257, 38016 Grenoble Cedex, France
*LETI-CEA (Div. Techn. Avancées), CENG, 85X, Grenoble, France

INTRODUCTION

Even if the oxygen implantation into Si is far from being the softest method for forming a buried oxide, SIMOX is still the most successful silicon on insulator (SOI) technology. Indeed, SIMOX is unique in providing SOI wafers whose reproducibility, uniformity and yield match the standards of the industrial process for integrated circuits. The rate of progress of this technology is very impressive. SIMOX was "rediscovered" about 10 years ago and soon integrated components with promising performance were demonstrated. In the next 3–4 years, most of the basic mechanisms involved in the oxide synthesis and recrystallization were understood enough to open wide perspectives of optimization. Major milestones were: (i) annealing at very high temperatures ($> 1300°C$), (ii) implantation at controlled temperatures around $600°C$, (iii) high current (100 mA) implantation using specially designed implanters, and (iv) multiple step implants and anneals. High performance VLSI circuits are currently being fabricated on SIMOX substrates.

This paper is focused on the properties of the SIMOX structure that are essential for the proper operation of integrated circuits. These properties are determined by direct inspection of the wafer or inferred from the characterization of test MOS devices[1] (Fig.1). The next section presents the status of the SIMOX technology in terms of quality of the Si film, buried oxide and Si–SiO_2 interfaces. Interface coupling and floating body effects are typical phenomena in SOI transistors and will be briefly discussed. Emphasis is put, in the last section, on reliability aspects which may be governed by the resistance of the buried oxide to hot carrier injection and radiation effects.

PROPERTIES OF SIMOX STRUCTURES

Fabrication Conditions

Device-grade SIMOX wafers are fabricated by implantation of high doses ($1.6 - 1.8 \times 10^{18} cm^{-2}$) of oxygen at about $200 keV$ energy and $600 - 650°C$ temperature with

The Physics and Chemistry of SiO_2 and the Si-SiO_2 Interface 2
Edited by C.R. Helms and B.E. Deal, Plenum Press, New York, 1993

100mA beam. The crystallinity of the Si overlay is restored by annealing in argon (+ 1% O_2) ambient at 1300 – 1350°C for 2 – 6 hours. The thicknesses of the Si film and buried oxide are typically 150–200nm and 380–400nm, respectively. Alternatively, the synthesis can be made in three successive steps of implantation ($0.5 - 0.6 \times 10^{18} cm^{-2}$) and annealing, which results in even a better material.

Silicon interstitials are generated during the formation and growth of the buried oxide (BOX). They migrate towards the surface sink and tend to reconstruct the surface by internal epitaxy. The use of a screen oxide during implantation is detrimental because it inhibits the production of free accommodation volume (by surface regrowth) and causes strain induced dislocations[2]. On the other hand, the screen oxide is an efficient barrier against contamination, channeling effects and surface sputtering. During annealing, oxygen precipitates are dissolved and the available oxygen atoms complete the formation of the BOX.

Other variants of the process have been explored with the aim of optimizing the SIMOX synthesis or generating novel structures and geometries for the buried insulator[3]. This is possible by scanning wide ranges of doses and energies or selecting other implanted species. In particular, the local formation of SIMOX islands in designed regions of the wafer offers the opportunity of mixing SOI and bulk Si components with special functions (power, detection, speed) on the same chip. Implanted oxides with 3–D topology can also be fabricated in order to isolate totally the Si island by the surrounding buried and lateral oxides.

Silicon Film

The quality of the Si film is reflected by excellent values of carrier Hall mobility ($> 850 cm^2/Vs$ for electrons) and lifetime ($> 100 \mu s$). The residual doping is in general n-type; very low levels ($2-5 \times 10^{15} cm^{-3}$) have been achieved after studying carefully and suppressing the various sources of contamination in the implanter, furnace and capping layer. In this respect, Si or SiO_2 shields were placed in the implanter in order to avoid co-implanted species, whereas annealing in nitrogen (which generates non-intentional donors) was prohibited. Oxygen itself acts as a contaminant, since it can be activated by annealing around 450 – 550°C (thermal donors) and 750°C (new donors). The unclarified structure and features of oxygen donors are far more complex than in bulk Si and depend closely on the synthesis conditions[4]. Fortunately in recent material,[2] the concentration of interstitial oxygen was reduced below $10^{18} cm^{-3}$. Moreover, thermal and new donors are deactivated by annealing at 650°C and 900°C, respectively[4].

Comparison between the electrical properties measured near the front and back interfaces demonstrates that the film is remarkably homogeneous in the vertical direction and free from oxygen precipitates. This was not always the case in the past. Implantation at lower temperatures and annealing below 1250°C used to result in a highly defective region close to the buried oxide, where the carrier mobility and lifetime were orders of magnitude lower[3].

The density of threading dislocations has also been reduced from $10^{10} cm^{-2}$ in early low-temperature-annealed SIMOX to less than $10^4 cm^{-2}$ in multiple implanted SIMOX. Dislocations are due to defects generated during the implantation, which extend from the surface to the BOX interface, and to strain accumulation in the film. The density of dislocations increases if (i) a screen oxide is used, (ii) the implant energy is reduced, (iii) the dose is increased, or (iv) the film is thinner. The influence of the screen oxide thickness was systematically investigated and discussed in an excellent paper by Margail et al[2]. Other subsisting defects are small ($< 0.1 \mu m$) stacking faults with concentrations below $< 10^4 cm^{-2}$. Presently, the lateral uniformity of 6″ SIMOX wafers ($\pm 10 nm$) is unmatched by any other SOI technology.

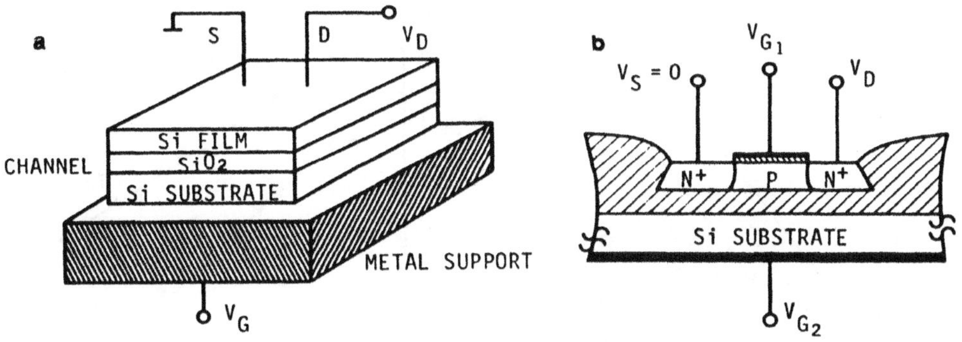

Figure 1. Schematic configurations of (a) pseudo-MOS transistor and (b) fully-processed MOSFET in SOI.

Interfaces

A natural consequence of the high crystal quality is that the top interface between Si film and thermal oxide presents the regular properties of MOS structures. No difference as compared to standard Si–SiO_2 interfaces in bulk Si is revealed by charge pumping, noise spectroscopy or MOSFET characterization techniques[1].

The quality of the buried interface between the Si film and implanted oxide is a key feature of SIMOX as it depends on implant and anneal conditions. Huge densities of back interface traps ($D_{it2} \simeq 10^{13} cm^{-2} eV^{-1}$) may be engineered in low temperature annealed SIMOX in order to shield the surface from the influence of the substrate bias. Although in device–grade SIMOX, atomically sharp interfaces have been achieved, D_{it2} is still larger than at the front interface ($D_{it2} \simeq 10^{11} - 10^{12} cm^{-2} eV^{-1}$)[5]. The surface generation velocity is also $2 - 6$ times higher at the buried interface. However, these densities of defects are not unreasonable and do not induce any significant mobility degradation at the back channel ($\mu_2 \simeq 600 cm^2/Vs$, in low doped films).

An interesting method of probing the back interface is to make use of the pseudo–MOS transistor (Ψ–MOSFET) which is inherent to any SOI material. The buried insulator acts as a gate oxide and the substrate is biased as a gate in order to activate an inversion or accumulation channel at the buried interface (Fig.1(a)). The Ψ–MOSFET is operated by applying source/drain pressure probes and, in principle, does not require any special preparation of the wafer. Typical $I_D(V_G, V_D)$ characteristics are obtained, from which the electron and hole mobilities, threshold voltages and subthreshold slope (*i.e.* D_{it2}) are easily determined. Details on the experimental set–up, theoretical background and parameter extraction procedure were discussed elsewhere[6].

Buried Oxide

Experimental data collected in the last years demonstrate that the physico–chemical structure and properties of the implanted oxide are far from being identical to those of a thermal oxide. Although the dielectric constant and breakdown field of the BOX are normal for SiO_2, variable transition layers have been observed by ellipsometry on each side of the oxide. Typical defects are listed below:

E' centers — Concentrations as large as $10^{18} cm^{-3}$ have been measured by Electron Spin Resonance[7]. They originate from oxygen vacancies, act as hole traps and govern the radiation–induced effects at low fields. Unlike the case of thermal oxides, E' centers spread through the whole volume of the buried oxide.

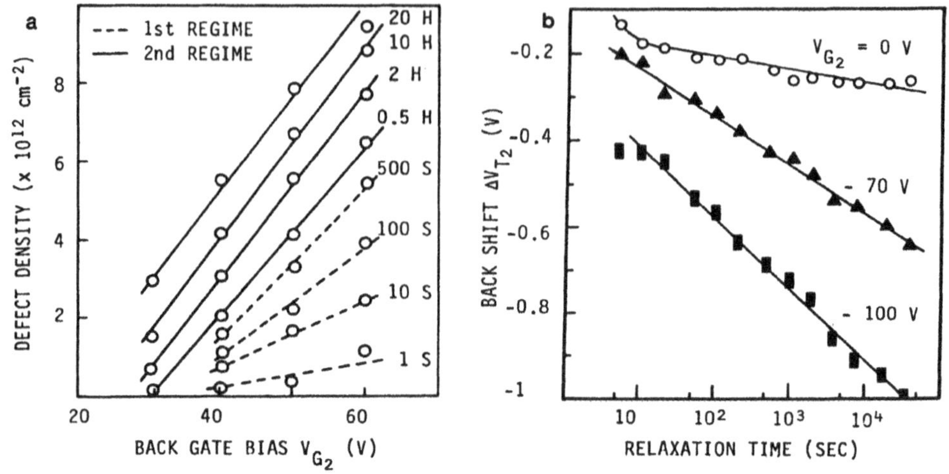

Figure 2. Direct degradation of the back interface in n–channel SIMOX MOSFET's. (a) Local defect density versus stress bias V_{G2} and duration ($V_D = 6V$, $L = 1\mu m$). (b) Bias-assisted threshold voltage relaxation ($T = 300K$).

Electron traps — An intrinsic feature of SIMOX, pointed out by Leray et al[8], is the large amount of electron trapping which is observed after radiation. Several groups have confirmed that many electron traps ($> 10^{12} cm^{-2}$) are present even before X–ray exposure[9]. A quantitative correlation between the concentrations of excess Si atoms and electron traps might help in clarifying their nature and origin.

Slow traps — Charge pumping experiments[10] revealed the presence of slow traps located near the film–buried oxide interface. Their concentration exceeds that of "fast" interface traps and their capture/emission time constants are in the range $0.1 - 1msec$.

Pipes — Silicon columna of $1 - 10\mu m$ diameter, which may extend from one interface to the other, have been observed by TEM, selective KOH etching or leakage current measurements through the buried oxide. These pipes represent an unbearable leakage path between integrated devices and substrate. They originate from large particles (1–10 μm) which may subsist accidentally on the surface and shadow the oxygen implantation[2]. However, the density of pipes has been reduced to less than $0.5cm^{-2}$.

Silicon islands — The aggressive interaction between implanted oxygen atoms and Si matrix as well as the need for an accommodation volume results in a more or less complete oxidation process. Spectroscopic ellipsometry suggests that the buried oxide is Si rich, amorphous Si defects being concentrated near the BOX–substrate interface[11]. Large Si islands may remain capped in the bottom of the buried oxide. Although it is not clear whether they affect or not the operation of integrated circuits, their number has been reduced by multiple implantation process or use of a screening oxide[2].

More information about the electrical behavior of these defects is found from radiation and hot carrier injection measurements.

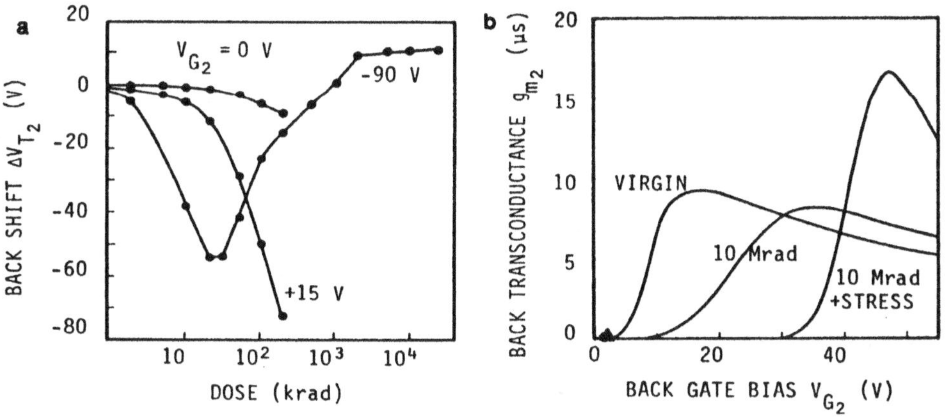

Figure 3. Radiation effects in n-channel SIMOX MOSFET's. (a) Back channel threshold voltage shift ΔV_{T2} versus radiation dose and bias. (b) Shift of the back channel transconductance after radiation (at $V_{G2} = -90V$) and subsequent stress ($V_D = 6V$, $V_{G2} = 60V$).

Direct injection of hot carriers into the buried oxide. A substantial electron trapping ($10^{12}cm^{-2}$ for $1.5MV/cm$) results from the injection of back channel hot electrons into the buried oxide[12]. This causes large variations in the back channel threshold and flat band voltages. The filling of pre-existing traps is field-dependent and the concentration of filled traps saturates to a value proportional to the back gate voltage V_{G2} (Fig.2(a)). In addition, new electron traps are generated by hot carriers.

A logarithmic detrapping time law has been observed (Fig.2(b)), which is typical for a broad distribution (in energy and/or depth) of emission time constants. The detrapping mechanism is field-dependent and has been attributed to either direct or phonon-assisted tunneling[12,13]. In the case of p-channel MOSFET's, a large electron trapping occurs even at zero field across the oxide.

Radiation-induced defects. SOI technologies have primarily been developed for their immunity to *transient* photocurrents induced during radiation exposure. SIMOX devices also hardened against *permanent* dose effects (up to 10Mrad) are now available[8]. The hole trapping is qualitatively similar in SIMOX oxides and in thick thermal oxides. Quantitatively, it is larger in SIMOX (due to more numerous E' centers) and saturates for doses between $2 - 10Mrad$. The recovery from radiation damage proceeds logarithmically with time. The positive charge is removed by thermal excitation, the activation energies being broadly distributed[14].

The trapping and detrapping of electrons is a more specific feature of SIMOX. For high doses and large negative substrate bias, the electron trapping may even overcome the effect of hole trapping and leads to an overall positive shift ($\Delta V_{T2} > 0$) of the back channel threshold voltage[9] (Fig.3(a)). The radiation-generated electron traps are also responsible for an enhanced hot carrier damage (Fig.3(b)). This suggests that positively charged E' centers are attracting the electrons and assist their recombination with holes. In comparison with the amount of electron and hole trapping, the effect of interface traps, which are also generated by radiation, is almost negligible immediately after exposure.

TYPICAL EFFECTS IN SIMOX TRANSISTORS

Interface Coupling

In thin and low doped SOI films, the two depletion regions associated with the front and back gates meet when appropriate voltages $V_{G1,2}$ are applied. In very thin films, it is even possible to fully deplete the whole film by acting on the front gate only, while maintaining $V_{G2} = 0$. In full depletion, the two interfaces (and channels) are electrically interacting. *Interface coupling* means that the properties of the front channel transistor are affected by the back gate bias and by the defects located at the buried interface. Typical consequences are:

- The front channel threshold voltage V_{T1} decreases when V_{G2} varies from accumulation to inversion.
- The subthreshold slope is a maximum and the swing can reach nearly ideal values (60 mV/decade) for depletion at the back interface[15].
- The transconductance curves $g_{m1}(V_{G1})$ are shifted and distorted as V_{G2} is increased. A plateau may emerge when the back channel is modulated by V_{G1} before the front channel activation[16].
- Inversion at the back interface is equivalent to a leakage current in the front channel MOSFET.

These effects are reversible and can equally be sensed on the back channel transistor. The intensity of the interface coupling actually depends on the film thickness, doping, and interface defects. Fully-depleted integrated circuits exhibit improved speed and are less sensitive to floating body effects. Care is needed, however, to minimize the influence of buried interface and oxide defects (generated by the fabrication process, radiation exposure or hot carrier injection) on the front channel properties. It will be shown in the last section that this indirect degradation, induced by interface coupling, represents a challenge for the reliability of p-channel MOSFET's.

Floating Body

Partially-depleted devices form the second family of SOI circuits. Thick Si films, free from interface coupling effects, are indeed very appropriate for radiation-hardened applications. Unfortunately, partially-depleted transistors are subjected to parasitic effects induced by their floating body. The absence of a body contact prevents the evacuation of majority carriers generated by impact ionization. Their accumulation near the source contact leads to an increase of the body potential and to a corresponding reduction in the threshold voltage V_{T1}. The drain current increases and, eventually, the body to source diode becomes forward biased, activating the intrinsic bipolar transistor. Major drawbacks of this positive feed-back are the onset of a "kink" in $I_D(V_D)$ characteristics and a hysteresis in $I_D(V_{G1})$ curves. The hysteresis is reflected by nearly vertical current transitions between the on- and off-states, which are different for increasing or decreasing V_{G1}. A (V_{G1}, V_D) domain is found where the gate control is lost and both the transconductance and the conductance take negative values[17].

The modeling of the various phenomena involved in the feed-back shows that the hysteresis can formally be described as a second-order phase transition and depends on the film doping, thickness, carrier lifetime, interface defects and V_{G2} bias[17]. The extent of floating body effects is also related to the ability of the diode to evacuate majority carriers.

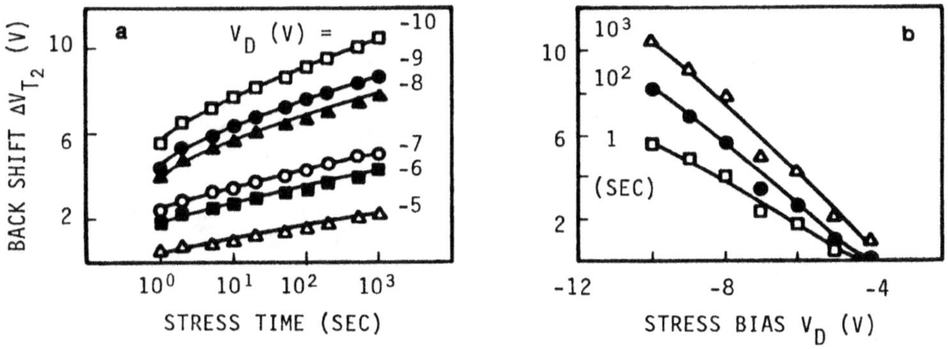

Figure 4. Back channel threshold voltage shift ΔV_{T2} during the stress of front channel p–MOSFET's ($L = 0.6\mu m$). Effects of (a) stress time and (b) drain voltage ($V_{G1} = -1.5V$, $V_{G2} = 0$).

Moreover, the floating body is responsible for detrimental transient effects[1]. The isolated film is unable to supply rapidly the majority carriers (in enhancement-mode MOSFET's) or the minority carriers (in accumulation-mode MOSFET's) needed to compensate for the deep-depletion regions which may be induced by the bias. Five-terminal MOSFET's, where the body can be biased independently or connected to the source, are especially fabricated for radiation applications. Although the floating body effects are suppressed, this solution implies a trade-off in terms of speed and integration density.

RELIABILITY ASPECTS

Aging of n–Channel MOSFET's

Carrier heating and injection into the gate oxide of submicron MOSFET's leads to interface defects (charge trapping and interface states) localized within the pinch off region. This damage is visible through the modification of the threshold voltage, subthreshold slope, mobility, series resistances and floating body effects[12]. The transistor aging depends on many structural and biasing parameters and an accurate comparison between the merits of bulk Si and SOI devices is impossible. However, there are strong arguments indicating the advantage of fully-depleted n–channel SIMOX MOSFET's, in particular after optimization of the LDD structure. In partially-depleted transistors the peak of the lateral field is larger and they behave more similarly to bulk Si MOSFET's.

Systematical measurements have recently demonstrated that defects are exclusively formed at the same interface where the electrons flow, for instance at the front interface in normal operation conditions. It was already mentioned that the buried interface can be severely degraded by direct injection of back channel electrons. Even though, no modification of the front channel properties was detected when the back interface was kept in accumulation, during the test, in order to mask the defects. In contrast, if the damaged back interface is depleted, the front channel characteristics suggest an *apparent* aging. This change must not be misinterpreted as a *physical* degradation of the front interface, because it merely results from the coupling of front channel electrons with back interface defects.

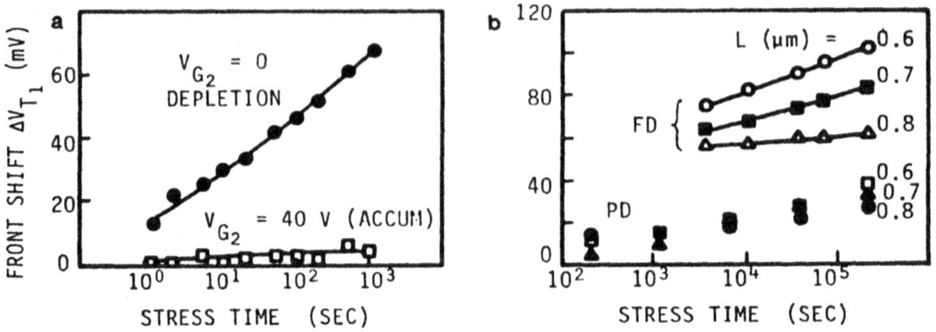

Figure 5. Front channel threshold voltage shift ΔV_{T1} induced by the trapping of electrons in the buried oxide during the stress of front channel p–MOSFET's. (a) Influence of V_{G2} bias (depletion or accumulation) during the test (L = 0.8μm). (b) Influence of the channel length for fully- and partially-depleted p–MOSFET's (stress at $V_D = -6.5V$, $V_{G1} = -2V$, $V_{G2} = 0$ and test at $V_{G2} = 0$).

Aging of p–Channel MOSFET's

Unlike the case of n–channel MOSFET's, the reliability of p–channel transistors is governed by the presence of many electron traps in the BOX. As far as the radiation hardness is concerned, this is not detrimental because electron traps can compensate the effect of hole trapping and help reducing ΔV_{T2} shifts[8,9]. The problem is that the electron trapping dramatically affects the hot carrier reliability of p–channel SIMOX MOSFET's. Indeed, electrons generated by impact ionization near the drain are driven by the electric field towards the back interface. In submicron MOSFET's, this systematically results in electron injection and trapping in the BOX. As a consequence, substantial variations of the *back* channel threshold voltage are monitored during the stress of the *front* channel transistor (Fig.4). This degradation happens in both fully-depleted and partially-depleted SIMOX MOSFET's, as long as the film is relatively thin ($\leq 0.2\mu m$, see also Fig.5(b)).

It was found that the threshold voltage shift increases logarithmically with stress time (Figs.4(a)) and is more accentuated in shorter MOSFET's. No saturation of the buried oxide aging was found for the times explored in this investigation ($10^5 - 10^6 sec$). There are arguments indicating that, unlike the case of the front thermal oxide interface, the defective region extends over almost the whole back channel interface. For a given duration of stress, ΔV_{T2} increases almost linearly with the stressing drain voltage (Fig.4(b)), depends on V_{G2} bias as well, but is rather independent on the front gate voltage V_{G1}.

The key point is that the electron trapping in the BOX makes a great impact on the front channel properties. Figure 5(a) demonstrates that the aging of the front channel transistor may be governed by the defects created at the back interface rather than by the standard degradation of the front interface. This again happens, via interface coupling, when the back interface is depleted. If the electron traps are screened by accumulating the back interface, only a very limited shift of the front gate threshold voltage ΔV_{T1} subsists which corresponds to the natural (physical) degradation of the top channel (Fig.5(a)). Another problem is that the electron trapping in the BOX is equivalent to $V_{G2} < 0$ and tends to drive the back interface from accumulation into depletion regime, where the influence of the defects is just a maximum. This interface coupling effect is enhanced in shorter channel transistors (Fig.5(b)).

CONCLUSIONS

Intensive research and industrial commitment have succeeded in achieving thin SIMOX films with excellent crystallinity and electrical properties. However, as the defect densities come under process control, special SOI effects induced by interface coupling and floating body become more severe. Also, the microstructure of the implanted oxide differs from that of thermal oxides and requires further investigation. Buried oxide defects and, in particular, electron traps play a prevailing role in reliability aspects. The hot carrier related degradation of p-channel MOSFET's appears to be a key challenge. These recent features should be carefully considered and appropriate technological solutions found, in order for SIMOX to fully materialize its exceptional potential.

Acknowledgements

Parts of this work have been made possible by grants from various organizations (GCIS, JESSI, ESPRIT). Our colleagues from the LPCS laboratory (Grenoble) as well as Drs. A-J. Auberton-Hervé, B. Giffard, G. Reimbold and J. Margail from LETI (Grenoble), G. Borel from Thomson-TMS (Grenoble), G. Campisi and H. Hughes from NRL (Washington D.C.), D. Ioannou from G. Mason University (Virginia) are thanked for interest expressed in this work.

References

1. S. Cristoloveanu, A review of the electrical properties of SIMOX substrates and their impact on device properties, *J. Electrochem. Soc.* 138:3131 (1991).
2. J. Margail, J.M. Lamure, and A.M. Papon, Defects in SIMOX structures: some process dependence, *Mat. Science and Eng.* B12:27 (1992).
3. S. Cristoloveanu, Advanced silicon on insulator materials: processing, characterization and devices, *in:* "Semiconductor Silicon," G. Harbeke and M.J. Schulz, eds., Springer Verlag, Berlin, 223 (1989).
4. S. Cristoloveanu, Oxygen-related activity and other specific electrical properties of SIMOX, *Vacuum* 42:371 (1991).
5. D.E. Ioannou, S. Cristoloveanu, C.N. Potamianos, X. Zhong, P.K. McLarty, and H.L. Hughes, Optimization of SIMOX for VLSI by electrical characterization, *IEEE Trans. Electron Dev.* 38:463 (1991).
6. S. Cristoloveanu and S. Williams, Point contact pseudo-MOSFET for in situ characterization of as-grown silicon on insulator wafers, *IEEE Electron Device Lett.* 13:102 (1992).
7. J.F. Conley, P.M. Lenahan, and P. Roiman, Electron spin resonance study of E' trapping centers in SIMOX buried oxides, *IEEE Trans. Nucl. Sci.* 38:1247 (1991).
8. J.L. Leray et al, From substrate to VLSI: investigation of hardened SIMOX without epitaxy for dose, dose rate and SEU phenomena, *IEEE Trans. Nucl. Sci.* 35:1355 (1988).
9. T. Ouisse, S. Cristoloveanu, and G. Borel, Electron trapping in irradiated SIMOX buried oxides, *IEEE Electron Device Lett.* 12:312 (1991).
10. T. Ouisse, S. Cristoloveanu, T. Elewa, H. Haddara, G. Borel, and D.E. Ioannou, Adaptation of the charge pumping technique to gated PIN diodes fabricated on silicon on insulator, *IEEE Trans. Electron Dev.* 38:1432 (1991).
11. P.J. McMarr, B.J. Mrstik, M.S. Barger, G. Bowder, and J.R. Blanco, Study of Si implanted with oxygen using spectroscopic ellipsometry, *J. Appl. Phys.* 67:7211 (1990).
12. T. Ouisse, S. Cristoloveanu, and G. Borel, Hot carrier induced degradation of the back interface in short channel silicon on insulator MOSFET's, *IEEE Electron Device Lett.* 12:290 (1991).
13. M. Simons, G. Campisi, and H.L. Hughes, Evaluation of pulsed radiation effects in buried oxide by fast CV measurements, *J. Appl. Phys.* 70:4784 (1991).
14. C.A. Pennise and H.E. Boesch Jr., Thermal annealing of trapped holes in SIMOX buried oxides, *IEEE Trans. Nucl. Sci.* 38:1240 (1991).
15. B.M. Mazhari, S. Cristoloveanu, D.E. Ioannou, and A.L. Caviglia, Properties of ultra thin wafer bonded silicon on insulator MOSFET's, *IEEE Trans. Electron Dev.* 38:1289 (1991).

16. T. Ouisse, S. Cristoloveanu, and G. Borel, Influence of series resistances and interface coupling on the transconductance of fully-depleted silicon on insulator MOSFET's, *Solid St. Electron.* 35:141 (1992).
17. T. Ouisse, G. Ghibaudo, J. Brini, S. Cristoloveanu, and G. Borel, Investigation of floating body effects in silicon on insulator metal-oxide-semiconductor field-effect transistors, *J. Appl. Phys.* 70:3912 (1991).

REOXIDIZED NITRIDED OXIDE GATE DIELECTRICS FOR ADVANCED CMOS*

Peter W. Wyatt[1] and Gregory J. Dunn[2]

[1]MIT Lincoln Laboratory, Lexington MA 01773
[2]U.S. Department of State, United States Embassy, Rome, Italy

ABSTRACT

Hot carrier damage limits the scaling of MOSFETs. Nitridation and reoxidation of the gate oxide can greatly reduce this damage and therefore allow a more favorable tradeoff among shorter channels, high operating voltage, and simpler fabrication processes. This paper reviews the nitridation-induced reliability improvements in the micron to deep submicron range for n-channel devices and the potential for p-channel improvements as well.

INTRODUCTION

Integrated circuit geometries are approaching half micron and below, resulting in increasingly high electric fields near transistor drains. These high fields accelerate charge carriers to high enough energy to produce damage in the gate oxide and at the oxide-silicon interface. Complex wafer fabrication processes are being explored[1] to reduce the field by drain engineering, but a better alternative may be to develop a gate dielectric which is more resistant to damage by the hot carriers. Reoxidized nitrided silicon dioxide (RNO) is receiving increasing attention as a candidate to replace the conventional oxide for this purpose. RNO offers significant improvements in charge-to-breakdown[2,3], n-MOSFET resistance to channel hot carrier stress[3–6], suppression of boron penetration from p-type polysilicon gates[7,8], resistance to latent electrostatic discharge damage[9], and suppression of neutral electron trap generation by x-ray lithography[6,10]. Although nitridation increases background electron trapping in the dielectric, the generation of *new* traps — bulk or interface — is suppressed. Furthermore, trapping in the bulk will diminish in importance as gate dielectric thickness approaches 5 nm for deep submicron devices since trapped electrons can easily escape by tunneling.

Nitridation is a high temperature process, and for submicron devices, exposure to

*This work was supported by the Department of the Air Force and the Office of Naval Research.

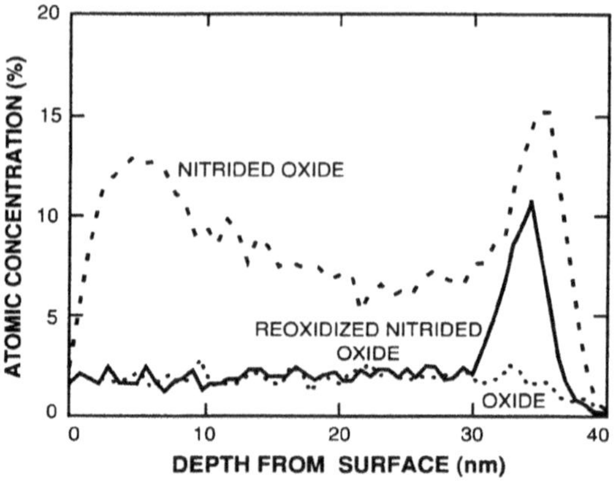

Figure 1. Typical Auger profiles of nitrogen concentration in nitrided oxide and RNO. The oxide curve indicates the noise level of the measurement.

high temperatures after the channel doping implants must be minimized in order to maintain the desired doping profiles. We have chosen to explore the maximum capabilities of this technology, and therefore have used furnace processing with temperature cycles that are not practical for conventional submicron, bulk silicon devices, though they might be quite acceptable for other high speed devices (*e.g.* fully depleted SOI) in which doping profiles are less critical. Several studies[2,11] have shown that rapid thermal nitridation and reoxidation at lower temperatures and much shorter times can produce some of the same advantages as the longer, more extensive furnace processes, and RTP is being seriously considered by a number of manufacturers for their 0.5-μm processes[12].

DEVICE FABRICATION

The transistors and capacitors discussed here were all fabricated with conventional poly gate, LOCOS technology, except for the nitridation and reoxidation steps. To simplify both processing and analysis no threshold adjust implants were used, so these are surface channel devices. The gate material was LPCVD polysilicon deposited at 620°C and doped with phosphorus by solid source diffusion at 925°C for 45 minutes. For n-channel transistors the substrate was (100), boron doped, 14-22 Ω·cm, and the source-drain implant was 6×10^{15} cm^{-2} arsenic at 100 keV, activated in dry O_2 at 950°C for 20 min. For p-channel the substrate was (100), phosphorus doped, 1-5 Ω·cm, and the source-drain implant was 2×10^{15} cm^{-2} boron at 25 keV, activated in dry O_2 at 850°C for 20 min. PECVD insulating glass was deposited from SiH_4 and N_2O and densified in nitrogen at 700°C for 30 min. 0.75 μm Al-Si-Cu was deposited by magnetron sputtering and patterned by plasma etching, followed by a 20-min, 455°C sinter in 3% H_2/N_2 forming gas.

The 37-nm gate oxide was grown in dry O_2 at 1000°C. Control device oxides were annealed in dry N_2 for 10 min at the oxidation temperature. Nitridation and reoxidation were accomplished *in situ* in pure NH_3 for 15 min and in dry O_2 for various times, respectively, at 1100°C. Gases were purged from the tube with N_2 before any

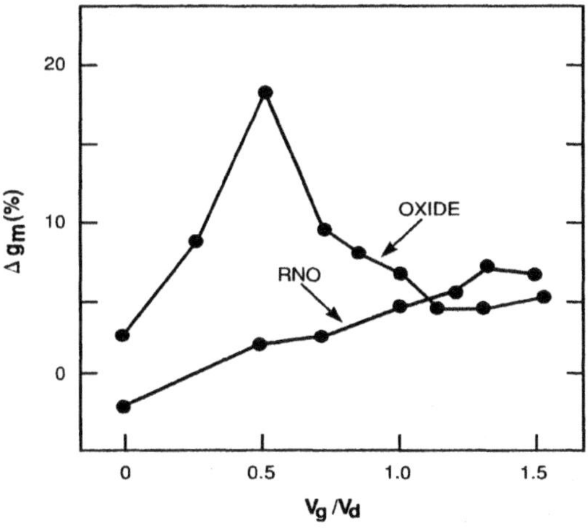

Figure 2. Decrease in transconductance due to 5000-second channel hot carrier stress in oxide-gate and RNO-gate (15-min reoxidation) transistors. The oxide channel length was 1.4 μm and stress voltage was $V_d = 7.0$V, while corresponding RNO values were 1.1 μm and 7.5V.

changes, during which period the temperature was ramped to its new value. Negligible change in thickness occurred during nitridation or reoxidation. These processes result in nitrogen profiles, measured by Auger spectroscopy with sputter etching, such as those illustrated in Fig. 1. Nitridation produces peaks in N concentration near both the free surface and the silicon interface, with lower concentration in the bulk. Reoxidation eliminates the surface N very quickly and the bulk N more slowly[13], while the interfacial nitrogen remains even after long reoxidation. This fabrication sequence results in fixed positive charge in RNO which is somewhat greater than in oxide, about 1×10^{11} cm^{-2}, and interface state density which is about the same as in oxide, 3×10^{10} per cm^2·eV[14].

INTERFACE STATES

There is broad agreement that the most important form of transistor damage produced by electrical stress in modern integrated circuits is the interface states in n-channel devices created by channel hot carriers. These states, located near the drain junction, reduce the transconductance and also cause a slight threshold voltage shift. They are formed in both p- and n-channel transistors[15], but they seriously affect device operation only in n-channel, at least for channel lengths of 0.5 μm and above. (Shorter channels are discussed below[16].) As channel lengths are reduced in order to increase device speed, n-channel degradation by interface state formation is becoming an increasingly important problem.

Heavy nitridation and reoxidation almost completely eliminates such interface state generation. Figure 2 compares the change in transconductance produced by channel hot carrier stress in oxide and RNO devices[4]. For oxide we observe the usual behavior of maximum degradation at $V_g \approx V_d/2$. The RNO devices, with shorter channel and

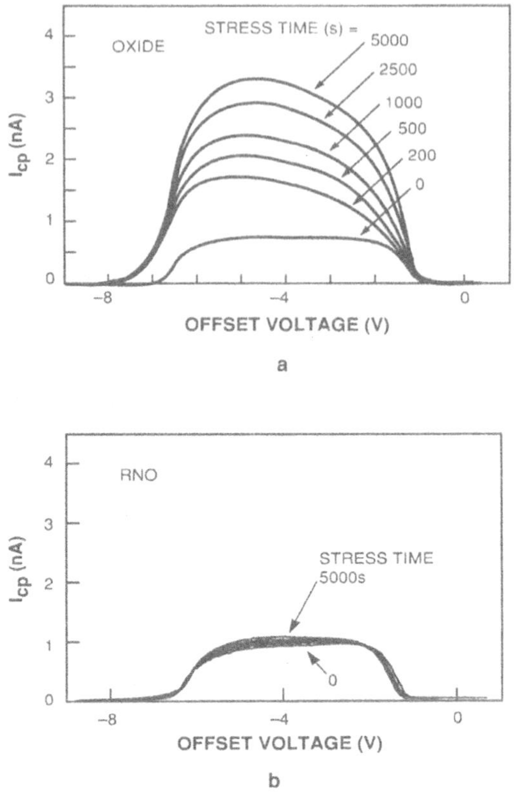

Figure 3. Charge pumping curves for (a) an oxide device with $L_{eff} = 1.4\mu m$ and (b) an RNO device with $L_{eff} = 1.1\mu m$, both stressed at $V_g = V_d/2$ with V_d=7.5 V.

stressed at higher drain voltage in order to produce readily measurable damage, show no such maximum in Δg_m, and instead have the greatest degradation for $V_g > V_d$. High gate voltage is the condition for maximum electron injection into the dielectric, and the observed change results from electron trapping.

Figure 3 demonstrates that interface states are not formed by such stress in the RNO devices. These are measurements of charge pumping current on devices like those of Fig. 2. The area under an I_{cp} curve is proportional to the number of interface states in the channel. The data show that the number of states before stress is about the same for the two dielectrics, and that stress at the condition which produces maximum g_m degradation in oxide results in negligible interface states in the RNO device. A similar measurement of an RNO device stressed at $V_g > V_d$ also shows negligible interface state generation.

Transconductance degradation limits the reliability of short channel transistors. N-channel lifetime is typically defined by a 10% reduction in g_m, determined by stressing at high drain voltage and extrapolating to operating voltage based on an inverse V_d dependence. A typical plot is shown in Fig. 4 for devices like those of Fig. 2. Each type of device was tested at its own worst case condition, $V_g = V_d/2$ for oxide and $V_g = 1.3V_d$ for RNO. The data for RNO are somewhat scattered because the degra-

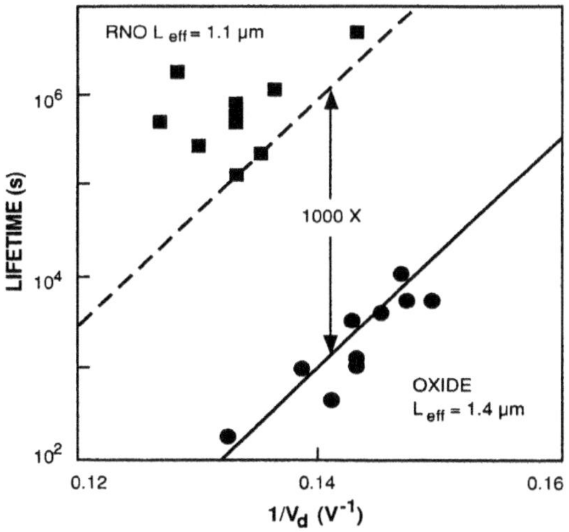

Figure 4. Stress time to 10% g_m decrease for n-channel transistors like those of Fig. 2. Each device type was tested at its own worst case condition, as described in the text, and the RNO data are extrapolated in time to 10% degradation.

dation is so small, which required extrapolation in time to reach the point of 10% degradation, but a factor of 1000× improvement is clearly demonstrated in this static stress experiment. It is well known that static stress is not completely representative of the damage encountered by transistors in CMOS circuits. In particular, the worst case condition for RNO devices does not occur in inverters and similar logic gates, though it may occur in some other circuits. In simulated inverter stress, we have shown[5] the lifetime advantage of RNO over oxide to be several orders of magnitude greater than the 1000× of Fig. 4.

As discussed in the Introduction, this work has used very heavily nitrided material in order to ascertain the limits of device performance. Many others[2,11] have investigated very light nitridations of thin oxides using rapid thermal nitridation. These reports have shown considerable reduction (though not elimination as found here) of interface state generation by hot carriers. Thus nitridation and reoxidation processes which are compatible with deep submicron devices can allow either simplified drain engineering (and hence better yield) or improved lifetime for a given device structure and operating voltage. These benefits are clear for n-channel devices and may apply for p-channel as well[16].

ELECTRON TRAPPING

The other important mechanism of transistor degradation is electron trapping in the gate dielectric. For oxide n-channel devices this is a small effect compared with interface state generation, while for p-channel it is the principal degradation mechanism at least for channel lengths down to 0.5 μm[15]. For oxide the trap densities and cross sections are so small that the p-channel rate of degradation is negligible compared with that of n-channel. Nitridation, however, introduces a high density[17] of large cross

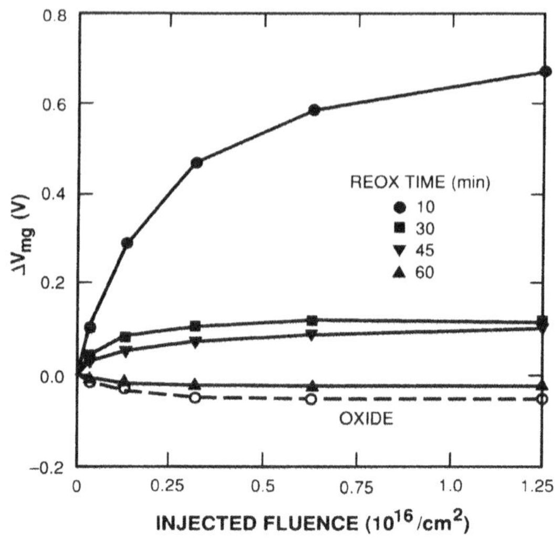

Figure 5. Capacitor midgap voltage shift *vs.* Fowler-Nordheim electron fluence in heavily nitrided and reoxidized, 37-nm RNO.

section[18,19] traps. Such traps are commonly studied by Fowler-Nordheim tunneling injection of electrons from the silicon at high field. Figure 5 shows the shift in midgap voltage of an MIS capacitor due to such injection as a function of reoxidation time for devices processed as described above, apparently indicating that the traps have been nearly eliminated.

Several authors[10,18–20] have shown, however, that such data can easily be misinterpreted due to field-enhanced emission of electrons from the traps. In transistor operation the field is considerably lower than in tunneling injection tests, so detrapping is suppressed. While Fig. 5 appears to show that 60 min of reoxidation produces a trap density in RNO almost as low as that in oxide, channel hot carrier stressing shows that in fact, far more electron trapping occurs in RNO transistors[10]. This trapping produces the small degradation in n-channel devices discussed in the previous section, and it also results in changes in p-channel characteristics.

The worst case stress condition for both oxide and RNO p-channel transistors is high drain voltage and gate voltage only slightly more negative than threshold. This condition maximizes the lateral field near the drain and produces a transverse field pushing electrons toward the gate. The resulting trapped electrons near the drain reduce $|V_T|$ in that region, thus effectively reducing the channel length. That *increases* g_m, which might be seen as a benefit, but it also decreases the punchthrough voltage and results in transistor leakage. Thus the time to 10% increase in transconductance is generally considered to represent p-channel lifetime. Negligible interface states are produced by such stress even in oxide, as shown by the charge pumping curves of Fig. 6. The main peak *decreases* in amplitude after stress, while a shoulder appears at more positive voltage. The shoulder results from the positive shift in threshold and flatband voltages near the drain. The area under the curve, which is proportional to the number of interface states, remains constant.

The lifetime of oxide and RNO p-channel transistors is plotted versus stress in Fig. 7, again with each type of device stressed at its worst-case gate voltage. It is evident

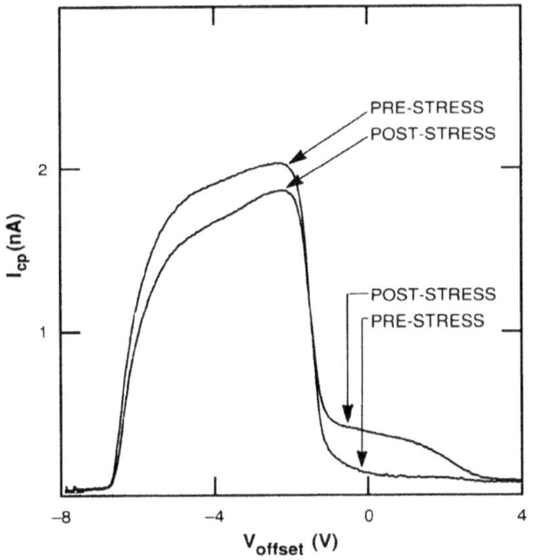

Figure 6. Typical charge pumping current for an oxide-gate, p-channel transistor, with 1.3 μm channel length. Stress was $V_d = -11V, V_g - V_T = V_d/6$ for sufficient time to produce about 10% increase in transconductance. Similar results were obtained with RNO transistors.

that even these relatively long reoxidation times have not eliminated the electron traps, and that the RNO p-channel devices have substantially shorter lifetime than the oxide ones. Note also, however, that all the p-channel lifetimes are much longer than those of oxide n-channel devices. Figure 8 combines the data of Figs. 4 and 7 to show that for RNO, p- and n-channel lifetimes are similar and about three orders of magnitude longer than that of n-channel oxide devices.

Few studies have been reported of p-channel lifetime in lightly nitrided RNO devices. Momose[21] reported enhanced electron trapping in surface-channel p-channel RNO devices with 6- and 10-nm dielectric and 1.0-μm channel length. The rapid thermal nitridation conditions were not stated, but the high fixed positive charge suggests a nitridation intermediate between ours and the very light nitridation of the following author. Hori[11] reported no such enhancement — oxide and RNO devices had the same lifetime under electron trapping conditions — for p-channel transistors with 8-nm dielectric and 0.35-μm channel length. These are probably buried channel devices, although that is not explicitly stated.

Tsuchiya, *et al.*[16] report a very important result for deep submicron, p-channel *oxide* gate transistors. They observe a *reduction* in transconductance and *negative* threshold shift in 0.25-μm gate length p-channel buried channel transistors with 3.5-nm gate dielectric after stress at $V_g \approx V_d/2$. This behavior is completely inconsistent with electron trapping. Based on charge pumping and other measurements the authors conclude that the degradation results from generation of interface states, just as is commonly observed for n-channel. This result suggests that for very thin gate dielectrics, from which trapped electrons can escape by tunneling, interface state generation may become the principal degradation mechanism for p-channel devices as well as n-channel, in which case nitridation and reoxidation would again improve the lifetime.

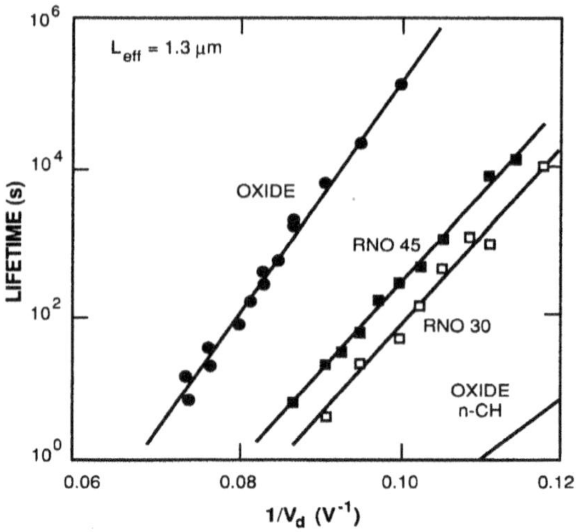

Figure 7. P-channel transistor lifetime (10% transconductance increase). The oxide devices had $V_g - V_T = V_d/6$ and the RNO devices, reoxidized for 30 and 45 minutes, had $V_g - V_T = V_d/8$. The oxide n-channel line is taken from Fig. 4.

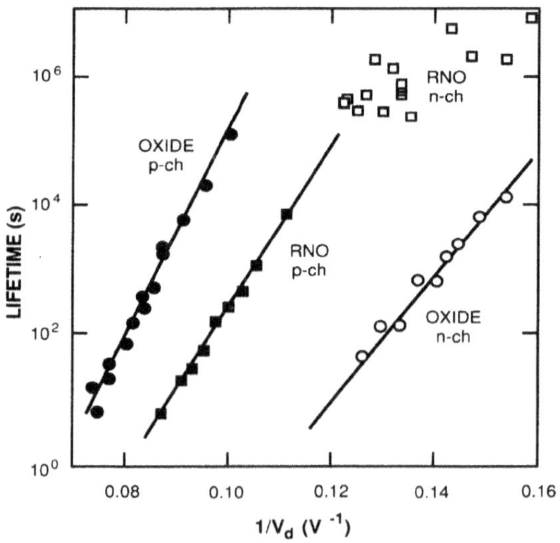

Figure 8. Lifetime comparisons of oxide and RNO n- and p-channel transistors, each stressed at its worst case condition.

SUMMARY

Reoxidized nitride oxide virtually eliminates the creation of interface states by electrical stress on transistors[4] and thus offers greatly improved n-channel reliability. It also allows more flexibility in wafer processing, including resistance to radiation damage in x-ray lithography[6,10], and reduces the need for complex drain engineering. The high temperatures used in nitridation and reoxidation in our work produce the best possible hot carrier resistance, but most of the improvement can be attained with much lower temperatures and very short times[2,11] which are compatible with deep submicron devices. The chief disadvantage of RNO is the high density of electron traps, which might be expected to affect device reliability. We have shown that for channel lengths in the 1 μm range electron trapping results in p- and n-channel RNO transistors of about equal reliability, both with lifetimes about 1000× better than n-channel oxide transistors. For the very thin gate dielectrics appropriate for quarter micron transistors, trapped electrons can tunnel out so that electron trapping is expected to become even less of a concern. Other studies[11,16] have supported this expectation and one[16] even indicates that for very small p-channel devices, interface states may be the principal degradation mechanism just as for n-channel. If further work confirms that observation then the advantage of RNO will be even greater.

REFERENCES

1. For example, T.N. Buti, S. Ogura, N. Rovedo, and K. Tobimatsu, A new asymmetrical halo source GOLD drain (HS-GOLD) deep sub-half-micrometer n-MOSFET design for reliability and performance, *IEEE Trans. Electron Dev.,* **ED-38**, 1757 (1991).

2. A.T. Wu, V. Murali, J. Nulman, B. Triplett, D.B. Fraser, and M. Garner, Gate bias polarity dependence of charge trapping and time-dependent dielectric breakdown in nitrided and reoxidized nitrided oxides, *IEEE Electron Dev. Let.,* **EDL-10**, 443 (1989).

3. W. Yang, R. Jayaraman, and C.G. Sodini, Optimization of low-pressure nitridation/reoxidation of SiO_2 for scaled MOS devices, *IEEE Trans. Electron Dev.,* **ED-35**, 935 (1988).

4. G.J. Dunn and S.A. Scott, Channel hot-carrier stressing of reoxidized nitrided silicon dioxide, *IEEE Trans. Electron Dev.,* **ED-37**, 1719 (1990).

5. B.S. Doyle and G.J. Dunn, Dynamic hot-carrier stressing of reoxidized nitrided oxide, *IEEE Electron Dev. Let.,* **EDL-12**, 63 (1991).

6. G.J. Dunn, Effect of synchrotron x-ray radiation on the channel hot-carrier reliability of reoxidized nitrided silicon dioxide, *IEEE Electron Dev. Let.,* **EDL-12**, 8 (1991).

7. T.Morimoto, H.S. Momose, Y. Ozawa, K. Yamabe, and H. Iwai, Effects of boron penetration and resultant limitations in ultra thin pure-oxide and nitrided-oxide gate-films, *IEEE Int. Electron Dev. Mtg. Technical Digest,* 429 (1990).

8. J.S. Cable, R.A. Mann, and J.C.S. Woo, Impurity barrier properties of reoxidized nitrided oxide films for use with p^+-doped polysilicon gates, *IEEE Electron Dev. Let.,* **EDL-12**, 128 (1991).

9. B.S. Doyle, K.R. Mistry and G.J. Dunn, Reoxidized nitrided oxides (RNO's) for latent ESD-resistant MOSFET devices, *IEEE Electron Dev. Let.,* **EDL-12**, 184 (1991).

10. G.J. Dunn and J.T. Krick, Channel hot-carrier stressing of reoxidized nitrided oxide p-MOSFET's, *IEEE Trans. Electron Dev.,* **ED-38**, 901 (1991).

11. T. Hori, Deep-submicron nitrided-oxide CMOS technology for 3.3-V operation, *IEEE Int. Electron Dev. Mtg. Technical Digest,* 837 (1990).

12. J. Yasaitis, private communication.

13. M.L. Naiman, C.T. Kirk, B.L. Emerson, J.B. Taitel, and S.D. Senturia, The constitution of nitrided oxides and reoxidized nitrided oxides on silicon, *J. Appl. Phys.,* **58**, 779 (1985).

14. G.J. Dunn and P.W. Wyatt, Reoxidized nitrided oxide for radiation-hardened MOS devices, *IEEE Trans. Nuclear Science,* **NS-36**, 2161 (1989).

15. P. Heremans, R. Bellens, G. Groeseneken, and H.E. Maes, Consistent model for the hot-carrier degradation in n-channel and p-channel MOSFET's, *IEEE Trans. Electron Dev.,* **ED-35**, 2194 (1988).

16. T. Tsuchiya, Y. Okazaki, M. Miyake, and T. Kobayashi, New hot-carrier degradation mode and lifetime prediction method in quarter-micrometer PMOSFET, *IEEE Trans. Electron Dev.,* **ED-39**, 404 (1992).

17. S.K. Lai, D.W. Dong, and A. Hartstein, Effects of ammonia anneal on electron trappings in silicon dioxide, *J. Electrochem. Soc.,* **129**, 2042 (1982).

18. S.-T. Chang, N.M. Johnson, and S.A. Lyon, Capture and tunnel emission of electrons by deep levels in ultrathin nitrided oxides on silicon, *Appl. Phys. Lett.,* **44**, 316, (1984).

19. F.L. Terry, Jr., P.W. Wyatt, M.L. Naiman, B.P. Mathur, C.T. Kirk, and S.D. Senturia, High-field electron capture and emission in nitrided oxides, *J. Appl. Phys.,* **57**, 2036 (1985).

20. D.J. DiMaria and J.H. Stathis, Trapping and trap creation studies on nitrided and reoxidized-nitrided silicon dioxide films on silicon, *J. Appl. Phys.,* **70**, 1500 (1991).

21. H.S. Momose, S. Kitagawa, K. Yamabe, and H. Iwai, Hot carrier related phenomena for n- and p-MOSFETs with nitrided gate oxide by RTP, *IEEE Int. Electron Dev. Mtg. Technical Digest,* 267 (1989).

INTERFACE PROPERTIES AND DEVICE RELIABILITY OF HIGH QUALITY PECVD OXIDE FOR MOS APPLICATIONS

L. K. Wang, C. C-H Hsu and W. Chang

IBM Research Division
T. J. Watson Research Center
Yorktown Heights, NY 10598, USA

INTRODUCTION

Recently PECVD oxide has been attracting much attention as gate dielectric material for silicon FET device applications because the high temperature oxidation process and the dopant segregation occurrence can be eliminated by depositing oxide directly on the device substrates in a low temperature environment [1-5]. In addition, this oxide can be the gate dielectric for Si-Ge alloy as well as III-V semiconductors such as GaAs and InP where high quality thermal oxide is not available. It has been reported that these deposited oxides can have similar defect densities, breakdown fields, and trapping property in the oxide as thermally grown oxides [1-4]. However, the reliability of the PECVD oxide as the gate dielectric for the deep submicron silicon gate FETs has not been reported. In this paper, we investigate the hot carrier induced degradation on the devices with PECVD and thermally grown gate oxides. To facilitate the comparison of hot-carrier reliability, CMOS devices and buried junction hot electron injector structure with both PECVD and thermally grown 7 nm gate oxides are fabricated using a 0.25 μm gate CMOS technology [6,7].

EXPERIMENT AND DEVICE CHARACTERISTICS

The CMOS devices are fabricated on a 2.2μm thick epi layer with a doping concentration around $1.5 \times 10^{16}/\ \text{cm}^{-3}$ which is grown on a p-type 0.005Ω-cm substrate. A retrograde N-well 0.25μm gate CMOS process is used in the device fabrication [6]. The dimension of polysilicon gate as low as 0.3μm are patterned in a I-line stepper using contrast enhancement lithography followed by a high selectivity reactive ion etching process. The arsenic implant doped N+ polysilicon gates are used for the n-channel FETs and boron doped P+ gates are used for p-channel devices. Both N+ and P+ poly gates are doped during the source/drain formation process. The N+ source/drain junctions are doped with arsenic at a junction depth around 0.1μm through a 900°C furnace annealing process. The boron doped P+ junctions at 0.12μm deep are formed by germinium amorphorization implant prior

The Physics and Chemistry of SiO_2 and the Si-SiO_2 Interface 2
Edited by C.R. Helms and B.E. Deal, Plenum Press, New York, 1993

to the BF_2 implant and subsequently driven in at 850°C nitrogen ambient. After the drive-in of source and drain junctions, a 350Å thick self aligned titanium silicide is formed by a rapid thermal annealing process on the source/drain and poly gates simultaneously at a final sheet resistance around 10 Ω/square. The fabricated devices are passivated with a LPCVD oxide and the Al-Cu-Si metal with a 100 nm titanium barrier layer is used as the final metallization step of this process. A 400°C forming gas annealing is followed as the post metallization sintering. To investigate interface trap generation, a buried junction injector structure [6] is also build to provide uniform hot electron injection from the substrate into the oxide. The buried junction injector is formed by fabricating a n-channel MOSFET in the p-well on a n-type epi grown on N+ substrate.

The thermal grown gate oxides at 7 nm thick are grown pyrogenically at 850°C on the control wafers. PECVD oxide at the same thickness are deposited on the experimental wafers in a PECVD system. The PECVD oxide films are deposited at very low substrate temperature (≤350°C) in a high vacuum system [1-3,5]. A base pressure at 1×10^{-4} torr is routinely operated before each deposition. The substrates are pre-cleaned in the dilute HF and rinsed in DI water before loaded into the system. A helium plasma treatment prior to the film deposition is applied to minimize the FET mobility degradation due to surface roughness [1,2]. The film deposition is operated in a gas flow composited with diluted SiH_4/helium, N_2O and helium carrying gas. At a low deposition rate (14Å/min.) gas phase nucleation is minimized to accomplish a true heterogeneous CVD process. A LPCVD polysilicon deposition is done immediately after the gate oxide process to prevent possible contaminations of the gate oxides.

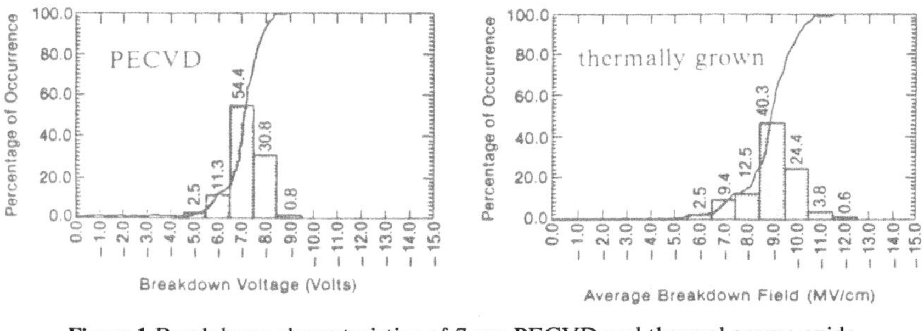

Figure 1 Breakdown characteristics of 7 nm PECVD and thermal grown oxide.

The as deposited PECVD oxide has a index of refraction at 1.463 +/- 0.002 which is the same as the thermally grown SiO_2. From the ion backscattering result, it also shows the same stoichiometric as SiO_2. These results suggest that the film should have the same chemical property as thermally grown SiO_2. However, the measured etch rate of the oxide film in diluted buffer HF is about 2.5 times higher than the thermally grown SiO_2. This may due to the fact that before any annealing these films exhibit high concentration of hydrogen. The thin oxide breakdown characteristics, as shown in Figure 1 are compared using the oxide capacitor structures with polysilicon electrode and thick oxide isolation. There is little difference in the oxide breakdown characteristics between the PECVD oxide and thermally grown oxides. The measured midgap interface trap density D_{it} is at $6 \times 10^{10} cm^{-2} eV^{-1}$, also the fix charge density D_f is at a value $6 \times 10^{11} cm^{-2}$. These are slightly higher than most of thermally grown oxides but still suitable for the device applications.

The $I_D - V_D$ characteristics of the 0.25μm n- and p-channel FETs are shown in Figures. 2(a) and (b) for both devices using PECVD gate oxide as well as thermally grown gate oxides. The FET parameters, such as threshold voltage and subthreshold slope, as listed in Table 1 are similar between the devices with PECVD and thermal gate oxides. The device transconductance from both n- and p-channel FETs with PECVD gate oxide are slightly reduced due to lower mobility. This suggests that the deposited oxide and silicon interface is still rougher than the thermal oxide/silicon interface. There is no indication of boron penetration from the boron doped P+ polysilicon gate in both types of oxides.

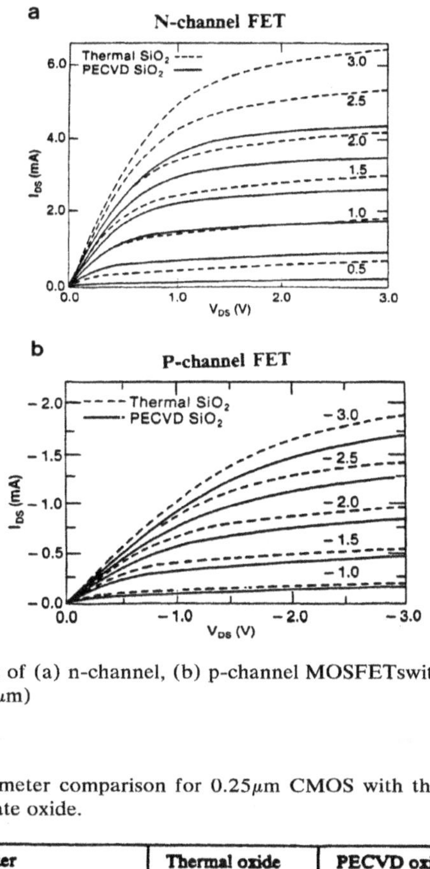

Figure 2 I-V characteristics of (a) n-channel, (b) p-channel MOSFETswith 7 nm PECVD and thermal gate oxides.(L_{eff}=0.25μm)

Table 1 Device parameter comparison for 0.25μm CMOS with thermally grown gate oxide and PECVD gate oxide.

Parameter	Thermal oxide		PECVD oxide	
N-channel FET:				
t_{ox}	73	Å	71	Å
sub V_t slope	81	mV/dec.	83	mV/dec.
G_m(Vd=2.5 V)	220	mS/mm	177	mS/mm
P-channel FET:				
t_{ox}	74	Å	71	Å
sub V_t slope	82	mV/dec.	83	mV/dec.
G_m(Vd=-2.5 V)	88	mS/mm	80	mS/mm

HOT CARRIER RELIABILITY

The oxides deposited by PECVD at low temperatures ($\lesssim$ 350°C) can be high quality and low defect density [1,2,3,5]. The CMOS device characteristics have also been shown to be only slightly worse than the devices with thermally grown gate oxide. There exists, however, a concern over the long-term reliability of the PECVD oxides under hot-carrier stress. It has been reported [4] that under high field tunneling stress the PECVD oxides is more reliable than thermal oxides. In this study, we have comprehensively investigated and compared the hot-carrier instability of CMOS devices with PECVD and thermal gate oxide by using (1) high field tunneling stress (as shown in Figure 3(a)) (2) hot electron injection from a buried junction injector (as shown in Figure 3(b)), and (3) channel hot carrier stress (as shown in Figure 3(c)).

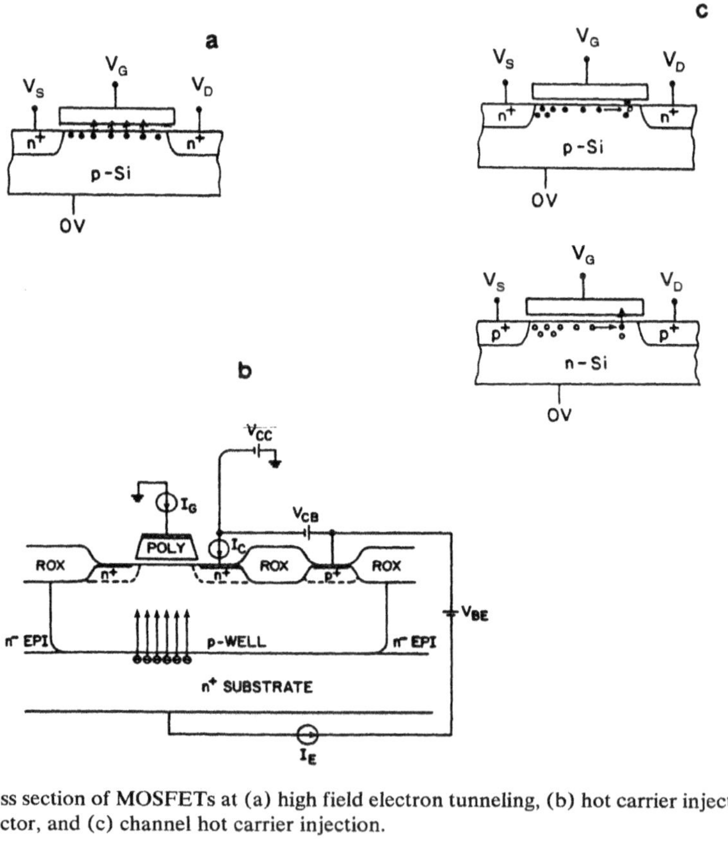

Figure 3 Cross section of MOSFETs at (a) high field electron tunneling, (b) hot carrier injection from a buried injector, and (c) channel hot carrier injection.

During high field tunneling stress, a constant voltage was applied to the gate of n-channel MOSFET. The similar tunneling characteristics of PECVD and thermal oxide are shown in Figure 4. Figure 5 shows a larger threshold voltage shift in the device with thermal gate oxide than with PECVD gate oxide. This finding is consistent with the result reported in [4] and suggests that the PECVD oxide film has a comparable high field characteristics with thermal oxide film.

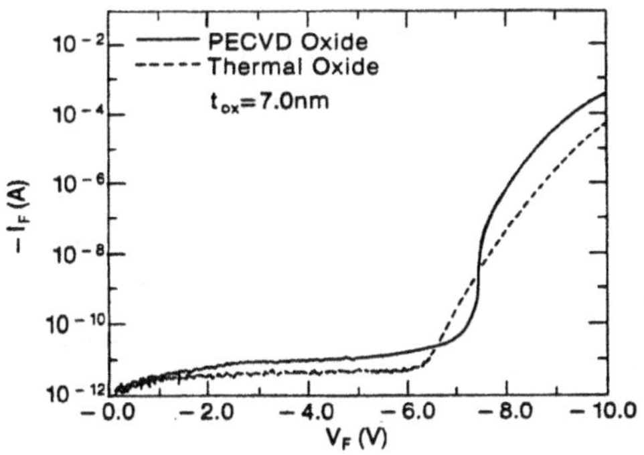

Figure 4 Electron tunneling characteristics in the MOS capacitors with PECVD and thermally grown oxide.

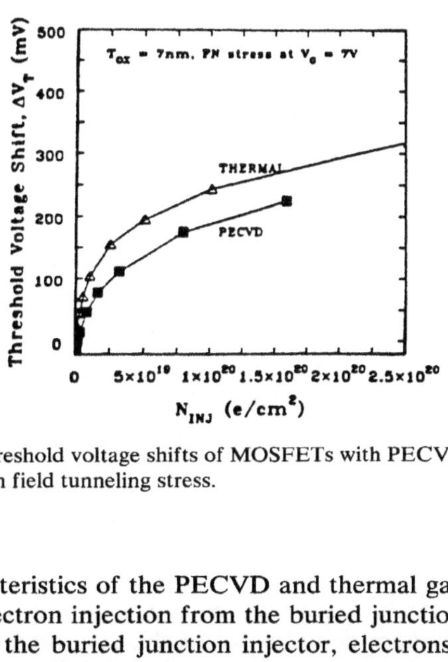

Figure 5 Comparison of threshold voltage shifts of MOSFETs with PECVD gate oxide and thermally grown gate oxide after high field tunneling stress.

Low field characteristics of the PECVD and thermal gate oxide are compared using substrate hot electron injection from the buried junction injector. During the stress of the oxide in the buried junction injector, electrons are emitted from the buried p-well/n-substrate junction and accelerated by the reverse bias between the inverted channel and p-well. The oxide field can be independently controlled by the voltage across the gate and the inverted channel and hence low oxide field during stress can be realized. In addition, uniform hot electron injection from buried junction injector will results in uniform degradation (electron trapping and/or trap generation). The number of generated interface traps can then be measured from the change of the subthreshold slope. The threshold voltage and subthreshold slope shifts of the devices with PECVD and thermal oxide after substrate hot electron injection are shown in Figure 6. In contrast to the results of high field stress, the threshold voltage and subthreshold slope shifts of the devices with PECVD oxide is illustrated to be larger than that of the devices with thermal oxide. The larger shift in subthreshold slope of the devices with PECVD oxide (about 5x higher than the devices with thermal oxide) clearly indicates that the number of the generated interface traps at the PECVD oxide/silicon interface is more than at the thermal oxide/silicon interface. More hydrogenated bonds at the PECVD oxide/silicon interface than at the thermal oxide/silicon interface due to the nature of the PECVD oxidation process can be accounted for the larger amount of generated interface traps which are dangling bonds left by released hydrogens [8]. The larger threshold voltage shifts, as shown in Figure 7(b), in the devices with PECVD oxide could also result from the greater amount of generated interface traps. From the 5x higher in interface traps and only 1.3x larger in the threshold voltage shift of PECVD oxide devices than thermal oxide devices, the less electron trapping in the PECVD oxide could be suggested. It will also be shown in the following that the electron trapping is less significant in the devices with PECVD oxide from the p-channel hot-carrier experiments.

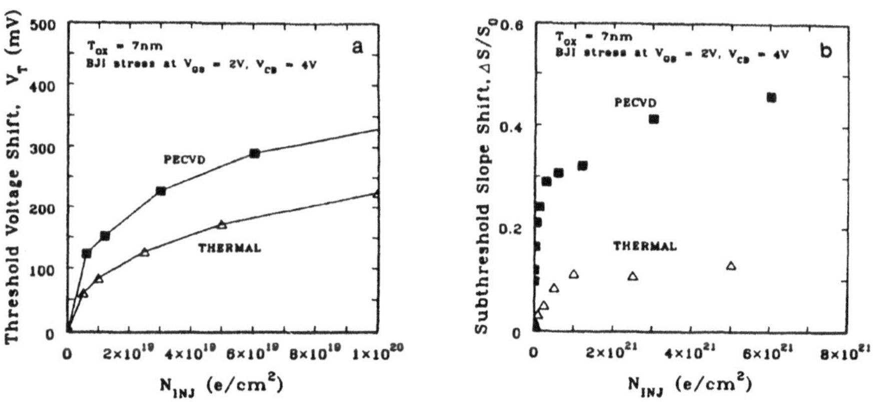

Figure 6 (a) Threshold voltage and, (b) subthreshold slope shifts of n-channel MOSFETs with PECVD and thermally grown gate oxide after hot carrier stress from a buried junction injector.

Channel-hot-carrier stress at the worst degradation condition has been commonly used as a technique to determine the hot-carrier limited lifetime [9,10] The worst stress condition for n-channel device is reported to be at the maximum substrate current ($V_G = 0.5\ V_D$) and the dominant degradation is found to be due to interface trap generation [9,10]. On the contrary, the worst stress condition for p-channel device is shown to be at the maximum gate current (low V_G) [11,12] and the dominant mechanism is reported to be due to electron trapping [11]. We have employed the maximum substrate current stress for n-channel devices and maximum gate current stress for p-channel devices to compare the hot-carrier instability of the PECVD and thermal gate oxides.

The hot-carrier instability in n-channel MOSFETs decreases the drain current of n-channel devices. The substrate current and drain current of the n-channel devices are measured during and after stress, respectively. The lifetime of n-channel MOSFETs with PECVD and thermal gate oxide are compared as a function of substrate current where the lifetime is defined as 20% decrease of reverse saturated drain current. The results, as shown in Figure 8, indicates the channel hot-carrier limited lifetime of n-channel MOSFETs with PECVD gate oxides is 4 to 6 times shorter than that of devices with thermal gate oxides for the same substrate current stress condition. The 4 to 6X reduction in lifetime is consistent with the 5X increase in the interface trap generation of the devices with PECVD oxide by comparing to the devices with thermal oxide. These results also confirmed that the major channel

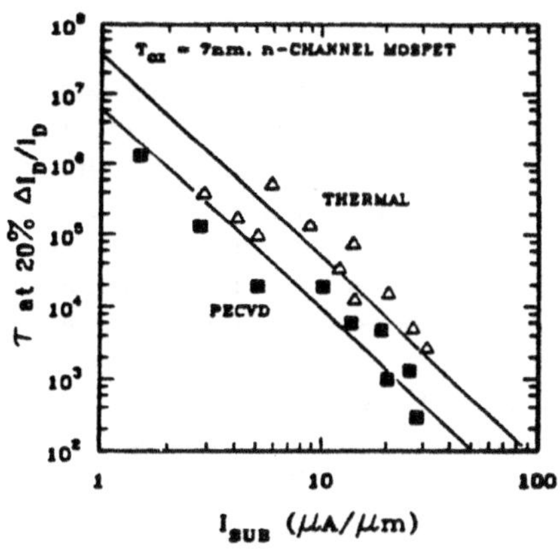

Figure 7 Channel hot carrier limited lifetime of n-channel MOSFETs with PECVD and thermally grown gate oxides.

hot carrier degradation mechanism in the n-channel MOSFETs is the interface trap generation.

The hot carrier instability in p-channel devices results in the negative charge trapping near the drain end. The trapped negative charges attract holes and form an extension of drain. The short channel effects are aggravated due to the shortening of channel after hot-carrier stress. The gate current and threshold voltage are measured during and after hot-carrier stress, respectively. As shown in Figure 9, the threshold voltage shifts measured as the function of the injected electron current during stress, the threshold voltage shifts of p-channel devices with PECVD oxides is less than that of p-channel devices with thermal oxides. The smaller threshold voltage shift at the same gate current in the devices with PECVD oxide suggests that the PECVD oxide could have less electron trapping than thermal oxide. The less electron trapping in the PECVD oxides than thermal oxide is consistent with the conclusion of hot electron injection using buried junction injector structure. Less dangling bonds due to more hydrogenation process during PECVD oxidation could be accounted for the less electron trapping in the PECVD oxide.

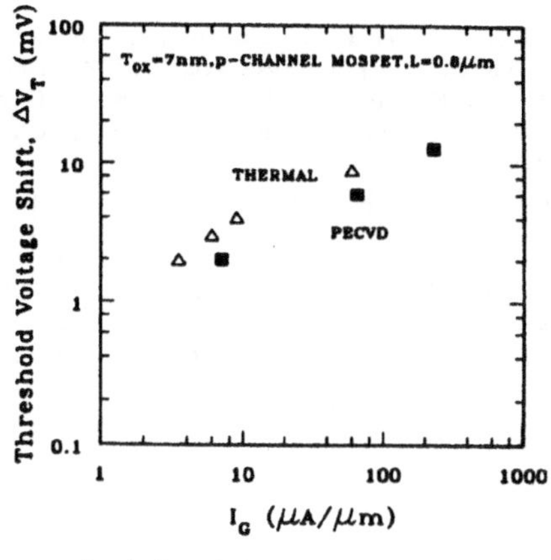

Figure 8 Comparison of hot carrier induced threshold voltage shifts of p-channel MOSFETs with PECVD and thermally grown gate oxides after 1000 seconds stress.

CONCLUSIONS

The hot carrier reliability of thin PECVD gate oxide (7 nm) used for 0.25μm gate CMOS devices are studied. The device characteristics of the fabricated FETs with PECVD gate oxides are close to the devices with thermal gate oxide. However, due to more hydrogenated bonds at the PECVD oxide/silicon interface and therefore more generation of interface traps during the hot carrier stress, the n-channel MOSFETs with PECVD gate oxides have larger degradation and slightly shorter lifetime than the n-channel devices with thermal gate oxide. On the contrary, electron trapping which dominates the degradation of p-channel devices, is reduced in the p-channel devices with PECVD oxides due to more hydrogenated dangling bonds. The experiments using hot electron injection from buried junction injector provide better understanding of hot electron instability mechanisms in the PECVD oxides.

ACKNOWLEDGMENTS

The authors would like to thank the personnel in the IBM Yorktown Silicon Facility for their process support. The authors also acknowledge D. Wen, W. Chang, A Bright, Thao N. Nguyen, Tak Ning, M. Wordeman, J. Batey and E. Tierney for helpful discussions of this work.

REFERENCES

1. J. Stasiak, J. Bateyia, E. Tierney and J. Li, "High-quality deposited gate oxide MOSFET and the importance of surface preparation" IEEE Electron Device Lett., Vol.10, pp.245-248, June 1989.
2. J. Batey, E.Tierney, "Low-Temperature Deposition of High-Quality Silicon Dioxide by Plasma-Enhanced Chemical Vapor Deposition," J. Appl. Phys. Vol.60, No.9, 3136(1986).
3. J. Batey, E.Tierney, and T.N. Nguyen, "Electrical Characteristics of Very Thin SiO_2 Deposited at Low Substrate Temperatures," IEEE Trans. on Electron Device Lett., Vol. EDL-8, No.4, 148(1987).
4. J.Lee, C. Hegarty, and C. Hu, "Electrical Characteristics of MOSFET's Using Low-Pressure Chemical-Vapor-Deposition," IEEE Electron Device Lett., Vol. EDL-7, No.9, 506(1986).
5. L.K. Wang, D.S. Wen, A.A. Bright, T.N. Nguyen and W. Chang, "Characteristics of CMOS devices fabricated using high quality thin PECVD gate oxide", 1989 IEDM Tech, Dig., pp.463.
6. C.C.-H. Hsu, D.S. Wen, M.R. Wordeman, Y. Taur, and T.H. Ning, "A comparative study of hot-carrier instabilities in P- and N-type poly gate MOSFETs," 1989 IEDM Tech, Dig., pp.75.
7. B. Davari et. al., "A high performance 0.25μm CMOS technology," 1989 IEDM Tech, Dig., pp.75.
8. C.T. Sah, S.S.-C. Pan, and C.C.-H. Hsu, "Hydrogenation and annealing kinetics of group-III acceptors in oxidized silicon," J. Appl. Phys., Vol. 57, 5148(1985).
9. C. Hu, S.C. Tam, F-C. Hsu, P-K. Ko, T-Y. Chan, and K.W. Terrill, "Hot-electron-induced MOSFET degradation - model, monitor, and improvement," IEEE Trans. Electron Devices, Vol. ED-32, 375(1985).
10. E. Takeda and N. Suzuki, "An empirical model for device degradation due to hot-carrier injection," IEEE Electron Device Lett., Vol. EDL-4, 111(1983).
11. C.C.-H. Hsu, L.K. Wang, J.Y.-C. Sun, M.R. Wordeman, and T.H. Ning, "Radiation damage and its effect on hot-carrier induced instability of 0.5μm CMOS devices patterned using synchrotron x-ray lithography," J. Electronic Materials, Vol. 19, No. 7, 721(1990).
12. K.K. Ng and G.W. Taylor, "Effects of hot-carrier trapping in n- and p-channel MOSFET's," IEEE Trans. Electron Devices, Vol. ED-30, 871(1983).

CHARGE TRAPPING IN AN OXIDE-NITRIDE-OXIDE GATE DIELECTRIC

Richard B. Klein,[1] Nelson S. Saks,[2] and Richard R. Siergiej[3]

[1]SFA, Inc.
Landover, MD 20785-5396

[2]Naval Research Laboratory
Washington, DC 20375-5000

[3]Lehigh University
Bethlehem, PA 18015-3185

ABSTRACT

The average densities and capture cross sections of electron (N_{et}, σ_e) and hole (N_{ht}, σ_h) traps in a composite oxide-nitride-oxide (ONO) gate dielectric were measured at 80 and 295 K using low-field substrate hot electron injection. We determine $N_{et} = 8\times10^{18}$ traps/cm^3 and $\sigma_e = 3\times10^{-14}$ cm^2 at 80 K at an applied electric field $\mathbf{E}_{app}$ of +2.0 MV/cm. At 295 K, we find $N_{et} = N_{ht} = 5.4\times10^{18}$ traps/cm^3 and $\sigma_e = \sigma_h = 1.2\times10^{-13}$ cm^2 at $\mathbf{E}_{app} = +1.0$ MV/cm. These results support amphoteric models of traps in ONO dielectrics.

INTRODUCTION

The incorporation of nitrogen into the SiO_2 layer of MOS devices to form an oxynitride or composite oxide-nitride-oxide (ONO) dielectric is typically found to improve several dielectric properties of the layer. Improved characteristics include reduced defect density (i.e., increased yield), reduced diffusion of dopants and contaminants, increased dielectric strength, and improved radiation hardness.[1-3] These properties make ONO dielectrics potentially advantageous in applications such as radiation-hardened electronics for infrared sensors at cryogenic temperatures. However, one drawback to nitridation can be the introduction of large numbers of electron traps[2,4-7] which may result in an unstable threshold voltage V_T. A wide range of electron trap densities N_{et} and capture cross sections ($10^{-18} < \sigma_e < 10^{-13}$ cm^2) has been reported,[2,5-12] which reflects the many different techniques used to fabricate and characterize the dielectric. For low-temperature, radiation-hard applications, we are interested in the properties of very thin (~10 nm or

The Physics and Chemistry of SiO_2 and the Si-SiO_2 Interface 2
Edited by C.R. Helms and B.E. Deal, Plenum Press, New York, 1993

less) composite dielectrics for gate insulators. Much of the previous research on deposited ONO was performed on relatively thicker dielectrics. In this work, we determine N_{et} and σ_e as well as the equivalent parameters for holes (N_{ht}, σ_h) in such a thin ONO MISFET gate dielectric with a deposited nitride layer using low-field substrate hot electron injection (SHE). We have also examined hole trapping in these dielectrics following irradiation.

SAMPLE PREPARATION

N-channel MIS transistors were fabricated using a conventional LOCOS (*loc*al *o*xidation of *s*ilicon) process. After definition and growth of the LOCOS field oxide and substrate threshold adjust implants, thin gate oxides were grown in O_2 and Ar at 900°C in an ultra-high purity oxidation system to a thickness of 7.4 nm. A 7.8 nm thick silicon nitride layer was then deposited over the SiO_2 using low-pressure chemical vapor deposition. The dielectric was then reoxidized in O_2 at 900°C for 1 h. This reoxidation process is estimated to have converted ~2 nm of the nitride surface to SiO_2. The effective dielectric thickness (i.e., thickness assuming that the dielectric is all SiO_2) of this composite ONO layer is 12 nm. Following reoxidation, a 350 nm polysilicon gate was deposited and doped by phosphorous implantation. Samples with a 19 nm pure SiO_2 dielectric (without a deposited nitride layer) were also fabricated in the same process lot for comparison.

EXPERIMENTAL TECHNIQUES

SHE injection[13] was used to measure N_{et} and σ_e. This technique is easy to use, provides uniform injection of electrons over the gate area, and utilizes relatively low electric fields across the dielectric to minimize field-induced detrapping of trapped charge. All SHE injections were performed with an applied electric field $\mathbf{E}_{app}$ = +1-2 MV/cm. To calculate N_{et} and σ_e we first determine the areal density of electrons injected N_{inj}. This is found by integrating the current that flows across the dielectric with respect to time, measured at the transistor gate. The injections were performed sequentially, in equal log-time steps. Threshold-voltage shifts ΔV_T were determined after each step either by fitting the MISFET I_D-V_G (source-to-drain current vs gate voltage) characteristics in the linear region (at drain voltage V_D = 50 mV) to a simple transconductance model which includes field-dependent mobility and series resistance effects,[14] or by simply measuring the shift in the voltage required to obtain a given subthreshold drain current (100 pA). Since the location of the traps is not known *a priori*, the traps are assumed to be uniformly distributed throughout the composite dielectric. (This assumption is plausible given voltage-shift vs dielectric thickness[11] and photoemission-current vs voltage[7] measurements showing a uniform charge distribution in nitrided SiO_2 films.) Changes in the amount of fixed charge in the dielectric were determined after each injection from ΔV_T.

Radiation experiments were also performed to characterize N_{ht} and σ_h for hole traps in the dielectric. Devices were irradiated at a ^{60}Co pool source with $\mathbf{E}_{app}$ = +2.0 MV/cm. The major advantage of irradiation over other techniques used to characterize hole traps (e.g., hole-injection) is that it is simple to implement. No special injecting structures are needed, and experiments can be performed on either *n*- or *p*-channel transistors. Results from radiation experiments, however, may be more difficult to interpret since radiation creates both holes *and* electrons in the insulator. Electron trapping is generally not important in pure SiO_2 dielectrics at room temperature (295 K) because of the relatively small probability of electron trapping, but may be of greater importance in dielectrics incorporating nitrogen. Here we will ignore the effects of electron trapping which accompany the hole trapping in these ONO dielectrics in order to obtain at least a rough characterization of the hole traps. Our primary interest is the characterization of traps in ONO dielectrics at 80 K. It was not possible to perform these experiments at 80 K, however, because the radiation-induced holes are essentially immobile at temperatures less than ~ 100 K for E_{app} < 3 MV/cm. We show that the trap characteristics presented here are relatively independent of temperature.

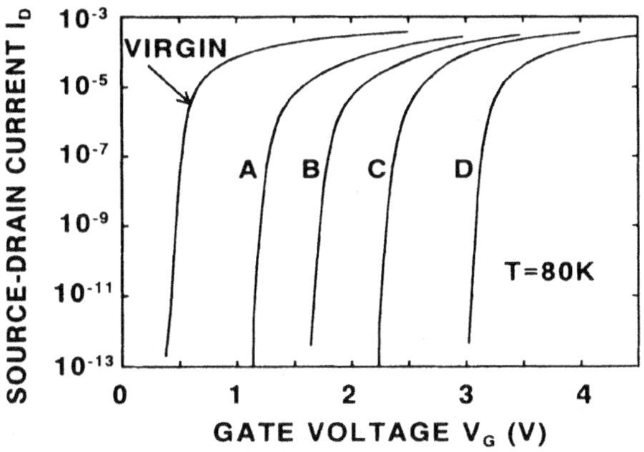

Figure 1. Source-to-drain current I_D as a function of gate voltage V_G. I_D-V_G characteristics shift to the right with increasing electron injection levels N_{inj}. Subthreshold slopes are nearly parallel to the virgin characteristic. *A*: $N_{inj}=2\times10^{13}$ *e*/cm^2. *B*: $N_{inj}=4\times10^{13}$ *e*/cm^2. *C*: $N_{inj}=8\times10^{13}$ *e*/cm^2. *D*: $N_{inj}=2\times10^{19}$ *e*/cm^2.

RESULTS AND DISCUSSION

Figure 1 shows the typical I_D-V_G response of a virgin MISFET to a series of SHE injections at 80 K. Large voltage shifts are seen at low injection levels, indicating a relatively large number of electron traps in the as-processed dielectric. Nearly parallel shifts of the subthreshold characteristic at injection levels of up to 2×10^{19} *e*/cm^2 suggest that the injection is (at least) approximately uniform and that the number of new interface traps created by the SHE injection is negligible. A more sensitive way to determine the density of interface traps D_{it} — the charge-pumping measurement technique[15] — was also used, and supports this conclusion. (For example, D_{it} increases from 9.0×10^9 to only 1.1×10^{10} traps/cm^2·eV at N_{inj} = 2×10^{15} *e*/cm^2.)

ΔV_T vs N_{inj}, calculated from the data shown in Figure 1, is shown in Figure 2. ΔV_T was determined from shifts in the I_D-V_G data as discussed above. If the electron trapping obeys first-order trapping kinetics, then the number of trapped electrons should increase as $1\text{-}\exp(-\sigma_e N_{inj})$,[16] assuming that the electron traps can be characterized by a single capture cross section σ_e. The best fit of an exponential of this form (solid line) to the data (triangles) is also shown in Figure 2. The good fit obtained indicates that first-order trapping kinetics are adequate to model the electron trapping, and that the trapping is dominated by a single electron trap level. From the data in Figure 2, assuming a uniform trap distribution and one electron per trap, we calculate N_{et} = 8×10^{18} traps/cm^3 and σ_e = 3×10^{-14} cm^2 at E_{app} = +2.0 MV/cm. For comparison, we also show in Figure 2 electron injection data from a MOSFET with a pure SiO_2 gate dielectric (circles). It is clear that no significant electron trapping occurs for N_{inj} < 10^{16} *e*/cm^2. From these results, we infer that the high average density of electron traps in the ONO dielectric is

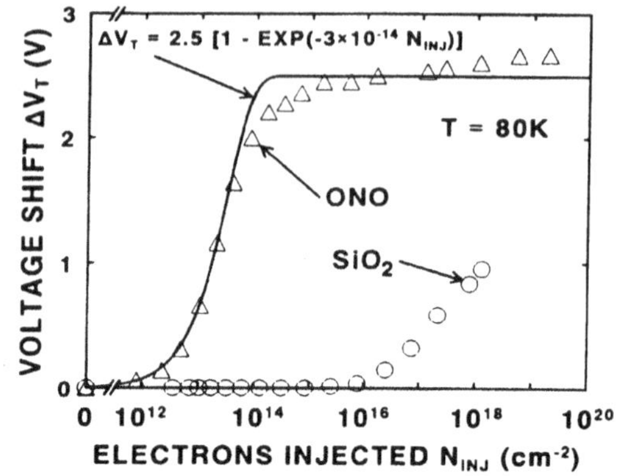

Figure 2. Threshold voltage shift ΔV_T as a function of the number of electrons injected N_{inj}. An electric field $\mathbf{E}_{app} = +2.0$ MV/cm is applied across the dielectric during injections. A good fit to an exponential of the form 1-exp($-\sigma_e N_{inj}$) is obtained. Also shown are data from a pure SiO_2 MOSFET (circles) performed at a slightly larger $\mathbf{E}_{app}$. Unlike the ONO dielectric, there is no electron trapping for $N_{inj} < 10^{16}$ e/cm^2.

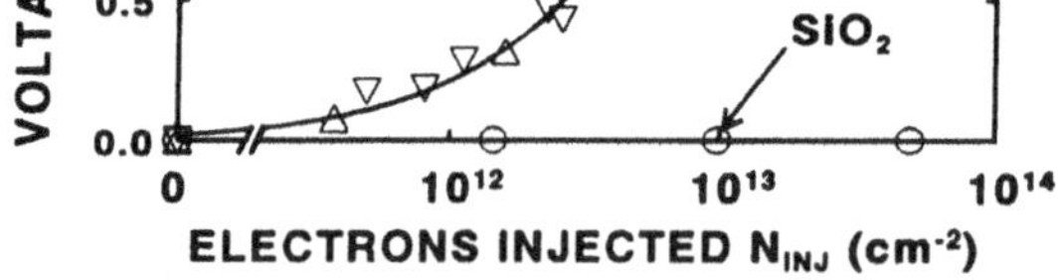

Figure 3. Threshold voltage shift ΔV_T as a function of the number of injected electrons N_{inj}. Line is least-squares fit of an exponential to the data. An electric field $\mathbf{E}_{app} = +1.0$ MV/cm was applied across the dielectric during injections. From these data, we calculate $\sigma_e = 1.2 \times 10^{-13}$ cm^2. Also shown are data from a pure SiO_2 MOSFET (circles) performed at a slightly larger $\mathbf{E}_{app}$. Unlike the ONO dielectric, there is no electron trapping for $N_{inj} < 10^{14}$ e/cm^2. Symbols: ▵ Virgin 100 × 100 μm^2 device. ▿ Virgin 10 × 100 μm^2 device.

related to defects introduced by the nitrogen. The monotonic increase in ΔV_T for $N_{inj} > 10^{16}$ e/cm^2 in the pure SiO_2 suggests that new traps are being created by the injected electrons. Likewise, the relatively small continued increase in ΔV_T at high N_{inj} in the ONO dielectric may be due to some creation of new electron traps.

Figure 3 shows the results of a similar SHE injection experiment performed at 295 K. Data are shown for two transistor sizes: length × width = 100 × 100 and 10 × 100 μm^2. Good agreement is obtained between these devices, showing that the injected currents scale linearly with area as expected. The excellent fit of the data to an exponential of the form 1-exp(-$\sigma_e N_{inj}$) again shows that the traps can be characterized by a single capture cross section σ_e. From these data, we calculate $N_{et} = 5.4\times10^{18}$ traps/cm^3 and $\sigma_e = 1.2\times10^{-13}$ cm^2 with E_{app} = +1.0 MV/cm. Electron injection experiments performed at 295 K on ONO MISFETs with E_{app} between 1 and 2 MV/cm show that σ_e decreases with increasing E_{app}, in qualitative agreement with the results of Ning.[16] Different σ_e values determined from the data in Figures 2 and 3 probably result primarily from this dependence on applied field and *not* from the temperature dependence, which is minimal.[16] For comparison, Figure 3 shows electron injection data from a MOSFET with a pure oxide gate dielectric (circles). Once again, no electron trapping is observed in the control oxide at these injection levels.

In order to characterize the hole traps, some devices were irradiated after electron injection at 295 K. Radiation creates electron-hole pairs in the gate dielectric. Assuming an effective electron-hole pair production rate of 7.6×10^{12} pairs/cm^3·rad (which is the value for SiO_2),[17] about 9.1×10^{12} pairs/cm^2·Mrad are produced in the ONO dielectric. Irradiation of virgin devices produces negligible shifts (~ -20 mV/Mrad), which suggests that either (*a*) no charge trapping occurs (either because there are no traps or all existing traps are filled with carriers) or (*b*) the electrons and holes are trapped in equal numbers.

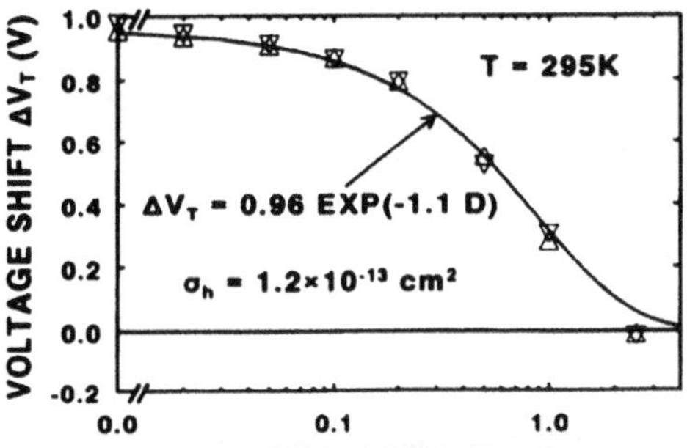

Figure 4. Threshold voltage shift ΔV_T as a function of total dose. Data are from two adjacent transistors that were previously injected with electrons. Line is least-squares fit of an exponential to the data. From these data, we estimate the capture cross section for holes $\sigma_h = 1.2\times10^{-13}$ cm^2. Symbols: ▵ Irradiated with gate voltage V_G = +2.4 V. ▿ Irradiated with V_G = -2.4 V.

In contrast, if irradiation is performed on a MISFET previously injected with a large number of electrons, then the rate of hole trapping is dramatically increased. For example, in a device with ΔV_T = +0.96 V after electron injection, ΔV_T decreases back towards zero with increasing irradiation as shown in Figure 4. Similar results are observed when the sample gates are biased either positively or negatively during irradiation (±2.4 V), which supports our assumption that the traps are approximately uniformly distributed throughout the dielectric. By fitting the data in Figure 4 to an exponential (again assuming first-order kinetics), we find that the cross section for capture of a hole $\sigma_h = 1.2\times10^{-13}$ cm^2. Within experimental error, this is about the same capture cross section obtained above for electron

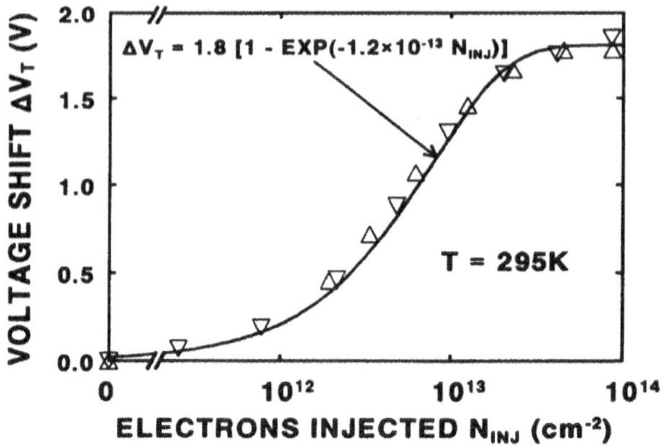

Figure 5. Threshold voltage shift ΔV_T as a function of the number of injected electrons N_{inj} at $\mathbf{E}_{\text{app}} = +1.0$ MV/cm. Line is least-squares fit of an exponential to the data. Data are from devices that were previously injected (see Figure 3) and irradiated (see Figure 4). From these data, we calculate $\sigma_e = 1.2\times10^{-13}$ cm^2. Symbols: ▵ Previously irradiated with gate voltage $V_G = +2.4$ V. ▿ Previously irradiated with $V_G = -2.4$ V.

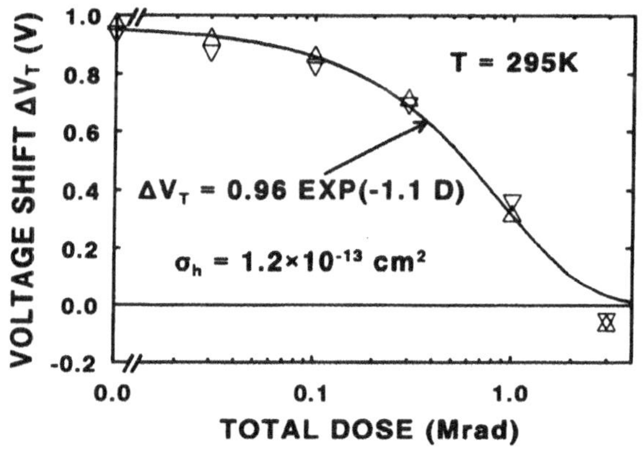

Figure 6. Threshold voltage shift ΔV_T as a function of total dose. Data are from two adjacent transistors that were previously injected with electrons (see Figure 3), irradiated (see Figure 4), and then re-injected with electrons (see Figure 5). Line is least-squares fit of an exponential to the data. From these data, we estimate the capture cross section for holes $\sigma_h = 1.2\times10^{-13}$ cm^2. Symbols: ▵ Previously irradiated with gate voltage $V_G = +2.4$ V. ▿ Previously irradiated with $V_G = -2.4$ V.

capture. This calculation of σ_h is only approximate, since it assumes negligible trapping of radiation-induced *electrons* as discussed above. The assumption is most accurate at low levels of irradiation, when all traps are filled with electrons. At high radiation levels, trapping of radiation-induced electrons increases substantially, and the assumption of simple first-order kinetics is invalid.

We find that data from these samples are reproducible when cycled (i.e., alternately fill traps with holes and electrons) by repeated electron injections followed by irradiations.

Some examples of this are shown in Figures 5 and 6. Note that the data for the second cycle (Figures 5 and 6) accurately reproduce the shifts obtained on the first cycle, which indicates that the shifts are due to the filling and emptying of existing traps, not creation of new traps.

Our data are consistent with two models that describe the traps in ONO dielectrics. Libsch and White[18] suggest that the traps in their ONO memory devices are initially neutral and amphoteric. Our finding that $N_{et} = N_{ht}$ and $\sigma_e = \sigma_h$ suggests that these traps are related and might arise from the same physical defect, i.e., an amphoteric trap. However, we find that $\sigma_e = \sigma_h = 1.2\times10^{-13}$ cm^2, which is a relatively large value generally considered to be characteristic of a Coulombic-attractive trap.[19] Although describing the traps as amphoteric with the same capture cross section when it is neutral as when it is Coulombic-attractive (cf. Figures 3 and 5 with Figures 4 and 6) appears unlikely, this idea cannot be dismissed on physical grounds.

Others[10,20-22] have noted results similar to those reported here for silicon-nitride films. They suggest that the amphoteric traps discussed above have a negative correlation energy (a "negative *U*"), as described by Ngai and Hsia.[23] In this model, each amphoteric trap initially contains an electron-hole pair, resulting in charge compensation[3,24] in the dielectric. (Although Dunn *et al.*[1] have shown that charge compensation does not occur in reoxidized nitrided oxides, their dielectrics are probably significantly different from the ONO dielectrics used here. Fabrication methods are known to greatly affect dielectric properties.) The large capture cross section measured for both electrons and holes is accounted for if we assume that the electric field produced by an electron-hole pair at a trap is sufficiently strong (Coulombic-attractive) for both electrons and holes. The negative-*U* center may be invoked to explain why we see no evidence of a neutral trap. Both the Libsch and negative-*U* models appear to explain the radiation hardness exhibited by our devices. In the Libsch model, it results from electron and hole trapping in equal numbers. In the negative-*U* model, charge compensation prevents accumulation of net radiation-induced charge.

SUMMARY

We find that there is a very large density of electron and hole traps with a large (Coulombic-attractive) capture cross section for both electrons and holes in the ONO dielectric studied. Results at 80 and 295 K show no significant dependence of trap parameters on temperature. Radiation experiments clearly show that the radiation hardness of this ONO dielectric is *not* due to a lack of intrinsic hole traps, but rather must be due to compensation (equal electron and hole trapping) or initially filled hole traps in the virgin dielectric. Although this conclusion seems to conflict with the conclusions of Dunn *et al.*,[1] we believe that it is a consequence of the different methods used to fabricate the dielectrics (reoxidized nitrided oxide vs oxide-nitride-oxide). As we have shown, significant charge trapping occurs during radiation experiments under certain conditions. We therefore conclude that although the use of ONO dielectrics has some advantages, significant charge trapping may lead to reliability problems in applications such as radiation-hardened electronics at cryogenic temperatures.

ACKNOWLEDGEMENTS

The authors thank Sukyoon Yoon of National Semiconductor Corporation (formally of Lehigh University) for fabricating the ONO dielectric and Marvin White of Lehigh for useful discussions.

REFERENCES

1. G.J. Dunn, R. Jayaraman, W. Yang, and C.G. Sodini, Radiation effects in low-pressure reoxidized nitrided oxide gate dielectrics, *Appl. Phys. Lett.* 52(20):1713 (1988).
2. Z.H. Liu, P.T. Lai, Y.C. Cheng, Characterization of charge trapping and high-field endurance for 15-nm thermally nitrided oxides, *IEEE Trans. Electron Devices* 38(2):344 (1991).
3. R. Sundaresan, M.M. Matloubian, and W.E. Bailey, Rapid-thermal nitridation of SiO_2 for radiation-hardened MOS gate dielectrics, *IEEE Trans. Nucl. Sci.* NS-33(6):1223 (1986).
4. Z.A. Weinberg, K.J. Stein, T.N. Nguyen, and J.Y. Sun, Ultrathin oxide-nitride-oxide films, *Appl. Phys. Lett.* 57(12):1248 (1990).
5. Z.H. Liu, P.S. Chen, Y.C. Cheng, and P.T. Lai, Charge trapping and interface state generation by avalanche hot-electron injection in rapid thermal NH_3 annealed and reoxidized SiO_2 films, *J. Electrochem. Soc.* 137(6):1871 (1990).
6. T.B. Hook, M.E. Johnson, and A.V. Ferris-Prabhu; Electron trapping, nitride conduction, and forward gain instability in a lateral *p-n-p* device; *IEEE Trans. Electron Devices* 37(3):755 (1990).
7. S.K. Lai, D.W. Dong, and A. Hartstein, Effects of ammonia anneal on electron trappings in silicon dioxide, *J. Electrochem. Soc.* 129(9):2042 (1982).
8. Y.C. Park, W.B. Jackson, N.M. Johnson, and S.B. Hagstrom, Spatial profiling of electron traps in silicon nitride thin films, *J. Appl. Phys.* 68(10):5212 (1990).
9. D.J. DiMaria and J.H. Stathis, Trapping and trap creation studies on nitrided and reoxidized-nitrided silicon films on silicon, *J. Appl. Phys.* 70(3):1500 (1991).
10. P.M. Lenahan, D.T. Krick, and J. Kanicki, The nature of the dominant deep trap in amorphous silicon nitride films: evidence for a negative correlation energy, *Appl. Surf. Sci.* 39:392 (1989).
11. M. Severi and M. Impronta, Charge trapping in thin nitrided SiO_2 films, *Appl. Phys. Lett.* 51(21):1702 (1987).
12. F.L. Terry, Jr., P.W. Wyatt, M.L. Naiman, B.P. Mathur, C.T. Kirk, and S.D. Senturia, High-field electron capture and emission in nitrided oxides, *J. Appl. Phys.* 57(6):203 (1985).
13. N.S. Saks, R.B. Klein, S. Yoon, and D.L. Griscom, Formation of interface traps in metal-oxide-semiconductor devices during isochronal annealing after irradiation at 78 K, *J. Appl. Phys.* 70(12):7434 (1991).
14. P.I. Suciu and R.L. Johnson, Experimental derivation of the source and drain resistance of MOS transistors, *IEEE Trans. Electron Devices* ED-27(9):1846 (1980).
15. G. Groeseneken, H.E. Maes, N. Beltran, and R.F. DeKeersmaecker, A reliable approach to charge-pumping measurements in MOS transistors, *IEEE Trans. Electron Devices* ED-31(1):42 (1984).
16. T.H. Ning, High-field capture of electrons by Coulomb-attractive centers in silicon dioxide, *J. Appl. Phys.* 47(7):3203 (1976).
17. H.E. Boesch and J.M. McGarrity, Charge yield and dose effects in MOS capacitors at 80 K, *IEEE Trans. Nucl. Sci.* NS-23(6):1520 (1976).
18. F.R. Libsch and M.H. White, Charge transport and storage of low programming voltage SONOS/MONOS memory devices, *Solid-State Electron.* 33(1):105 (1990).
19. T.P. Ma, Process-induced radiation effects, in: "Ionizing Radiation Effects in MOS Devices and Circuits," T.P. Ma and P.V. Dressendorfer, eds., John-Wiley and Sons, New York (1989).
20. S.E. Curry, P.M. Lenahan, D.T. Krick, J. Kanicki, C.T. Kirk, Evidence for a negative electron-electron correlation energy in the dominant deep trapping center in silicon nitride films, *Appl. Phys. Lett.* 56(14):1359 (1990).
21. W.L. Warren, J. Kanicki, F.C. Rong, and E.H. Poindexter, Paramagnetic point defects in amorphous silicon dioxide and amorphous silicon nitride thin films II. *a*-SiN_x:H, *J. Electrochem. Soc.* 139(3):880 (1992).
22. W.L. Warren, J. Kanicki, F.C. Rong, E.H. Poindexter, and P.J. McWhorter, Charge trapping centers in *n*-rich silicon nitride thin films, *Appl. Phys. Lett.*, to be published.
23. K.L. Ngai and Y. Hsia, Empirical study of the metal-oxide-semiconductor device characteristics deduced from a microscopic model of memory traps, *Appl. Phys. Lett.* 41(2):159 (1982).
24. G.Q. Lo, W.C. Ting, D.K. Shih, and D.L. Kwong, Charge trapping properties in thin oxynitride gate dielectrics prepared by rapid thermal processing, *Appl. Phys. Lett.* 56(10):979 (1990).

INTERFACE TRAP DENSITY REDUCTION AND OXIDE PROFILING FOR MOS CAPACITORS WITH FLUORINATED GATE OXIDE DIELECTRICS

Dimitrios N. Kouvatsos,[1] Ralph J. Jaccodine[1] and Fred A. Stevie[2]

[1]Sherman Fairchild Center for Solid State Studies
Lehigh University, Bethlehem PA 18015
[2]AT&T Bell Laboratories, 555 Union Boulevard, Allentown, PA 18103

ABSTRACT

The effect of fluorine incorporation on the as-grown Si-SiO_2 interface state density of the metal/SiO_2/Si system was investigated using MOS capacitors with fluorinated oxide dielectrics grown by NF_3-enhanced oxidation as test structures. A clear reduction of the interface trap density was shown for NF_3 additions in the ppm range as compared to dry oxides. The fluorine incorporation in the oxide was investigated by means of SIMS profiling. A tendency of fluorine to be partially immobilized in the oxide at the time of its incorporation was observed.

INTRODUCTION

Halogenic additions introduced in the oxidizing gas stream during dry silicon oxidation result in significant improvement in the electronic properties of the oxide and of the underlying silicon. Chlorine-related effects on the Si-SiO_2 system and on MOS devices such as oxidation rate enhancement[1], mobile ion gettering[2,3], improved breakdown strength[4] and higher minority carrier lifetime[5,6] have been investigated since the early 1970s and shown to yield MOS devices with improved electrical properties and better threshold stability; chlorinated oxidation is now used industrywide.

Within this broader context of halogenic oxidation as a means of growing thermal silicon dioxide films with improved properties to be used in VLSI fabrication, the influence of fluorine on the properties of the Si-SiO_2 system has recently been of great interest. Fluorine-related oxidation rate enhancement[7,8,9] is much more pronounced than the corresponding chlorine effect; moreover, fluorine has been shown to result in a marked oxide stress relaxation[10]. Effects of fluorine on the electrical properties of gate oxides constitute a subject of considerable significance. Fluorine effected suppression of the hot-electron induced generation of interface traps[11,12] and enhancement of hot electron immunity[13,14] has been reported. Moreover, fluorine effects on dielectric breakdown strength[15], radiation response[16] and junction leakage currents[12] have been investigated.

Interface trap density reduction is a matter of great significance and with important consequences for device stability and for the operation of MOS integrated circuits. Previous investigations have focused on fluorine-induced suppression of the generation of radiation-induced or hot electron-induced interface traps. In this work we have investigated the influence of fluorine on as-grown interface traps, using MOS capacitors with fluorinated oxide dielectrics grown by

NF_3-enhanced oxidation of silicon as test structures. Moreover, we have determined the fluorine profiles in the oxides as they vary with different processing conditions in an attempt to shed light on the mechanism of fluorine incorporation in the oxide.

EXPERIMENTAL

We have used *n*-type epitaxial silicon wafers of 12 mil thickness, (100) orientation and with a resistivity of 1-2 $\Omega \cdot cm$ as substrates. After a standard RCA cleaning procedure was performed the wafers were oxidized in a quartz hot double wall oxidation furnace. Wafers were inserted very slowly into the furnace tube while nitrogen was flowing and then the oxidation was started by switching over to the oxidant supply. The oxidizing gas flow consisted of 1.65 liters/min of oxygen along with a much smaller flow of NF_3 gas, monitored by a microflowmeter, as a fluorine source. Nitrogen was flowing between the furnace tube walls throughout the oxidation process in order to sweep away water molecules indiffusing from the atmosphere.

Oxide films were grown at an oxidation temperature of 900°C in dry O_2 or in dry O_2 with NF_3 added at several different concentration values in the 30 to 150 ppm range during either the entire oxidation time or a part of it; in the latter case, several values for the proportion of the total oxidation time during which NF_3 was flowing were used. Wafers for MOS structure fabrication and electrical testing as well as samples for profiling and kinetics studies were oxidized.

Aluminum evaporation, with the use of suitable masks, was subsequently performed in order to form the metal gates of the MOS capacitor structures. After the back side oxide was stripped, substrate contacts were formed by another evaporation step depositing aluminum on the back side. Electrical characterization using computer-controlled capacitance-voltage (C-V) and charge-voltage (Q-V) measurement setups was then carried out and the interface state density distribution across the silicon energy band gap was extracted. The fluorine profiles in the oxides were determined by means of secondary ion mass spectrometry (SIMS) performed on samples oxidized in the same processing runs with the MOS capacitor wafers. The oxide thickness and refractive index values were determined by a Rudolph Research Auto EL-II ellipsometer using a helium-neon laser.

RESULTS AND DISCUSSION

Fluorine effect on interface trap density

Plots of the as-grown interface trap density against the surface potential for an energy range covering most of the silicon band gap are shown in figure 1 for MOS capacitors with oxide dielectrics grown with either dry or fluorinated oxidation with 30 ppm NF_3 added to the oxidizing gas stream. Similar plots are shown in figure 2 for a larger surface potential range covering almost the entire bandgap. The U-shaped curve is clearly wider for the fluorinated oxidation case, indicating a lower value of the interface state density for a given value of the surface potential, and thus for a given energy value in the silicon band gap, in this case as compared to dry oxidation. Moreover, the average interface trap density over most of the silicon energy band gap (approximately -0.4V to +0.4V of surface potential sweep) is plotted against the NF_3 concentration in the oxidant gas in figure 3; NF_3 has been added during the entire oxidation time. A clear reduction is effected with small NF_3 additions of 30 to 50 ppm, but higher NF_3 concentrations do not result in significant further reduction.

We believe that this fluorine-induced reduction of the interface trap density can be attributed to passivation of dangling bonds (in particular, P_b centers) at the $Si\text{-}SiO_2$ interface as well as to removal of Si-Si and Si-O weak bonds, which are all defects introducing energy levels within the silicon band gap[17], by way of fluorine incorporation and Si-F bond formation. Furthermore, fluorine bonding and incorporation in the oxide network in the area of the oxide near the $Si\text{-}SiO_2$ interface

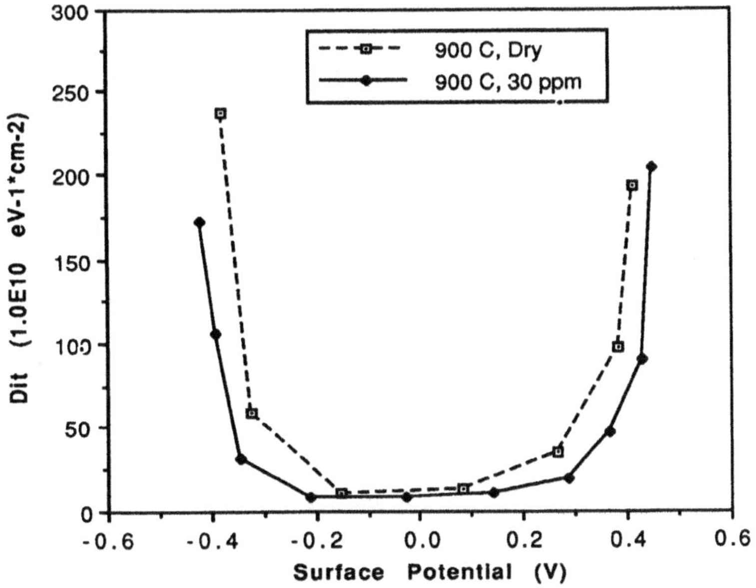

Figure 1. Interface trap density against surface potential, -0.4V to 0.4V potential sweep, dry or continuous fluorinated oxidation with 30 ppm NF_3

would break Si-O-Si rings and induce interfacial strain relaxation; this would reduce the migration towards the interface of non-bridging oxygen defects generating interface traps according to the bond strain gradient model[16, 18]. Such a mechanism, which would be much more important in the presence of radiation or hot electron injection, is consistent with the observed fluorine-induced oxide stress relaxation[10].

Fluorine incorporation and profiling in the oxide

The fluorine profiling in the oxide has been studied for both continuous fluorinated oxidation, where the exposure to the NF_3 fluorine source lasts for the entire duration of the oxidation, and for "pulsed" fluorinated oxidation, where NF_3 gas is only allowed in the furnace tube for specific time intervals. In the former case the NF_3 concentration is the only variable, while in the latter the proportion of the total time that corresponds to the fluorinated oxidation step is also varied. In all cases of "pulsed" oxidation the NF_3-enhanced oxidation step is performed after an initial dry oxidation step lasting for 30% of the total oxidation time; dry oxidation for the remainder of the oxidation time follows the fluorinated step. As evidenced by the SIMS profiling plots obtained, this method can provide more insight on the mechanism of fluorine incorporation in the growing oxide.

The fluorine profile in the oxide for the case of a dry oxidation is shown in figure 4; a small fluorine concentration, with a peak near the Si-SiO_2 interface which is probably due to the HF dip performed as a step of the RCA cleaning procedure, can be observed. The fluorine profiles that result from a continuous NF_3-enhanced oxidation are shown in figures 5 and 6 for NF_3 concentrations of 30 ppm and 150 ppm respectively. It can be seen that, apart from a small peak in the interface area, the fluorine concentration is uniform throughout the oxide film; its level is higher for the case of the larger NF_3 concentration, which seems to have reached the point where the amount of fluorine that is incorporated in the oxide saturates[10].

Fluorine profiles resulting from "pulsed" fluorinated oxidation are distinctly different, as can be observed from figures 7, 8 and 9 which correspond to NF_3-enhanced oxidation steps of 10% of

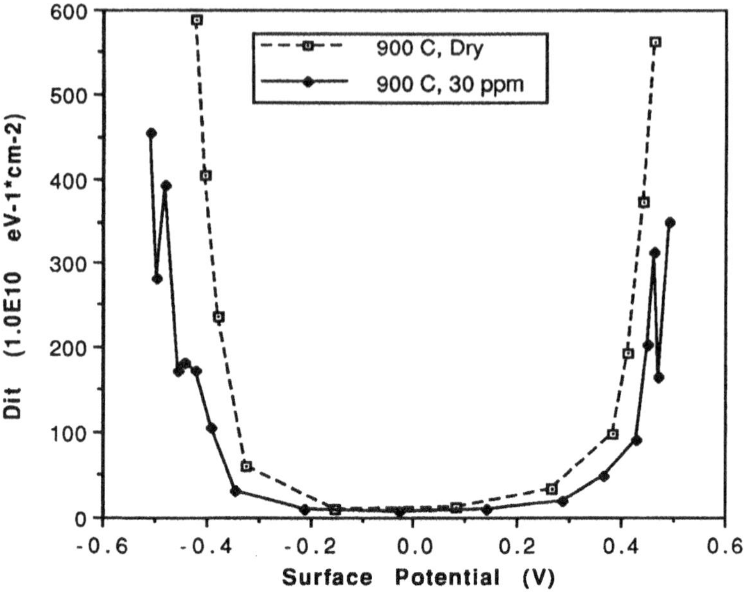

Figure 2. Interface trap density against surface potential, -0.5V to 0.5V potential sweep, dry or continuous fluorinated oxidation with 30 ppm NF_3

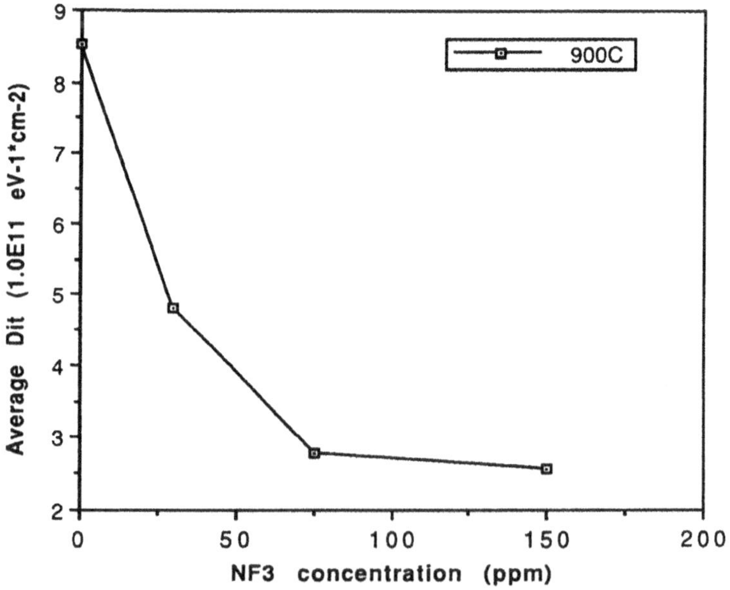

Figure 3. Average interface trap density against NF_3 concentration, continuous fluorinated oxidation

oxidation time with 50 ppm NF_3, 20% of time with 100 ppm NF_3 and 40% of time with 100 ppm NF_3, respectively. The fluorine profiles can be seen to have two peaks, one in the interface area and one in the area roughly corresponding to the oxide grown during the NF_3-enhanced oxidation step. It

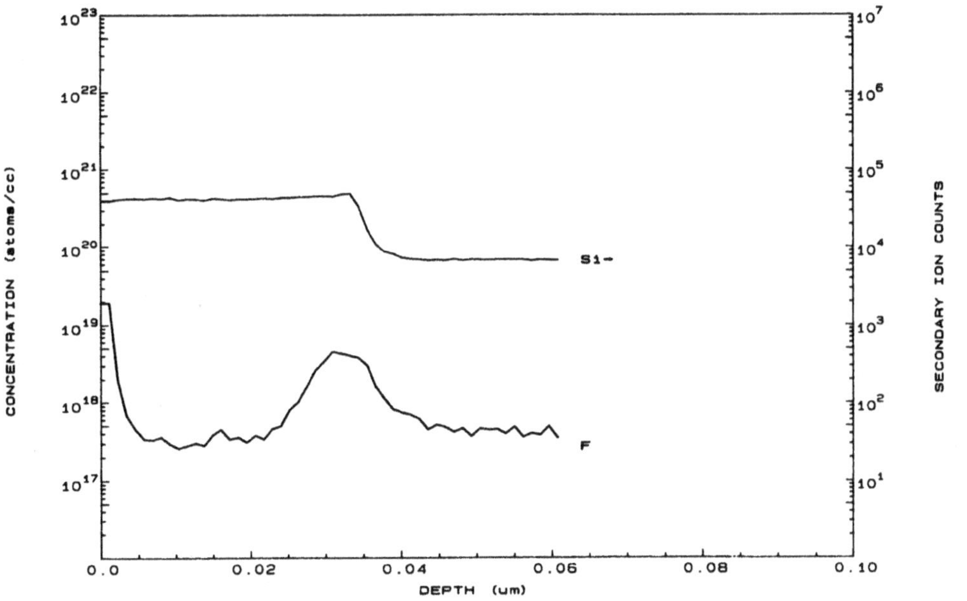

Figure 4. Fluorine concentration against depth in the oxide, dry oxidation

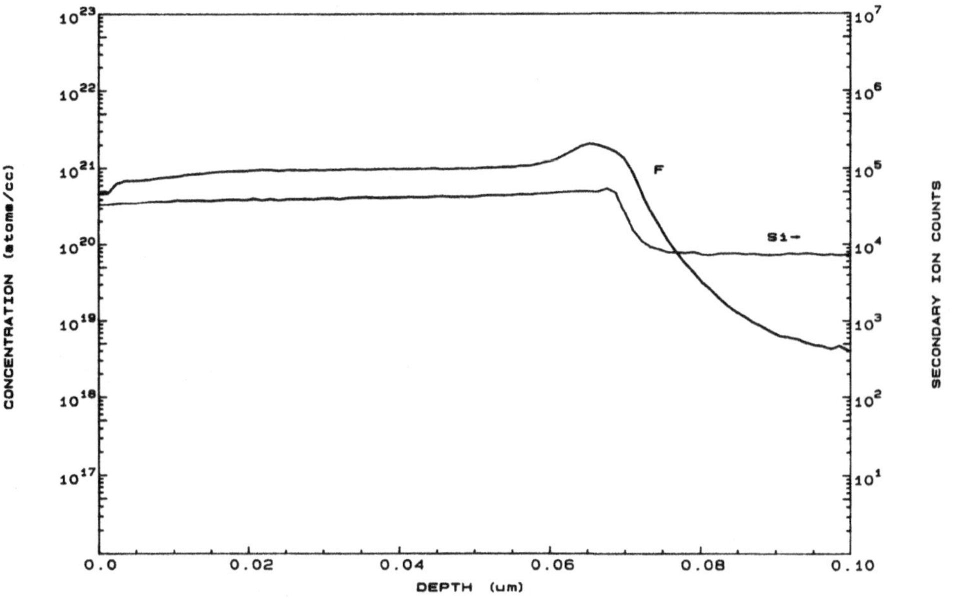

Figure 5. Fluorine concentration against depth in the oxide, continuous fluorinated oxidation with 30 ppm NF_3

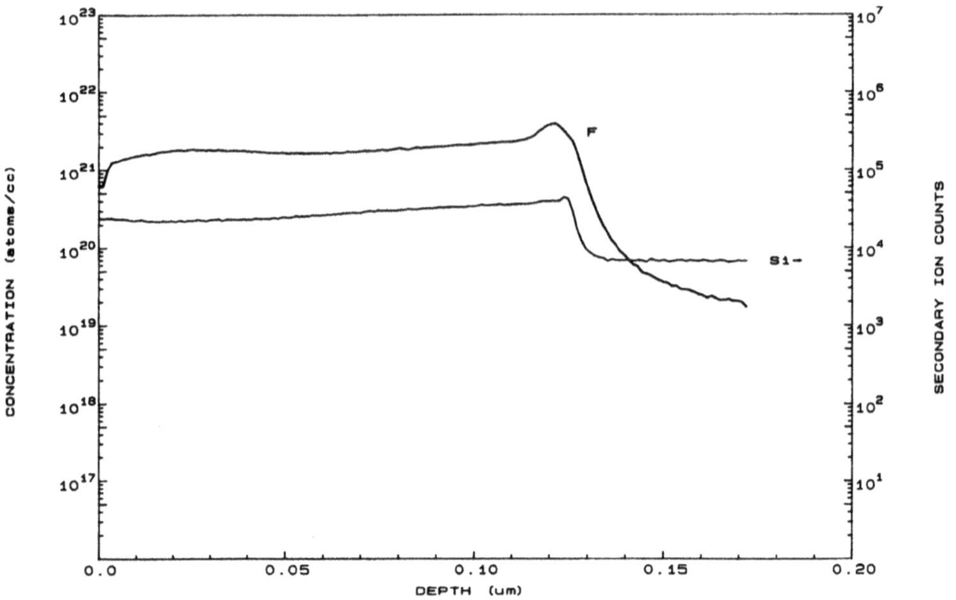

Figure 6: Fluorine concentration against depth in the oxide, continuous fluorinated oxidation with 150 ppm NF_3

can be clearly observed that this latter peak becomes more pronounced as the duration of the fluorinated oxidation step or the NF_3 concentration increase; if the fluorinated step becomes large enough, the continuous NF_3-enhanced oxidation case is approached, as can be seen from figure 9.

The above observations are consistent with a mechanism in which fluorine bonds in the oxide matrix at the time of its incorporation during fluorinated oxidation; we have direct evidence from X-ray photoelectron spectroscopy (XPS) analysis to the effect that fluorine is bonded to silicon atoms in the oxide[19]. During NF_3-enhanced oxidation and, in the case of "pulsed" oxidation, during the subsequent dry oxidation step, some of this fluorine is freed, probably being replaced by indiffusing oxygen atoms, and diffuses away. Fluorine atoms moving towards the oxidizing interface are incorporated in the growing oxide network there and result in a fluorine peak in the interfacial area; interfacial fluorine passivates dangling bonds at the Si-SiO_2 interface, as discussed above.

CONCLUSIONS

A fluorine-induced reduction in the as-grown interface trap density in MOS structures having oxides grown by NF_3-enhanced oxidation of silicon was demonstrated. Furthermore, fluorine profiles in the oxides indicate that fluorine tends to bond and become incorporated in the oxide network and to remain partially fixed in the oxide area grown at the time of its incorporation.

ACKNOWLEDGEMENTS

The authors would like to thank T-C. Lin for providing a Q-V measurement setup as well as Drs. D.R. Young and D. Xie for useful discussions. This work was supported by ARO through contract No. DAAL03-88-K-0095 and by DNA through contract No. DNA001-88-C-0122-P1.

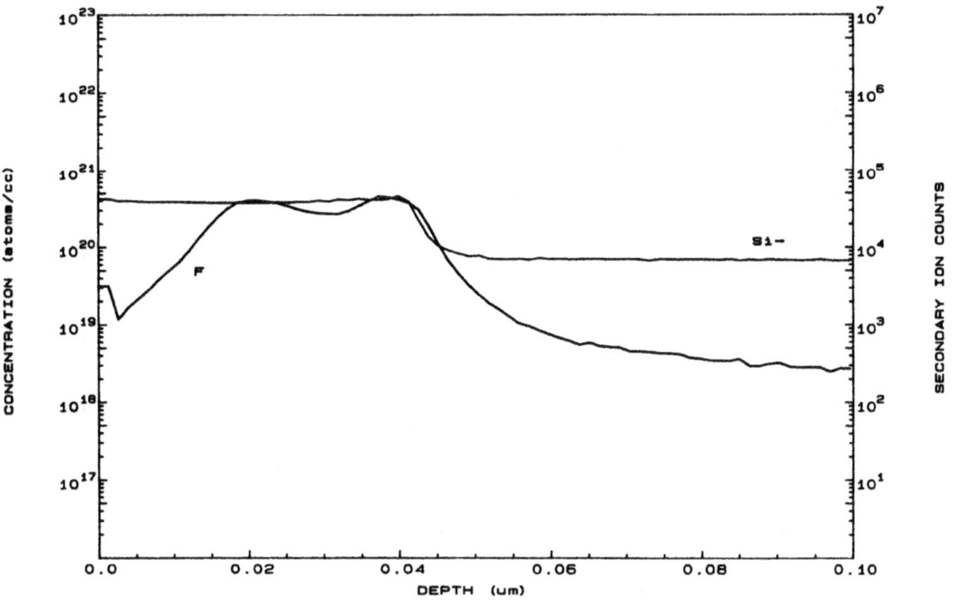

Figure 7. Fluorine concentration against depth in the oxide, pulsed fluorinated oxidation with 50 ppm NF_3 for 10% of the oxidation time

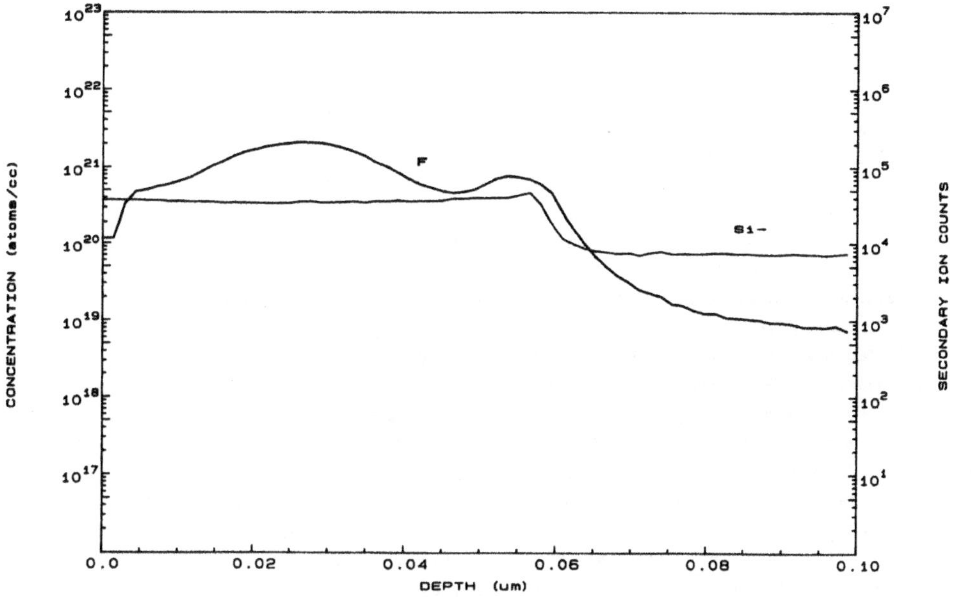

Figure 8. Fluorine concentration against depth in the oxide, pulsed fluorinated oxidation with 100 ppm NF_3 for 20% of the oxidation time

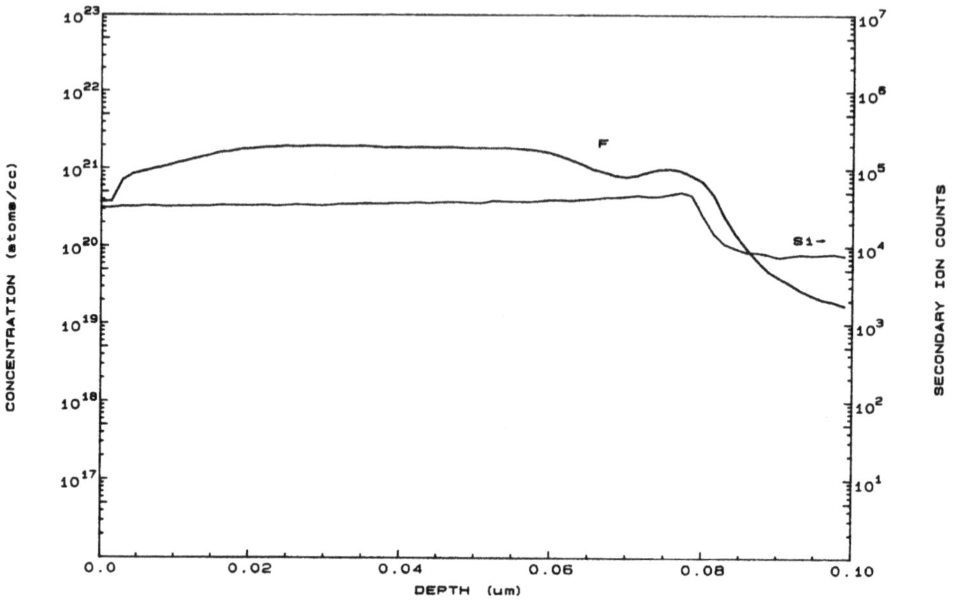

Figure 9. Fluorine concentration against depth in the oxide, pulsed fluorinated oxidation with 100 ppm NF_3 for 40% of the oxidation time

REFERENCES

1. R.J. Kriegler, Y.C. Cheng and D.R. Colton, *J. Electrochem. Soc.*, **119**:388 (1972).
2. R.J. Kriegler, *Appl. Phys. Lett.*, **20**:449 (1972).
3. M. Chen and J.W. Hile, *J. Electrochem. Soc.*, **119**:223 (1972).
4. C.M. Osburn, *J. Electrochem. Soc.*, **121**:809 (1974).
5. P.H. Roginson and F.P. Heiman, *J. Electrochem. Soc.*, **118**:141 (1971).
6. D.R. Young and C.M. Osburn, *J. Electrochem. Soc.*, **120**:1578 (1973).
7. P.F. Schmidt, R.J. Jaccodine, C.H. Wolowodiuk and T. Kook, *Mat. Lett.*, **3**:235 (1985).
8. M. Morita, T. Kubo, T. Ishihara and M. Hirose, *Appl. Phys. Lett.*, **45**:1312 (1984).
9. U.S. Kim, C.H. Wolowodiuk, R.J. Jaccodine, F. Stevie and P. Kahora, *J. Electrochem. Soc.*, **137**:2291 (1990).
10. D. Kouvatsos, J.G. Huang and R.J. Jaccodine, *J. Electrochem. Soc.*, **138**:1752 (1991).
11. Y. Nishioka, E.F. da Silva, Y. Wang and T-P. Ma, *IEEE Electron Device Lett.*, **9**:38 (1988).
12. K. Ohyu, T. Itoga, Y. Nishioka and N. Natsuaki, *Japan. J. Appl. Phys.*, **28**:1041 (1989).
13. P.J. Wright and K.C. Saraswat, *IEEE Trans. Electron Devices*, **ED-36**:879 (1989).
14. K.P. MacWilliams, L.F. Halle and T.C. Zietlow, *IEEE Electron Device Lett.*, **11**:3 (1990).
15. Y. Nishioka, K. Ohyu, Y. Ohji, N. Natuaki, K. Mukai and T-P. Ma, *IEEE Electron Device Lett.*, **10**:1141 (1989).
16. E.F. da Silva, Y. Nishioka and T-P. Ma, *IEEE Trans. Nucl. Sci.*, **NS-34**:1190 (1987).
17. T. Sakurai and T. Sugano, *J. Appl. Phys.*, **52**:2889 (1981).
18. F.J. Grunthaner, P.J. Grunthaner and J. Maserjian, *IEEE Trans. Nucl. Sci.*, **NS-29**:1462 (1982).
19. D. Kouvatsos, F.P. McCluskey, R.J. Jaccodine and F.A. Stevie, submitted to *Appl. Phys. Lett.*.

PHYSICS OF EXTREME QUANTUM CONFINEMENT EXEMPLIFIED BY $Si/Si0_2$ SYSTEM

Raphael Tsu

Electrical Engineering Department
Microelectronics Division
University of North Carolina at Charlotte
Charlotte, NC 28223

INTRODUCTION

Since the introduction of superlattices[1] and quantum well structures,[2] quantum confinement in man-made structures has been an important area of research in semiconductor physics and material science, as well as many new functional devices[3] utilizing the discrete nature of these artificially created energy states. Before explaining the physics of extreme quantum confinement and its manifestation in the $Si/a\text{-}Si0_2$ system, we shall first examine what assumptions are involved in conventional quantum confinement. The following assumptions and situations are applicable in conventional quantum confinement.

1. Band structure of bulk, with dielectric constant and effective mass approximations, is valid.
2. Band-edge discontinuities characterize well and barrier regions.
3. Conventional scattering mechanisms apply: mobility and mean free path.
4. Mostly lattice-matched heterostructures are employed.
5. Strain layered heterojunctions are included which broaden the class of heterostructures.

On the other hand, in extreme quantum confinement, the bulk band structure and effective mass approximations may not apply. The dielectric constant is usually reduced. Charge accumulation in the confined region results in a drastic change of energy eigenstates. In what follows some salient features, particularly in reference to the recent observation of tunneling via nanoscale silicon particles embedded in an $a\text{-}Si0_2$ matrix are discussed.

The Physics and Chemistry of SiO_2 and the $Si\text{-}SiO_2$ Interface 2
Edited by C.R. Helms and B.E. Deal, Plenum Press, New York, 1993

IMPROVEMENT IN QUANTUM CONFINEMENT

In quantum dots, quantum confinements in three dimensions, the presence of localized states such as impurities merely redefines the eigenstates of the quantum box. In other words, impurity scattering satisfying time reversal, may be "rediagonalized" away. Traditional phonon scattering cannot be rediagonalized, however, they are much reduced as the energy states are pushed farther and farther apart as the dimension shrinks. Reduction in scattering leads to a better definition of quantum states. What we discuss here is also present even in quantum wire and quantum well to a lesser degree. The reduction in the phase space in initial and final states is the cause of reduction in scattering. The overall reduction in many scattering mechanisms improve confinement, except perhaps interfacial defects now dominate over other deleterious effects.

REDUCTION OF DIELECTRIC CONSTANT

The dielectric constant is lowered when the separation of energy states is pushed apart. In the case of quantum well such as GaAs/AlGaAs structures, the dielectric constant is reduced 20% when the well width is 25 A.[4] For quantum dot, the reduction is even more dramatic. The upper bound of the dielectric constant for a sphere of silicon of 25 A diameter is only 6, a reduction of 50%. Recently, an extremely low index of refraction of a porous silicon sample was obtained by fitting the angle resolved ellipsometry at a wavelength of 5145 A.[6] The determined index of refraction is 1.8 which is less than half the value for bulk silicon of 4.22 at this wavelength. This translates to the value of dielectric constant only 20% of the bulk value, which is surprisingly low. As we know, that the binding energy of a hydrogenic impurity is inversely proportional to the squared of the dielectric constant. A reduction by a factor of two gives rise to quadruple increase in the binding energy only from the point of view of the dielectric constant. A drastic increase in the binding energy can lead to a room temperature freeze out of carriers. Recently, the luminescence in porous silicon under anodic etching in HF has been attributed to quantum confinement.[7,8] It is quite reasonable to attribute the self-limiting action of highly preferential anodic etching to the reduction in the carrier concentration by extreme quantum confinement. Even without taking into account this reduction of dielectric constant, quantum confinement alone can increase the binding energy of shallow impurity states.[9]

COMPLEXITY OF QUANTUM STATES

Lowering of the dielectric constant makes the calculation of the electrostatic solution far more complex. In the course of trying to better understand the resonant tunneling results of Ye et al,[10] Babic et al[11] calculated the one- and two-electron states of a spherical silicon embedded in an amorphous SiO_2 matrix assuming a constant dielectric constant. The sign of the image force depends on whether the dielectric constant inside the silicon is greater or smaller than that of SiO_2. Imagine the complexity as the dielectric constant changes from greater to smaller values! Although the difference in transverse and longitudinal masses were taken into account in an earlier calculation,[12] it was decided to use only an effective isotropic mass. The difference between one-and two-electron systems may be used to define an effective capacitance which is much smaller than the voltage independent capacitance calculated from electrostatics used by Likharev.[13]

EFFECTS OF CHARGE ACCUMULATION

Charge accumulation in a silicon nanocrystal (a nanoscale microcrystal) manifests in what is known as "Coulomb Blockade," in addition to the presence of a helium-like, and Li-like atoms, etc. Electrons accumulated inside the silicon particle raise the energy levels, which is the origin of the so-called Coulomb Blockade. However, if the electrons are trapped inside the oxide near the $Si/Si0_2$ interface, the effective barrier will be further raised from the usual 3.2eV allowing the application of even larger voltage to the silicon particle surrounded by an amorphous $Si0_2$ barriers. From the resonant tunneling work in Ref. 10, it is estimated that extremely high electric field, in the range of 10^7 V/cm, can exist in the Si-nanocrystal. Such high fields further cause wave function localization known as Stark ladder localization. Perhaps the observed sharp structure by Ye et al[10] is due to this Stark localization[14] in addition to quantum confinement.

A $Si0_2$ - Si - $Si0_2$ STRAIN LAYERED COMPOSITE BARRIER

Similar to the strain layered superlattices when thicknesses involved having stored energy below that which creates defects and dislocations,[15] heterojunctions with lattice mismatched systems may produce a defect free interface. An alternate layer of $Si0_2/Si/Si0_2$ may serve as a better barrier for Si confinement. Since a single monolayer of $Si0_2$ is not sufficient serving as a barrier, it is proposed to build alternate strain layers expitaxially. Such a strain layered barrier should be superior to the usual a-$Si0_2$ barrier, apart from the fact that an epitaxially grown silicon quantum well is now possible.

In summary, the $Si0_2S/Si$ system can play a major role in devices operating in the extreme quantum confinement regime. In this regime, contrary to intuition, many parameters that characterize the quality actually improve, because most of the usual scattering mechanisms in solids have energies far below the eigenstates. It is a regime between atomic and solid state physics. In particular, the success of superlattices and quantum well structures with III-V compounds can be extended to the $Si/Si0_2$ system utilizing the strain layer $Si0_2/Si/Si0_2$ as barriers in an epitaxially grown silicon quantum well. This system should be compatible with silicon processing technology. The increase in the binding energy of shallow impurity states due to the reduction of dielectric constant should be an important field of research in its own right.

ACKNOWLEDGEMENT

Work reported is supported by the Army Research Office and the Office of Naval Research.

REFERENCES

1. L. Esaki and R. Tsu, IBM J. Res. and Devel. 14, 61, (1970).

2. R. Tsu and L. Esaki, Appl. Phys. Letts. 22, 562, (1973).

3. F. Capasso, K. Mohammed and A. Y. Cho, IEEE Quantum Electronics, QE 22, 1853 (1986).

4. R. Tsu and L. Ioriatti, Superlattice and Microstructures, 1, No. 4, 295 (1985).

5. L. Ioriatti and R. Tsu, to be published.

6. J. F. Harvey, H. Shen, R. A. Lux, M. Dutta, J. Pamulapati and R. Tsu, Mat. Res. Soc. Symp. Proc. 256, 175 (1992).

7. L. T. Canham, Appl. Phys. Letts. 57, 1046 (1990).

8. R. Tsu, Hishen and M. Dutta, Appl. Phys. Letts. 60, 112 (1992).

9. L. Ioriatti and R. Tsu, Surf. Sci. 174, 420 (1986).

10. Q. Y. Ye, R. Tsu and E. H. Nicollian, Phys. Rev. B 44, 1806 (1991).

11. D. Babic, R. Tsu and R. F. Greene, Phys. Rev. B 15, June, 1992.

12. R. Tsu, SPIE Proc. 1361, 231 (1990).

13. K. I. Likharev, in "Granular Nanoelectronics," ed. by D. Ferry (Plenum New York, 1991).

14. R. Tsu and L. Esaki, Phys. Rev. B 43, 5204 (1991).

15. J. W. Matthews and A. E. Blakeslee, J. Vac. Sci. Tech. 14, 989 (1977).

THE INTEGRITY OF VERY THIN SILICON FILMS DEPOSITED ON SIO_2

Mark Chonko,[1] David Vandenberg,[2] and Douglas Keitz[2]

[1] Advanced Product Research and Development Laboratory
Motorola, Inc. 3501 Ed Bluestein Blvd., Austin TX 78721

[2] MOS 11 Wafer Fabrication Facility, Motorola, Inc.
6501 William Cannon Drive West, Austin, TX 78735

ABSTRACT

The physical integrity of very thin LPCVD silicon films has been studied as a function of deposition temperature and pressure. Integrity is defined as the ability of the film to protect an underlying gate oxide from an HF etch. Films deposited under certain common conditions are quite porous and cannot protect the gate oxide. A pressure dependence was seen for deposition temperatures greater than 560°C. Films deposited in the amorphous phase do provide protection over a broad range of pressures and therefore must be continuous.

INTRODUCTION

LPCVD silicon (poly) films are widely used in integrated circuits as gate electrodes and interconnects. Very thin films (≈500Å) are also used in resistors and transistors as well as to protect gate oxides during a variety of processes prior to the gate electrode deposition.[1] The structure and morphology of thicker (0.5-1.0 μm) films have been characterized by a variety of techniques [2-4] but very thin films have received little attention[5]. If these films are not continuous, they cannot completely protect an underlying gate oxide. We have studied the ability of thin silicon films (≈ 500Å) to protect gate oxides from a wet oxide etch as function of deposition pressure and temperature. Films deposited between 590°C and 625° C do not insure protection against an oxide etch while those deposited at or below 560°C and above 650°C do provide protection over a broad range of deposition pressures.

EXPERIMENTAL

MOS capacitors were fabricated on p-type <100> wafers. The gate oxide thickness was 400Å. A two step silicon deposition process was employed for the gate electrodes. The poly-1 layer was deposited under a various temperature and pressure conditions to a thickness of 500Å, determined ellipsometrically. Depositions were performed in a conventional hot-walled, horizontal reactor with a flat temperature profile and gas injectors using 100% silane. Deposition temperatures ranged from 525°C to 650°C, pressures between 200 and 500 mTorr were used.

Following the first poly deposition, the wafers were etched for 2 minutes in 10:1 buffered HF @ 30°C. Unetched control wafers were included with each poly-1 deposition. After the BOE etch, 4000Å of poly-2 was deposited and phosphorus diffused with PH_3. Electrodes were then patterned with photoresist and plasma etched. Oxide breakdown statistics were gathered using a LOMAC automated tester.

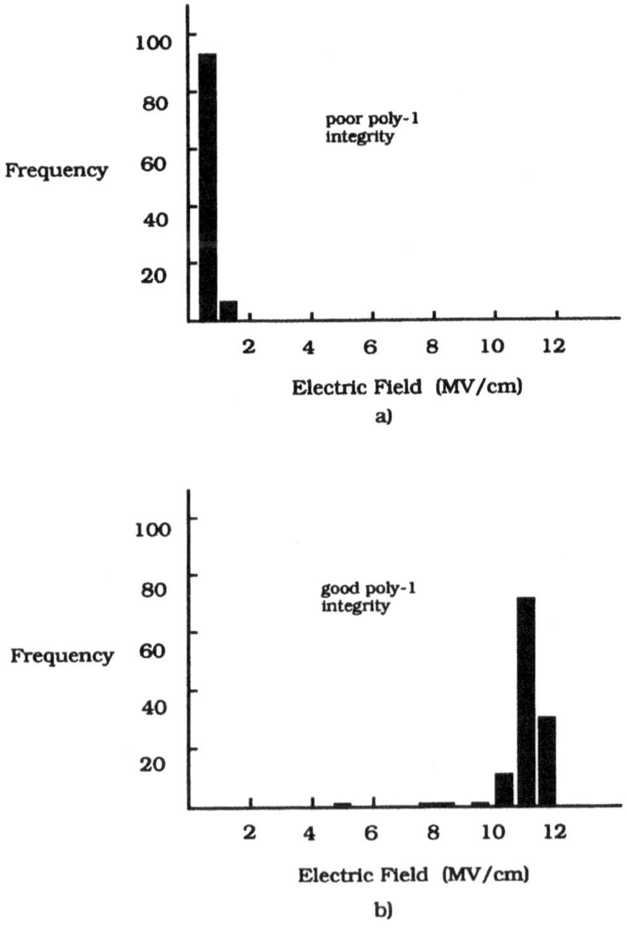

Figure 1. Breakdown field distributions for wafers with a) good poly-1 integrity and b) poor poly-1 integrity. Both were etched for 2 minutes in 10:1 BOE prior to poly-2 deposition.

RESULTS

The integrity of the poly-1 layer was determined by the gate oxide breakdown statistics. Shorted capacitors resulted from the oxide etch when the integrity of poly-1 was poor. Figure 1 shows an example of the breakdown field distributions for a poly-1 film with poor integrity and a film with good integrity.

Both samples were subjected to the 2 minute buffered HF treatment. Due to the extended etch, a clear, unambiguous signal is observed. Excellent breakdown distributions were obtained for all of the control samples. Figure 2 shows the results for wafers prepared over a wide range of deposition conditions. The values shown represent the average oxide breakdown field. Good poly-1 integrity is obtained for all films deposited at or below 560°C regardless of deposition pressure. As the temperature increases a dependence on deposition pressure is observed. Pressures above 200-250 mTorr result in poor thin poly integrity between 590°C and 625°C.

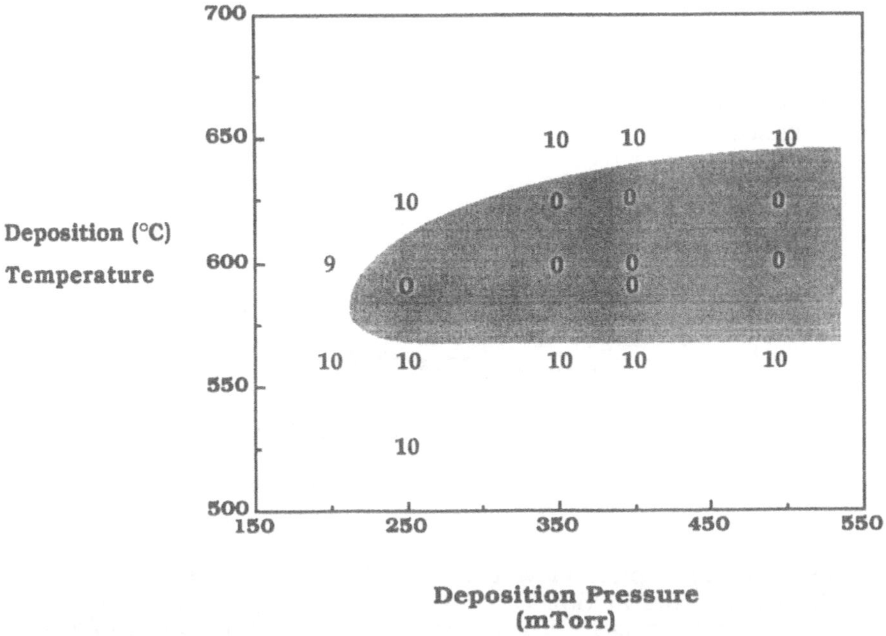

Figure 2. Poly-1 integrity as measured by the average oxide breakdown field under various deposition conditions.

The dependence of oxide breakdown on the thickness of poly-1 deposited under worst case conditions (590°C, 400 mTorr) was examined. Poly-1 films up to 1200Å were not sufficient to prevent acid from reaching the gate oxide. In fact, the critical thickness is somewhere between 1200Å and 1900Å as shown in Figure 3. This implies a network of voids which can be penetrated by acid for even relatively thick films.

DISCUSSION

The conditions under which poor thin poly integrity occurs are similar to those reported to result in a mixed amorphous/crystalline material, as determined by X-ray diffraction,[1,6] however the poor integrity is not a direct result of this mixed phase. Several samples with poor integrity were deposited under conditions reported to produce crystalline films. Additionally, electron diffraction patterns of 500Å films deposited at 590°C, 260 mTorr did not show any evidence of an amorphous component. It has been reported that in the mixed phase regime the film structure is layered with the crystalline portion near the the substrate with the amorphous portion above, near the free surface.[6] This implies that in the transition zone very thin films are crystalline while thicker films are of mixed phase.

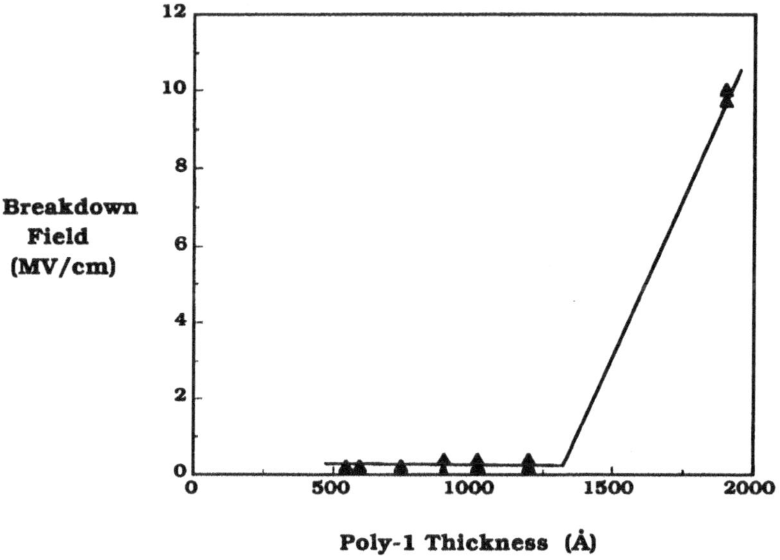

Figure 3. Thickness dependence of poly-1 integrity deposited at 590°C, 400 mTorr.

Figure 4a is a plan view transmission electron micrograph of a 500Å film deposited at 590°C, 260 mTorr. The porous nature of this film is readily apparent. Contrasted is the uniform morphology of a 560°C film, shown in Figure 4b. Apparently films deposited under amorphous conditions nucleate and grow uniformly while those grown at intermediate (580-600°C) temperatures temperatures exhibit a tendency to form islands during the early stages of the deposition. The small amount of space that exists between these islands is enough to permit the BOE to penetrate to the gate oxide. Decreasing the silane pressure during the deposition may increase the surface diffusion coefficient of silicon[4] which would account for the pressure dependence seen in Figure 2. At higher temperatures (≈625°C) the surface mobility of silicon may increase, allowing the voids to be filled. Since the amorphous films form uniformly, no pathway for the etchant exists and the oxide is protected.

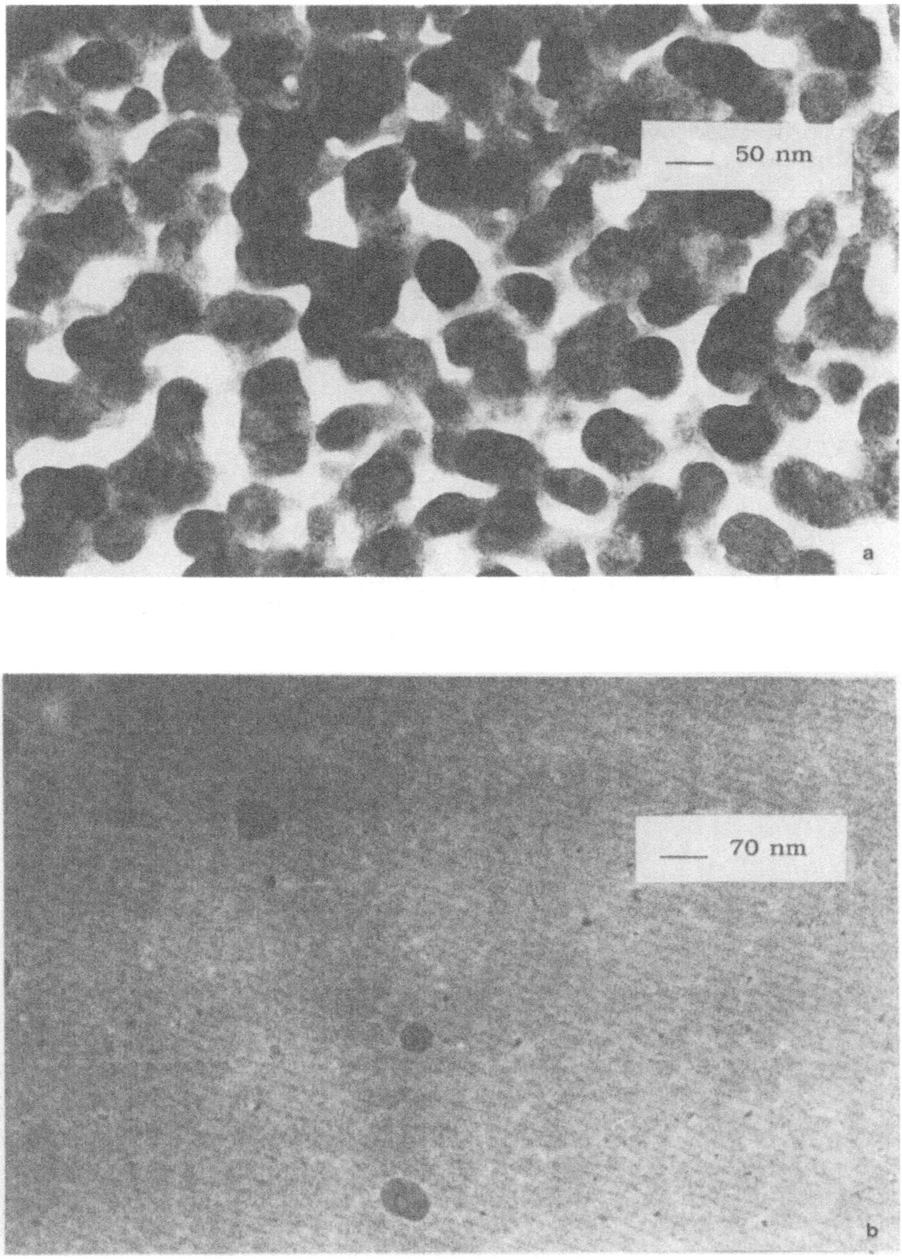

Figure 4. Plan view transmission electron micrograph of poly-1 deposited at a) 590°C, 260 mTorr and b) 560°C, 200 mTorr.

SUMMARY

We have investigated the ability of very thin LPCVD silicon films to protect an underlying gate oxide from a wet etch and have found that under certain common deposition conditions these films are quite porous and cannot provide that protection. A dependence on deposition pressure was seen for films deposited at temperatures greater than 560°C. Films deposited in the amorphous phase do provide protection over a broad range of pressures and therefore must be continuous.

ACKNOWLEDGEMENTS

The authors would like to thank Phil Tobin for valuable discussions, Nancy Dickerson and Joan Kelly for the wafer processing. The TEM support of Peter Fejes and Vidya Kaushik is also gratefully acknowledged.

REFERENCES

1. R. Szeto et al, *ECS Extended Abstracts*, **82-2**, 277 (1982).
2. T.I. Kamins et al, *J. Electrochem. Soc.*, **125**, 927 (1978).
3. G. Harbeke et al, *ibid*, **131**, 675, (1984).
4. P. Jaoubert et al, *ibid*, **134**, 2541 (1987).
5. J. Franka et al, *Semiconductor International*, **May 1991**, 195.
6. Bisaro et al, *J. App. Physics*, **59**, 1167 (1986).

RESEARCHES OF SiO_2 ON InP AND GaAs MOS STRUCTURES

Y.K. Su, C.J. Hwang, and J.D. Lin

Department of Electrical Engineering
National Cheng Kung University
Tainan, Taiwan, R.O.C.

ABSTRACT

High quality SiO_2 layers have been grown on InP and GaAs substrates by direct photo-enhanced chemical vapor deposition (photo-CVD) using monosilane (SiH_4) and oxygen (O_2) as gas sources under the irradiation of deuterium Lamp. The oxide films were evaluated by ellipsometer, Auger electron spectroscopy, Fourier transform infrared spectrum and X-ray photoelectron spectroscopy. The refractive index of photo-oxide is 1.462 when the substrate prepared at 250 °C and gas flow rate ratio (O_2/SiH_4) is 5:1. The high frequency (1MHZ) capacitance capacitance-voltage(C-V) was measured on InP MOS diode and a counterclockwise hysteresis was observed. Deep level transient spectroscopy(DLTS) system was applied to measure the interface state of the SiO_2-GaAs interface.

Introduction

Dielectric thin film deposition is an important processing in semiconductor device fabrication. In III-V compound semiconductor, due to its high saturation velocity compared to silicon, much attention has been focuse on InP and GaAs for high speed device applications. In those semiconductors, the heat treatment temperature should be low to avoid shift in its stoichiometry at elevated temperature owing to evaporation or oxidation of one of its constituents, especially group V element. A low temperature deposition technique is necessary for the formation of the insulator in fabrication III-V MIS diodes. Up to now, anodic oxidation, plasma-enhanced chemical vapor deposition (PECVD) and low temperature chemical vapor deposition (LTCVD) have been the major process employed to form insulating layers on III-V semiconductors. Recently, low damage, low pressure and low temperature photo-chemical vapor deposition (photo-CVD) has attracted much attention in the semiconductor industry[1-7]. Amorphous insulator growth by photo-CVD at temperature of 200°K or below it has been

The Physics and Chemistry of SiO_2 and the Si-SiO_2 Interface 2
Edited by C.R. Helms and B.E. Deal, Plenum Press, New York, 1993

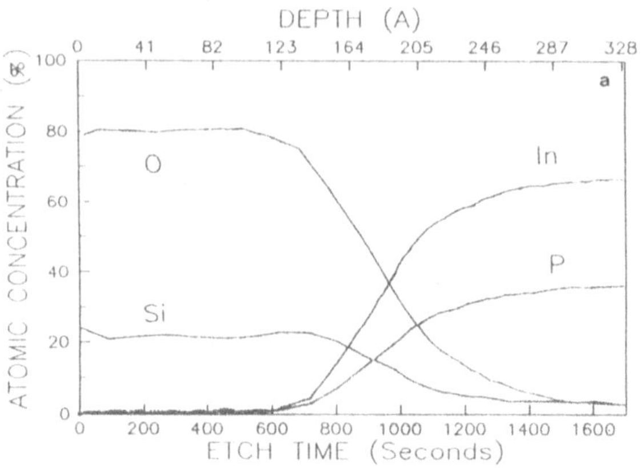

Figure 1.(a)The AES depth profile of the SiO_2/InP structure prepared by direct photo-CVD.

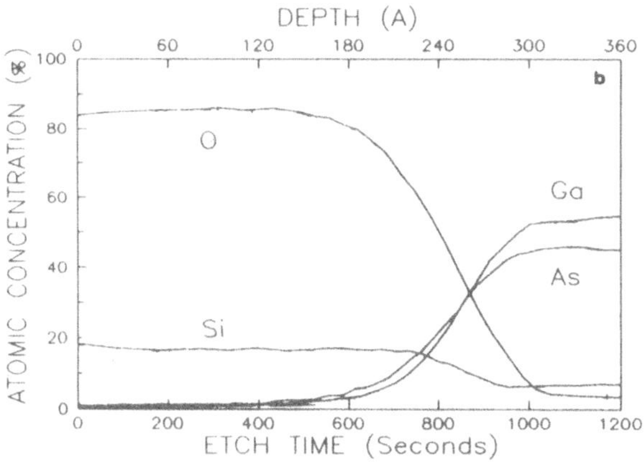

Figure 1.(b) The AES depth profile of the SiO_2/GaAs structure prepared by direct photo-CVD.

reported[5-7]. Therefore, this technique is suitable to fabricate MIS devices on compound semiconductors which must process at low temperature (Tsub<350°K). In this process, silicon dioxide (SiO_2) films have been deposited on indium phosphide (InP) and gallium arsenide (GaAs) by direct photo-chemical vapor deposition using deuterium (D_2) lamp without introducing mercury vapor into reaction chamber at low temperature (<250°K) and low pressure (<2 Torr). The amorphous photo-silicon dioxide films were examined using Fourier transform infrared spectrum (FTIR), Auger electron spectroscopy (AES) and X-ray photoelectron spectroscopy (XPS). The metal oxide semiconductor (MOS) characteristics were measured by high interface state densities were also determined.

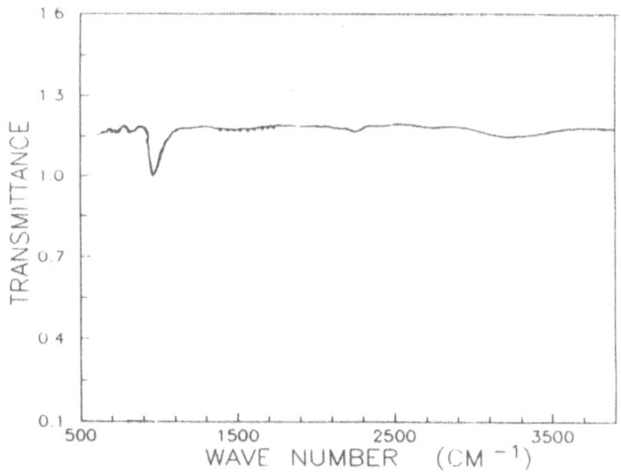

Figure 2.(a)The FTIR spectrum of the photo-SiO_2/InP sample at room temperature and atmosphere.

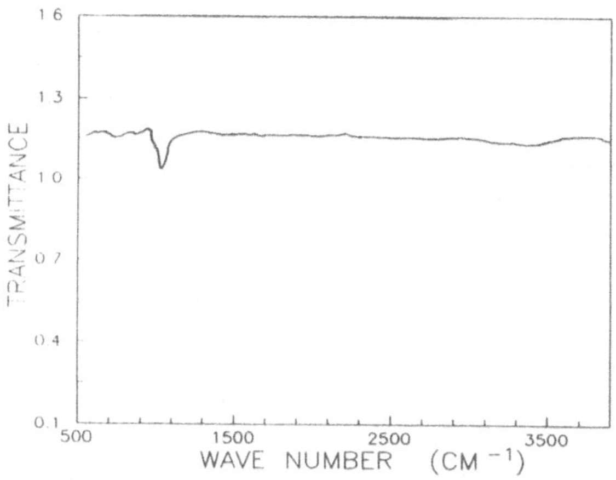

Figure 2.(b)The FTIR spectrum of photo-SiO_2/GaAs sample at room temperature and atmosphere.

Experimental

Metal oxide semiconductor (MOS) capacitor structure was fabricated on n+-type (100) oriented InP and GaAs wafer. Prior to oxide deposition, the wafer was clean by 2-propanol, acetone and methanol successively in ultrasonic vibrator 10 minutes respectively. Following this process, the samples is rinsed in DI water for 10 minutes. The InP sample was then etched in NRL solution [HCl + HF + H_2O(1:1:4): H_2O_2 in the ratio of 12:1], and rinsed in DI water for 10 min again. The GaAs sample was etched in the solution (H_2SO4:H_2O_2=4:1:1) and then rinsed in DI water for 10 min. The sample was then dried by N_2 and loaded into a low pressure direct photo-CVD reaction chamber for silicon dioxide deposition. Low temperature oxide was grown at a temperature in the range of 150-250°K at a pressure of 1-3 Torr with 30 W deuterium lamp irradiation. We believe that the pressure in the reactor is an important parameter in obtaining high quality SiO_2. Higher pressure usually leads to larger deposition rates but tends to favor homogeneous reactions as evidenced by dust formation and leaky films[8]. the thickness and refractive index of the photo-oxide were measured by Rudolph "Auto EL-III" ellipsometer. It was evaluated by FTIR to study its chemical bond and by AES and XPS to analyze its elemental composition. MOS capacitor structure was defined using standard photolithography techniques.

Results and Discussion

To evaluate the properties of photo-oxide and compound semiconductors interface and to understand the physical and chemical process involved in the oxide growth, compositional profile from the surface of the oxide to the bulk of the semiconductor throughout the interface have been studied using AES combined with Ar^+-ion sputtering. Figure 1 (a) and (b) describe the AES depth profile of the various constituent concentration of Si and O, obtained at different depth during Ar^+ (1.5 KeV) sputtering of a photo oxide on InP and GaAs, respectively. The grown oxide appears to be stoichiometric with uniform composition of silicon and oxygen atoms throughout the surface to the oxide-semiconductor interface. The transition region, evaluated by the variation of oxygen concentration (15%-85%), is 90Å for SiO_2/InP interface and 70 Å for SiO_2/GaAs interface. The exact constitutions in this region are unknown but some native oxide should inhabit it. The larger transition region for SiO_2/InP structure maybe due to phosphorus out-diffusion from the surface of InP during oxide deposition or to absorbed unreacted oxygen atoms diffusion into the InP. The penetration increases the oxygen concentration depth and/or fill the In dangling bonds to form In native oxide (e.g. In_2O_3 and $In_x(OH)_y$). In order to study the bonding between Si, O and H in the oxide, infrared spectra were measured using Fourier Transform Infrared (FTIR) spectroscopy as shown in Figs.2(a) and (b). The dominant absorption peaks are near 1060-1080 cm^{-1} This is the same as that by thermally grown oxide [8]. This indicates that the Si-O bond of the photo oxide is as strong as in the thermally grown oxide. The signals of Si-H bond in the range near 2260 cm^{-1} are not found in Figs.2(a) and (b). This suggests that the hydrogen incorporated in this oxide is few. The other absorption mode characterized the Si-O bonding near 800 and 880 cm^{-1} are not found which implies the oxide is purely bonded by Si-O bonds. Figures. 3(a) and (b) show the AES SiO_2/InP and SiO_2/GaAs in the form of front view with semiconductor as backside. The montages provide an excellent vehicle to follow the propagation of both the semiconductor and oxide element in the interface region. The Auger peaks near 64 and 78 eV are the characteristics of oxidized silicon peak of silicon dioxide and near 115 eV are for phosphorus[9] In Fig.3(a), the concave silicon peaks near 64 and 78 eV become flat step by step and, at the same time, the ridgy peak near 115 eV arises gradually and in Fig.3(b) the silicon peaks also vanish in several steps. This transitions are on behalf of interface obviously and the interface between InP, GaAs and oxide is sharp. The depth scale was calibrated according to sputtering rate and is 30Å per step. Looking

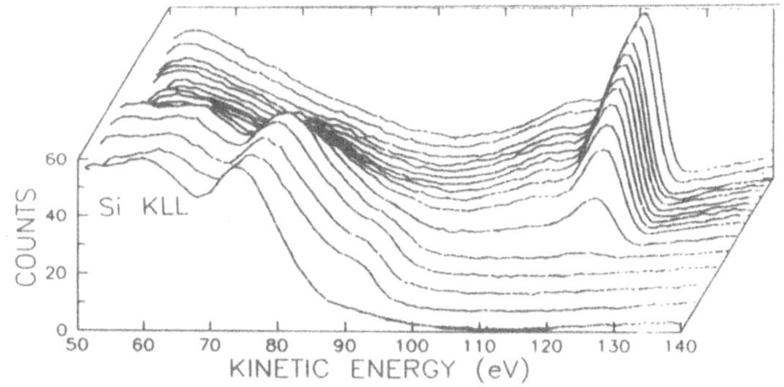

Figure 3 (a)The front wivew of Suger montage diagram of SiO_2/InP structure with InP as backside.

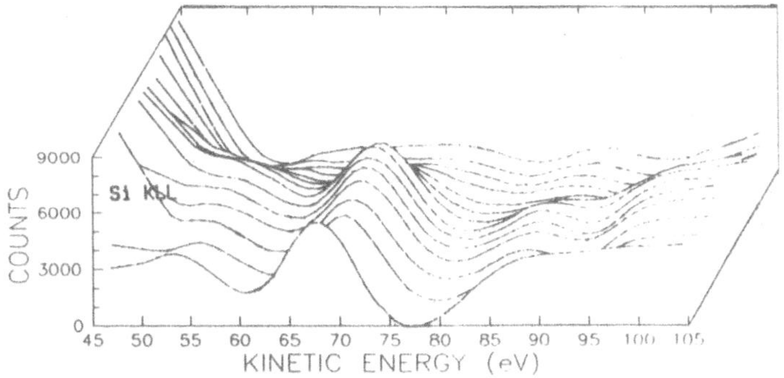

Figure 3.(b)The front view of Auger Montage diagram of SiO_2/GaAs structure with GaAs as backside.

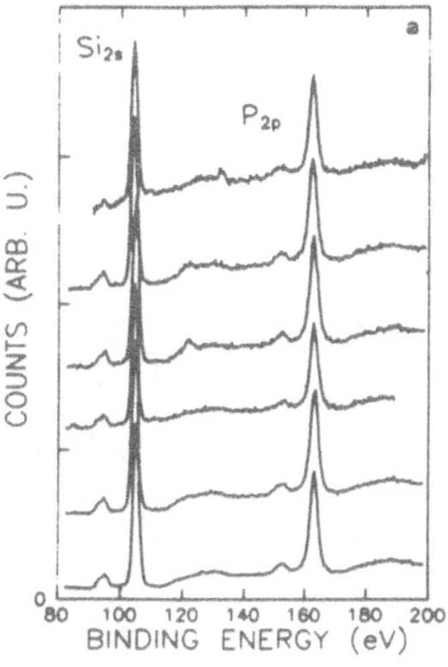

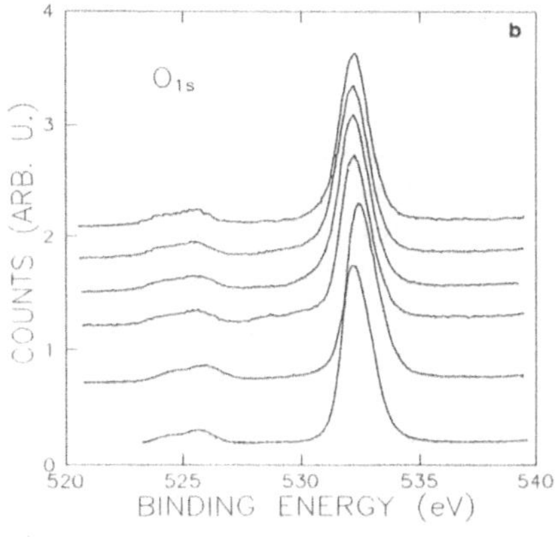

Figure 4.(a)The chemical XPS depth profile of Si peak of the SiO_2 film on InP at different depth from oxide surfce.

Figure 4.(b)The chemical XPS depth profile of O peak of SiO_2 film on InP at the same depth with Fig.4(a).

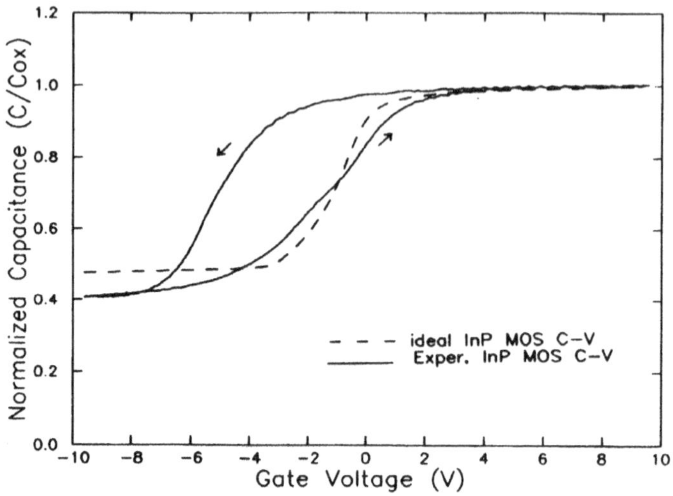

Figure 5.The experimental (solid line)and theoretical (dash line0 1 MHz C-V data of InP MOS dilde prepared at 250óJ

up Fig.3(a), the variation accomplished in three steps which is equivalent to about 90 Å. The XPS chemical depth profile using etching solution of $HF:H_2O$=1:20 was measured and Figs. 4(a) and (b) show the XPS lines of Si_{2s}, Si_{2p} and O_{1s} peaks for SiO_2/InP sample at different depth. The phosphorus didn't be detected until the oxide thickness is smaller than 90 Å. The value is consistent with the AES depth profile transition region data. The bonding energy of silicon and oxygen calibrated by carbon energy value is 103.3 and 522.4 eV. Checked with XPS handbook, the films can be assigned to be silicon dioxide (SiO_2) and is stoichiometric from surface to interface throughout the bulk layer due to the invariation of the signal of Si and O at each depth. The typical high frequency, 1 MHz, capacitance-voltage (C-V) measurement of the n-InP MOS diode is presented in Fig.5 which has both experimental (solid line) and theoretical (dash line) curves. The counterclockwise and distort hysteresis was observed. it intimates that interface traps capture electron when the trap energy level located in the upper half of the band gap (acceptor-like) and capture hole when the trap level is below the Ei (donor-like). The interface trap densities (D_{it}) was obtained by utilizing Terman's method. A U-shape distribution and the minimum interface trap densities of $2x10^{11}$ $cm^{-2}eV^{-1}$ were obtained. To determine the spatial distribution of trap density and energy distribution of trap level in MOS diode, the DLTS (deep level transient spectroscopy) was applied. the correlation signal ($C=C(t_1)-C(t_2)$) was measured from 180 °K to 300 °K at different sampling times (t_1,t_2) with the same time ratio ($t_2=2t_1$). Figure 6 is the DLTS spectrum of the GaAs MOS diode with different pulse voltage and the maximum peaks shifted by varying the pulse voltage (V_b). This shift results from the interface traps [10]. After Arrhenius plot of the thermal emission rate for the peak, the energy level (ET) and interface state densities (N_{ss}) could be obtained. the interface state density distribution is plotted in Fig. 7 which shows that the N_{ss} are all smaller than 1×10^{13} cm^{-2} eV^{-1} except that at $E_t=E_c$-0.44eV there is a high density of N_{ss}. Its value is about $6.5\times10^{13}cm^{-2}eV^{-1}$. this high density of N_{ss} was suggested

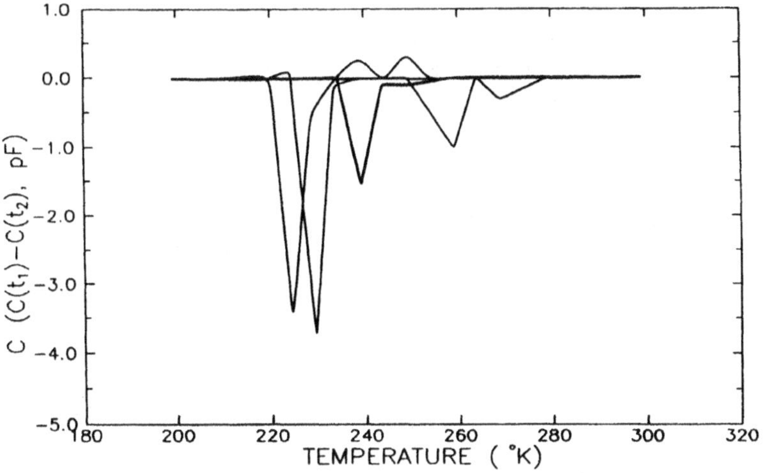

Figure 6. The DLTS spectrum of the n+-GaAs MOS diode prepared by photo-CVD. The temperature of the maximum peak were different at different pulse voltage due to interface state densities.

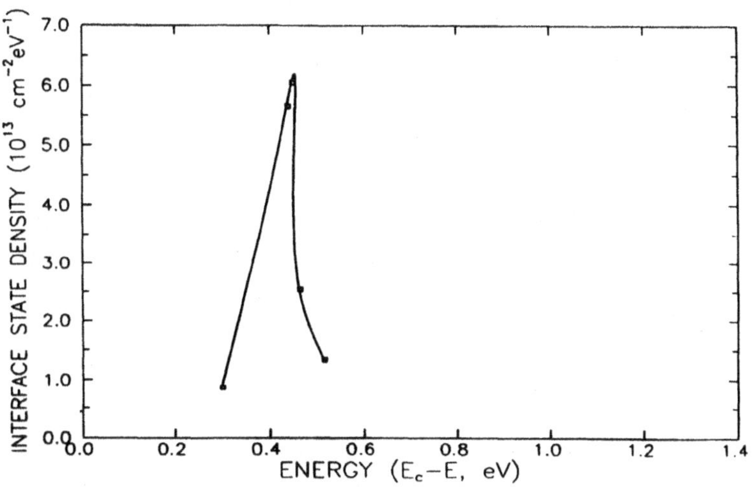

Figure 7. The interface state density distribution derived from the DLTS spectrum

to be due to absorbed unreactive oxygen generated by photolysis diffuse into GaAs and/or inhabit at the Sio2/GaAs interface. Conclusion SiO_2 was successfully deposited on InP and GaAs substrate at low temperature using direct photo-CVD by deuterium lamp without introducing mercury vapor. FTIR absorptionshows that the grown film has strong Si-O bonds with few Si-H bonds. AES depth profile and montage measurements indicate that the composition is stoichiometric and uniform throughout the SiO_2 film and a sharp interface is observed between oxide and semiconductors. The chemical XPS depth profile shows that the constituents of the semiconductors outdiffusion into the oxide is few. the C-V characteristics of the InP MOS structure were measured. A counterclockwise hysteresis was observed. The minimum interface trap density is 2×10^{11} $cm^{-1}eV^{-1}$. DLTS techinque was applied to GaAs MOS diode to measure the interface state between SiO_2 and GaAs. The N_{ss} is 6.5×10^{13} $cm^{-2}eV^{-1}$ at E_c-0.44eV. All these results shows that the MOS structure prepared by direct photo-CVD has good quality and interface to fabricate MOSFET.

ACKNOWLEDGEMENTS

The authors would like to Dr. M. Yokoyama for his system assistance and discussions. The authors would acknowledge the Dr. F. M. Pan for the AES and XPS measurements and discussions. This research was supported by National Science Council, R.O.C., under Contract No. NSC-80-0404-E006-36.

REFERENCES

1.J.W.Peter, IEEE Electron Devices Meeting Digest (IEEE, New York, 1981), p.240

2.Y.Mishina, M. Hirose, Y. Osaka and Y. Ashida, J. Appl. Phys,55, 1234(1984)

3.C.J. Hwang and Y.K. Su, J. Appl. Phys., 67,3350(1990)

4.J.Y. Chen, R.C. Henderson, J.T. Hall and U.W. Peters, J. Electrochem. Soc.131,2145(1984)

5.K.Tamagawa, T. Hayashi and S.Komiya, Jap. J.Appl. Phys., 25, pp.728(1986)

6.S.Nishida, H. Tasaki, M.Konagai and K.Takahashi, J. Appl. Phys., 58, 1427(1985)

7.S.Nishida, T. Shiimoto, A. Yamada, S.Karasawa, m. Konagai and K.takahashi, Appl. Phys. Lett., 49,79(1986)

8.L.E.Davis, N.C.MacDonald, P.W.Palmberg, G.E. Riach and R.E. Walker, Handbook of Auger Electron Spectroscopy, 2nd ed. (Physical Electrons, Eden Paririe, Nm, 1976)

9.K.Yamaski, M. Yoshida and T.Sugano, J. Appl. Phys., 18,113(1979)

VII. DEFECTS AND HOT-CARRIER INDUCED DAMAGE IN $Si\text{-}SiO_2$ SYSTEMS

INTRODUCTION

E. H. Poindexter
U.S. Army Electronics Technology and Devices Laboratory
Forth Monmouth, New Jersey 07703

N. S. Saks
Naval Research Laboratory
Washington, D.C. 20375

M. Schulz
Universität Erlangen–Nürnberg
D-8520 Erlangen, Germany

M. A. Stroscio
U. S. Army Research Office
Research Triangle Park, North Carolina 27709

This chapter contains papers on defects and hot-carrier induced damage in $Si\text{-}SiO_2$ systems, as well as the related SiN_x system. Topics discussed include the generation of random telegraph signals or switching (RTS) via single interface defects, determination of the free energy associated with single-electron Coulomb blockage in submicron MOSFETs, studies of P_b centers by non-standard electron paramagnetic resonance (EPR) approaches, the influence of crystal orientation and processing conditions on the energy distribution of traps, charge trapping and degradation in thin dielectric layers, and optically-induced nitrogen dangling bonds in amorphous SiN_x.

In the area of RTS in ultrasmall MOSFETs, several key issues have been clarified in recent experiments. Uren and Cobden have examined the entropy associated with carrier capture. Effective mobility changes due to the trapping/detrapping of a single carrier are well substantiated. A novel application of substrate bias by Schulz and Pappas has revealed the free energy of the single-electron Coulomb blockage to be of order 200 meV. The free energy associated with this single-electron blockage can be explained in terms of a "breathing" of the depletion region which arises from the 3-dimensional nature of the blockage. Important unsettled aspects are the absence of an entirely satisfactory model for the RTS noise generation, especially noise amplitude, and the complete absence of any physico-chemical or structural model for the traps.

Stathis, Rigo, and Trimaille have studied the hyperfine structure of P_b centers through the use of electrically detected magnetic resonance (EDMR) in actual devices, which provides evidence for the high sensitivity of the method. Defects can be clearly identified from the g-value and hyperfine splitting. At present, however, the key limitation of EDMR is the nonquantifiable meaning of the signal amplitude.

The influence of crystal orientation and processing conditions on the energy distribution of traps at the $Si\text{-}SiO_2$ interface is now known from a variety of experimental approaches reported by Bjorkman, Ma, Yasuda, and Lucovsky. Based on the surprising behavior of the (110) surface, it seems that future research on this case could be very meaningful. Heyns, v. Schwerin, Verhaverbeke, and Kelleher examine charge trapping and degradation of thin dielectric films, and point to the need for adequate models of traps and injection mechanisms.

EPR studies of optically-induced nitrogen dangling orbitals in amorphous SiN_x suggest that the creation of the N dangling orbital is somehow related to hydrogen evolution from the films, and that the N orbital is an electrically active defect. Future research in this area could be profitably correlated with more extensive studies in realistic device structures.

GENERATION OF RANDOM TELEGRAPH NOISE BY SINGLE Si/SiO_2 INTERFACIAL DEFECTS

M.J. Uren[1] and D.H. Cobden[2]

[1]Defence Research Agency
RSRE Malvern
Worcestershire WR14 3PS
England
[2]Cavendish Laboratory
Madingley Road
Cambridge CB3 0HE
England

ABSTRACT

Submicron silicon MOS structures frequently show discrete two-level fluctuations in their conductivity associated with the change in occupancy of a single SiO_2 defect. This article will review the progress that has been made in understanding the complex kinetics and scattering behaviour of these traps.

INTRODUCTION

Most modern electronic devices rely for their effective function on the near perfection of certain solid state interfaces. In particular, the Si/SiO_2 interface which is at the heart of the silicon MOSFET, can readily be made with defect densities of $10^{10}cm^{-2}eV^{-1}$ or less. The Si/SiO_2 defects were traditionally studied in large capacitor structures where very many were present. An entirely new approach to the study of these defects became possible with the pioneering work of Ralls *et al*[1] who were investigating the physics of operation of submicron MOSFETs. They observed in all their devices, discrete fluctuations in the conductance, which jumped between two well defined levels. In a classic paper they mapped out most of the basic characteristics of the phenomenon which will form the subject of this review. What they discovered was that if the device was small enough, a situation could be reached where there was only a single active defect in the device. Similar signals (Random Telegraph Signals, RTS) had earlier been seen in bipolar devices[2] and JFETs[3] in the form of burst noise and were known to be related to bulk silicon point and extended defects

RTSs in a wide variety of systems have been previously reviewed[4,5,6,7]. Here we will talk only about defects in the thermally grown oxide on silicon which is the system that has been most intensively studied, and where the simple and well modelled device structure allows the most progress towards an understanding of the microscopic defect processes.

The Physics and Chemistry of SiO_2 and the Si-SiO_2 Interface 2
Edited by C.R. Helms and B.E. Deal, Plenum Press, New York, 1993

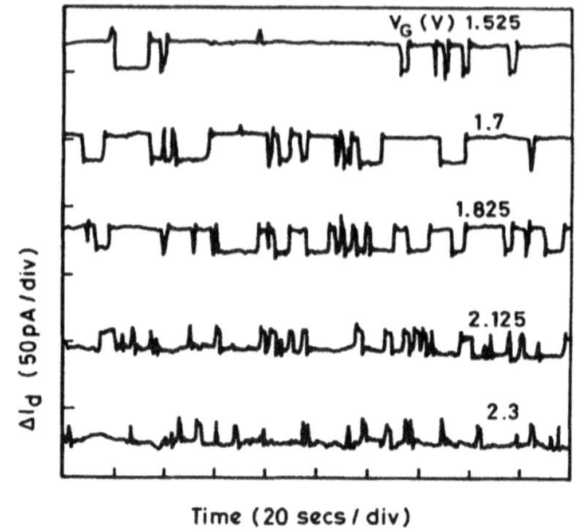

Figure 1. Gate voltage dependence of a RTS measured in a 0.75 by 0.5μm n-channel MOSFET[4].

RANDOM TELEGRAPH SIGNALS IN MOSFETS

In the MOSFET, RTSs are normally observed by measuring the conductance of the inversion layer between source and drain contacts in a submicron device. The fluctuation in occupancy of a single defect level in particle equilibrium with the channel free carriers results in a change in the conductance. The RTSs themselves are characterised by the "capture" and "emission" times, which reflect kinetics, and the amplitude of the conductance step, which reflects the interaction of the defect with the inversion layer carriers.

Current DLTS has also been used to monitor single defects in small MOSFETs by filling all the traps and then monitoring the emission transient as all the filled traps over the accessible range in energy emit, each generating a step in the device conductance[8]. This technique has proved particularly useful for surveys of devices to count the number of defects present.

Capture Kinetics

RTSs are observed in both n- and p-channel MOSFETs where the free carriers are electrons and holes respectively, and are found to display similar behaviour. Figure 1 shows an example of an RTS measured in an n-channel MOSFET at room temperature for a range in gate voltages. The two-level signal is always found to show exponentially distributed times in the up and down states, indicating that the capture/emission probability is time-independent (history-independent). Only a very small voltage is applied to the drain during the measurement and there is no transport through the gate oxide, so the change in defect occupancy occurs in equilibrium. Assuming the process involves the exchange of a carrier from the inversion layer with the defect, the principle of detailed balance can be applied,

$$\frac{\tau_E}{(\tau_E+\tau_C)} = \frac{1}{\{1+g\exp[(E_T-E_F)/k_BT]\}} \tag{1}$$

where τ_E is the emission time for electrons, τ_C is the capture time, g is the ratio of degeneracy change on emission, E_T is the free energy level of the trap and E_F is the Fermi energy. The capture rate may be factorised thus:

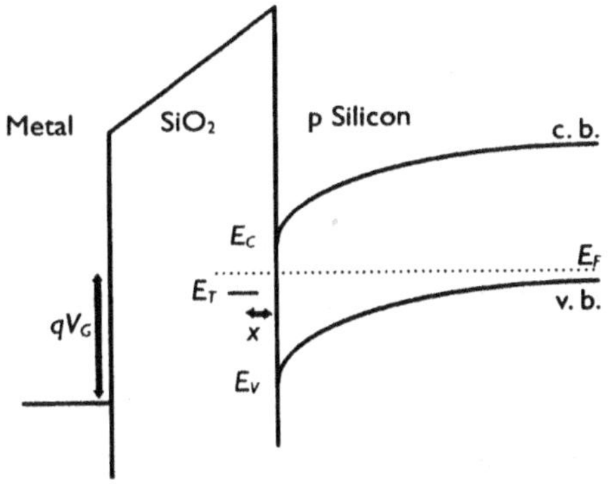

Figure 2. Band diagram of a n-channel MOSFET perpendicular to the surface.

$$1/\tau_C = \sigma v_{TH} n, \tag{2}$$

where σ is defined as the trap capture cross-section, v_{TH} is the free electron thermal velocity and n is the electron number density. Assuming that σ is constant we can identify the up time in figure 1 with τ_C as it falls with increasing gate voltage and hence E_F. At low temperature a similar identification was confirmed[9]. The inversion layer electrons could be heated above the lattice temperature using a high source-drain electric field, and τ_E was found to be independent of electron temperature while τ_C was not, as expected.

Figure 2 shows the energy band diagram for an n-channel MOSFET. By measuring the ratio of τ_E to τ_C as gate voltage V_G is varied, one can monitor the variation of E_T-E_F using equation 1. The electrostatics of this structure are easily modelled, and so the variation in band bending at the inversion layer, $d(E_C\text{-}E_F)/dV_G$ can be calculated from the known gate voltage and source-drain conductance. Since the potential varies linearly with position through the oxide, the distance of the trap into the oxide[1,4] can be deduced from the difference between $d(E_C\text{-}E_F)/dV_G$ and $d(E_C\text{-}E_T)/dV_G$. In strong inversion, where there is a sufficiently high carrier concentration for screening to be principally due to the inversion layer, this procedure always gives a location 0.5-2nm inside the oxide. In contrast, at low carrier concentration (in weak inversion) the locations often appear to be anomalously deep (up to 20nm) into the oxide. A very wide range in gate voltage dependences is observed, with various rates of fall in τ_C with increasing V_G, and with τ_E showing between no dependence and an equal and opposite dependence to τ_C.

This brings us to the crucial question of the number of electrons involved in the trapping process. Equation 1 holds if only a single electron is involved. If the defect were a "negative U" centre, and so liable to trap two (or more) electrons simultaneously, the trap level would not need to shift as fast with V_G to generate the same change in the time constants. Analysing the data for any simple RTS observed to date would indicate a location inside the silicon[4]. As we shall see later, the large activation energies (more than half the bandgap[10]) and the wide range of electrical characteristics of the defects are wholly consistent with an oxide defect but not with a defect in bulk silicon. It therefore appears that the defects responsible for the great majority of RTSs are oxide defects which trap a single electron.

The drain voltage can be used to determine the location of the trap down the channel[11,12]. The trap level shifts relative to E_F as the drain voltage is varied by an amount depending on distance between source and drain, which can be modelled. Hence from a model of the change in Fermi level down the channel, the location can be found.

The temperature dependence of the RTSs has been extensively investigated and

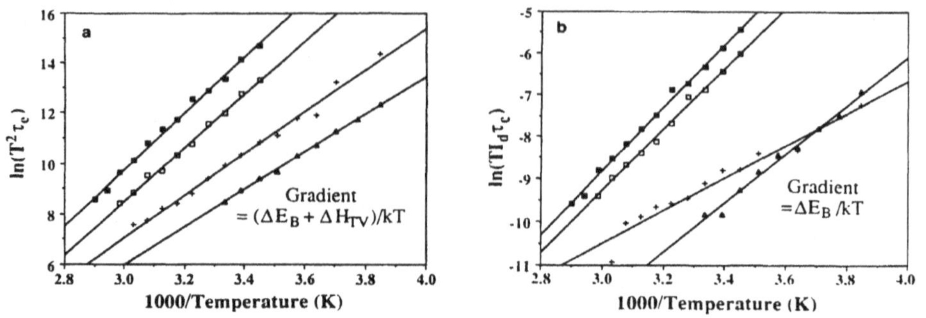

Figure 3. Arrhenius plots for four RTSs in p-channel MOSFETs. (a) emission time, (b) capture time.

reveals some important information about the defects responsible. The principle feature is that all defects show strongly activated capture and emission times. Some typical temperature dependences are plotted in figure 3 for RTSs in p-channel transistors. For capture, $TI_D\tau_C$ is plotted (I_D is the drain current), since this is proportional to the capture cross-section[4]. It is clear that the capture time (cross-section) is activated, ie

$$\sigma = \sigma_0 \exp(-E_B/k_B T). \quad (3)$$

Values of E_B are different for every trap, lying in the range 50meV to 600meV, with values of σ_0 between 10^{-24} and 10^{-15}cm^2 at room temperature. This observation of a featureless distribution in properties is quite consistent with the location of the trap being in the amorphous SiO_2, so that every defect is located in a different environment and thus has its own unique parameters.

The activated capture process is normally explained in terms of a lattice relaxation model of the defect which is very well known in the study of bulk defects in semiconductors[14]. The application of the idea to the capture of a single electron into a near-interfacial defect in SiO_2 is illustrated in figure 4 which shows a configuration coordinate diagram for the defect[15]. The diagram plots the variation of the sum of the electronic and the local phonon energy versus the generalised coordinate Q for one of the vibrational normal modes of the defect. The two minima correspond to the occupied and empty states of the defect. The effect of the electron-lattice interaction is to produce a lattice distortion so that the empty and occupied states lie at different values of Q. Tunnelling between these two configurations involves nuclear motion and so is relatively unlikely. The preferred route is for the lattice to thermally distort until there is overlap between the empty and filled states of the defect, at which point the electron can make

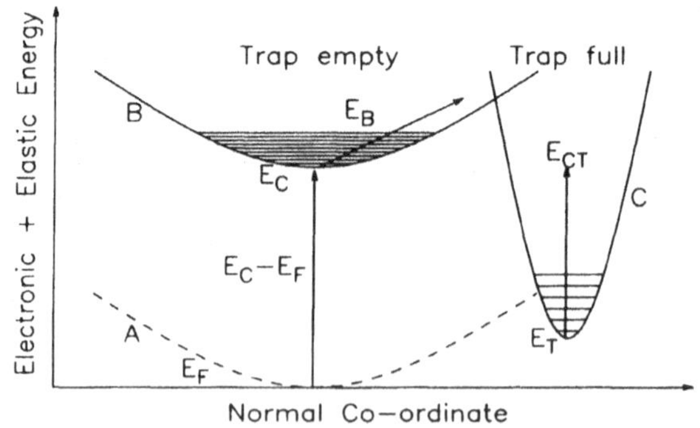

Figure 4. Configuration coordinate diagram for an oxide defect. The different curvatures of the empty and full states is indicative of some lattice softening on emission.

a transition from the inversion layer to the defect. E_B then represents the thermal activation energy to reach the crossover, as indicated on the diagram, and σ_0 depends on the matrix element for electron tunnelling to the defect and probably on the the phonon frequency.

An alternative explanation has been put forward by Schulz and Karmann[10] who argue that a purely electronic model based on thermionic emission over a barrier can account for the activated capture cross-section. If the defects are truly located at the measured distances from the interface then the probability of inelastic tunnelling through the barrier into a state below the silicon bandedge must be especially low.

As was first pointed out by Engstrom and Alm[16], thermal measurements that rely on detailed balance via Equation 1 do not directly measure the eigenenergies of defects (enthalpies), but rather their Gibbs free energies[17]. The free energy difference between the conduction band and the trap level $\Delta E_{CT}=E_C-E_T$ is related to the corresponding enthalpy and entropy differences by $\Delta E_{CT}=\Delta H_{CT}-T\Delta S_{CT}$. The activation energy measured in an Arrhenius plot such as that in figure 3 for the emission times equals the enthalpy change, while the entropy change appears exponentially in the prefactor. From the combination of the prefactors for the emission and capture activation data, ΔS_{CT} can be extracted. Directly analogous arguments apply to p-channel devices. The values obtained for emission of electrons in n-channel devices[13] or holes in p-channel devices[18] are surprisingly large, lying in the range 5-12k_B in both cases, implying that E_T is strongly temperature dependent. The entropy change is the result of a change in the density of accessible states for the entire system on emission of the carrier. The most obvious contribution to the entropy change is a change in the ground state degeneracy. In a disordered system, it is very unlikely that this will amount to more than 4 yielding an entropy change of $\Delta S=k_B\log_e(4)$. A second contribution which in the past has been used to attempt to explain the observed ΔS's is a strong electron-phonon interaction leading to changes in phonon mode frequencies on emission. However, to explain a 10k_B entropy change would require an infeasibly large mode softening[7]. Also to explain the positive sign for carrier emission in both n- and p-channel devices there would have to be a softening of the lattice, yielding an increase in the density of phonon modes on emission. This requires that the defects studied in the n-channel MOSFET are distinct from those studied in the p-channel MOSFET since in the first case the lattice would have to soften and in the second harden on emission of an electron. The two cases involve defect levels near the centre of the 9eV oxide energy gap which happen to line up with the upper and lower halves of the 1.1eV silicon bandgap. It is very unlikely that these should be of a different nature.

Rather than describing the trap energy level temperature dependence in terms of an entropy, Schulz and Karmann[10] have proposed a model that considers the effect the Coulomb potential of the charged defect has on the inversion layer. They define a local Fermi energy that is a function of the *average* occupation of the defect, which has the consequence that the capture time is in turn a function of the average occupation.

We have recently suggested that there is an additional and potentially large contribution to the entropy which has been ignored and can explain both the size and sign of the observed entropy changes[19]. The entropy of a free particle (Fermi-Dirac) gas is well understood[20], but it has nevertheless always been assumed, without justification, that the entropy change of the free electron gas in a semiconductor on addition of an electron would have an insignificant effect on defect thermodynamics. The inversion layer was assumed to act as an ideal particle reservoir, which is not modified in any way when an electron was added or removed. In fact the inversion layer in a screening region around the defect is significantly affected by the change in the Coulomb potential of the defect caused by electron capture[21] and this should properly be accounted for when considering the thermodynamics of the capture process. An estimate of the size of the entropy change of the nondegenerate free-electron gas on adding an electron can easily be made. The

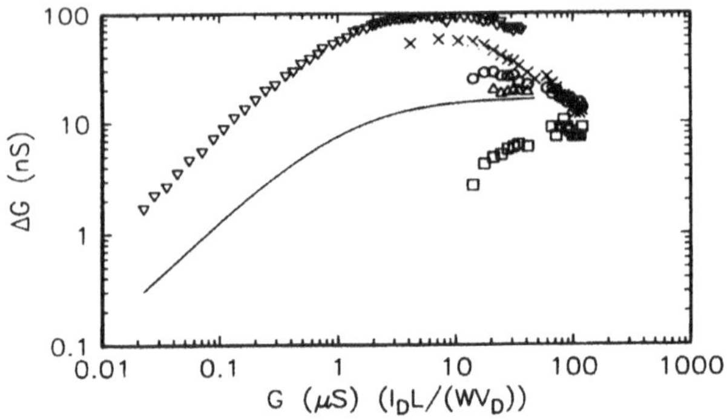

Figure 5. Amplitude of conductance change for five different RTSs in a single n-channel MOSFET[31].

result is only weakly sensitive to the details of the electron distribution in the inversion layer (by a maximum of $\pm\frac{1}{2}k_B$) and is given by:

$$\Delta S_{GAS} \approx -k_B[2 - \ln(\rho / N_{EFF})] \tag{4}$$

where $\rho = N/V$ is the volume number density of free carriers and N_{EFF} is the effective level density, which in an ideal gas is equal to $g^{-1}(h^2/2\pi m^* k_B T)^{3/2}$. Importantly, the entropy change on adding a hole to a hole gas has the same sign as that on adding an electron to an electron gas, in accordance with the observations. Equation 4 predicts an entropy change that rises with falling carrier density to reach a maximum at around the gate threshold voltage, where the screening from the metallic gate starts to become important; at lower V_G the contribution should then fall. Typical values of entropy change for carrier emission with a free carrier concentration of 10^{12}cm^{-2}, are $5.3k_B$ for n-channel and $4.4k_B$ for p-channel devices. Thus the free carrier entropy can explain large values for the entropy change on capture and does not require that the oxide defects in the two halves of the silicon bandgap be distinct. Note also that this contribution will affect any thermally based measurement of deep levels in semiconductors and can be used to explain various anomalies in the interpretation of data on such other defect systems as the Au deep acceptor in Si[22].

Amplitudes of RTSs

The capture of an electron into an oxide defect may affect the conductance of a MOSFET in two ways. Firstly, capture involves the removal of an electron from the channel which thus reduces the number of electrons available to carry current. Secondly, the trapping of a charge changes the charge state of the defect and therefore its scattering cross-section and consequently changes the effective mobility μ. Averaging the effect over the entire device, yields the observed change in conductivity σ:

$$\frac{\Delta\sigma}{\sigma} = \frac{\Delta N_{INV}}{N_{INV}} + \frac{\Delta\mu}{\mu} \tag{5}$$

where N_{INV} is the free carrier number density per unit area. Many workers have now calculated the change in number density, which is a simple function of the electrostatics of the device[23,4,25,24]. At high carrier concentrations, the defect charge is entirely screened by the image charge in the inversion layer, thus the change in the number of channel carriers is just 1, giving $\Delta N_{INV}/N_{INV} \propto 1/N_{INV}$. Below the threshold voltage, there is insufficient inversion charge to screen, so the image charge resides mainly on the gate. The screening length becomes roughly constant, leading to $\Delta N_{INV}/N_{INV}$=constant.

At room temperature, the conductance is universally found to fall on capture with

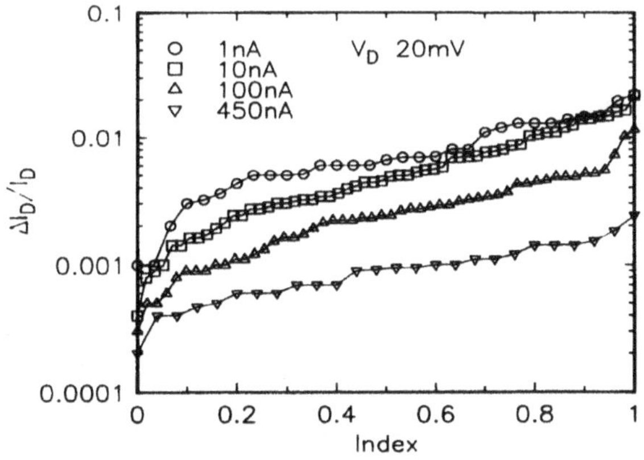

Figure 6. Each point shows the amplitude of an RTS measured in 1.5 by 1.5 micron n-channel MOSFETs ranked in order of increasing size at the four device conductances indicated.

broadly the amplitude dependence on gate voltage described above[4,12]. This can be seen in figure 5 where the effective change in the number of electrons in the channel deduced from the RTS step height is plotted against conductance for five defects in the same device. The full line shows the prediction of the simple number-density-change model assuming constant mobility. Above threshold, the curves tend towards the prediction, but they lie within a wide range up to an order of magnitude above and below the line. Figure 6 shows the distribution in amplitudes in many devices from the same wafer, measured at four carrier concentrations lying both above and below threshold. There is no preferred amplitude but a distribution is seen which is roughly log-normal. Attempts have been made to parameterise the mobility change and relate it to the scattering from a screened Coulombic scatterer, with contributions above and below the line being related to scattering from donor or acceptor states respectively[26]. However, there is no sign of a resulting bimodal distribution. At present, there is no clear explanation available as to why there is such a broad distribution of scattering amplitudes. One possibility is that the local-interference mechanism of scattering from complex defects, which has been invoked to explain the noise amplitude in dirty metals[27], may play some part.

At low temperatures, some RTSs show the inverse of the "normal" behaviour, ie the conductance rises on electron capture[1]. This was observed particularly in early devices that showed some process damage and was less evident in other samples[28]. It is clear that at low temperatures the scattering term can be dominant, and multiple contact submicron width devices can show surprising effects such as the reversal of the "polarity" of an RTS on changing the choice of voltage probes in a four-terminal configuration. This is a clear demonstration of the quantum-interference phenomenon of universal conductance fluctuations which dominates the conductance of submicron structures at liquid helium temperatures[29].

Complex RTSs

Most RTSs behave as we have discussed above but a few per cent show more complex behaviour[30]. Many papers have been published concentrating on particular examples, but the problem is that a different model is required to explain each complex RTS.

Figure 7 shows an example of a complex signal where one two-level RTS turns on and off, with a transition to a third level[31]. Signals have been described with three levels that appear to represent capture of two electrons into the defect[24,30]. In DLTS, a "giant step" was seen which showed several ledges, presumably implying the capture of

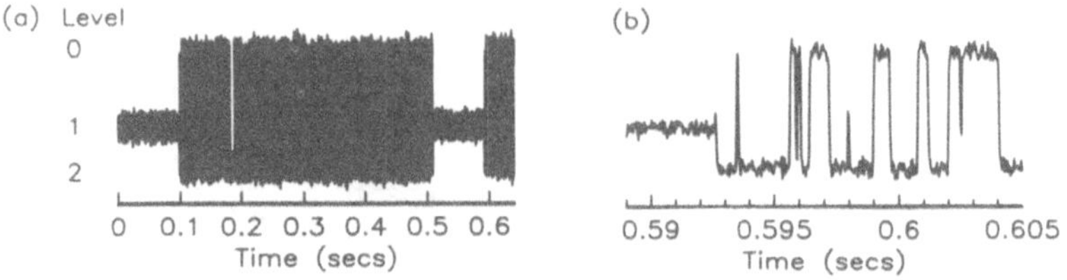

Figure 7. Complex RTS observed in a 0.7 by 0.6μm n-channel MOSFET [31]. (b) is an expanded view.

several electrons into a single defect complex[32]. The probability of two randomly distributed defects being close enough together to interact via a simple Coulomb interaction can be estimated and is too small to explain the observed incidence of these signals[30]. It is necessary to postulate some preferential clustering of point defects in the oxide which can then interact via their Coulomb fields or a strain field[34,33]. Or alternatively, the signals may come from complex extended defects that have multiple metastable configurations and charge states. Such states have been observed in computer simulations[35] of SiO_2. However, it is very difficult to make general statements in the face of the wide variety of signals and the plethora of possible models that can explain each signal.

RTS with No Change in Charge State

Some RTSs have been observed that display an anomalous variation in RTS amplitude with gate voltage. They have been interpreted using models involving no change in charge state[31], so the signal is due purely to a change in the scattering cross-section of the defect accompanying some internal rearrangement. One particular signal observed by Schulz and Karmann[36] in a p-channel device at a temperature around 100K showed a very weak temperature dependence, and had a small amplitude that showed the reverse gate voltage dependence to normal, ie the time for the low conductance level fell with increasing gate voltage. The gate voltage dependence of the time constants appeared to show peak-like structure. They attributed the signal to an "attractive" centre that can bind a hole in a shallow level at the interface from which it can tunnel to and from the defect. Thus the hole is never free and the conductance modulation results from a change in scattering cross-section. We have recently observed similar but more pronounced effects at lower temperatures, which are in agreement with a qualitatively similar model, involving modulation of the scattering by a single two-level-system with no carrier capture involved.

MOS TUNNEL DIODES

The first observations of RTSs in the tunnel current through silicon MOS were made by Farmer *et al*[33] using 1μm^2 area devices with a thin tunnel oxide of approximately 2nm thickness[6]. In this situation the current carrying electrons in the oxide are far from equilibrium and Equation 1 does not apply. The energy levels responsible for the RTSs can lie anywhere up to about 2 eV above the silicon conduction band edge, as opposed to the situation in the MOSFET where they must be close to the silicon band edges.

The RTSs in the two situations show many similarities. The amplitude of the tunnelling RTSs are distributed over many orders of magnitude, are strongly gate voltage dependent and can either increase or decrease with temperature. The up and down times are all found to be exponentially distributed, and display a strongly activated behaviour in general, with activation energies observed in the range 7meV to 442meV over the temperature range 50K to 270K. In many cases, complex signals were observed that were sometimes not stable to wide variations in temperature or gate voltage. Farmer *et al*[33,6] saw several different types of behaviour of multilevel signals, including three level signals where transitions between two particular levels could only take place via the third level, a four level signal where both the amplitude and duty cycle of one RTS was modified by another, and a large transition that decomposed into smaller steps as T was increased.

The large signals were up to 2000 times greater in amplitude than the model for a single Coulomb centre in a homogeneous barrier predicts, which they felt ruled out single electron trapping. Large scale inhomogeneities were not present based on the behaviour of the current-voltage characteristics. They developed a model based on reconfiguration of bistable oxide defects with charge trapping following as a *consequence* of the reconfiguration. In this model, the defect can trap either one or many electrons in one of its configurations and a different number in another configuration. The voltage dependence was ascribed to the variation in the relative energy levels of the two configurations on the assumption that they are located at different positions through the oxide barrier. The temperature dependence would arise through the activation energy necessary to surmount the lattice distortion barrier. In some cases a curvature in the Arrhenius plots was seen that was ascribed to a local heating of the defect by the current passing through the device.

Andersson *et al*[37] reported detailed results on one RTS in a tunnel diode biassed in accumulation that showed rich and complex behaviour. The temperature dependence changed from weak to strongly activated at about 130K. They developed a model based on trap-assisted tunnelling through two different states of the same defect. When the ground state of the defect is occupied, the efficient trap-assisted tunnelling via the excited state of the same defect through the oxide is inhibited. Capture into the ground state has to be as much as 10^8 times slower than capture into the excited state; this can be explained by a lattice relaxation process which is absent for the excited state. Using this model, they can quantitatively explain the gate voltage dependence of the amplitudes, which ranged up to 16% of the underlying current flow.

RELATIONSHIP OF RTSs TO ENSEMBLE MEASUREMENTS

One technologically important manifestation of the traps responsible for RTSs in small devices is that in large devices the superposition of many RTSs causes the 1/f noise. This has been proven in both tunnel junctions[38] and in MOSFETs[23] by observing the transition from the RTS noise due to a few defects, to a smooth 1/f spectrum as more defects are added, either by increasing temperature, or by looking at larger and larger devices.

The states responsible for the RTSs can also be observed in large capacitor structures using either DLTS[39] or conductance[40] measurements. These defects are quite distinct from the "fast" states normally measured using capacitance-voltage techniques which have a single well-defined capture cross-section and small or absent capture activation energy. They are in fact the "slow" states and are characterised by a wide range in capture cross-section, entirely consistent with the RTS measurements.

CONCLUSIONS

A great deal is now known about the defects that cause RTSs in MOS structures from a phenomenological point of view. We can determine that their location is in the oxide within a nm or two of the Si/SiO_2 interface; we know that they show a very wide distribution in their characteristics, such as activation energies and capture cross-sections and we can relate them to the behaviour of larger devices. However, it is not yet possible to identify the chemical nature of the defects, in the absence of any clear "fingerprint" that is characteristic of *all* defects. The challenge for the future is to find a universal model that encompasses all the observed phenomena and whose predictions can be compared with the results of other experimental techniques.

REFERENCES

1. K S Ralls, W J Skocpol, L D Jackel, R E Howard, L A Fetter, R W Epworth, D M Tennant, *Phys Rev Lett*, 52;228(1984).
2. K F Knott, *Solid State Elec.*, 33; 1347 (1990).
3. K Kandiah and F B Whiting, *Solid State Elec.*, 21; 1079 (1978).
4. M J Kirton and M J Uren, *Adv. in Phys.*, 38; 367 (1989).
5. M J Kirton, M J Uren, S Collins, M Schulz, A Karmann and K Scheffer, *Semicond. Sci. Technol.*, 4; 1116 (1989).
6. K R Farmer and R A Buhrman, *Semicond. Sci. Technol.*, 4; 1084 (1989).
7. K R Farmer, in: "Insulating films on semiconductors 1991," ed. W Eccleston and M Uren, Adam Hilger, Bristol (1991).
8. A Karwath and M Schulz, *Appl. Phys. Lett.*, 52; 634 (1988).
9. L D Jackel, W J Skocpol, R E Howard, L A Fetter, R W Epworth, D M Tennant, in: "Proc. 17th Int. Conf. on Physics of Semiconductors," ed. D J Chadi and W A Harrison, Springer, New York (1985).
10. M Schulz and A Karmann,*Physica Scripta*, T35; 273 (1991).
11. P Restle, *Appl. Phys. Lett.*, 53; 1862 (1988).
12. P Restle and A Gnudi, *IBM J. Res. Develop.*, 34; 227 (1990).
13. M J Kirton and M J Uren, *Appl. Phys. Lett.*, 48; 1270 (1986).
14. C H Henry and D V Lang, *Phys. Rev. B*, 15; 989 (1977).
15. M Jaros, "Deep levels in Semiconductors," Adam Hilger, Bristol (1982).
16. O Engstrom and A Alm, *J. Appl. Phys.*, 54; 5240 (1984).
17. O Engstrom and H Grimmeiss, *Semicond. Sci. Technol.*, 4; 1106 (1989).
18. D H Cobden, M J Uren and M J Kirton, *Appl. Phys. Lett.*, 56; 1245 (1990).
19. D H Cobden and M J Uren, submitted for publication.
20. H Eyring, D Henderson, B J Stover and E M Eyring, "Statistical mechanics and dynamics," Wiley, New York (1989).
21. M Schulz, in: "Insulating films on semiconductors 1991," ed. W Eccleston and M Uren, Adam Hilger, Bristol (1991).
22. D V Lang, H G Grimmeiss, E Meijer and M Jaros, *Phys. Rev. B*, 22; 3917 (1980).
23. M J Uren, D J Day and M J Kirton, *Appl. Phys. Lett.*, 47; 1195 (1985).
24. H Nakamura, N Yasuda, K Taniguchi, C Hamaguchi and A Toriumi, *Jap. J. Appl. Phys.*,28; L2057 (1989).
25. O Roux, B Dierickx, E Simoen, C Claeys, G Ghibaudo and J Brim, *Micoelectronic Engineering*,15; 547 (1991).
26. K K Hung, P K Ko, C Hu and Y C Cheng, *IEEE Elec. Dev. Lett.*, 11; 90 (1990).
27. J Pelz and J Clarke, *Phys. Rev. B*, 36; 4479 (1987).
28. W J Skocpol, private communication.
29. W J Skocpol, in: "Physics and fabrication of microstructures," ed. Kelly and Weissbuch, Springer Verlag (1986).
30. M J Uren, M J Kirton and S Collins, *Phys. Rev. B*, 37; 8346 (1988).
31. M J Uren and M J Kirton, *Appl. Surf. Sci.*, 39; 479 (1989).
32. A Karwath and M Schulz, in: "The physics and chemistry of SiO_2 and the SiO_2 interface," ed. C R Helms and B E Deal, Plenum, NewYork (1988).
33. K R Farmer, C T Rogers and R A Buhrman, *Phys. Rev. Lett.*, 58; 2255 (1987).
34. A Ohata, A Toriumi, M Iwase and K Natore, *J. Appl. Phys.*, 68; 200 (1990).
35. L Guttman and S M Rahman, *Phys. Rev. B*, 33;1506 (1986).
36. M Schulz and A Karmann, *Appl. Phys. A*,52; 104 (1991).
37. M O Andersson, Z Xiao and O Engstrom, *Phys. Rev. B*, 41; 9836 (1990).
38. C T Rogers and R A Buhrman, *Phys. Rev. Lett.*, 53; 1272 (1984).
39. H Lakhdari, D Vuillaume and J C Bourgoin, *Phys. Rev. B*, 38; 13124 (1988).
40. M J Uren, S Collins and M J Kirton, *Appl. Phys. Lett.*, 54; 1448 (1989).

SINGLE ELECTRON TRANSFER FROM THE CHANNEL OF A SUB-μm MOSFET TO AN INDIVIDUAL INTERFACE TRAP

M. Schulz, A. Pappas, and J. Vennemann

Institute of Applied Physics, University of Erlangen
Staudtstr. 7, W-8520 Erlangen, Germany

INTRODUCTION

Random telegraph switching (RTS) of the source-drain current in μm-sized MOSTFETs provides a unique opportunity to study the trapping of mobile charge carriers in the channel into individual defect centers at the SiO_2-Si interface[1,2]. The present understanding of RTS was reviewed in several recent papers [3-5]. The interface states causing the RTS reside in the oxide a few nanometers away from the interface to silicon. Most frequently only a single electron or hole is captured and re-emitted by these states from or to the MOS channel. It has been shown that the 1/f-noise is generated through a superposition of RTSs of many individual interface defects[6,4].

A rather complex dependence on the gate bias voltage is frequently observed for the trapping and emission rate constants[1,4]. The transfer rates may be strongly activated for Coulomb-repulsive centers, i.e. acceptors near an n-type channel or donors near a p-type channel[8,9]. For Coulomb-attractive centers, a negligible temperature activation is observed and the transfer of charge carriers takes place by tunneling[9]. The energy position of the level is lowered with increasing gate bias voltage by the voltage drop across the oxide in the region between the channel and the trap location[3,4].

In the present paper, we report on RTS rate constants measured for two different defect types as a function of the gate bias voltage, of the temperature, and for the first time of the substrate bias voltage. We can show that the trapping rates are determined by the Coulomb interaction of the trap charge with mobile carriers in the channel and on the gate. The key information for the interpretation is obtained by the substrate bias voltage effect because the voltage drop in the oxide and thus the level lowering by the gate bias can be varied with maintaining a constant carrier concentration.

The measurements are performed for two specific interface trap levels classified as type A and type B. The two interface traps differ in the functional dependence of the rate constants on gate bias voltage.

The Physics and Chemistry of SiO_2 and the Si-SiO_2 Interface 2
Edited by C.R. Helms and B.E. Deal, Plenum Press, New York, 1993

EXPERIMENTAL

The experiments are performed on two different n-channel MOSFETs which were kindly provided to us by the Siemens Research Lab, Munich. The technical data of the devices are: oxide thickness $t_{ox}=17.7/10$nm, channel length $l=0.55/0.7\mu$m, and channel width $w=0.7/0.7\mu$m for the two devices in which the levels A/B, respectively, are observed. The effective channel conductance is varied by the gate bias voltage. The channel conductivity is measured by the drain current I_D observed for a fixed, sufficiently small source-drain voltage $V_D=10$mV$<$kT to maintain an approximately uniform channel. This drain current is used as a measure of the carrier density in the channel. The magnitude $I_D=1$nA relates to a carrier density of approximately $n_s=1.3\cdot10^9$cm^{-2}.

The RTS is recorded as a voltage drop between source and drain for a constant current by a transient recorder. The mean switching times τ_e and τ_c, or the rates $1/\tau_e$ and $1/\tau_c$, respectively, are determined from averages of up to 300 switching events.

RESULTS

The capture and emission times are depicted as a function of the channel conductance in Figs. 1 and 2 for the defects A and B, respectively. Parameter is the substrate bias voltage in both figures.

The defect A (Fig. 1) is characterized by an emission time which steeply increases over more than four orders of magnitude. The capture time decreases for a low channel conductance and increases again after a turnaround at approximately threshold (thr) conditions for the MOSFET channel. This turnaround and the increase of the capture time with increasing carrier density in the band is in strong contrast to the normal trapping behavior as described by the Shockley-Read-Haynes (SRH) statistics[10] for bulk trap levels for which the trapping rate increases proportional to the free carrier density in the bands.

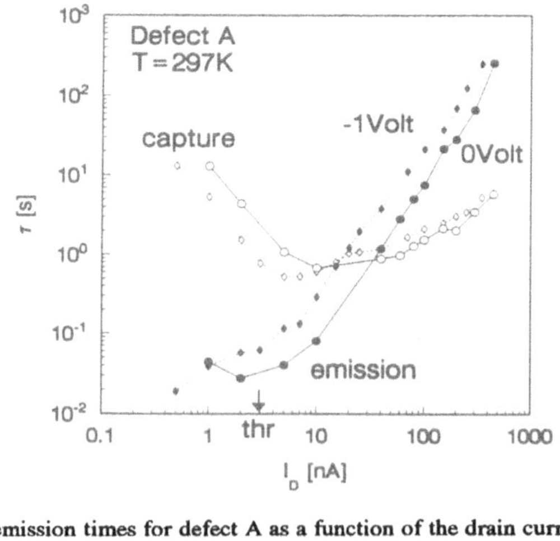

Figure 1. Capture and emission times for defect A as a function of the drain current I_D at $V_D=10$mV as varied by the gate voltage. The continuous and dashed curves are for the substrate bias voltages $V_s=0$V and $V_s=1$V, respectively. The arrow (thr) on the horizontal axis marks the threshold for channel formation at $I_D=3$nA.

The defect B (Fig. 2) is characterized by an approximately constant emission time and a capture time which decreases steeply with increasing channel conductance. This trapping behavior resembles that of SRH statistics; however, we will see below that the defect occupation increases proportional to a power $\alpha \geq 2$ of the free carrier density.

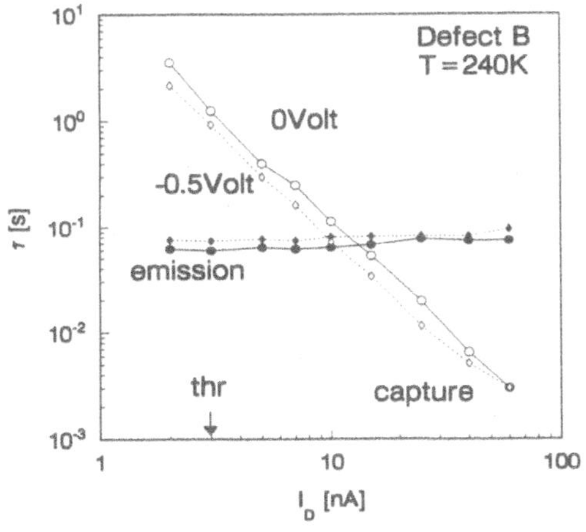

Figure 2. Capture and emission times for defect B as a function of the drain current I_D at $V_D = 10mV$ as varied by the gate voltage. The continuous and dashed curves are for the substrate bias voltages $V_S = 0V$ and $V_S = -0.5V$, respectively. The arrow (thr) on the horizontal axis marks the threshold for channel formation at $I_D = 3nA$.

For both interface defect types A and B, the emission time increases when a reverse substrate bias voltage is applied. The reverse substrate bias reduces the capture time in the sub-threshold and increases the capture time above threshold for defect A while for defect B, the capture time is reduced in the full measurement range.

Several defects of the typical dependence types A and type B have been observed and published earlier[3,4,9]. The magnitudes of the time constants vary for both types from milliseconds to hundreds of seconds over the full measurement range. The typical functional dependencies shown in Figs. 1 and 2 for the types A and B, respectively, remain similar and may be used to classify the defects.

Due to the turnaround of the capture time, the defect type A remains in the time constant measurement window for a wide range of carrier density magnitudes in the channel. The defect A can therefore be monitored with a variation of the channel conductance over four orders of magnitude. For defect B for which the trapping time steeply decreases, this range is only two orders of magnitude wide.

Both defects A and B show a strong temperature activation of the capture and emission times. The activation enthalpy H is depicted in Fig. 3. The temperature shift for the defect A has been reported in an earlier publication[7]. From the temperature shift of the capture and emission times, we deduce the activation enthalpy H for defect A as a function of the channel conductance as depicted by the circles in Fig. 3. The scatter of the data points indicates the error in this evaluation.

The smooth curves for the defect A in Fig. 3 are drawn to demonstrate the continuous increase of the activation with increasing channel conductance. The temperature activation reaches a magnitude of $H=1.2eV$ for the emission time constant. The activation enthalpy for the capture time is approximately $\Delta H=0.24eV$ lower than the values for emission. Because of the high temperature activation, the defect A is only observed in a narrow temperature range $T=280\text{-}300K$.

For defect B, we observe activation enthalpies $H=0.22eV$ and $H=0.63eV$ for capture and emission, respectively, independent of the channel conductance. Two values as deduced from an Arrhenius plot with a better accuracy as for level A are shown in Fig. 3 by the full and open triangles for emission and capture, respectively. It is noted that the enthalpy difference $\Delta H=0.40eV$ for capture and emission is higher for defect B than the equivalent value of defect A.

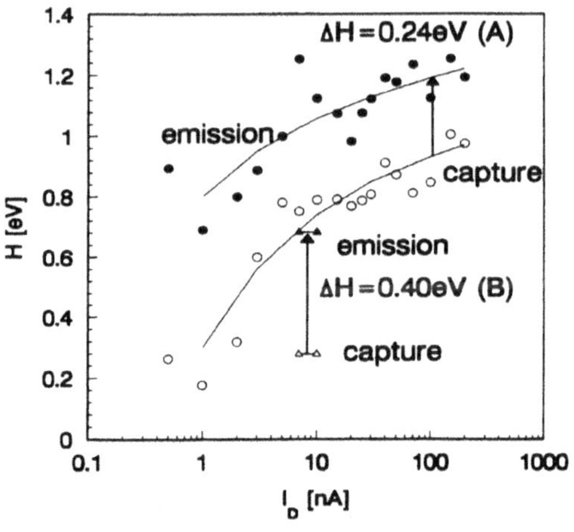

Figure 3. Activation enthalpy H vs. drain current I_D at $V_D=10mV$ for defect A (circles) and defect B (triangles). Full and open data points are for emission and capture, respectively.

The two defects A and B also remarkably differ in the magnitude of the RTS switching step. The conductance change ΔG for the two RTS switching states is depicted in Fig. 4 for the two defects A and B as a function of the channel conductance. Parameter is the reverse substrate bias voltage. The switching step maximum exceeds the magnitude of the conductance change estimate of one electron in the channel with a constant mobility as indicated by the horizontal line. The increased magnitude of the switching step is commonly interpreted by a combined effect of a carrier and a mobility change during charge trapping. The switching step of defect A, however, exceeds the charge change of one mobile carrier by more than one order of magnitude. Such a high conductance change cannot be explained by the creation and the annihilation of an additional scattering center in a uniform channel.

The application of a reverse substrate bias voltage has a negligible effect on the switching step of level B. The level A shows a 10% splitting induced by a 1 Volt reverse substrate bias.

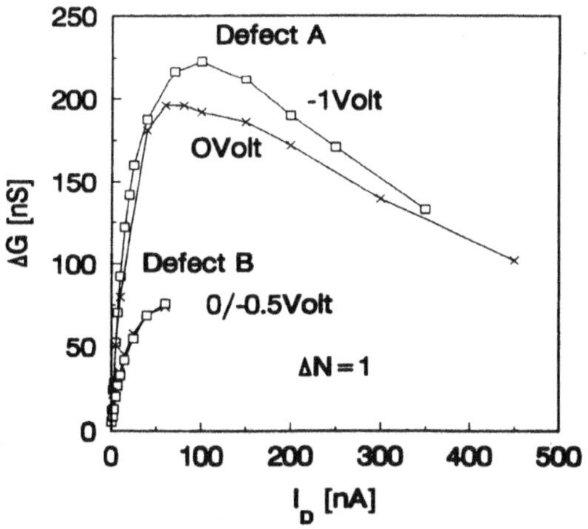

Figure 4. Channel conductance change ΔG of the switching step in RTS vs. drain current I_D at V_D=10mV. The defect type and the substrate bias are as indicated. The horizontal line $\Delta N=1$ marks an estimate of the conductance change for one electron and constant mobility in a uniform channel.

DISCUSSION

For the evaluation, of the measured time constants, we make use that the interface defect causing the RTS is in equilibrium with the electrons in the channel. The average occupation of the defect can therefore be described by a Fermi energy. The ratio of the emission to the capture time constant yields the energy E_T of the occupied defect with respect to the Fermi energy in units of the thermal energy kT where k is the Boltzmann's constant.

$$\ln\left(\frac{\tau_e}{\tau_c}\right) = \frac{E_T - E_F}{kT} \qquad (1)$$

It is noted, however, that the energy E_T as deduced from the time constant ratio is not necessarily a level position; it may contain a free energy contribution due to a configuration change of the system during transfer. The relation eq. (1) is useful to test the trap energy because the transfer probability prefactor cancels in the time constant ratio.

The time constant ratios are evaluated for the defects A and B and depicted in Figs. 5 and 6, respectively in double-logarithmic scales. The two defects A and B differ in the slope (power law) of the time constant ratio which also represents an occupation factor of the defect. The defect A showing a power of $\alpha \approx 3/2$ is approximately occupied proportional to the volume carrier density n in the channel. The volume carrier density increases slightly stronger than the area density n_s with increasing channel conductance because of the channel narrowing. The occupation of level A with a power $\alpha \approx 3/2$ is therefore in accordance with a volume carrier density n in the channel. Other defects observed for this type show similar power laws. The level B, however, is occupied more rapidly with a power law $\alpha \approx 2$. Other defects observed for this type are occupied even more rapidly with $\alpha > 2$. Defects of this type are therefore characterized by an occupation which increases with a superlinear power law of the volume carrier density.

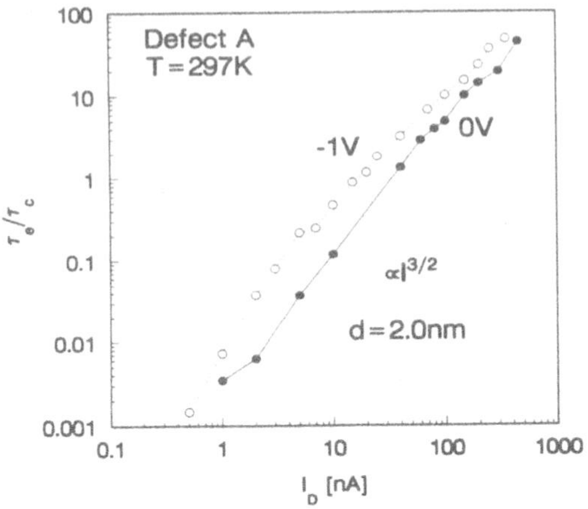

Figure 5. Ratio of emission and capture time for defect A vs. drain current I_D at V_D=10mV. The continuous and dashed lines are for substrate voltages V_S=0V and V_S= -1V, respectively. The dotted line indicates the slope.

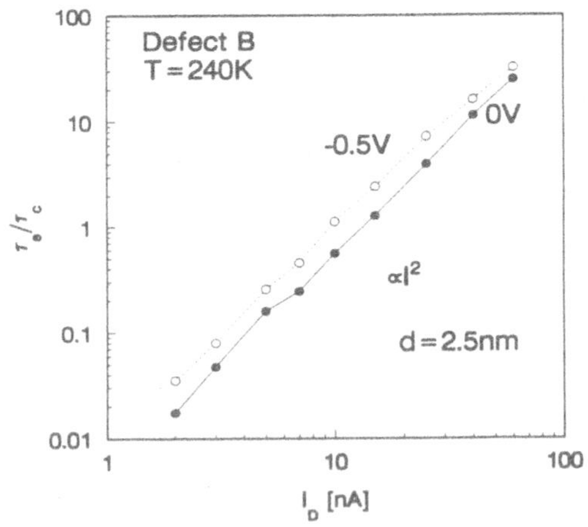

Figure 6. Ratio of emission and capture time for defect B vs. drain current I_D at V_D=10mV. The continuous and dashed lines are for substrate voltages V_S=0V and V_S= -1V, respectively. The dotted line indicates the slope.

It is noted that in both figures the dotted curve which connects the data points (open circles) for a reverse substrate bias is shifted with respect to the continuous curve representing the data (full circles) without a substrate bias. From the shift we conclude, that the energy position of the defect has been shifted by a constant energy after application of the substrate bias. The shift is only weakly dependent on the carrier density in the channel in contrast to the time constants themselves and to the ratio of the time constants which strongly vary with the channel conductance. If we assume that the shift is due to the voltage drop in the oxide in the region between the defect position and the channel, we can determine the depth d of the defect in the oxide from the interface by

$$d = \frac{t_{ox} kT}{\Delta V_G q}\left[\ln(\tau_e/\tau_c)_{V_{sub}} - \ln(\tau_e/\tau_c)_0\right] \qquad (2)$$

where ΔV_G is the gate voltage shift with substrate bias for the same channel conductance, t_{ox} is the oxide thickness, kT/q is the thermal voltage. The average depths d = 2.0nm(A) and d = 2.4nm determined for the levels A and B, respectively, are given in the insert of the figures.

The depth values for the defects A and B, respectively, as deduced from the energy shift are plotted in Fig. 7 as a function of the drain current I_D. A large scatter of the data is observed. The average values as given in the Figs. 5 and 6, respectively, for the levels A and B are indicated by the horizontal lines. The depth values show a trend to decrease for high values of I_D. This trend may be caused by the oversimplified approximation of eq. (2) which ignores the variation of the channel width and any twodimensional effects in the channel. These effects will be further studied in the future.

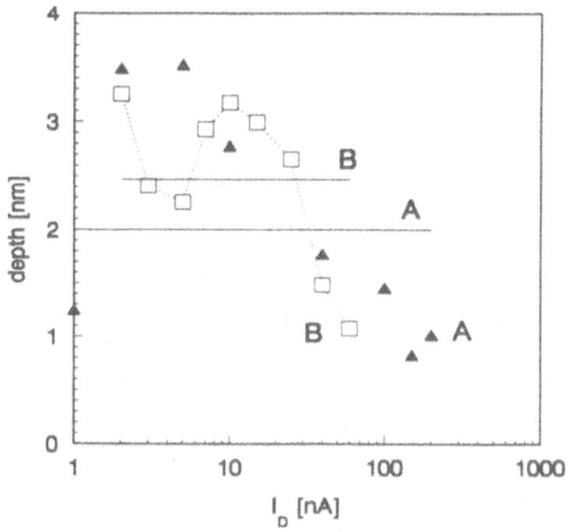

Figure 7. Depth d of the interface defects A (triangles) and B (squares) as a function of the drain current I_D for V_D=10mV. The data is evaluated from Figs. 5 and 6 by eq. (2). The horizontal lines are average values for the two defects as indicated.

Both defects A and B are located approximately d ≈ 2nm deep in the oxide. This depth in the oxide is a reasonable distance for which defect levels can be reached with a time constant of the order of seconds.

In the original data Figs. 1 and 2 where the time constants are depicted as a function of the channel conductance $I_D(V_G)$, the gate voltage dependence of the trap energy cannot be simply extracted. For instance, a lowering of the trap energy with gate bias voltage should be visible in an increased emission time constant according to the SRH statistics[10] by

$$\frac{1}{\tau_e} = e_e \cdot \exp(-E_T(V_G)/kT) \tag{3}$$

where e_n is a constant emission prefactor.

For the trap level B, we find an approximately constant emission rate. The slope is less than the lowering expected from the gate bias change with a depth d=2nm in the oxide. The emission rate for the defect A varies so steeply that the estimated position of the defect in the oxide is closer to the gate than to the semiconductor interface.

We conclude that the time constants of the two defect levels are a function of the carrier concentration in the channel in addition to the dependence of the electric field in the oxide. The two parameters may be independently varied by the additional application of a substrate bias. The real oxide field dependence of the trap energy is only visible when the carrier density in the channel is maintained constant.

The energy lowering due to the voltage drops in the oxide is estimated from the defect position and the gate bias voltage applied to be less than $\Delta E(V_G) < 50$meV leading to roughly the same variation of the emission time between minimum and maximum of one order of magnitude. $\tau_{e,max}/\tau_{e,min} \approx 10$for the defects A and B in contrast to the experimental observation in Figs. 1 and 2. The effect of the carrier density in the channel on the trap energy therefore compensates the field effect for defect A and induces a large energy change for defect B. In order to obtain a variation of the emission time of four orders of magnitude as observed in Fig. 1 for defect A, a trap energy change of at least $\Delta E_T(n_s) \approx 200$meV is estimated from eq. (3). This high value is also visible in the measured temperature activation of defect A depicted in Fig. 3.

In contrast to the electric field dependence, this energy contribution related to the carrier density is a configurational energy change occurring when the electron transfers into the defect. Configurational energy contributions were already postulated in earlier papers[2,3]. Such energies are known for defects which show a lattice relaxation associated with a charge change. Such a lattice relaxation is in principle possible for defects in the oxide and cannot be excluded; the lattice relaxation must be screened by the mobile carriers in the channel to explain the large variation with carrier density in this experiment.

We explain our data by a much simpler and rather obvious configurational change in the system. When an electron is moved from the substrate or source into the defect it charges the MOS capacitor. For weak channels and in depletion, the charge change at the interface is a full electronic charge. For strong inversion, the trapped charge is fully compensated by the mobile charge carriers in the channel which screen the charge change.

We have calculated the change in free energy induced by a single electron trapped at a distance d=2nm in the oxide from the Si interface. The estimation is performed by a numerical finite element calculation of the charge and potential distributions in the 3-dimensional system. The computational method has been previously published[11]. The result is shown in Fig. 8.

A simple estimate of the energy ΔE is obtained by the image charge Q_i of the trap on the gate at bias voltage V_G which leads to a Coulomb energy on the capacitor of

$$\Delta E = Q_i \cdot V_G \approx 0.1 \ldots 0.4 \text{ eV}$$

which amounts to several hundred millivolts. Figure 8 shows the computed Coulomb energy change ΔE as a function of carrier density in the channel. In the subthreshold region, the energy weakly increases due to the gate voltage increase. When the channel is formed, the image charge on the gate is rapidly reduced by screening thus leading to a large change in the free energy $\Delta E \approx 300\text{mV}$.

The computation is performed for a uniform channel leading to a similar behavior of all the defects. The differing properties may be caused by potential fluctuations in the channel. Potential fluctuations are known to be present in the MOS interface and have been considered for RTS in an earlier paper[3].

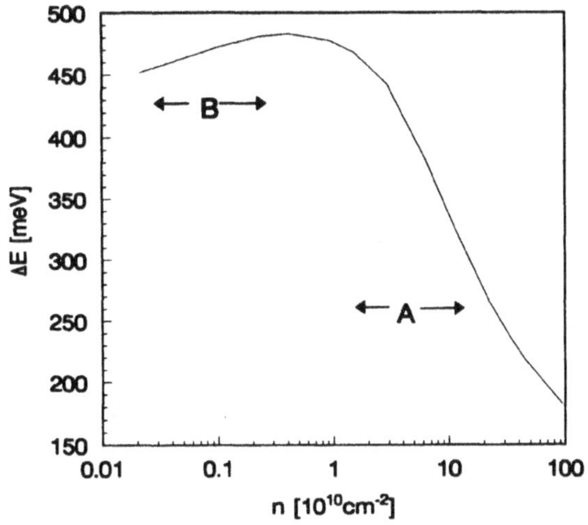

Figure 8. Energy change ΔE of the MOS capacitor when an electron is moved to a trap position d=2nm in the oxide from the interface as a function of the mobile carrier density in the channel. The ranges A and B indicate the conditions of defects A and B, respectively.

The typical behavior of the defects type A and type B can be explained, when we assume that the defect A is located near a potential minimum in the channel containing a large density of mobile carriers. The defect type B is assumed to be located near a potential maximum where the carrier density is reduced and only partial screening of the defect is possible; however, the potential fluctuation in the channel is screened thus leading to a steeply increasing occupation of the defect with carrier density in the channel and an independence or only weak dependence of the emission time of carrier density.

The magnitude of the switching step is defined by the modulation of the current flow in the channel. It is obvious that a defect positioned in front of a puddle of carriers which concentrates the current flow, causes a higher modulation than a defect in a position near a potential maximum near an area which does not carry a large fraction of the channel current in accordance with the observation in Fig. 4.

CONCLUSIONS

The full details of the modelling will be described elsewhere[12]. In the present paper we like to point out that the defects in the oxide causing the RTS can be classified into two types A and B. Type A shows a turnaround in the capture time and a strongly increasing emission time with increasing average carrier density in the channel. Type B shows an almost constant or only weakly increasing emission time with increasing carrier density in the channel. The difference of the defects is due to a change in the Coulomb energy of the capacitor system which may reach large values of several hundred millivolts for the defect type A by screening. For the defect type B, where the screening is incomplete, the Coulomb energy is almost fixed. The different regions for the defects type A and type B are indicated in the energy diagram of Fig. 8.

The key information on the energy changes involved in the charge transfer was obtained by applying a substrate bias which makes possible to differentiate the field lowering in the oxide from the larger free energy change change of the Coulomb energy in the capacitor system.

ACKNOWLEDGEMENT

The authors are grateful to Dr. K. Hofmann at the Siemens Research Lab, Munich who provided the sub-μm MOSFETs investigated in this study.

REFERENCES

1. K. S. Ralls, W. J. Skocpol, L. D. Jackel, R. D. Howard, L. A. Fetter, R. W. Epworth and D. M. Tennant, Discrete resistance switching in submicrometer silicon inversion layers: Individual interface traps and low-frequency (1/f?) noise, Phys. Rev. Lett. 52:228 (1984)
2. M. J. Kirton and M. J. Uren, Capture and emission kinetics of individual Si:SiO_2 interface states, Appl. Phys. Lett. 48:1270 (1986)
3. M. J. Kirton, M. J. Uren, S. Collins, M. Schulz, A. Karmann, and K. Scheffer, Individual defects at the Si:SiO_2 interface, Semicond. Sci. Technol. 4:1116 (1989)
4. M. Schulz, Single, individual traps in MOSFETs, Physica Scripta T35:273 (1991)
5. R. Farmer, Discrete conductance fluctuations and related phenomena in Metal-Oxide-Silicon device structures, in: "Insulating Films on Semiconductors, INFOS 91" eds. W.Eccleston and M. J. Uren, Adam Hilger, Bristol 1 (1991)
6. M. J. Kirton and M. J. Uren, Noise in solid-state microstructures: A new perspective on individual defects, interface states and low-frequency (1/f) noise, Adv. Phys. 38:367 (1989)
7. M. Schulz and A. Pappas, Telegraph noise of individual defects in the MOS interface in: "Noise in Physical Systems and 1/f Fluctuations" eds. T. Musha, S. Sato, and M. Yamamoto, Ohmsha Ltd., 256 (1991)
8. A. Karmann and M. Schulz, Characterization of individual defects in MOSFETs, in: "Insulating Films on Semiconductors, INFOS 89", eds. F.Koch and A. Spitzer, North-Holland Publ., Amsterdam 500 (1989)
9. M. Schulz and A. Karmann, Individual, attractive defect centers in the SiO_2-Si interface on μm-sized MOSFETs, Appl. Phys. A-52:104 (1991)
10. W. Shockley and W. T. Read Jr., Statistics of the Recombinations of holes and electrons, Phys. Rev. 87:835 (1952)
11. M. Schulz, Modelling of individual interface states in MOSFETs, in: "Insulating Films on Semiconductors 1991", W. Eccleston and M. J. Uren eds., Adam Hilger, Bristol 127 (1991)
12. C. Kroneder, J. Vennemann, A. Pappas, and M. Schulz to be published

DEFECT STRUCTURE AND GENERATION MECHANISMS AT THE Si/SiO$_2$ INTERFACE

J.H. Stathis

IBM Research Division
T.J.Watson Research Center
Yorktown Heights, NY 10598

ABSTRACT

We first review recent work elucidating the oxide structure around the P_b center and illustrating important chemical differences among defects at the (100) interface. Then we present new results concerning the identification of defects generated by hot electrons in the gate oxide of devices. Using electrically-detected magnetic resonance to study interface degradation in MOSFETs, we observe the generation of P_{b0} centers by hot-electron stress at fields above the electron heating threshold. In contrast, no P_{b0} centers are created by electron-hole recombination near the interface, even though this process generates interface states.

INTRODUCTION

The goal of the work described in this paper is to increase our fundamental understanding of the structure and formation mechanisms of defects at the Si/SiO$_2$ interface, and to relate this microscopic understanding of interface states to issues of device reliability. As technology moves toward thinner gate oxides the importance of this interface becomes ever greater. The corresponding tendency toward higher oxide fields in the design of CMOS and EEPROM devices demands a microscopic understanding of the role of hot electrons in causing interface degradation.

A thorough understanding of defects at the the Si/SiO$_2$ interface requires first of all a detailed knowledge of their microscopic structure. This includes not only the "core" structure of the defect (e.g. the wave function character of the dangling bond) but also the arrangement of neighboring atoms in the environment of the defect. In particular, we would like to understand the structure of the oxide close to the interface, immediately adjacent to the defect. This is the important region for device reliability issues, since it the interaction region for hydrogenous species such as H_2O, OH, H_2, H^0, or H^+ which may be involved in interface degradation. It is also the region containing the greatest concentration of hole traps, and so is key to a microscopic understanding of radiation and hot-electron effects.

The Physics and Chemistry of SiO$_2$ and the Si-SiO$_2$ Interface 2
Edited by C.R. Helms and B.E. Deal, Plenum Press, New York, 1993

Our most detailed knowledge of the structure of defects at the Si/SiO_2 interface, including details of the ^{17}O hyperfine structure, concerns the (111) interface. A single defect, P_b, has been identified and extensively studied at this interface.[1] On (100) silicon, two related defects called P_{b0} and P_{b1} are observed. While the structure of the $P_b(111)$ center is quite well established, the precise origins of the P_{b0} and P_{b1} resonances on (100) silicon are still controversial. While it is important to study the (111) interface as a model system, current experiments must focus on the more complex but technologically more relevant (100) interface. The emergence of the powerful new technique of electrically-detected magnetic resonance allows spin-resonance measurements to be made on individual fully-processed MOSFETs, leading to important new revelations concerning defect generation mechanisms in devices.

EXPERIMENTAL METHODS

The samples used in these studies comprised both bare oxides and fully-processed MOSFETs, and will be described more fully in the appropriate sections below. Bare oxides grown on high-resistivity wafers were used for standard electron paramagnetic resonance (EPR) measurements, which require large sample areas. MOSFETs were used for electrically-detected magnetic resonance (EDMR) measurements and hot-electron degradation studies.

The work described in this paper includes standard EPR measurements at both X-band (~9.5 GHz) and S-band (~4 GHz), electrically-detected magnetic resonance (EDMR, also referred to as spin-dependent recombination or SDR), oxide stressing using both Fowler-Nordheim (FN) tunneling and optically-induced substrate hot-electron (SHE) injection, and two- and three-terminal electrical device characterization techniques (capacitance-voltage (CV) and current-voltage (IV) measurements).

^{17}O HYPERFINE MEASUREMENT

In spite of much study, certain aspects of the structure of P_b centers remain unknown. Recently, we have studied the position of oxygen atoms surrounding the Si dangling bond using ^{17}O EPR hyperfine measurements.[2-4] To enhance the concentration of ^{17}O ($I=5/2$), oxides were grown in 55.65% isotopically enriched oxygen. By deconvolution of the ^{17}O-broadened spectrum, we are able to ascertain that the unpaired election on the P_b center interacts weakly with only a single oxygen atom.

Experimental Details

The samples consisted of stacks of (111)-oriented wafers polished on both sides to a thickness of 5-6 mil. Prior to oxidation, the wafers were cleaved into 2mm×20mm strips, then cleaned using standard procedures. The samples were then oxidized in atmospheric-pressure dry oxygen of normal isotopic ratio at 800°C for approximately 5.25 hours, to a thickness of approximately 140 Å. Following this, all samples received further oxidation at 800°C and 38 mbar for approximately 5 hours in either normal oxygen or in oxygen that was isotopically enriched to 55.65% ^{17}O, adding about 40 Å to the oxide thickness. The interfacial region is formed during the second oxidation step, with the isotopic make-up of the oxide at the interface equal to that of the gas.[5]

Results

The P_b resonance in the partially ^{17}O-enriched sample is broadened slightly compared to the normal linewidth,[6] reflecting a small interaction with neighboring oxygen atoms. To determine the number of oxygen neighbors which give rise to this hyperfine broadening, we make use of the fact that the ^{17}O enrichment of the interface is only about 50%. Therefore some percentage of the dangling bonds must see only ^{16}O (I=0) neighbors. If N oxygen atoms comprise the oxide side of the P_b center, then for an ^{17}O enrichment ϵ the fraction of sites with *only* ^{16}O nuclei will be $(1-\epsilon)^N$. This gives the fractional intensity of the expected central (non hyperfine-broadened) line. The spectrum in the ^{17}O-enriched sample consists of a central line with this intensity, with the remainder of the signal in a a broader background arising from those sites with one or more ^{17}O nuclei. To separate these two components we performed a simple fourier-transform deconvolution and digital filtering, using the unenriched spectrum as the deconvolution function. The result is shown in Figure 1. We obtain a sharply peaked spectrum, with broad wings. We find that the deconvolved line can be very accurately described as a sum of two lines: a broad gaussian and a narrower lorentzian.

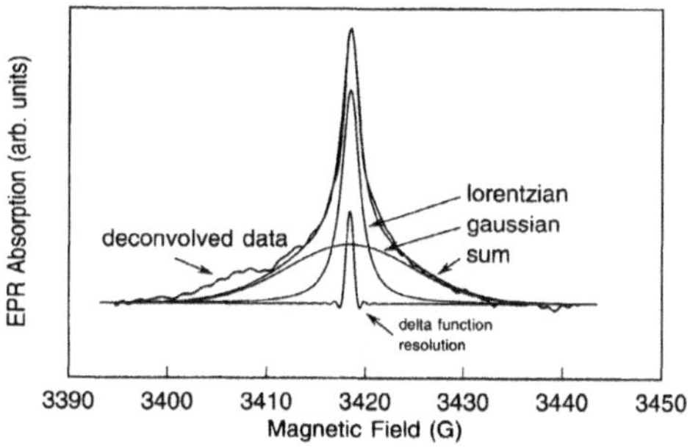

Figure 1. Deconvolution of the EPR spectrum in ^{17}O-enriched oxide. The fit to a sum of lorentzian and gaussian components is shown. The resolution of our deconvolution procedure, given our particular choice of digital filtering, is indicated.

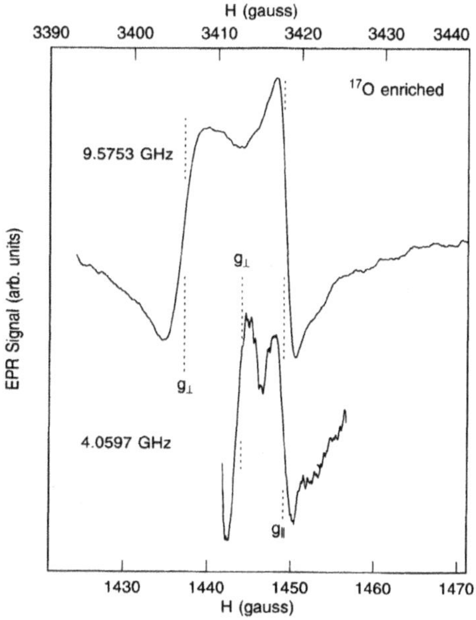

Figure 2. Comparison of 10 GHz and 4 GHz spectra of P_b centers. The samples were grown in ^{17}O-enriched oxygen. The field axes are arranged so that the spectra are aligned at $g_{\|}=2.0016$.

We interpret the gaussian component as the envelope of the hyperfine lines for those sites with one or more ^{17}O nuclei near the defect. Comparing the integral of this line with that of the central line (the lorentzian component) we find that the hyperfine-broadened component of the spectrum is 56% of the total. This is in precise agreement with the predicted value if each P_b center interacts predominantly with only one oxygen atom. Because the hyperfine structure appears as only a broad gaussian line, not as individual resolved lines, the position of this single oxygen must vary randomly from one defect site to the next. According to our results, the mean hyperfine coupling A_{iso} of the dangling bond to the oxygen is about 2.5 gauss, corresponding to only about 0.15% wave function amplitude on the oxygen 2s orbital.

A comparison of 4 GHz and 10 GHz measurements of the P_b center in an ^{17}O-enriched sample is shown in Figure 2. Each curve is the *difference* between spectra taken with the magnetic field parallel to [111] direction and with the field normal to this direction. The spectrum thus appears as a superposition of the resonance with H||[111] and an *inverted* resonance with H⊥[111]. This convenience allows the resonances from both orientations to be displayed simultaneously and automatically subtracts orientation-independent background signals, e.g. from the cavity or from the scribed edges of the sample. For ^{17}O-enriched samples, the linewidth for H||[111] is independent of frequency, as expected for hyperfine broadening. The H⊥[111] linewidth, however, is considerably larger at X-band compared to S-band, confirming the idea[7] that at X-band (and higher frequencies) this linewidth is determined by g variations. At S-band the linewidth apparently becomes independent of orientation. This is consistent with a weak overlap with the oxygen 2s orbital or with core polarization. However, using the peak-to-peak width as an upper limit to any observable anisotropy, the spectrum is also consistent with up to 5% wave function amplitude on the oxygen 2p orbital.

The fact that the every P_b center interacts with exactly one oxygen atom may imply that the structure of the oxide around the defect is not purely random, and should be regarded as *part of* the structure of the P_b center. This could have important implications for interface degradation mechanisms.

(100) VS (111) INTERFACES

The P_b center that is best understood is the one at the (111)-oriented interface; on the other hand, the (100) interface is more important technologically, and there are fundamental differences in the structure and chemistry of defects on these two interfaces.[8, 9] Figure 3 shows the effect of various annealing treatments on P_b centers at (111) and (100) interfaces.

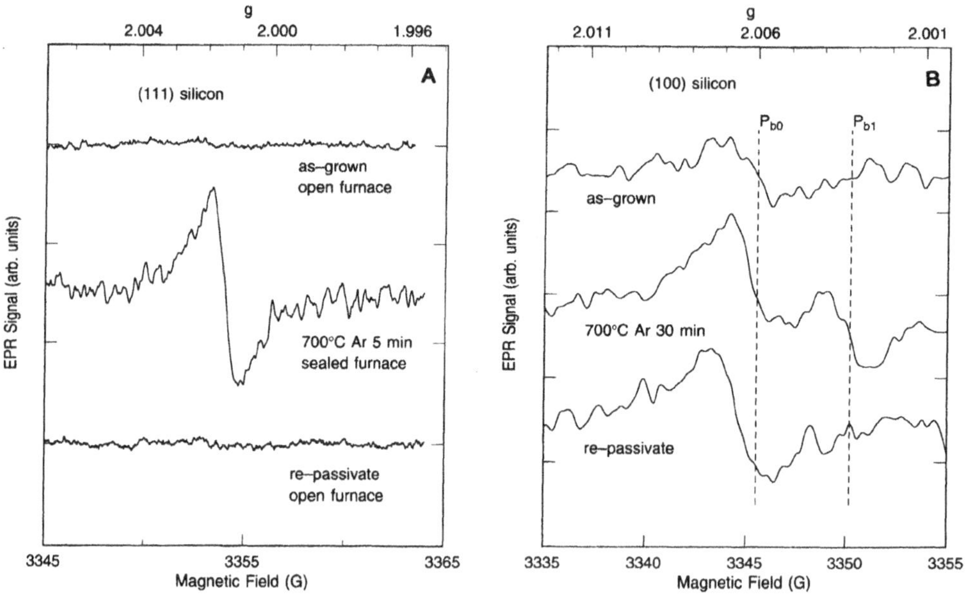

Figure 3. Effect of thermal treatments on P_b signals. (A) the (111) interface, (B) the (100) interface. SAF refers to a silicide-anneal furnace, which is sealed to prevent entry of ambient atmosphere. The top traces are the as-grown oxides. The middle traces show the effect of annealing in an oxygen- and water-free ambient. The bottom traces show the effect of re-passivating the interface by exposure to trace water vapor as the samples are pulled from the furnace.

Experimental Details

The samples consisted of approximately 70Å SiO_2 grown in a standard furnace at 800°C in dry O_2 + 4% HCl, on 50-90 Ω-cm p-type Czochralski silicon wafers polished on both sides. The wafers received a post-oxidation anneal (POA) in argon at 700°C for 5-30 minutes in a sealed oxygen-free silicide-anneal furnace (SAF), and then a subsequent thermal treatment in the standard furnace. This treatment consisted of simply ramping the wafers into the furnace in argon at approximately 760°C and then pulling them out after 30 seconds. EPR measurements were made after each step, as shown in Figure 3.

Results

We observe that, for this particular annealing sequence, P_{b1} behaves in a manner analogous to the P_b center on (111), but the P_{b0} center behaves entirely differently. It is present in the as-grown oxide, and it cannot be passivated by the same thermal treatment that passivates $P_b(111)$ and $P_{b1}(100)$ centers. A continuous slight growth of the P_{b0} center is observed during the thermal cycle. It is likely that the relevant ingredient in the annealing of $P_{b1}(100)$ and $P_b(111)$ is water vapor, although more experiments are needed to prove this specific identification. The different behavior of P_{b0} vs the other P_b centers may thus be a matter of kinetics rather than energetics. For example, the P_{b0} center may lie deeper in the silicon, two or three atomic layers away from the interface. In this position it would be less susceptible to passivation by water vapor, since water will diffuse into the silicon less readily than through SiO_2. This is consistent with recent electronic structure calculations.[10]

HOT-ELECTRON EFFECTS

For many years there have existed two fields of study relating to Si/SiO_2 interface degradation: the radiation effects field which focuses on the damaging influence of ionizing radiation (e.g. electron beam or gamma ray),[11] and the electrical device community which focuses on the effects of hot carriers in the SiO_2 conduction band.[12]

Radiation effects are believed to result principally from radiation-induced holes in the oxide.[11] Two processes described in the literature are believed to account for nearly all radiation-induced interface states: (1) electron-hole recombination near the interface[11, 13] and (2) a multi-step process wherein holes in the SiO_2 bulk interact with hydrogen to form H^+ ions, which then transport to the interface where they react to form interface states.[14, 15] A third process,[16] accounting for a small fraction of the radiation-induced interface states, is believed to result from radiolytic H^0.

Interface states are also generated in MOSFETs by electron transport through the oxide. Either Fowler-Nordheim tunneling injection or substrate hot-carrier injection over the Si-SiO_2 barrier can be used. It is experimentally observed that bulk trap and interface state generation occur when the average electron energy exceeds a threshold of about 2 eV. On oxides thicker than 100Å this corresponds to an electric field of about 3.5 MV/cm.[17] Monte Carlo calculations[18] and experiments[19] have previously shown that the average electron energy in SiO_2 never exceeds ~6 eV at oxide fields up to 16 MV/cm due to stabilization by acoustic phonon (non-polar) scattering. Since this is well below the threshold for impact ionization (9 eV) no connection to radiation effects or holes could be seen. On the other hand, both radiation damage and trap creation have been associated with the generation and reaction of mobile hydrogen atoms or ions in the SiO_2.[15-17] This opens the possibility that the *underlying microscopic* mechanism for interface state creation may be closely related in the two cases.

Recent calculations[20] have revealed the possibility of electron-hole pair formation by impact ionization of hot electrons at high electric fields in SiO_2. A small portion of the electron distribution can obtain energies sufficient to generate electron-hole pairs in the oxide. This impact ionization quantitatively accounts for positive charge formation in thick oxides (>500Å) at high fields (>7 MV/cm) and low injected-charge fluences (<.001 C/cm^2)[21] as some of the generated holes become trapped at near-interfacial sites. Some of these trapped holes are then annihilated by subsequently-injected electrons, creating interface states.[22, 23] The thickness, field, and fluence dependance of these processes are more fully described elsewhere in this volume,[24] and have been used here to distinguish interface defects generated by electron-hole recombination from those generated by the trap creation process.

Nearly all studies of interface state generation mechanisms in MOSFETs have relied solely on electrical measurements. Such measurements provide data concerning the number density, location, capture cross section, and energy level of the defects, but cannot provide a microscopic identification. No study of defect generation mechanisms can be complete without such microscopic information. Using an electron spin resonance technique in conjunction with electrical characterization we have for the first time identified the defect generated by the trap-creation process due to electrons with energies over 2 eV.

Experimental Details

The samples used in this study were poly-silicon-gate *n*-channel MOSFETs on *p*-type (100) silicon. Two oxides of different thickness (t_{ox}) have been studied so far. The samples were stressed at constant average oxide field, while recording and integrating the current to obtain the injected electron fluence. The 245Å oxide was stressed at 4 MV/cm using SHE to study trap creation, and the 670Å oxide was FN stressed at 9 MV/cm to observe effects due to impact ionization and electron-hole recombination. The stress was interrupted periodically for CV and EDMR measurements. The number of interface states was determined from the broadening of the high-frequency (1MHz) CV characteristic of the device.

The EDMR Technique

We have used an electrically detected magnetic resonance (EDMR) technique to identify the defect responsible for the interface states generated by high-field stress. In this technique, originally described by Vranch *et al.*,[25] the FET is configured as a gated diode.[26] The source and drain *n*-diffusions of the MOSFET are shorted together, and are slightly forward biased (0.2V) with respect to the *p*-type substrate. The current flowing from source/drain to substrate is measured as function of gate voltage. If interface states are present, an increase in current is observed when the silicon surface is in depletion. This excess current results from recombination through the interface defects. Under the combined application of a DC magnetic field and a microwave field at the Larmor frequency of an electron localized on the defect, a small resonant enhancement of this current can be observed. This spin-dependent recombination (SDR) signal[27] is in most respects identical to a standard electron paramagnetic resonance (EPR) signal, giving the same spectroscopic quantities about point defects. The advantage lies in the vastly reduced sample size necessary for measurement. The technique is ideally suited for study of *fully-processed* MOSFETs under realistic operating conditions. It is thus an extremely powerful technique, which has just begun to see use in studies of MOSFET degradation phenomena.[28]

Results

Figure 4 shows the EDMR spectrum obtained in the sample stressed at 4 MV/cm. The resonance is characterized by a g value of 2.0066 for the magnetic field along the [100] direction, consistent with the P_{b0} center,[8] and a slight asymmetry, being broader on the high-field side. The same lineshape is observed independent of oxide thickness, field, or fluence.

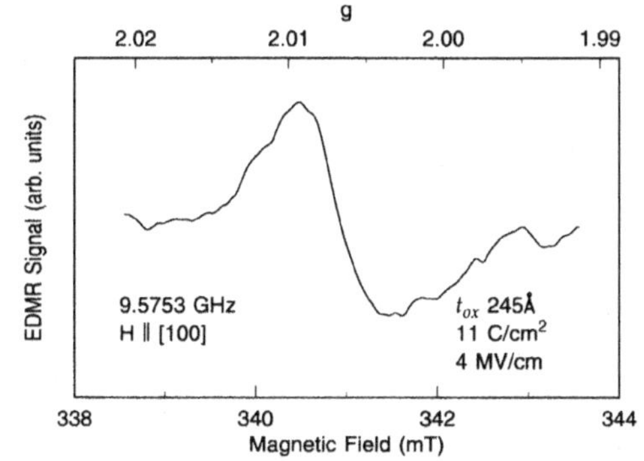

Figure 4. EDMR spectrum of interface defect created by hot electrons in SiO_2.

The interface state density and EDMR amplitude as a function of fluence are shown in Figure 5 for two combinations of oxide thickness and field. Note the arbitrary normalization of the EDMR signal. In contrast to normal EPR, this technique does not possess a means of calibration, hence it cannot provide an absolute measure of the defect density.

At 4 MV/cm we see that the initial growth of the spin signal closely parallels the interface state density, then saturates while the interface states continue to increase. At 9 MV/cm in the thicker oxide, two interface state generation processes are evident.[21, 24] At this combination of field and thickness, some electrons acquire sufficient energy to cause impact ionization.[20] The initial rise in the interface state density at 10^{-4}C/cm^2 is due to electrons recombining with trapped holes. In this sample there is is some overshoot in the apparent interface state density because of lateral non-uniformity in the hole trapping, which causes additional broadening of the CV characteristic. Note that there is no EDMR signal associated with these recombination-induced interface states. Trap creation by hot electrons appears at ~ 10^{-3}C/cm in this sample. (This figure should be compared to Figure 6 in the paper by DiMaria *et al.*[24] in this volume.) The EDMR signal appears only after a fluence sufficient to produce trap creation. It can therefore be said with certainty that the P_{b0} center is generated by hot electrons in SiO_2.

Also plotted in Figure 5 is the interfacial contribution to the recombination current, extracted from the gated diode characteristic and arbitrarily normalized to overlay the EDMR data. The EDMR amplitude is proportional to the recombination current, as expected.

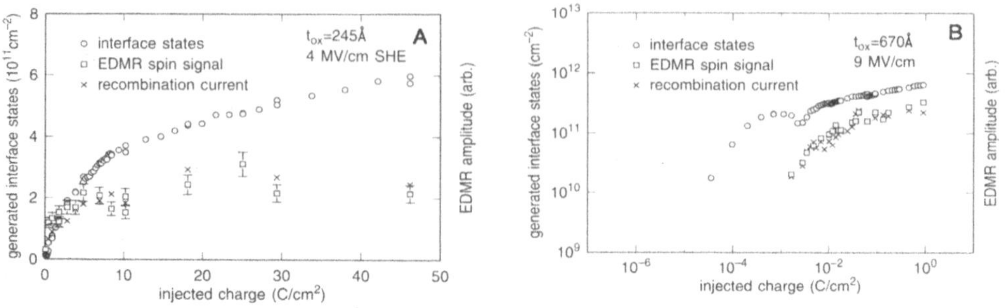

Figure 5. Interface state density and EDMR amplitude vs fluence. (A) t_{ox}=245Å, 4 MV/cm SHE stress, (B) t_{ox}=670Å, 9 MV/cm FN stress.

DISCUSSION

Previous investigators have observed P_b centers generated by electron injection into SiO_2, using standard EPR. These studies have used (111)-oriented wafers, and have not attempted to study the dependance on field and electron fluence in detail as we have done here. Nonetheless, it is useful to compare our results with these previous studies. Mikawa and Lenahan[29] used UV internal photoemission with a corona bias of 2 MV/cm on thick oxides (1750Å and 2700Å). They observed generation of P_b centers. The field they used is below the electron-heating/trap-creation threshold, but DiMaria[30] has shown that there is some interface state generation at this field associated with background trapping. Warren and Lenahan[31] used corona charging at higher field (>8 MV/cm) on both thin (230Å) and thick (1150-2700Å) oxides. They observed production of E′ centers (trapped holes) only in the thick oxides. They correctly interpreted this as reflecting the absence of impact ionization in thin oxides. An earlier report by these same authors[32] stated that they also observed P_b centers after Fowler-Nordheim stress of a 1750Å oxide. Trombetta *et al.*[33] injected a 1500Å CVD oxide at fields of field 4-6 MV/cm to a fluence of 6.4×10^{-3}C/cm². They observed neither E′ nor P_b centers, in spite of a high interface state density, $\sim 5 \times 10^{12}$cm². Although E′ production would not be expected at these fields, which are too low for impact ionization,[20] the lack of P_b centers is in contrast to the other results.

The present work shows that, on (100) silicon, P_{b0} centers are generated by hot electrons in SiO_2, with a field and fluence dependance that correlates well with the trap-creation process which has been studied extensively by DiMaria.[17, 30] Quantitatively, we cannot say whether the P_{b0} center accounts for most of the interface states, or only a fraction. At 4 MV/cm (Figure 5a) the EDMR signal appears to saturate before the interface state density. This could be an artifact of the EDMR technique, or it may signify the creation of another defect which is not detected by EDMR.

EDMR spectra essentially identical to ours were also observed by Krick *et al.* on *n*-channel MOSFETs subjected to channel hot-carrier stress at low gate bias.[28] For the stress conditions they used, the dominant localized degradation mechanism is believed[34] to involve hot hole injection near the drain. From the data in Figure 5b, however, we see that no EDMR signal is associated with holes in the oxide. Hot electrons are therefore probably also involved in the degradation.

Krick *et al.* also suggested that the asymmetry in the line shape may arise from E' centers (trapped holes). The fact that we observe the same asymmetry in the sample stressed at 4 MV/cm, where no hole generation by bandgap ionization is possible, rules out this interpretation.

We must wonder why only one defect is observed, whereas two different P_b varieties are known to exist at the (100) Si/SiO_2 interface.[8] It is possible that P_{b0} centers are preferentially generated because of the different chemical behavior of the P_{b0} and P_{b1} centers, as described above. But a variety of electrical results[35] clearly show that the spectrum of radiation-induced interface states on (100) silicon is too complex and variable to be accounted for by a single species of defect. Yet only P_{b0} centers are detected by the EDMR technique, whether the interface states are generated by hot electrons, channel hot carriers,[28] or gamma irradiation,[36] and no spin signal is observed from defects generated by electron-hole recombination near the interface. One conventional electron paramagnetic resonance study[37] of irradiated (100) interfaces found a weak P_{b1} component in addition to the P_{b0} center. One must consider the possibility that the SDR technique may be insensitive to the P_{b1} center. Recent results[38-40] have shown the existence of a least two different defects with capture cross sections differing by more than an order of magnitude. Since the magnitude of the recombination current in the gated diode depends on the cross section,[26] EDMR using this detection scheme will be most sensitive to those defects with the largest capture cross section. This provides a plausible explanation for the lack of any other resonances, but a quantitative theory relating the magnitude of the SDR signal to the microscopic parameters (capture cross section, spin-lattice relaxation time, etc) of a defect is lacking at this point.[41, 42] These results may also have important implications for the interpretation of charge-pumping measurements, which are also sensitive to the capture cross section of the interface states.

SUMMARY

The mechanisms by which interface states are generated by hot electrons in SiO_2 and by recombination of electrons with interfacial trapped holes are fundamentally different. The former process produces P_{b0} centers, while the latter does not. In the absence of any detectable spin resonance signal, the mechanism for interface state creation by electron-hole annihilation remains mysterious. Hopefully, the knowledge of the oxide environment of the P_b center may allow theorists to shed some light on this issue.

The fact that the P_{b0} center is produced by both hot-electron stress and radiation damage suggests some common element in the damage mechanism. This is consistent with the idea that both involve hydrogen atoms or ions which react with an Si-H bond to form H_2, leaving a dangling bond at the interface.[43] Whether the absence of a P_{b1} signal reflects the different chemistry of this defect compared to the P_{b0} center, or is because the SDR technique is less sensitive to P_{b1} centers, awaits further study.

ACKNOWLEDGEMENTS

I wish to thank Serge Rigo and Isabelle Trimaille for growing the oxides for the ^{17}O-enrichment experiment. Without their expertise this experiment would never have come to fruition. Mark Crowder generously provided the use of his S-band spectrometer, and Chris D'Emic provided invaluable help with sample preparation. I also thank Joe Boulais for help in making the sample holder for spin-dependent recombination, and Keith Milkove for the use of a wire-bonder. Discussions with Ed Cartier, Doug Buchanan, Doug Arnold, and Leonello Dori have been extremely helpful throughout. In particular, I am indebted to Dan DiMaria for guidance in understanding MOSFET electrical characterization, and for motivating and encouraging this work.

REFERENCES

[1] E. H. Poindexter and P. J. Caplan, Prog. Surface Science **14**, 210 (1983).

[2] J. H. Stathis, S. Rigo, and I. Trimaille, Solid State Commun. **79**, 119 (1991).

[3] J. H. Stathis, S. Rigo, I. Trimaille, and M. Crowder, Materials Science Forum **83-87**, 1421 (1992).

[4] A. Stesmans and K. Vanheusden, Phys. Rev. B **44**, 11353 (1991).

[5] F. Rochet, B. Agius, and S. Rigo, J. Electrochem. Soc. **131**, 914 (1984).

[6] K. L. Brower, Z. Phys. Chem. Neue Folge **151**, 177 (1987).

[7] K. L. Brower, Phys. Rev. B **33**, 4471 (1986).

[8] E. H. Poindexter, P. J. Caplan, B. E. Deal, and R. R. Razouk, J. Appl. Phys. **52**, 879 (1981).

[9] J. H. Stathis and L. Dori, Appl. Phys. Lett. **58**, 1641 (1991).

[10] A. H. Edwards, in *The Physics and Chemistry of* SiO_2 and the SiO_2 Interface, edited by C. R. Helms and B. E. Deal (Plenum, New York, 1988), p. 271.

[11] T. R. Oldham, F. B. McLean, H. E. Boesh, and J. M. McGarrity, Semicond. Sci. Technol. **4**, 986 (1989).

[12] D. J. DiMaria, in *Insulating Films on Semiconductors*, edited by W. Eccleston and M. Uren (Hilger, Bristol, 1991), p. 65. and references contained therein.

[13] P. U. Kenkare and S. A. Lyon, Appl. Rev. Lett. **55**, 2328 (1989).

[14] P. S. Winokur, H. E. Boesch, J. M. McGarrity, and F. B. McLean, J. Appl. Phys. **50**, 3492 (1979).

[15] N. S. Saks and D. B. Brown, IEEE Trans. Nucl. Sci. **36**, 1848 (1989).

[16] N. S. Saks, R. B. Klein, and D. L. Griscom, IEEE Trans. Nucl. Sci. **35**, 1234 (1988).

[17] D. J. DiMaria and J. W. Stasiak, J. Appl. Phys. **65**, 2342 (1989).

[18] M. V. Fischetti, D. J. DiMaria, S. D. Brorson, T. H. Theis, and J. R. Kirtley, Phys. Rev. B **31**, 8124 (1985).

[19] S. D. Brorson, D. J. DiMaria, M. V. Fischetti, F. L. Pesavento, P. M. Solomon, and D. W. Dong, J. Appl. Phys. **58**, 1302 (1985).

[20] D. Arnold, E. Cartier, and D. J. DiMaria, Phys. Rev. B **45**, 1477 (1992).

[21] D. J. DiMaria, D. Arnold, and E. Cartier, Appl. Phys. Lett. **60**, 2118 (1992).

[22] S. K. Lai, J. Appl. Phys. **54**, 2540 (1983).

[23] D. A. Buchanan and D. J. DiMaria, J. Appl. Phys. **67**, 7439 (1990).

[24] D. J. DiMaria, D. Arnold, and E. Cartier, this volume.

[25] R. L. Vranch, B. Henderson, and M. Pepper, Appl. Phys. Lett. **53**, 1161 (1988).

[26] A. S. Grove, *Physics and Technology of Semiconductor Devices* (Wiley, New York, 1967).

[27] D. J. Lepine, Phys. Rev. B **6**, 436 (1972).

[28] J. T. Krick, P. M. Lenahan, and G. J. Dunn, Appl. Phys. Lett. **59**, 3437 (1991).

[29] R. E. Mikawa and P. M. Lenahan, J. Appl. Phys. **59**, 2054 (1986).

[30] D. J. DiMaria, J. Appl. Phys. **68**, 5234 (1990).

[31] W. L. Warren and P. M. Lenahan, J. Appl. Phys. **62**, 4305 (1987).

[32] W. L. Warren and P. M. Lenahan, Appl. Phys. Lett. **49**, 1297 (1986).

[33] L. P. Trombetta, G. J. Gerardi, D. J. DiMaria, and E. Tierney, J. Appl. Phys. **64**, 2343 (1988).

[34] P. Heremans, R. Bellens, G. Groeseneken, and H. E. Maes, IEEE Trans. Electron Devices **35**, 2194 (1988).

[35] T. P. Ma, Semicond. Sci. Technol. **4**, 1061 (1989).

[36] M. A. Jupina and P. M. Lenahan, IEEE Trans. Nucl. Sci. **37**, 1650 (1990).

[37] Y. Y. Kim and P. M. Lenahan, J. Appl. Phys. **64**, 3551 (1988).

[38] D. Vuillaume, D. Goguenheim, and G. Vincent, Appl. Phys. Lett. **57**, 1206 (1990).

[39] N. Haneji, L. Vishnubhotla, and T. P. Ma, Appl. Phys. Lett. **59**, 3416 (1991).

[40] M. J. Uren, K. M. Brunson, and A. M. Hodge, Appl. Phys. Lett **60**, 625 (1992).

[41] D. Kaplan, I. Solomon, and N. F. Mott, J. Physique Lett. **39**, L51 (1978).

[42] F. C. Rong, W. R. Buchwald, E. H. Poindexter, W. L. Warren, and D. J. Keeble, Solid State Electron. **34**, 835 (1991).

[43] K. L. Brower and S. M. Meyers, Appl. Phys. Lett **57**, 162 (1990).

THE INFLUENCE OF CRYSTAL ORIENTATION AND PROCESSING CONDITIONS ON THE ENERGY DISTRIBUTION OF TRAPS AT THE Si-SiO_2 INTERFACE

C.H. Bjorkman, Y. Ma, T. Yasuda, and G. Lucovsky

Departments of Physics and Materials Science & Engineering
North Carolina State University
Raleigh, N.C. 27695-8202

ABSTRACT

We have studied the local atomic strain as a function of crystal orientation and processing conditions, and its relationship to the energy distribution of interface traps in the Si bandgap. For a given oxidation temperature (up to 1050°C), the strain in SiO_2 films grown on (100), (110), and (111) Si was reduced by a rapid thermal anneal, RTA, at 1100°C. This was accompanied by a reduction in the midgap interface state density at midgap. Moreover, for oxides grown on Si(100) surfaces, an increase in oxidation temperature promotes strain relaxation and reduction of interface states at energies above 0.4 eV in the Si bandgap, but not at energies between 0.2 and 0.4 eV.

INTRODUCTION

Even though thermally grown silicon dioxide has been used as a gate dielectric material for more than 30 years, there are several questions related to the properties of the oxide material and the Si-SiO_2 interface that are as yet unresolved. One of the more important questions concerns an understanding of factors that contribute to the formation of electronically active defects at the Si-SiO_2 interface. Previous studies have shown that for thermally grown oxides there is a correlation between interface state density at midgap and oxide stress as obtained directly from laser beam deflection measurements, and also strain as obtained from analysis of infrared (IR) spectroscopic data.[1-3] As a continuation of this research, we have studied the local atomic strain as a function of crystal orientation and processing conditions, and its relationship to the energy distribution of interface traps relative to the band edges of the Si as identified from the analysis of C-V data from MOS structures.

EXPERIMENTAL PROCEDURES

P-type (100), (110), and (111) silicon wafers with resistivities of 5-30 Ω-cm were

The Physics and Chemistry of SiO_2 and the Si-SiO_2 Interface 2
Edited by C.R. Helms and B.E. Deal, Plenum Press, New York, 1993

thermally oxidized at 850°C and 1050°C for 250 and 11 minutes, respectively. This corresponds to thicknesses in the range of 300-400 Å. Half of the samples were then subjected to a rapid thermal anneal, RTA, at 1100°C for 100 s in an argon ambient. Al was evaporated onto the samples, followed by a post-metallization forming gas anneal. The interface state density, D_{it}, was determined using the combined high-frequency and quasi-static C-V method. IR spectroscopy was used to determine the dominant SiO_2 bond-stretching frequency.

EXPERIMENTAL RESULTS AND DISCUSSION

The bond-stretching frequency increased as the growth temperature increased, and showed additional increases for samples subjected to RTA. In previous work,[4] it has been shown this corresponds to an increase in the Si-O-Si bond angle. This demonstrates that higher oxidation and/or rapid thermal annealing temperatures promote increased relaxation of strained bonds in the SiO_2. Figures 1 and 2 also show that films grown on (110) substrates exhibit the highest bond-stretching frequencies and therefore the lowest level of strain when grown at either 850°C or 1050°C, or after a RTA. Note that the film thicknesses vary for films grown on substrates of different orientations, because of differences in the thermal oxidation rates. These differences have been partially explained by taking into account the density of surface atoms, C_{Si}, for the different substrate orientations:[5]

$$C_{Si}(110) > C_{Si}(111) > C_{Si}(100)\ . \tag{1}$$

It was found that the oxidation rate followed the above relationship for thicknesses up to about 300 Å, i.e. the (110) orientation exhibited the highest rate and (100) the lowest. For thicker films however, a crossover was observed with the (111) orientation yielding the fastest oxidation rate. It should be pointed out that the crossover thickness is temperature dependent, which was illustrated in this experiment with the (111) orientation yielding the thickest film, 430 Å, at 850°C while the (110) orientation exhibited the fastest oxidation rate at 1050°C. Previous work has shown that the thickness-averaged strain, on (100) substrates, varies as a function of film thickness.[6]

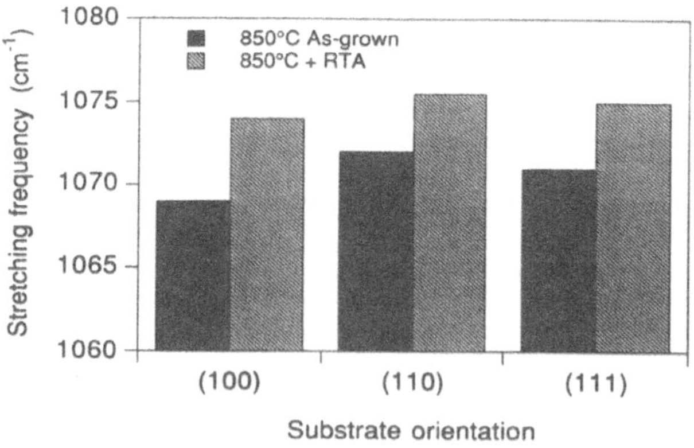

Figure 1. Bond stretching frequencies for oxides grown on different substrate orientations at 850°C and after subsequent RTA.

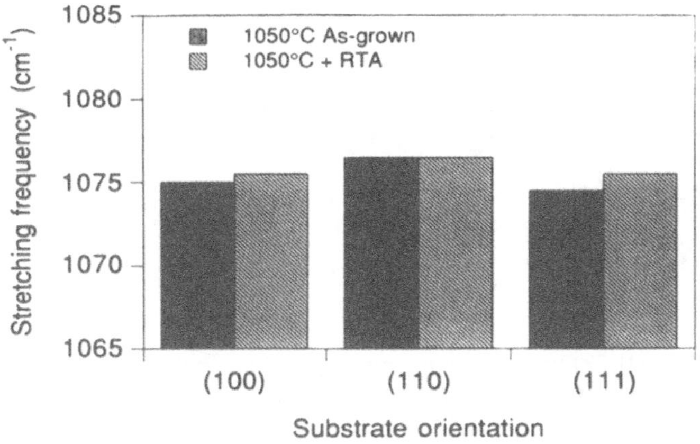

Figure 2. Bond stretching frequencies for oxides grown on different substrate orientations at 1050°C and after subsequent RTA.

From Ref. 6, one can make the assumption that up to 1.5 cm^{-1} of the observed difference between films grown on (110) and (100) substrates is due to this thickness dependence. However, the remaining difference, < 2 cm^{-1}, follows the relationship in Eqn. (1), i.e. the (110) orientation exhibits the highest stretching frequency (lowest strain) followed by (111) and (100).

The interface state density at midgap decreased as the oxidation temperature increased for the (100) substrate orientation, while the opposite was observed for the (110) orientation. Moreover, for films grown on (111) Si, D_{it} at midgap did not exhibit any significant change as the oxidation temperature was increased. For a given oxidation temperature (up to 1050°C), the interface state density was reduced by a RTA at 1100°C. As Figs. 1 and 2 illustrate, oxidation at higher temperatures and/or RTA result in lower levels of strain for films grown on all three substrates.

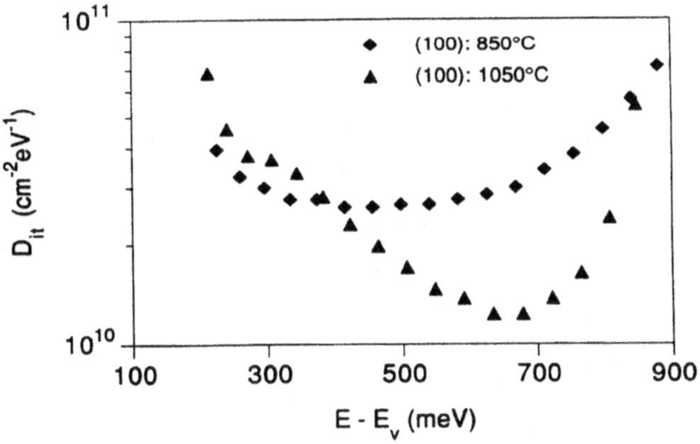

Figure 3. Energy distributions of interface traps for SiO_2 films grown at 850°C and 1050°C on (100) Si.

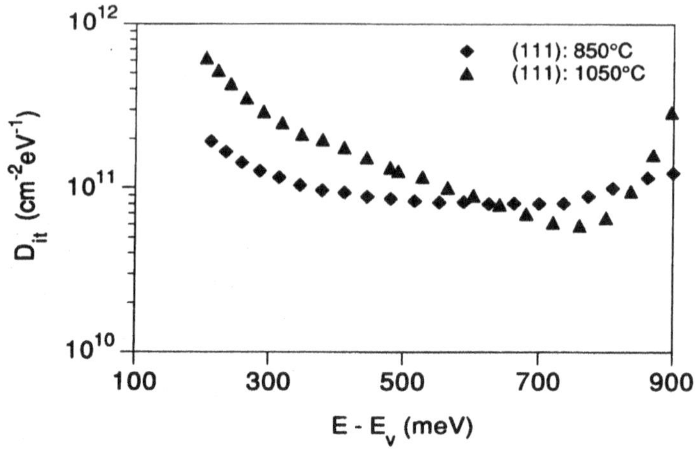

Figure 4. Energy distributions of interface traps for SiO_2 films grown at 850°C and 1050°C on (111) Si.

However, the previously reported correlation for (100) substrates between strain and interface state density at midgap,[1] apparently does not apply to as-grown films on (110) substrates.

When the energy distribution of the interface traps was compared for films grown at different temperatures, the difference in D_{it} at midgap did not correlate with relative differences in D_{it} elsewhere in the bandgap. For example, Fig. 3 shows the energy distribution of interface traps for films grown at 850°C and 1050°C on (100) Si. At midgap, as well as above midgap the film grown at the higher temperature exhibited a lower D_{it}, but the reverse effect was observed in the region 0.3 ± 0.1 eV above the valence band edge, i.e., the relative defect state density increased significantly in the film grown at 1050°C. As Fig. 4 shows, a similar relationship was observed for oxides grown on (111) substrates, while the oxides grown on (110) substrates at 850°C surprisingly exhibited a lower density of interface states than films grown at 1050°C throughout the bandgap (see Fig. 5.).

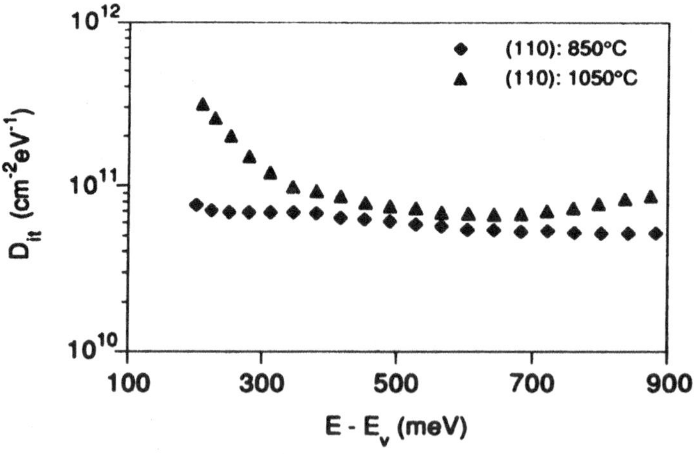

Figure 5. Energy distributions of interface traps for SiO_2 films grown at 850°C and 1050°C on (110) Si.

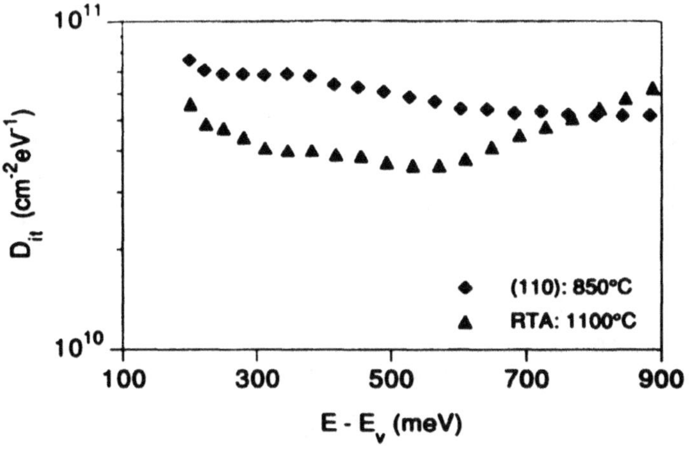

Figure 6(a). Energy distributions of interface traps for SiO_2 films grown on (110) Si at 850°C and after RTA.

However, the densities of states for the (110) and (111) orientations were between five and ten times higher than that of the (100). It is interesting to note that the region where differences are observed between films grown at different temperatures is the same region where P_b centers have been observed in electron paramagnetic resonance (EPR) and deep level transient spectroscopy (DLTS) studies,[7-8] i.e., up to about 0.4 eV from the valence band edge.

The density of interface traps was reduced significantly throughout the Si bandgap by RTA for SiO_2 films grown on all three types of substrates. Figures 6(a) and 6(b) illustrate this effect for the (110) orientation. The energy distribution of traps after RTA is relatively flat without any distinct features between 0.3 and 0.8 eV above the valence band edge, except in the case of the (100) orientation, where a peak at ~0.35 eV above the valence band edge emerged, see Fig. 7.

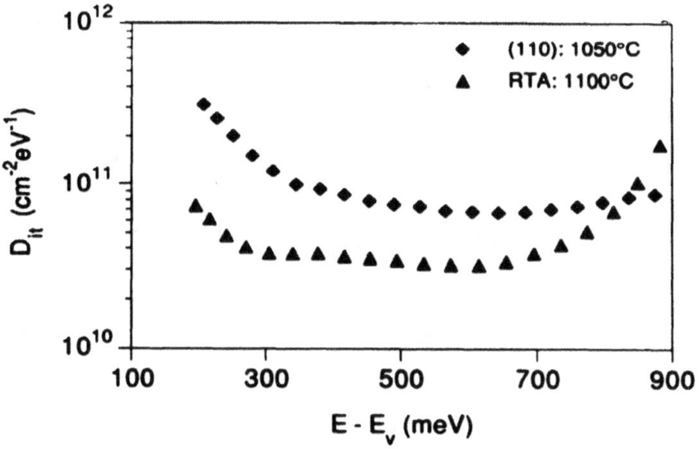

Figure 6(b). Energy distributions of interface traps for SiO_2 grown on (110) Si at 1050°C and after RTA.

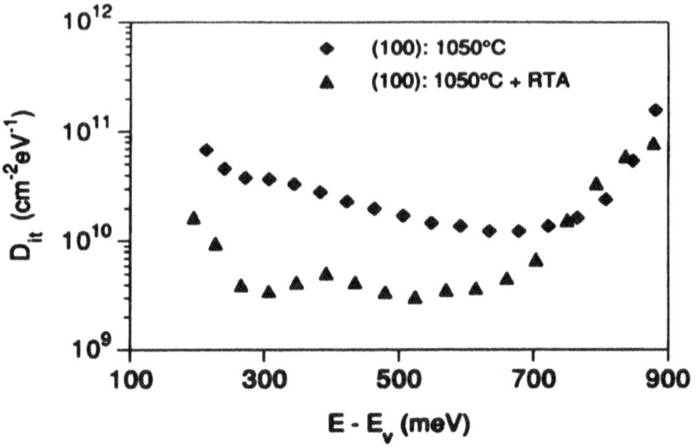

Figure 7. Energy distributions of interface traps for SiO_2 grown on (110) Si at 1050°C and after RTA.

At this stage, the energy distribution of interface traps of thermally grown films was compared to that of Remote Plasma-Enhanced CVD (Remote PECVD) films. The growth conditions for the Remote PECVD process have been reported elsewhere,[9] but it should be pointed out that the initial 5 - 6 Å of oxide are grown by exposure to plasma generated O-atoms, and the remainder is created by deposition using N_2O as the source gas.

A striking similarity was observed between thermal oxides on (100) Si substrates which had been subjected to RTA and 150 Å gate quality Remote PECVD films deposited on the same Si (100) substrates. For both of these films, the distribution of interface traps was relatively flat between 0.2 and 0.7 eV above the valence band edge, except for a peak at approximately 0.35 eV. This can be seen by comparing the lower curve in Fig. 7 with the curves in Fig. 8, which also illustrate that annealing of a Remote PECVD film does not change the shape of the energy distribution of interface states. Depositing SiO_2 films by Remote PECVD on (110) and (111) substrates yields the curves shown in Fig. 9.

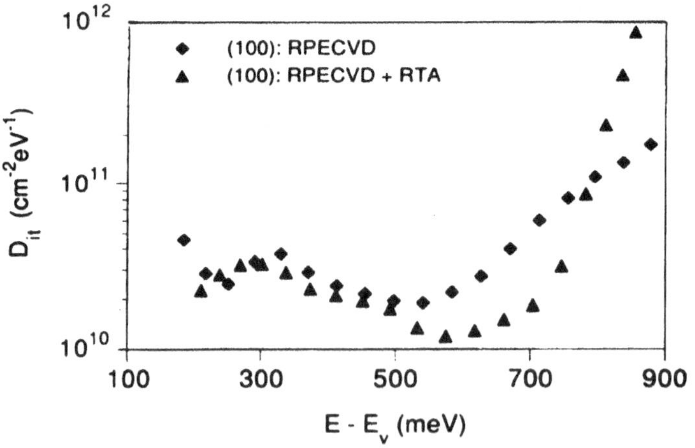

Figure 8. Energy distributions of interface traps for SiO_2 grown on (110) Si at 1050°C and after RTA.

Again the densities of states are higher than that of the (100) orientation, but the shapes of the distribution are similar. One should note that the (111) exhibits a sharper peak around 0.3 eV while the (110) has a higher density of states around midgap. As previously mentioned, thermal oxides subjected to a rapid thermal anneal exhibit very low levels of strain. This is also true for deposited films, since the deposition takes place at the SiO_2 surface and not at the interface where the 120% molar volume difference between Si and SiO_2 is the source of intrinsic growth stress for thermal oxides. This indicates that the shape of the energy distribution of interface traps shown in Figs. 7-9, is representative of a low strain Si-SiO_2 interface.

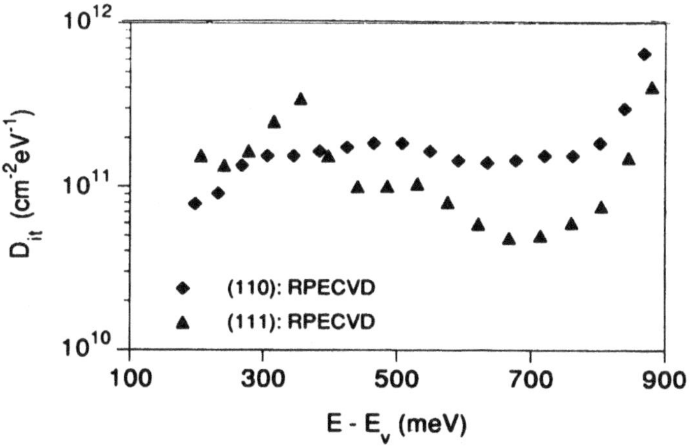

Figure 9. Energy distributions of interface traps for SiO_2 grown on (110) Si at 1050°C and after RTA.

CONCLUSION

In summary, we have studied the energy distribution of interface traps in the Si bandgap as a function of crystal orientation and processing conditions during SiO_2 formation. Increased oxidation temperatures resulted in higher infrared bond-stretching frequencies (lower strain) on (100), (110), and (111) Si. Although lower levels of strain in as-grown SiO_2 films were accompanied by lower interface state densities on (100) substrates, the opposite was observed for the (110) substrate orientation. However, there was a correlation between relaxation of strain by RTA and reduction in D_{it} at midgap for all three substrate orientations. For thermally grown oxide on (100) Si, an increased oxidation temperature promotes a reduction of interface states at energies above 0.4 eV in the Si bandgap, but not at lower energies between 0.2 and 0.4 eV above the valence band edge. Rapid thermal annealing further reduces the interface traps throughout the bandgap, but a peak at ~0.35 eV remains. Films deposited by Remote PECVD on Si (100) exhibit an energy distribution of interface traps similar to that of a thermally grown oxide which has been subjected to rapid thermal annealing indicating that this shape of the energy distribution of interface traps is representative of a low strain Si-SiO_2 interface.

REFERENCES

1. C.H. Bjorkman, J.T. Fitch, and G. Lucovsky, Appl. Phys. Lett. 56, 1983 (1990).
2. C.H. Bjorkman, J.T. Fitch, and G. Lucovsky, in *Proc. of the 20th Int. Conf. on the Phys. of Semicond.*, ed. by E.M. Anastassakis and J.D. Joannopoulos (World Scientific Publishing, Singapore, 1990), p. 419.
3. C.H. Bjorkman, D.R. Lee, and G. Lucovsky, MRS Symp. Proc. 202, 271 (1991).
4. G. Lucovsky, M.J. Mantini, J.K. Srivastava, and E.A. Irene, J. Vac. Sci. Technol. B5, 530 (1987).
5. S.C. Vitkavage, E.A. Irene, and H.Z. Massoud, J. Appl. Phys. 68, 5262 (1990).
6. J.T. Fitch, E. Kobeda, G. Lucovsky, and E.A. Irene, J. Vac. Sci. Technol. B7, 153 (1989).
7. E.H. Poindexter, P.J. Caplan, B.E. Deal, and R.R. Razouk, J. Appl. Phys. 52, 879 (1981).
8. G.J. Gerardi, E.H. Poindexter, P.J. Caplan, and N.M. Johnson, Appl. Phys. Lett. 49, 348 (1986).
9. Y. Ma, T. Yasuda, S. Habermehl, and G. Lucovsky, J. Vac. Sci. Technol. (in press).

CHARGE TRAPPING AND DEFECT GENERATION IN THERMAL OXIDE LAYERS

M.M. Heyns, A.v. Schwerin[a], A. Kelleher[b],
M. Kubota[c], and S. Verhaverbeke

IMEC, Kapeldreef 75
B-3001 Leuven, Belgium

INTRODUCTION

Thin thermal oxide layers are an important part of MOS-technologies because they are always related with the active parts of the device. An important reliability issue for these layers is the charge injection in the gate oxide as a result of the high fields in small geometry transistors. This causes charge build-up in the SiO_2 layer and degradation of the Si/SiO_2 interface. After a brief introduction to the charge injection techniques two aspects of charge trapping are discussed in this paper which have received much attention over the last few years : the oxide field dependence of the charge trapping and interface degradation and the effect of small amounts of chlorine and fluorine on the trapping properties of oxide layers.

CHARGE INJECTION TECHNIQUES

Most studies on the trapping properties of thermal oxide layers[1,2,3] have used the avalanche injection technique[4] to introduce electrons or holes into the oxide layer. Although much important information has been obtained with this technique it has the drawback that the oxide field during injection can not be controlled. This limitation can be overcome when homogeneous injection in MOS-transistors is used. This technique is schematically illustrated in figure 1. Minority carriers generated in the Si-substrate by optical means or from an (underlying) diode[5,6] are accelerated towards the Si/SiO_2 interface by a substrate bias while the source and drain of the transistor are grounded. Under these conditions the oxide field is determined by the gate voltage. Part of the carriers gain sufficient energy to overcome the barrier at the Si/SiO_2 interface and are injected into the oxide layer. The technique works relatively easy for electron injection but hole injection is more difficult. This is due to the shorter inelastic scattering length of holes in the Si-substrate and the higher energy barrier they have to overcome at the Si/SiO_2 interface. The hole injection efficiency can, however, be increased by providing a large transverse field in the silicon[7] so that measurable injection currents can be obtained.

---

[a] now with Siemens, Corporate Research and Development, Otto-Hahn-Ring 6, D-8000 Munchen 83, Germany

[b] now with National Microelectronics Research Centre, University College, Lee Maltings, Prospect Row, Cork, Ireland

[c] on leave from Sony Corporation, Atsugi Technology Center, ULSI R&D Group, 4-14-1, Asahi-cho, Atsugi-shi, Kanagawa-ken, 243 Japan

The Physics and Chemistry of SiO_2 and the Si-SiO_2 Interface 2
Edited by C.R. Helms and B.E. Deal, Plenum Press, New York, 1993

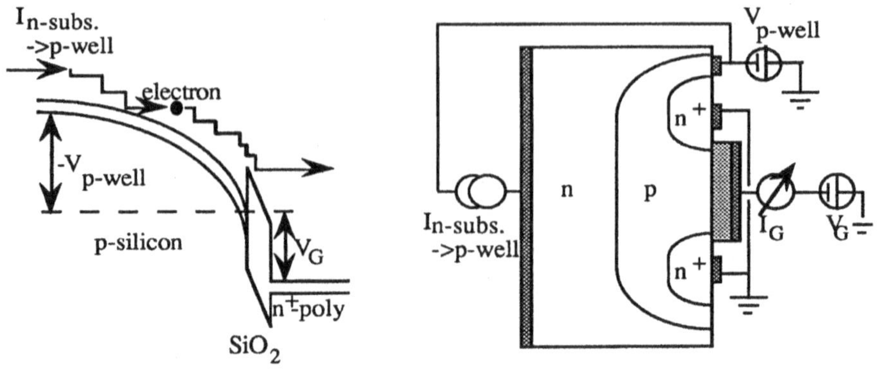

Figure 1. Schematic representation of the electron injection process in the Si-SiO_2-poly-Si band diagram (left part) and of the measurement set-up (right part).

Charge injection on capacitor structures can also be performed by applying fields large enough to cause Fowler-Nordheim tunneling of electrons through the triangular barrier at the interface into the oxide conduction band[8]. Due to the fast detrapping at these high fields the technique can not be used to study the characteristics of trapping centres in the oxide layer. However, the technique is very well suited to study degradation phenomena occurring at high fields and prior to breakdown and can provide valuable information on the characteristics of tunnel dielectrics operating under these conditions.

OXIDE FIELD DEPENDENCE OF ELECTRON AND HOLE TRAPPING

Experimental conditions

Electron and hole injection experiments were performed on poly-Si gated transistors with a typical gate oxide thickness of 20 to 30 nm. Transistors with various lengths and widths were used. The interface state density (D_{it}) was measured using the charge pumping technique[9]. The density of trapped oxide charge was obtained from the gate voltage shift for a fixed drain current level in deep subthreshold at a fixed low drain voltage. As the location of the charge is not known only an effective density of trapped charge can be given with a centroid assumed to be located at the Si/SiO_2 interface. It was demonstrated[7] that the effect of the interface state generation on this measurement can be neglected. More experimental details can be found in refs. 7 and 10.

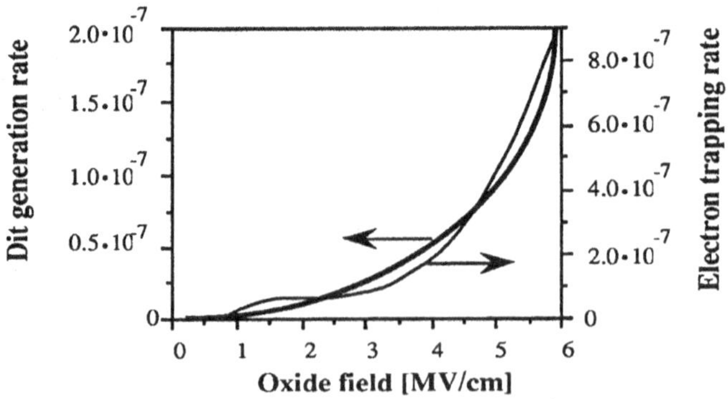

Figure 2. Oxide field dependence of the interface state generation rate (left scale) and the electron trap generation rate (right scale) during electron injection.

Defect generation during electron injection

During electron injection it is observed that electron traps are generated in the oxide layer. When the generation rate of the charged trap centres is calculated from the linear part of the trapped charge vs. injected charge curves (with the charge trapping at the lowest measured field subtracted from the measurements as background trapping in pre-existing electron traps) the result shown in figure 2 is found. The generation efficiency increases with increasing oxide field. Using thin oxide layers it was demonstrated that this correlates with a threshold in the electron energy for damage generation around 2.3 eV above the oxide conduction band[11].

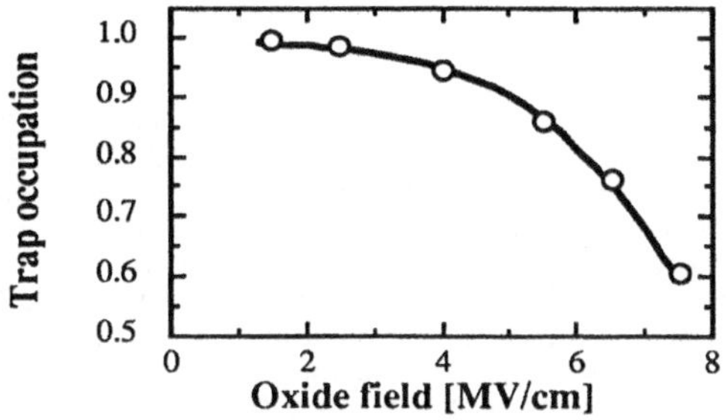

Figure 3. Normalized equilibrium trap occupation level as a function of the oxide field.

On some series of samples a different behaviour was found and it was observed that a subsequent injection of a relatively small amount of electrons at varying oxide field (E_{ox}), changes the apparent amount of trapped electrons as detected in an electrical measurement[12]. This means that the fraction of present traps which is actually charged with trapped electrons is in equilibrium with the field which is applied during electron injection. The equilibrium level is reached after switching to the respective field value upon injection of approximately 2 to 4.10^{16} e/cm^2. Switching of the electric field without simultaneous injection of electrons does not change the trap occupation considerably within the same time frame. The effect is reversible. The occupation of traps with electrons is a decreasing function of E_{ox}, as shown in figure 3. As a consequence, for these samples evidence for the field enhancement of the trap generation process at $E_{ox} > 4$ MV/cm is found only when a short low-field injection is carried out after the stressing. This is necessary in order to fill eventually generated, but empty, traps with electrons which enables their detection in an electrical measurement.

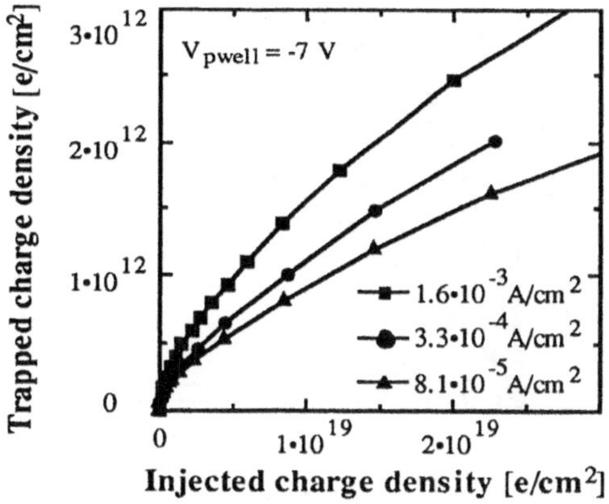

Figure 4. Trapped electron density as a function of the density of electrons injected at an oxide field of 3.5 MV/cm, a p-well bias of -7 V and different gate current densities (j_g).

When the current flow through the oxide is kept fixed by adjusting the carrier supply from the n-substrate, it was observed that the p-well bias (V_{pwell}) does not affect the electron trapping. Thus, the energy of the electrons at the moment of injection (which is controlled by V_{pwell}) has apparently no effect on the electron trap generation. This is most likely due to a fast thermalization of the injected electrons in SiO_2. This finding is apparently in contrast to previously reported results on the substrate bias dependence of electron trapping[11]. However, under normal experimental conditions an increase in substrate voltage will also cause an increase in oxide current density or, with other words, the injection of the same amount of electrons takes less time for higher substrate voltages.

The effect of this is illustrated in figure 4, which shows the result of the complementary experiment performed at an oxide field of 3.5 MV/cm. In this experiment V_{pwell} was kept fixed but the oxide current density (j_g) was set to three different values by adjusting the substrate/p-well forward bias, which actually changes the carrier supply to the accelerating space charge region. It is found that the electron trapping at this oxide field increases with increasing oxide current density. This means that the less time it takes to inject a certain amount of electrons, the more electrons are trapped. This effect disappears for E_{ox} well above 4 MV/cm, both before and after trap filling. The j_g-dependence was found likewise in experiments where optically stimulated substrate hot electron injection on NMOS samples was used and where j_g was changed by varying the light intensity in order to change the carrier supply to the accelerating space charge region in the Si. The knowledge of the current density dependence of electron trapping is important because j_g depends on V_{pwell} as well as on E_{ox} in the injection experiment, if no special precautions are taken. Therefore, the j_g-dependence of the electron trapping at low oxide fields can be easily misinterpreted as a dependence on E_{ox} or on V_{pwell}.

The observed j_g-dependence can *not* be explained by a simple dynamic trapping-detrapping argument. This was verified in an experiment where between every two subsequent characterization measurements electrons were injected at a high current density for part of the time, whereas the electric field was applied as long as in a corresponding low j_g experiment. Thus, for part of the time between two subsequent characterization steps, the electric field was applied without electron injection. In this case no considerable difference in the amount of trapped electrons was found compared to the standard high j_g measurement. A conclusive model for the origin of this effect is still missing but as a possible model it can be suggested that a shallow trap is transformed into a deep one when an electron is trapped and relaxes into a shallow one after detrapping of the electron. A key parameter in this model would be the time constant of the latter relaxation process, which is in competition with the mean capture time (defined by the capture cross section times the current density).

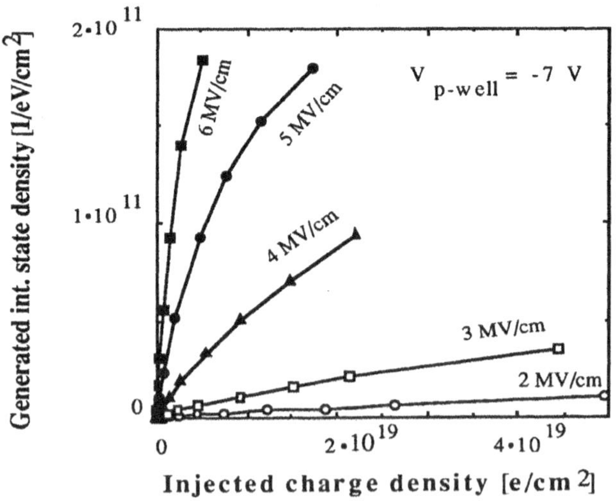

Figure 5. Increase in Si/SiO_2 interface states as a function of the total density of electrons injected at different constant oxide fields between 2 and 6 MV/cm.

No evidence was found for a current density dependence of the D_{it} generation. Also no increase in D_{it} generation with increasing energy of the incoming electrons (as varied by V_{pwell}) was found. This is in contradiction with results obtained from photoinjection experiments[13]. However, in these experiments not only electrons are involved, but also photons with energies well above 4 eV, which may influence the D_{it}-generation. The D_{it} generation rate strongly increases with increasing E_{ox}, as shown in figure 5. Extracting the generation *rate* from the starting slope of the curves in figure 5 shows that D_{it}-generation is almost exponentially dependent on the magnitude of E_{ox} (figure 2). The shape of the interface state density distribution is independent of E_{ox} and shows a larger density in the upper half of the bandgap similar to results found in irradiation and high-field injection experiments.

The similar field dependence of the electron trap and D_{it} generation suggests a common origin for both mechanisms. A correlation between electron trapping and D_{it} generation has already been suggested[14] and could occur by the release of hydrogen from water-related electron traps when an electron is captured at this defect. The hydrogen can diffuse towards the Si/SiO_2 interface where it can generate a dangling silicon bond, acting as an interface state. Another possible common origin for the electron trap and D_{it} generation is the injection of holes from the anode[15]. The importance of holes in the D_{it} generation has already been extensively discussed[16,17], while the generation of electron traps by hole injection[18] or by the recombination of electrons with trapped holes[19] has also been reported. Within this model the electron trap and D_{it} generation are not necessarily directly correlated but have the same origin, i.e. the injection of holes from the anode. The average electron energy above the oxide conduction band at an oxide field of 4 MV/cm is approximately 2.5 to 3 eV[15], which is comparable to the threshold energy for damage generation, reported as 2.3 eV[11]. Summed with the conduction band offset between the SiO_2 layer and the poly-Si electrode this gives a threshold energy for hole injection between 5.5 and 6.0 eV. This is larger than the energy barrier for holes at the poly-Si electrode/SiO_2 layer interface but smaller than the suggested 7.5 eV threshold for the generation of surface plasmons[15].

Hole trapping and interface state generation

Using homogeneous injection of holes in transistor structures it was demonstrated[20] that the trapping of holes does not depend on the substrate bias (and therefore on the energy of the injected holes) during injection. In contrast with the results for electron injection apparently no strong field dependence is present for the hole trapping. Only a small trend towards a decreased trapping rate at higher fields and low fluences could be detected, which can be related to either a field dependent capture cross section or to detrapping from shallow hole traps at higher fields. The possibility of electron injection from the cathode at the higher fields, which recombine with the trapped holes, could be excluded[20]. When the injection is prolonged to larger injected hole densities a clear saturation level is observed for all oxide fields. This level is independent of the oxide field during injection and no evidence could be found for the generation of hole traps during hole injection.

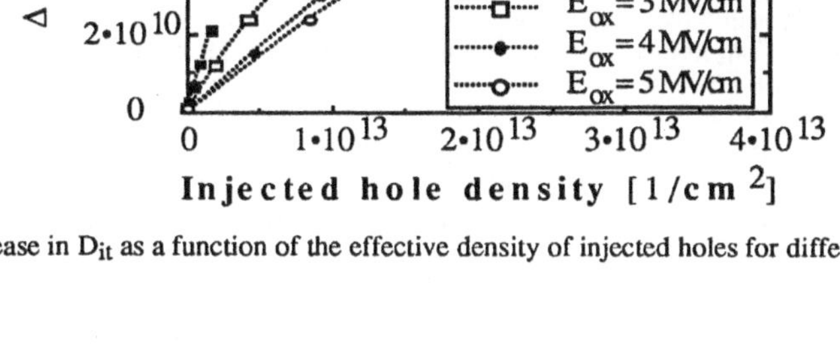

Figure 6. Increase in D_{it} as a function of the effective density of injected holes for different oxide fields.

The D_{it} generation during irradiation, high-field stressing or hot electron injection is still a point of controversy in literature. One of the suggested models assumes the D_{it} generation to be a two-step process[16]. The first step is the trapping of holes without any D_{it} generation. In the second step electrons recombine with the trapped holes causing the creation of interface states. On the other hand, experiments on hot carrier injection in transistor structures observed the maximum D_{it} generation under conditions where electrons and holes are simultaneously injected in the gate oxide layer[21] and no evidence could be found for a two-step mechanism[22,23]. However, in most experiments on transistor structures it is very difficult to investigate the effects of hole injection while completely avoiding the injection of (a small number of) electrons. As the capture efficiency for a trapped hole to capture an electron is very high[24] neutralization of the trapped holes occurs very efficiently and the second step of the two-step process may pass unnoticed. This will lead to the generation of an interface state and a re-structured hole trap which can act as a slow state[25]. The positive charge in these states gives rise to the positive charge observed after the hole injection which can be neutralized without the generation of interface states[22,23].

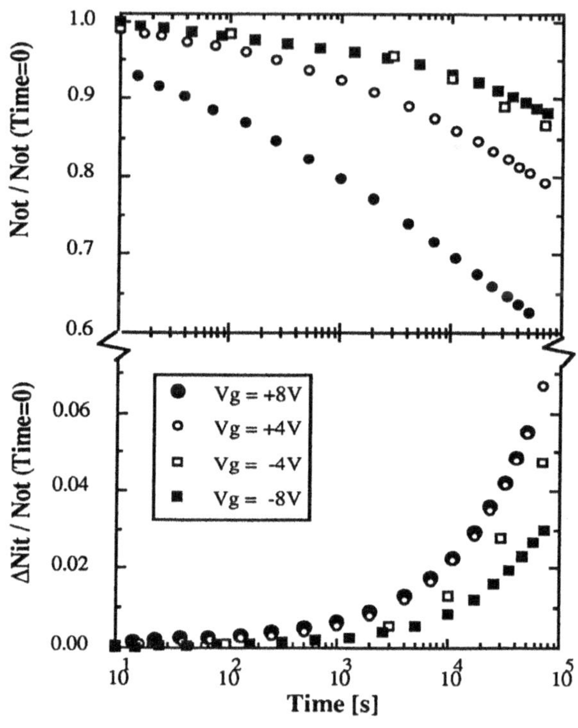

Figure 7. Change in the density of trapped holes (upper part of the figure) and the interface state density (lower part of the figure) as a function of time after hole injection. The values are normalized to the effective density of trapped holes measured directly (at time 0) after the hole injection (Not(Time=0)). A constant voltage (V_g) was applied to the gate after the hole injection. Note the different scales for detrapping and interface state build-up.

The homogeneous hole injection technique was used to study the E_{ox} dependence of the hole trapping and D_{it} generation *during* hole injection. The result is presented in figure 6 for E_{ox} between 2 and 5 MV/cm. From the starting slope of the curves in this figure the direct D_{it} generation rate can be extracted. It is found to be decreasing with increasing oxide field approximately like $1/(E_{ox})^2$, with a generation efficiency of about 1 interface state per 100 injected holes at E_{ox} = 2 MV/cm. This result is in good agreement with results obtained with gate-controlled diodes[26]. When the D_{it} values were re-measured one week after the hole injection (with the devices stored at room temperature with the gate floating) a strong D_{it} increase was observed together with a decrease in the trapped hole density. However, no one-to-one correlation between hole detrapping and D_{it} build-up after hole injection is found. This can be seen from the time and field dependence of the two phenomena as shown in figure 7. The hole detrapping after hole injection depends on both the magnitude and the polarity of the applied gate voltage. The decrease in the trapped charge density as a function of time is qualitatively consistent with a simple model of direct tunneling of the charge carriers between the SiO_2 and the Si substrate. In contrast with this the simultaneous D_{it} build-up is apparently only dependent on the field polarity. These observations can not be easily explained in the simple two-step model[16] where trapped holes are transformed into interface states since under this assumption a one-to-one correlation would be expected.

EFFECT OF CHLORINE AND FLUORINE ON THE CHARGE TRAPPING

The importance of chlorine in the fabrication of thin oxide layers has been considerably changed over the last few years. While initially the passivation of mobile ions[27] and the annihilation of stacking faults were targetted[28], recent work has focussed on the improved degradation resistance and radiation hardness which can be reached when minute amounts of TCA (1-1-1 trichloroethane) are added to the oxidation ambient[29]. In order to clarify the role of Cl in thin thermal oxides its effect on the trapping characteristics was investigated. Chlorine was introduced in the oxide layer by ion implantation and by adding small amounts of TCA to the oxidation ambient.

The effect of fluorine on the oxide properties has also received a great deal of interest over the last years. Fluorine can be introduced in the gate oxide in several ways. It has been suggested that some cleaning procedures leave the Si-surface covered with F which can be incorporated into the oxide layer upon oxidation[30]. Ion implantation in the Si-substrate prior to the gate oxidation[31], short purges of NF_3 during the oxidation[32] or ion implantation in the poly-Si electrode and subsequent diffusion[33,34] can be used to control the F-content in the oxide. It was also reported that a tungsten polycide process can introduce large quantities of F in the gate oxide[35]. In this study F was introduced by ion implantation in the poly-Si layer.

Experimental conditions

The experiments on Cl implantation were performed oxide layers which were grown in dry O_2 to a thickness of 12 nm. Immediately after the oxidation a 10 nm poly-Si layer was deposited to protect the oxide layer. Chlorine was implanted at an energy of 20 KeV with three different doses : 10^{11}, 10^{12} and 10^{13} cm^{-2}. The projected range based on the LSS-theory was 2.2 nm so that the maximum of the implanted profile is expected to be located near the Si/SiO_2 interface. The implantation was followed by a two-step annealing of 750°C for 120 min and 950°C for 60 min in N_2. The first step allows the reaction of Cl with defect sites in the oxide before being outdiffused to either the anneal ambient or the Si-substrate during the second step[36], which is needed to anneal the implantation damage. After deposition of the remaining part of the poly-Si gate electrode and solid source doping at 900°C, a drive-in was performed at 975°C during 20 min. Capacitor structures were defined by conventional wet lithography.

In order to test the effect of Cl-addition to the oxidation ambient, oxide layers with a thickness of 15 nm were grown at 900°C, 950°C and 1000°C with various amounts of TCA added to the oxidation ambient either at the start or at the end of the oxidation cycle. After deposition of the poly-Si electrode capacitors were formed as described above.

The effect of fluorine on the oxide trapping properties was investigated on 15 nm oxides which were grown in dry O_2 at 900°C. The poly-Si electrode was doped by ion implantation of phosphorus and annealed at 850°C. Fluorine was implanted in the poly-Si gate with a dose of either $5 \cdot 10^{14}$, $2 \cdot 10^{15}$ or $1 \cdot 10^{16}$ cm^{-2} at an implantation energy of 50 keV. Subsequent annealing was performed at 900°C for 30 min to introduce the F in the gate oxide. After this procedure capacitors were formed by standard lithography and wet etching.

The electron and hole trapping properties were investigated using avalanche injection. The data were fitted with a first order kinetic trapping model[37] which describes the trapping properties with a capture cross section and an effective density of traps.

Trapping properties of Cl-implanted oxides

The results of the hole injection experiments on Cl-implanted oxides are given in table 1. In all samples two types of traps were observed[38]. The relatively large total trap density is due to the fact that the highly-doped n-type substrates necessary for avalanche injection were formed by ion implantation[39]. The density of the trap with a capture cross section of 4-$6 \cdot 10^{-14}$ cm^2 is decreased when Cl is implanted. The decrease is only a fraction of the density of implanted ions. Therefore, either only a limited number of defects can be annealed by Cl or the diffusion of Cl towards the defects and the reaction at these defects is the limiting step. The density of the other hole trap is not affected by the Cl implantation. No electron trapping could be observed on the samples (detection limit $\approx 10^{10}$ cm^{-2}) both during injection at room temperature and at 140°C, which was used to lower the positive charge build-up by slow trapping instabilities[25]. This indicates that the implanted Cl does not generate electron traps.

It must be remarked that the Cl content used in these experiments is extremely small. A dose of $10^{13}/cm^2$ Cl ions, which is the highest implantation dose used, corresponds roughly to a HCl content of 0.01% in the oxidation ambient. For typical concentrations of 2% HCl in the oxidation ambient, peak concentrations of Cl as high as $2 \cdot 10^{21}$ cm^{-3} were measured[40] and concentrations per unit area, averaged over the total oxide thickness were in the range[41] of 10^{15}-10^{16} cm^{-2}. During the first annealing cycle (750°C for 120 min) the Cl diffuses over a distance of approximately 20 nm. During the second cycle the diffusion distance is of the order of 70 nm and all of the non-bonded Cl is diffused out[42,43]. The observed effect on the hole trapping is, therefore, due to the part of the Cl which becomes bonded on an oxide defect acting as a hole trap and, thereby, neutralises this hole trap.

Table 1. Capture cross sections (σ_c) and effective concentrations (N_{eff}) of the hole traps as a function of the implanted Cl dose.

impl. dose (cm^{-2})	Trap 1 σ_{c1} (cm^2)	Trap 1 N_{eff1} (cm^{-2})	Trap 2 σ_{c2} (cm^2)	Trap 2 N_{eff2} (cm^{-2})	Tot. density N_{efft} (cm^{-2})
no implantation	$6 \cdot 10^{-14}$	$3.3 \cdot 10^{12}$	$6 \cdot 10^{-15}$	$0.7 \cdot 10^{12}$	$4.0 \cdot 10^{12}$
$1 \cdot 10^{11}$	$4 \cdot 10^{-14}$	$3.0 \cdot 10^{12}$	$6 \cdot 10^{-15}$	$0.9 \cdot 10^{12}$	$3.9 \cdot 10^{12}$
$1 \cdot 10^{12}$	$5 \cdot 10^{-14}$	$2.9 \cdot 10^{12}$	$6 \cdot 10^{-15}$	$0.7 \cdot 10^{12}$	$3.6 \cdot 10^{12}$
$1 \cdot 10^{13}$	$6 \cdot 10^{-14}$	$2.6 \cdot 10^{12}$	$7 \cdot 10^{-15}$	$0.7 \cdot 10^{12}$	$3.3 \cdot 10^{12}$

Hole trapping properties of TCA oxides

A detailed discussion of the electrical properties of TCA oxides will be given elsewhere[44]. In general it was observed that when TCA is added to the oxidation ambient, even for relatively short times, the total hole trapping is increased. The effect becomes more pronounced for higher oxidation temperatures and less pronounced when the TCA is added at the end of the oxidation cycle[44]. The effect of the oxidation temperature and TCA-time on the hole trapping is illustrated in figure 8. Comparing the results obtained by Cl implantation and by TCA addition to the oxidation ambient it can be suggested that the introduction of Cl in an oxide layer can reduce the total density of hole traps. This probably occurs by the neutralization of dangling Si-bonds, acting as hole traps, through the formation of Si-Cl bonds. When Cl is introduced by adding TCA to the oxidation ambient this effect may be counteracted by the hydrogen (or water) introduced simultaneously. Furthermore, in the implantation experiments the Cl is introduced at a moment that the oxide network is already formed. Most of the Cl diffuses out during the subsequent temperature treatments and only the Cl which can form stable bonds remains in the SiO_2 layer. When TCA is added during the oxidation it will change the oxide structure and it is well known that the Cl is incorporated mainly near the Si-SiO_2 interface, i.e. the region where hole traps are normally found.

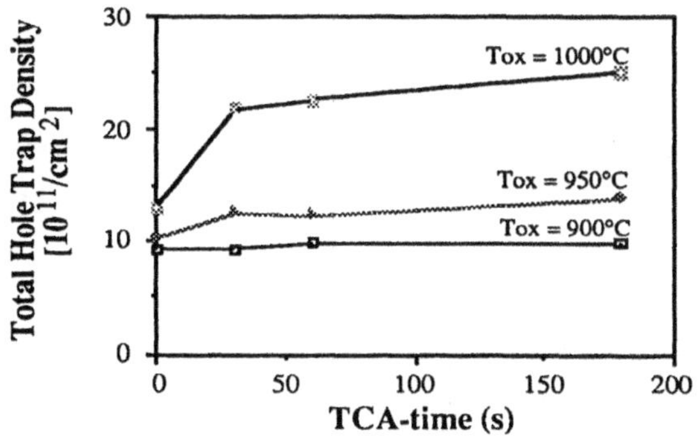

Figure 8. Total hole trap density as a function of the oxidation temperature and of the TCA-time at the start of the oxidation.

Electrical properties of F-implanted oxide layers

A detailed investigation of the characteristics of F-implanted oxides will be the subject of another publication[45]. A summary of the hole trapping properties is given in table 2. A good fitting of the experimental data could only be obtained by assuming three traps in the first order kinetic model. It is clearly observed that the implantation of F increases the total number of hole traps up to the implantation dose of 10^{16} cm^{-2}. At this dose the total number of hole traps slightly decreases again. This goes together with a decrease in the capture cross section of the observed traps. As a result, the initial hole trapping probability (σ.N) is decreased to below the level of the unimplanted sample. No strong electron trapping could be observed in the F-implanted samples.

Table 2. Capture cross sections (σ_c) and effective concentrations (N_{eff}) of the hole traps as a function of the implanted F dose.

impl. dose	Trap 1		Trap 2		Trap 3		Tot. dens.
(cm^{-2})	σ_{c1} (cm^2)	N_{eff1} (cm^{-2})	σ_{c2} (cm^2)	N_{eff2} (cm^{-2})	σ_{c3} (cm^2)	N_{eff3} (cm^{-2})	N_{efft} (cm^{-2})
no implantation	$4 \cdot 10^{-14}$	$4.5 \cdot 10^{11}$	$6 \cdot 10^{-15}$	$4.0 \cdot 10^{11}$	$1 \cdot 10^{-15}$	$3.7 \cdot 10^{11}$	$1.22 \cdot 10^{12}$
$5 \cdot 10^{14}$	$4 \cdot 10^{-14}$	$9.5 \cdot 10^{11}$	$7 \cdot 10^{-15}$	$6.8 \cdot 10^{11}$	$1 \cdot 10^{-15}$	$4.9 \cdot 10^{11}$	$2.12 \cdot 10^{12}$
$2 \cdot 10^{15}$	$5 \cdot 10^{-14}$	$9.8 \cdot 10^{11}$	$9 \cdot 10^{-15}$	$8.0 \cdot 10^{11}$	$1 \cdot 10^{-15}$	$5.8 \cdot 10^{11}$	$2.36 \cdot 10^{12}$
$1 \cdot 10^{16}$	$2 \cdot 10^{-14}$	$7.1 \cdot 10^{11}$	$4 \cdot 10^{-15}$	$6.4 \cdot 10^{11}$	$6 \cdot 10^{-16}$	$7.7 \cdot 10^{11}$	$2.12 \cdot 10^{12}$

CONCLUSIONS

The E_{ox} dependence of the trapping and D_{it} generation during electron and hole injection was investigated using homogeneous injection in transistors. A strong field dependence is found during electron injection. The hole trapping apparently does not depend on E_{ox}. Holes can efficiently generate interface states but no direct correlation between hole trapping and D_{it} generation seems to exist. These observations have important consequences on the interpretation of accelerated life-time test. The implantation of Cl in the oxide layer decreases the number of hole traps. The introduction of Cl in the oxidation ambient increases the hole trap density. The introduction of F by ion implantation and subsequent annealing generally increases the hole trap density. However, at sufficiently high implantation doses a turn-around is observed and the initial hole trapping probability can be decreased.

REFERENCES

1. D.J. DiMaria in "The Physics of SiO_2 and its interfaces", Ed. S.T. Pantelides (Pergamon, New York, 1978), p.160
2. R.F. De Keersmaecker in "Insulating Films On Semiconductors", Eds. J.F. Verweij and D.R. Wolters, (North-Holland, Amsterdam), p. 85 (1983)
3. P. Balk in "Solid State Devices 1983", Ed. E. H. Rhoderick, The Institute of Physics Conf. Ser. No. 69, p. 63 (1984)
4. E.H. Nicollian and C.N. Berglund, J. Appl. Phys. 41, p. 3052 (1970)
5. J.F. Verwey, J. Appl. Phys. 44, p. 2681 (1973)
6. T.H. Ning and H.N. Yu, J. Appl. Phys. 45, p. 5373 (1974)
7. A.v. Schwerin, M.M. Heyns and W. Weber, J. Appl. Phys. 67, p. 7595 (1990)
8. M. Lenzlinger and E.H. Snow, J. Appl. Phys. 40, p. 278 (1969)
9. G. Groeseneken, H.E. Maes, N. Beltran and R.F. De Keersmaecker, IEEE Trans. Electron. Dev. ED-31, p. 42 (1984)
10. M.M. Heyns, D. Krishna Rao and R.F. DeKeersmaecker, Appl.Surf.Sci. 39, p. 327 (1989)
11. D.J. DiMaria and J.W. Stasiak, J. Appl. Phys. 65, p. 2342 (1989)
12. A.v. Schwerin and M.M. Heyns in "Proceedings of Insulating Films on Semiconductors 91 (INFOS91)", Eds. W. Eccleston and M. Uren (Adam Hilger, Bristol), p. 263
13. V. Zekeryia and T.P. Ma, Appl. Phys. Lett. 43, p. 95 (1983)
14. L. Do Thanh, M. Aslam and P. Balk, Solid State Electron. 29, p. 829 (1986)
15. M.V. Fischetti, Phys. Rev. B 31, p. 2099 (1985)
16. S. K. Lai, J. Appl. Phys. 54, p. 2540 (1983)
17. S.J. Wang, J.M. Sung and S.A. Lyon in "The Physics and Technology of Amorphous SiO_2", Ed. R. A. Devine (Plenum, New York, 1988) p. 465 (1988)
18. S. Ogawa, N. Shiono and M. Shimaya, Appl. Phys. Lett. 56, p. 1329 (1990)
19. I.C. Chen, S. Holland and C. Hu, J. Appl. Phys. 61, p.4544 (1987)
20. A.v. Schwerin and M.M. Heyns in "Proceedings of Insulating Films on Semiconductors 91 (INFOS91)", Eds. W. Eccleston and M. Uren (Adam Hilger, Bristol), p. 283
21. N.S. Saks, P.L. Heremans, L. Van den hove, H.E. Maes, R.F. De Keersmaecker and G.J. Declerck, IEEE Trans. Electron. Dev. ED-33, p. 1529 (1986)
22. P. Heremans, R. Bellens, G. Groeseneken and H. Maes, IEEE Trans. Electron. Devices 35, 2194 (1988)
23. D. Krishna Rao, M.M. Heyns and R.F. De Keersmaecker in "Proc. of ESSDERC 88", Eds. J.-P. Nougier and D. Gasquet, (Les Editions de Physique, France), p. 669 (1988)
24. D.J. DiMaria, Z.A. Weinberg and J.M. Aitken, J. Appl. Phys. 48, p. 898 (1977)
25. M.M. Heyns and R.F. De Keersmaecker in "The Physics and Technology of Amorphous SiO_2", Ed. R.A. Devine (Plenum, New York, 1988), p. 411 (1988)
26. P. Heremans, G. Groeseneken and H.E. Maes, paper presented at IEE "Colloquium on hot carrier degradation in short channel MOS", London, Jan. 1987
27. E.J. Janssens and G.J. Declerck, J. Electrochem. Soc. 125, p. 1696-1703 (1978)
28. C.L. Claeys, E.E. Laes, G.J. Declerck and R.J. Van Overstraeten, paper 224 presented at the Electrochem. Soc. Meeting, Philadelphia (1977)
29. Y. Wang, Y. Nishioka, T.P. Ma and R.C. Barker, Appl. Phys. Lett. 52, p. 573 (1988)
30. Y. Nishioka, E.F. da Silva Jr., Y. Wang and T.P. Ma, IEEE Electron. Dev. Lett. 9, p. 38 (1988)
31. X.-W. Wang, A. Balasinski and T.P. Ma, J. Electrochem. Soc. Vol. 139, p. 238 (1992)
32. E.F. da Silva Jr, Y. Nishioka and T.P. Ma, IEEE Trans. on Nuclear Science, NS-34, p. 1190 (1987)
33. P.J. Wright and K.C. Saraswat, IEEE Trans. on Electron. Dev., ED-36, p. 879 (1989)
34. Y. Nishioka, K. Ohyu, Y. Ohji, N. Natuaki, K. Mukai and T.P. Ma, IEEE Electron. Dev. Lett. 10, p. 141 (1989)
35. M.D. Deal, D. Pramanik, A.N. Saxena and K.C. Saraswat, Proceedings of the 1985 VLSI Multilayer Interconnections Conference, Santa Clara CA, p. 324 (1985)
36. Z.M. Ling, L.H. Dupas and K.M. De Meyer, Abs. No 146, Electrochem. Soc. Spring Meeting, Atlanta (1988)
37. T.H. Ning and H.N. Yu, J. Appl. Phys. 45, p. 5373 (1974)
38. S. Verhaverbeke, M.M. Heyns and R.F. De Keersmaecker in "Proceedings of Insulating Films on Semiconductors 91 (INFOS91)", Eds. W. Eccleston and M. Uren (Adam Hilger, Bristol), p. 179
39. S. Verhaverbeke, M. Meuris, M.M. Heyns, R.F. De Keersmaecker, M. Murrel and C.J. Sofield, paper presented at the ECS Fall Meeting, Phoenix, Arizona, 1991, to be published in the proceedings
40. B.E. Deal, A. Hurrle and M.J. Schulz, J. Electrochem. Soc. 125, p. 2024-2027 (1978)
41. S.R. Butler, F.J. Feigl, A. Rothatgi, H.W. Kraner and K.W. Jones, paper 77 presented at the Electrochem. Soc. Meeting, Philadelphia (1977)
42. G. Greeuw and H. Hasper in "Proceedings of Insulating Films on Semiconductors 81 (INFOS81)", Eds. M. Schulz and G. Pensl (Springer, Berlin), p. 203 (1981)
43. A.S. Vengurlekar, K.V. Ramanathan and V.T. Karulkar, J. Electrochem. Soc. 132, p. 1172 (1985)
44. A. Kelleher, M. Kubota, S. Verhaverbeke, B. Lane and M.M. Heyns, to be published
45. M. Kubota, A. Kelleher, S. Verhaverbeke and M.M. Heyns, to be published

OPTICALLY INDUCED NITROGEN DANGLING BONDS IN AMORPHOUS HYDROGENATED SILICON NITRIDE THIN FILMS

W. L. Warren,[1] J. Kanicki,[2] P. J. McWhorter,[1] and E. H. Poindexter[3]

[1]Sandia National Laboratories
Albuquerque, New Mexico 87185
[2]IBM Research Division, T. J. Watson Research Center
Yorktown Heights, New York 10598
[3]U. S. Army Electronics Technology and Devices Laboratory
Fort Monmouth, New Jersey 07703

ABSTRACT

Using X-band and Q-band electron paramagnetic resonance (EPR) microwave frequencies, we have confirmed a model for the ultraviolet (UV) induced nitrogen dangling bond in N-rich amorphous hydrogenated silicon nitride thin films. We also report for the first time that the UV-induced N dangling bonds can be photo-bleached (light induced annealing) by sub-bandgap light. Since the photo-bleaching phenomenon is reversible, i. e., these defect centers can be reversibly photo-created or photo-bleached—a process requiring short or long-wave UV, respectively—without any change in the net space charge density of the films, it is suggested that an optical rearrangement of spin state and charge state of positive, negative, and neutral nitrogen sites occurs. This study has also shown that the N dangling bond is an electrically active point defect in these thin films.

INTRODUCTION

The principal deep trap centers in amorphous hydrogenated silicon nitride (a-SiN_x:H) thin films have been predicted to be silicon and nitrogen dangling bonds.[1] The Si dangling bond has been widely studied,[2-4] and is believed to be an amphoteric trap site,[3] behavior consistent with a mid-gap energy level.[5,6] To date, the focus of most electron paramagnetic resonance (EPR) studies in a-SiN_x:H has been on the Si dangling bond, primarily because other EPR-active centers evaded detection. Many properties of the Si dangling bond in a-SiN_x:H have been recently reviewed.[7]

Recently, another intrinsic defect center was discovered and attributed to a N dangling bond.[8] This defect is created in stoichiometric and N-rich films by a fairly-high-temperature post-deposition anneal (generally higher than the deposition temperature) followed by ultraviolet (UV) illumination; the Si dangling bond is no longer detected following this anneal/UV sequence. In this present study, we confirm and extend a model of the N dangling bond in a-$SiN_{1.6}$:H thin films, as well as investigate some of its physical and electronic properties.

EXPERIMENTAL DETAILS

The a-$SiH_{1.6}$:H thin films were deposited by plasma-enhanced chemical vapor deposition (PECVD) from an ammonia and silane mixture in a 12:1 gas flow ratio at a substrate temperature of 400 C. The resulting films were N-rich with a N/Si atomic ratio of 1.6. The films were deposited on either crystalline Si substrates that had a thin (200A) thermal oxide previously grown, or on fused silica substrates, to a thickness of 0.4 and 1 μm, respectively. Further details regarding the samples can be found elsewhere.[9] The films were given a post-deposition anneal (T_a) at either 600 or 700 C for 30 or 10 m, respectively; dry N_2 was flowing in the tube during the anneal.

The EPR measurements were performed using either a Bruker ESP-300 or an E-Line Varian spectrometer at room temperature; a non-saturating microwave power of 5 mW, and a 3 G modulation amplitude were used. One-MHz capacitance-vs.-voltage (*CV*) measurements were made with a Princeton Applied Research (*CV*) meter and a Hg probe. To inject charge carriers into the nitride thin films, a corona discharge apparatus was used. Optical illumination of the a-$SiN_{1.6}$:H thin films was performed using a Oriel 100W Hg arc lamp at 300K. The thin films were illuminated with monochromatic light of different photon energies by use of narrow-band (10 nm) interference filters.

RESULTS AND DISCUSSION

Fig. 1(a) illustrates the X-band (9.5 GHz) EPR spectrum of the nitrogen dangling bond which was generated by subjecting the N-rich a-$SiN_{1.6}$:H films to a two-step process:[8,10] (1) a fairly high temperature post-deposition anneal (T_a = 600 C) followed by (2) broad-band UV-illumination. (Before UV-illumination of the annealed samples, no EPR signals were detected.) Since essentially 100% of nitrogen nuclei (^{14}N) have a nuclear spin (*I*) of one, the N dangling bond is easily identified by its three-line EPR spectrum. It should be stressed that following this anneal/UV sequence, the N dangling bond is the dominant EPR-active defect center; the Si dangling bond is no longer detected in these films.

To confirm that this defect center is due to a hyperfine interaction with a nucleus with a nuclear spin of unity, Q-band (34 GHz) EPR measurements were performed; the results are

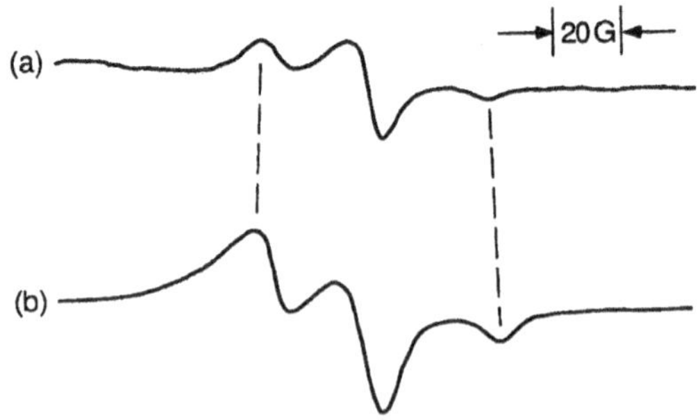

Fig. 1. EPR spectra of the N dangling bond at (a) X-band, and (b) Q-band microwave frequencies.

illustrated in Fig. 1(b). As shown in Fig. 1, the hyperfine splitting between the transitions corresponding to the nuclear spin quantum numbers of −1, 0, and 1 is independent of the microwave excitation frequency. This shows that this spectrum is due to an unpaired electron on a nucleus with $I = 1$, for which ^{14}N is the only reasonable possibility. The relatively small differences in line shape between the X-band and Q-band EPR spectra are due to g-tensor anisotropy.

Previous analysis of this line shape indicated that the unpaired electron is in a pure $2p$ orbital on a N nucleus,[8] in good agreement with that first predicted by Robertson.[5] Nitrogen radicals with pure $2p$ character always appear to be two-coordinated;[11,12] that is, the dangling N bond is bonded to two other atoms. In the absence of other hyperfine interactions from nearest neighbors, it is reasonable to ascribe this defect center to the following configuration:

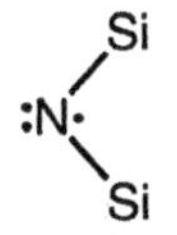

where • denotes the unpaired electron, and **:** denotes the lone pair electrons.

In light of the chemical identity of this paramagnetic defect, it is of interest to explore some of its associated properties. As mentioned earlier, the generation of N dangling bonds requires a two-step process. We now consider the second step (involving optical illumination). Fig. 2 shows the normalized N dangling-bond concentration vs. photon energy for a-$SiN_{1.6}$:H thin films given a 600 C post-deposition anneal. (The a-$SiN_{1.6}$:H films in this particular experiment were deposited on fused quartz to rule out the possibility that the UV-light is photo-injecting charge carriers from the underlying substrate.) The normalization was done with respect to the broad-band (unfiltered) illumination level. The N dangling bonds show a strong dependence upon photon energy. A minimum photon energy dependence to created N dangling bonds closely parallels the photon energy dependence of the optical absorption coefficient in these films.[13] Therefore, it appears that electron/hole pair creation is required to induce this paramagnetic center; i.e., the capture of electron/hole pairs by charged states and/or the capture of these photo-excited carriers by these states creates the paramagnetic N dangling bonds. The large photon energy required to create the N dangling bond is also consistent with the energy level of the N dangling bond being near the valence band edge, which has been both theoretically[5] predicted and experimentally verified.

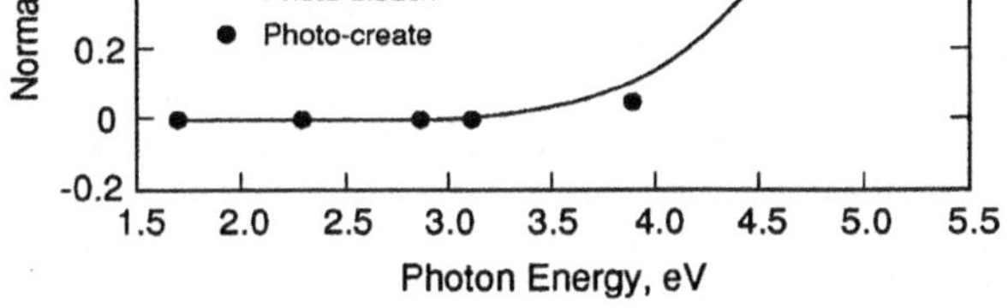

Fig. 2. Room-temperature dependence for photo-creation and photo-bleaching of the N dangling bond. The densities represent the saturated values. The dashed line represents the saturated broad-band density, which is the normalization value.

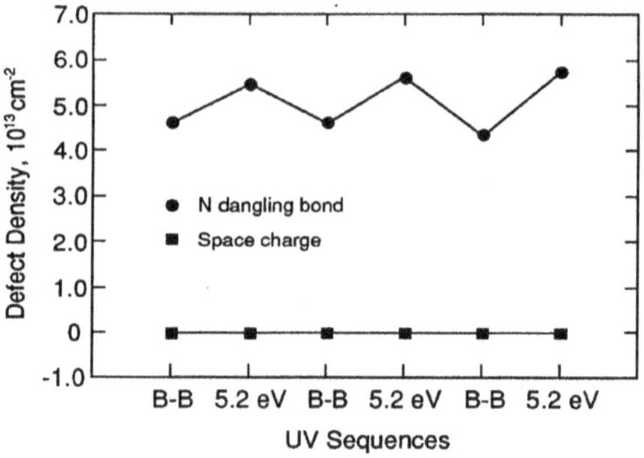

Fig. 3. Room-temperature N dangling-bond and space-charge densities in the nitride films subjected to various UV-illumination sequences with broad-band (B-B) and monochromatic (5.2 eV) light. Illumination time was constant at 30 m.

Fig. 2 also shows another phenomenon; 5.2 eV monochromatic light creates more spins than broad-band (unfiltered) UV illumination. Since monochromatic light is achieved by filtering the broad-band UV, there must be components in the broad-band light that reduce, or photo-bleach, the N dangling bond. To investigate this photo-bleaching mechanism further, the films were first illuminated with broad-band light to saturation level, followed by re-illumination with monochromatic light of various wavelengths to photo-bleach the paramagnetic N centers. The results are illustrated in Fig. 2. The photo-bleaching effect occurs over a wide range of photon energies, but the greatest extent of bleaching occurs around 2.85 eV; at this photon energy about 35% of the N dangling bonds are photo-bleached during 30 m illumination at room temperature.

Fig. 3 shows that the N dangling bond can be repeatedly photo-created and photo-bleached by different illumination conditions (5.2 eV and broad-band UV light, respectively). The illuminations were performed until the effects were saturated (about 30 m for each sequence). The photo-bleached N dangling bonds can be restored to their pre-bleached density by re-illuminating with 5.2 eV light. It appears from this result that the photo-bleached N dangling bonds are not permanently annihilated; rather, the host centers simply return to their original diamagnetic state.

To elucidate the above mechanism, the net space charge in the a-$SiN_{1.6}$:H films was monitored during the photo-creation/photo-bleaching sequences; the data are also shown in Fig. 3. The space charge density of the a-$SiN_{1.6}$:H films was monitored by the mid-gap voltage shift in the *CV* measurements. Fig. 3 shows that the net space charge in the films does not change even though the N dangling-bond concentration changes by 1×10^{13} cm^{-2} in the aforementioned photo-creation/photo-bleaching sequences. (The mid-gap voltage remained essentially fixed at −3.2V during these illumination sequences.) The net space charge density in the a-$SiN_{1.6}$:H thin films is independent of the N dangling-bond concentration. Since previous works have shown that the *paramagnetic* N dangling bond is electrically neutral, the photo-bleached, *diamagnetic* N sites should have an associated charge state. Yet, the *CV* measurements indicate that there is no change in the *net* space charge during photo-bleaching; therefore, charge compensation must occur. Based on this, paramagnetic N dangling bonds become compensated charged diamagnetic sites through an optical rearrangement of charge, thereby explaining the decrease in the spin density without any change in the net space charge density. One possibility is that the charged diamagnetic sites are simply positively or negatively charged N sites.

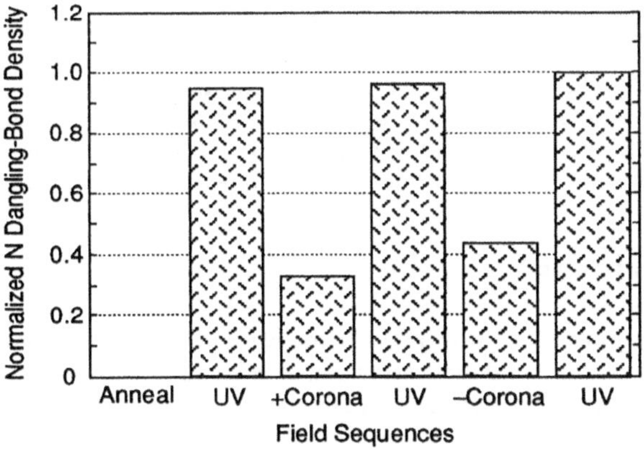

Fig. 4. Room temperature electric-field and UV-response of the N dangling bond.

At this juncture, it is interesting to consider whether the N dangling bond is an electrically active point defect; that is, can it capture and electron or a hole. Fig. 4 shows the effect of charge injection (electric field stimulation) on the N dangling-bond concentration. Again, to create the N dangling bonds, the a-$SiN_{1.6}$:H films were given a post-deposition anneal (T_a = 700C for 10 m) and then UV-illuminated with broad-band UV light. If the a-$SiN_{1.6}$:H films are subsequently biased positively with positive corona ions, the density of N dangling bonds is substantially reduced. It is reasonable to assume that the neutral paramagnetic N dangling bond has captured (or emitted) a charge carrier, becoming diamagnetic. Re-illumination restores the initial spin density; evidently, optical rearrangement of spin and charge occurs to recreated the neutral paramagnetic centers. Next, subjecting the films to a negative electric field reduces the UV-induced N dangling-bond density. It appears that the paramagnetic neutral center has captured (or emitted) a charge carrier, becoming charged and diamagnetic. Last, re-illumination again restores the initial N dangling-bond density. Unfortunately, the associated *CV* measurements were not reliable; hence, the net space charge could not be mapped out during the corona charge experiments. Therefore, we cannot determine if there is a one-to-one correlation between the changes in the spin and space charge densities during the charge injection experiments.

Although these results could be interpreted in a number of ways, the fundamental point is that the paramagnetic N dangling bond is electrically active; it readily responds to an electric field (charge injection). Also, since the process is reversible, this experiment is consistent with the aforementioned mechanism for photo-bleaching, and may support the notion that the N dangling bonds are created, following the post deposition anneal, by an optically induced rearrangement of charges in existing charged diamagnetic defects.

Last, it is interesting to note that many features reported here for the N dangling bond in a-$SiN_{1.6}$:H films given a post-deposition anneal are identical to that observed for the Si dangling bond in as-deposited films (films that were not given a post-deposition anneal). For instance, the Si dangling bond can be photo-created[3,4,7,13] and photo-bleached;[4,14] it is electrically neutral when paramagnetic;[3,7,13] and it is also an electrically active point defect.[3,7]

CONCLUSIONS

In summary, we have confirmed that the N dangling bond is two-coordinated in a-$SiN_{1.6}$:H thin films. It has been shown that the photo-creation of the N dangling bonds is strongly dependent on the photon energy, and that these centers can also be photo-bleached. By monitoring the EPR and *CV* response of the films subjected to optical illumination, it has been suggested that photo-bleaching occurs by an optically induced rearrangement of charges between positive, negative, and neutral N sites. Last, it has been demonstrated that the N dangling bond is an electrically active point defect in a-$SiN_{1.6}$:H thin films.

ACKNOWLEDGMENTS

The authors are grateful to J. Robertson for many useful conversations regarding this work. We thank B. Poindexter for word processing and page layout. This work was supported by the U. S. Department of Energy under contract DE-AC04-76-DP0078.

REFERENCES

1. J. Robertson and M. J. Powell, Gap states in silicon nitride, *Appl. Phys. Lett.* 44:415 (1984).
2. S. Fujita and A. Sasaki, Dangling bonds in memory-quality silicon nitride films, *J. Electrochem. Soc.* 132:398 (1985).
3. D. T. Krick, P. M. Lenahan, and J. Kanicki, Nature of the dominant deep trap in amorphous silicon nitride, *Phys. Rev. B* 38:8226 (1988).
4. M. S. Crowder, E. D. Tober, and J. Kanicki, Photobleaching of light-induced paramagnetic defects in amorphous silicon nitride films, *Appl. Phys. Lett.* 57:1995 (1990).
5. J. Robertson, Electronic structure of silicon nitride, *Phil. Mag. B* 63:47 (1991).
6. W. L. Warren, J. Kanicki, J. Robertson, and P. M. Lenahan, Energy level of the nitrogen dangling bond in amorphous silicon nitride, *Appl. Phys. Lett.* 59:1699 (1991).
7. W. L. Warren, J. Kanicki, F. C. Rong, and E. H. Poindexter, Paramagnetic point defects in amorphous silicon dioxide and silicon nitride thin films II. a-SiN_x:H, *J. Electrochem. Soc.* 139:880 (1992).
8. W. L. Warren, P. M. Lenahan, and S. E. Curry, First observation of paramagnetic nitrogen dangling-bond centers in silicon nitride, *Phys. Rev. Lett.* 65:207 (1990).
9. J. Kanicki, Role of hydrogen in silicon nitride films prepared by various deposition techniques, in "Amorphous Silicon Technologies," A. Madan, M. J. Thompson, P. C. Taylor, P. G. LeComber, and Y. Hamakawa, eds., Materials Research Society, Pittsburgh (1988).
10. W. L. Warren, P. M. Lenahan, and J. Kanicki, Electrically neutral nitrogen dangling-bond defects in amorphous hydrogenated silicon nitride thin films, *J. Appl. Phys.* 70:2220 (1991).
11. P. W. Atkins and M. C. R. Symons, "The Structure of Inorganic Radicals," Elsevier, Amsterdam (1967).
12. T. E. Tsai, D. L. Griscom, and E. J. Friebele, Characterization of a nitrogen center in high-purity SiO_2:OH glass, *Phys. Rev. B* 38:2140 (1988).
13. J. Kanicki, M. Sankaran, A. Gelatos, M. S. Crowder, and E. D. Tober, Stretched exponential illumination time dependence of positive charge and spin generation in amorphous silicon nitride, *Appl. Phys. Lett.* 57:698 (1990).
14. C. H. Seager and J. Kanicki, Photodarkening and bleaching in amorphous silicon nitride, *Appl. Phys. Lett.* 57:1378 (1990).

VIII. RADIATION AND HYDROGEN INDUCED EFFECTS IN SILICON-SILICON DIOXIDE SYSTEMS

INTRODUCTION

M. M. Heyns
Interuniversity Microelectronics Centre (IMEC)
Kapeldreef 75, B-3001, Seuven, Belgium

D. J. DiMaria
IBM Research Division, T.J. Watson Research Center
Yorktown Heights, NY 10598

In this chapter, the contributing authors treat the question of defect creation by the deposition of energy into the silicon dioxide layer. This energy is supplied by means of high-energy radiation/particles or through large-applied electric fields. For either means of energy deposition, two distinct classes of defects can be produced. One type is believed to be due to the presence of holes and/or their subsequent annihilation by electrons. The other is thought to be related to the release of a hydrogenic species, its motion to, and its subsequent reaction at an interfacial region.

The first section of the chapter treats the high field experiments. In these studies, the hot electron energetics in the oxide are shown to be now well understood. At high fields, both modes of defect generation are demonstrated to be present. The high energy tail of the hot-electron distribution is shown to produce holes on thick oxide layers at high fields, mostly through impact ionization. Hydrogen release by hot carriers with energy exceeding 2 eV is directly shown in one article. The complicated behavior of destructive breakdown of silicon dioxide is demonstrated in another article to be correlated with the net sum of both defect production modes.

For the radiation-exposed-oxide studies, there still remains a controversy over which mode dominates. Several of the authors in this chapter present results supporting the hydrogen-release mechanism for defect production. Specific impurity and bonding configurations, which can account for many features of the experimental data, are described and modeled in several articles. Finally, the concept of a positively-charged "cracking site" produced by the radiation is discussed and shown to be consistent with these hydrogenic models.

IMPACT IONIZATION AND DEGRADATION IN SILICON DIOXIDE FILMS ON SILICON

D.J. DiMaria,
D. Arnold and
E. Cartier

IBM Research Division
T.J.Watson Research Center
Yorktown Heights, NY 10598

ABSTRACT

Degradation of silicon-dioxide films is shown to occur primarily near interfaces with contacting metals or semiconductors. This deterioration is shown to be caused by two mechanisms triggered by electron heating in the oxide conduction band. These mechanisms are trap creation and bandgap ionization by carriers with energies exceeding 2 and 9 eV, respectively. In this study, both are discussed with particular emphasis on the bandgap ionization process which has been a controversial issue for many years. A procedure by which each can be separately studied is demonstrated.

INTRODUCTION

For many years, there has been a controversy concerning the degradation and ultimate destruction of silicon dioxide (SiO_2) films in electronic devices. In this study, it will be shown that degradation is caused in the interfacial regions by two separate mechanisms related to the energetics of conduction-band electrons in the oxide. The first, trap creation, has been treated previously[1-10] and will be reviewed. The second, which is related to impact ionization,[11-16] will be discussed here and a procedure will be developed to separate it from trap creation. Finally, the processing dependence of each will be discussed and related to breakdown of the oxide films from the continual build-up of defect sites near the interfaces.

RESULTS

Degradation in silicon dioxide films (SiO_2) at electric fields lower than 7 MV/cm where carriers must be injected over the interfacial energy barriers is now widely accepted as primarily due to the phenomenon of "trap creation".[1-10] This occurs when any electron entering an interfacial region has an energy in excess of 2 eV. An electron can enter from a contact with this energy (for example, from an avalanching Si junction or a hot carrier in the inversion layer of the channel of a device), or it can gain its energy in the oxide from the voltage dropped across this layer. For the latter case, the hot electron distribution in the oxide conduction band is determined by the known transport properties of SiO_2 where energy gained from the applied field is dissipated by both polar and non-polar phonon scattering.[17-21] Electron heating in oxide films thicker than 10.0 nm starts at fields of $\simeq$ 1.5 MV/cm, but the average energy of the distribution does not exceed 2 eV until fields of 3-4 MV/cm are reached. When these hot carriers enter the opposite interfacial region near the anode, a few can also lose energy by releasing mobile species from oxide defect sites near this interface. These species are believed to be a form of hydrogen (H or H^+) which then diffuses through the oxide bulk to create electron traps and interface-states, mostly near the SiO_2/cathode interface.[3-9] Some of these interfacial sites can be donor-like and positively-charged when empty. At lower electric fields ($\lesssim$ 1.5 MV/cm), bulk electron trapping in sites incorporated into the oxide film during fabrication (related to OH and H_2O) can also be observed and studied separately.[3,10] This phenomenon can also produce interface-states under certain processing conditions. However as oxide layers have been made thinner to accommodate device scaling over the past 15 years, bulk trapping in as-fabricated films has because less important. These phenomena are schematically summarized in Fig. 1a. With current polycrystalline silicon (poly-Si) gate technologies, these modes of oxide deterioration are not readily apparent until > 0.001 Coul/cm^2 of hot electrons has traversed the film.

At fields above 7 MV/cm, the degradation of SiO_2 films thicker than 10.0 nm is more complicated. Another degradation mode must also be operational since positive charge formation and interface-state generation are observed at injected-charge-fluences much less than 0.001 Coul/cm^2 (references 11-16). This phenomenon has recently been shown

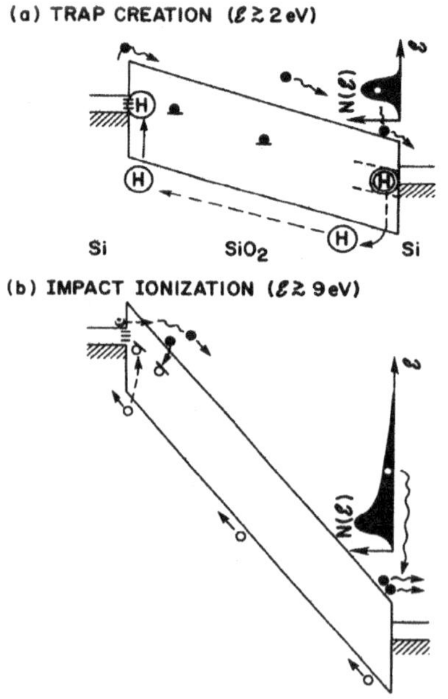

Fig. 1. Schematic energy-band diagram showing (a) trap creation near the cathode caused by mobile hydrogen release from near the anode and (b) defect generation near cathode caused by free-electron/trapped-hole recombination where holes were generated in the oxide bulk by impact ionization.

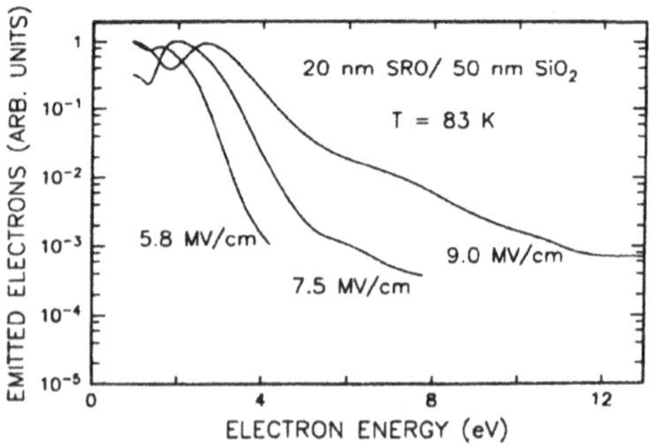

Fig. 2. Hot-electron energy-distributions (determined from vacuum emission experiments) as a function of kinetic energy with respect to the bottom of the SiO_2 conduction band for anode fields from 5.8 to 9.0 MV/cm. For this oxide thickness ($\approx$ 50.0 nm), high energy tails due to acoustic phonon runaway are seen above 7 MV/cm and increase rapidly with field.

to be triggered by impact ionization when any hot electron gains an energy that exceeds the oxide bandgap of 9 eV.[22,23] The bandgap ionization process generates electron-hole pairs. The holes move dispersively towards the cathode-oxide interface where most exit out through the cathodic contact.[24,28] A few of these positively-charged carriers can be trapped in energetically "deep" sites in the oxide forbidden gap near this interface.[29,30] Subsequently, some of these fixed positive charges can be annihilated by the injected electrons.[30-35] The energy release and/or lattice relaxation that occurs during this annihilation produces interface states.[34,35] This phenomenon is schematically depicted in Fig. 1b. Although deterioration near the anode can simultaneously occur for both trap creation and the impact ionization process, only the oxide deterioration near the cathode (Si-substrate for our studies) will be treated here.

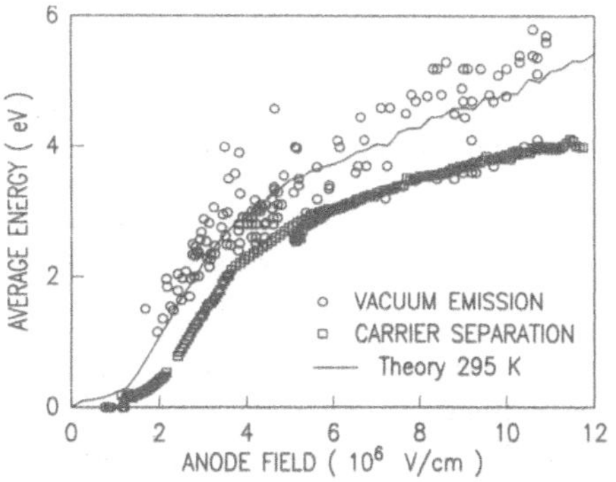

Fig. 3. Average energy of the hot electron distribution in SiO_2 as a function of the anode field from experimental data (vacuum emission and carrier separation) and theory (Monte-Carlo simulation).

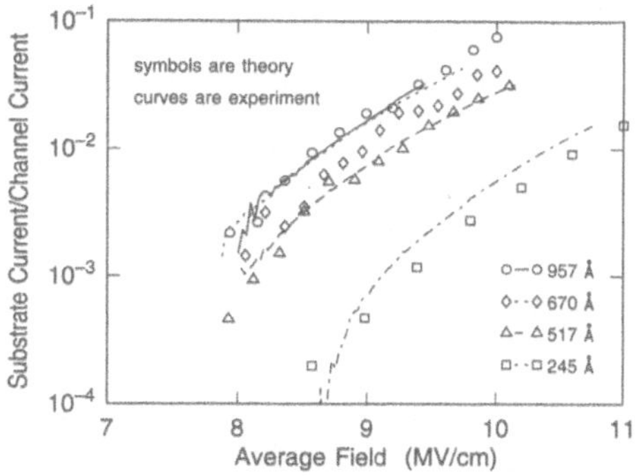

Fig. 4. The ratio of the number of holes collected by the Si substrate to the number of injected electrons from the channel of n-FETs as a function of the average oxide field for different oxide thickness. The curves are data from carrier separation experiments, and the symbols are from the new modified Monte-Carlo simulations described in reference 22.

Knowledge of the energy distributions of hot electrons in SiO_2 is necessary for the understanding of the electric field ranges where one or both of these phenomena are occurring. Figures 2 and 3 show measured distributions and average energies of these, respectively. Only the vacuum emission technique can be used to measure the distributions.[18] Two other techniques (carrier separation and electroluminescense) were additionally used to measure the average energies.[17-19] Previously, trap creation has been shown to track the average carrier energy.[1-8] No impact ionization would be predicted from these data in Fig. 3, since average energies of over 9 eV are not observed over this field range. However, the actual distributions in Fig. 2 do show the rapid development of a high energy tail at fields exceeding 7 MV/cm for films of 50.0 nm or thicker.[18] This tail requires 20-30 nm to reach an approximately steady-state condition as compared to about 2-3 nm for the main portion of the distribution. For oxides thinner than 30-40 nm, higher fields are required to observe the same number of carriers in the high energy tail, relative to the main portion of the distribution. For example with a 10.0 nm thick film, the average field would be expected to exceed 12 MV/cm before the tail would appear and bandgap ionization could occur.

Although previous theoretical Monte-Carlo calculations could predict the average electron energy[19] (as shown in Fig. 3), only recently has the existence of the high energy tails been theoretically confirmed.[22,23] Of importance to these new calculations, were the zero-field soft-X-ray-induced transmission-experiments of Cartier, McFeely, and co-workers.[20,21] These experiments were used to determine absolute scattering rates for hot electrons in SiO_2 with energies exceeding 8 eV. At these high energies, phonon scattering was found to be much weaker than initially predicted by first order perturbation theory. It has been shown by means of Monte Carlo transport simulations that the observed absolute values of the scattering rates lead to electron runaway from the acoustic phonons and to significant impact ionization at high fields.[22] (Recently other groups obtained similar results from Monte-Carlo calculations using input for the ionization rates based on the exact dielectric function of SiO_2 as described in reference 23.) In the modified model, the acoustic phonon scattering rate peaks at an energy of 5-6 eV and then decreases, rather than increasing indefinitely with electron energy raised to the 3/2 power as originally proposed by Fischetti.[19] These new simulations can now accurately predict the measured high energy tails of the hot electron distribution observed at high fields.[22,23] Consistent with the experimental results, it has been shown theoretically that the acoustic-phonon-runaway takes several tens of nanometers to develop. Therefore, bandgap ionization will begin to occur at $\approx$ 7 MV/cm only with oxide films of thickness (l_{ox}) $\gtrsim$ 30-40 nm, with this threshold field increasing rapidly on thinner oxides.

From this understanding of the high energy tails, other phenomena thought to be caused by band-gap ionization such as trapped-hole build-up and subsequent interface state generation can be addressed. Determination of the impact ionization probabilities over the field and oxide thickness ranges of interest are crucial. A measure of the hole current flow from the SiO_2 layer into the cathode would provide a direct measure of the ionization probabilities. These data can be obtained from carrier separation experiments on n-channel field-effect-transistors (n-FETs).[24-28] This is shown in Fig. 4 where the ratio of hole currents observed in the p-Si substrate (cathode) to the injected electron current (from the inversion layer) of n-FETs at high fields are plotted. Also, the recent Monte-Carlo simulations based on the soft-x-ray-induced transmission experiments are compared and shown to give quantitative agreement with this device data. The threshold field for the ionization process and oxide thickness dependence of these data clearly image the behavior expected due to the high energy tails previously discussed.

Using the ionization probabilities in Fig. 4, the quantities of interest in device degradation (trapped positive charge, trapped negative charge, and interface-state generation) can now be predicted and compared to the observed values. Figure 5 shows measured and predicted positive-charge build-up in the oxide layer. The experimental data shown for the charge build-up were obtained using capacitance-voltage (C-V) characterization[3] on the same device structures used for the ionization data in Fig. 4. Most of these positively-charged oxide-sites are due to the interface trapping of a small fraction of the transported holes generated by bulk-oxide impact-ionization as verified by the simulations shown. The calculations plotted in Fig. 5 were obtained from the solution of the first order rate equation where both hole trapping and annihilation by injected electrons are taken into account with no adjustable parameters. This equation which describes the positive charge build-up is

$$qdp/dt = \sigma_p \alpha J_e(N_p - p) - \sigma_e J_e p$$

where q is the magnitude of charge on an electron, J_e is the electron current density, p is the number of trapped holes per unit volume, N_p is the total number of hole traps per unit volume in the as-fabricated oxide layer, σ_p is the hole capture cross section for these energetically "deep" interfacial oxide sites, σ_e is the free-electron/trapped-hole annihilation-cross-section, and α is the bandgap ionization probability obtained from Fig. 4. The hole capture and annihilation cross-sections were taken from values in the literature which were checked independently on our devices. This value of the "effective" hole capture cross-section, which is not strongly field-dependent, is 1×10^{-14} cm^2 (reference 29). The annihilation cross-section, which is strongly field-dependent, has values near 1×10^{-16} cm^2 (references 30 and 31). The magnitude of total number of hole

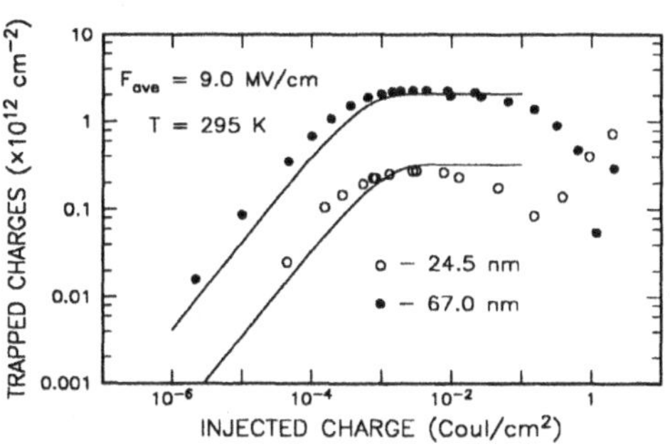

Fig. 5. Number of trapped charges as a function of injected-electron-fluence at an average field of 9 MV/cm for two different oxide thickness on devices similar to those used to obtain the data in Fig. 4. Solid lines are from the solution of the first order rate equation described in the text which uses the ionization rates from Fig. 4 as input. The decrease from the steady-state positive charge level and the eventual change in the sign of the net trapped charge from positive to negative (indicated by the inflection in the data between 0.1 and 1.0 Coul/cm^2) is due to electron trap creation and charging.

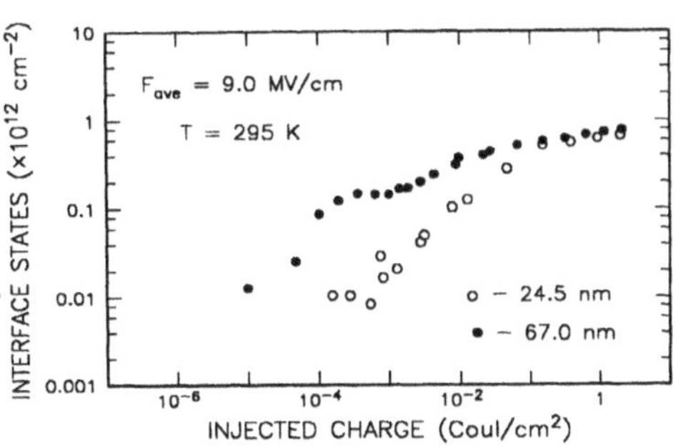

Fig. 6. Number of interface states generated due to hole trapping, recombination with injected electrons, and trap creation. These data where obtained simultaneously with the trapped charge data in Fig. 5 and show the different defect-generation regimes as a function of injected-charge-fluence.

traps per unit area ($N_p l_{ox} \approx 4\times10^{12}$ cm^{-2}) was taken from values in the literature for similarly processed structures where holes were injected into the oxide from the Si substrate.[32] This value was confirmed on the devices used here from the field dependence of the steady-state level of positive charging. Centroid locating techniques were used to show that most of this positive charge is trapped close to the SiO_2/Si-substrate interface.[33]

Figure 6 shows data (from C-V characterization[3]) for the resulting interface-states produced by the on-going electron-hole annihilation process. These data initially track the positive charge build-up as expected with a conversion efficiency of about one generated interface state per ten annihilated holes. This interface-state creation mode has been previously discussed in the literature,[1,32,34,35] and the rates determined here are consistent with those expected for high-temperature poly-Si-gate processing conditions. However, differences in the number of these sites due to variations in the conversion efficiency are more sensitive to processing changes than either the ionization rates or the number of trapped positive charges. Therefore, the generation rates for interface-state production are more difficult to quantitatively predict. The deviations in the data from the calculated predictions at injected-electron levels larger than 0.001 Coul/cm^2 for both the number of trapped holes (Fig. 5) and the number of generated interface-states (Fig. 6) are due to the onset of trap creation. Separating these two phenomena and relating the continuous build-up of charges and interface-states to destructive breakdown will be discussed later.

Experiments (such as those shown in Fig. 5) measuring the trapped hole build-up can also be used to study the dependency of the impact-ionization probability on electric field, oxide thickness, temperature, etc.. Figures 7 and 8 show the variations in the amount of positive charge observed when the electric field or the oxide thickness, respectively, is varied. These data reflect the dependency of the ionization rates in Fig. 4 on the same variables and are predictable from the analysis previously discussed. In Fig. 8, data for the 9.3 nm thick oxide devices was obtained at 11 MV/cm instead of 9 MV/cm as done for the thicker oxide structures. These data are shown because the trapped-positive-charge build-up at 9 MV/cm for these devices was below our detectability limit of about 1×10^{10} cm^{-2}. However, trap creation effects were still observed above 0.001 Coul/cm^2.

Figure 9 shows a processing dependence study of the positive charge build-up for device structures whose oxide layers were subjected to various high-temperature gaseous-annealing-treatments. The devices that had the reoxidation-nitridation processing (designated by NH_3/O_2 and performed with two sequential anneals using these gases) showed no net positive charge build-up, even though the ionization rate for hole generation in these devices was larger than that in the unannealed control-structure. The negative-charge build-up observed is due to trapping on nitrogen-related-sites in the as-fabricated film. The lack of measurable hole trapping is consistent with previous studies where holes were introduced into the oxide layer by radiation or injection from the contacts,[36,37] rather than generated internally using gap ionization. These two experimental observations for the magnitude of hole trapping and ionization rate lead to the conclusion that the number of "deep" interface-hole-traps has been significantly

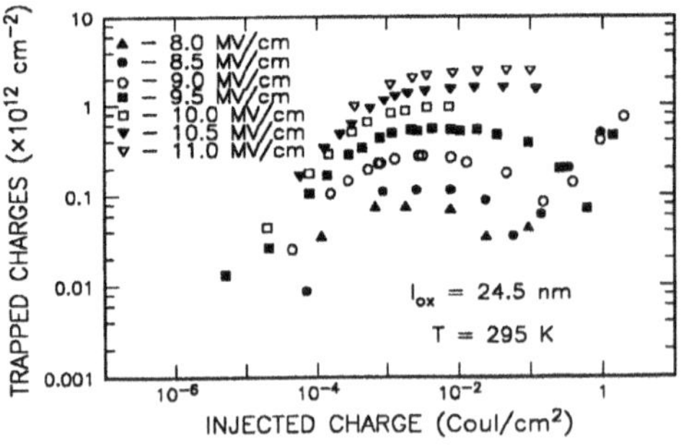

Fig. 7. Number of trapped charges as a function of injected-electron-fluence on devices with an oxide thickness of 24.5 nm for various average electric fields from 8.0 to 11.0 MV/cm.

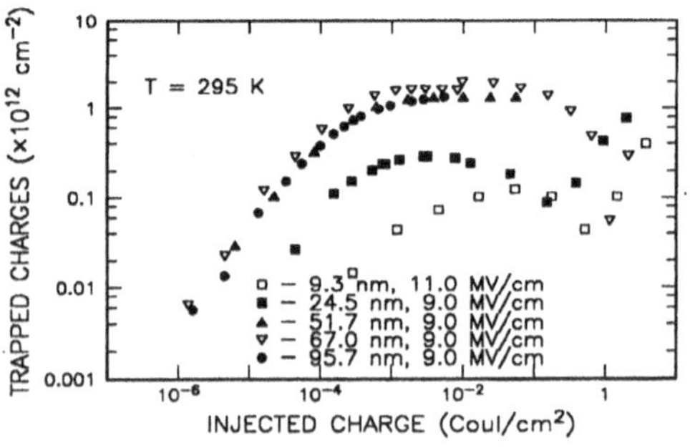

Fig. 8. Number of trapped charges as a function of injected-electron-fluence at 9.0 MV/cm on devices with oxide layers of varying thickness from 9.3 to 95.7 nm. For the 9.3 nm oxide devices, data shown is for 11 MV/cm since no measurable positive charges were observed at 9.0 MV/cm.

reduced by the NH_3/O_2 annealing step. The devices which had the high temperature forming gas anneal (indicated by N_2-H_2) are known to have significantly enhanced defect-generation-rates due to trap creation.[5,38] However, these annealing treatments in forming gas or nitrogen do not change the hole generation rates, but increase the number of interfacial hole traps as shown.

Deviations from the steady-state positive-charge levels in Figs. 5, 7, and 8 above 0.001 Coul/cm² are due to the onset of electron-trap-creation caused by hot electrons with energies exceeding 2 eV.[2-4] The subsequent charge trapping in these sites, which are distributed away from the cathode/SiO_2 interface, increases until destructive breakdown of the film occurs. (Areal negative-charge-densities of $> 1\times10^{13}$ cm^{-2} can be obtained.) Also, background trapping in bulk-oxide-sites of the as-fabricated film can contribute (particularly at very low fields) to the total negative charge build-up. Occupancy of these

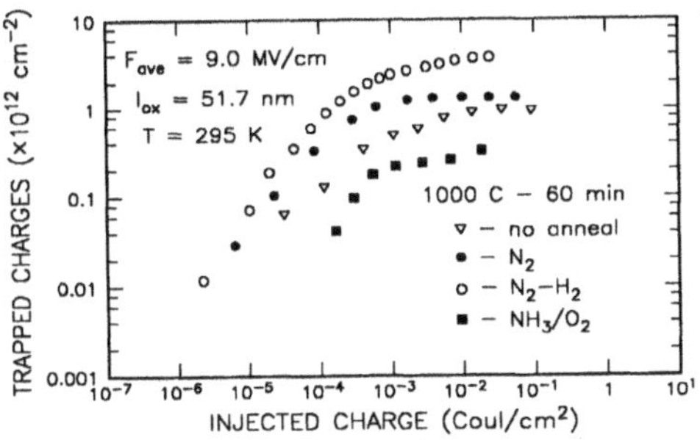

Fig. 9. Number of trapped charges as a function of injected-electron-fluence at 9.0 MV/cm on devices were the oxide has been annealed at 1000 °C for 60 min. in various gaseous ambients, including nitrogen, forming gas, and ammonia (followed by oxygen). The charge state for this latter case (NH_3/O_2) is negative (only trapped electrons), while it is positive for other three cases shown.

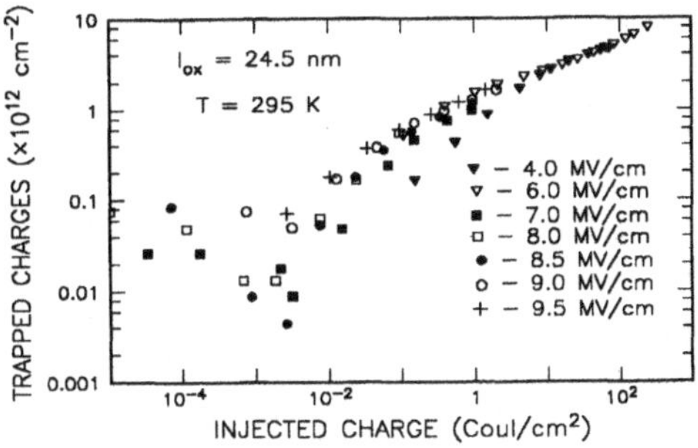

Fig. 10. Number of trapped electrons (from the trap creation process and background trapping in as-fabricated sites) as a function of injected-electron-fluence for various average fields from 4.0 to 9.5 MV/cm. For fields $\geq$ 7.0 MV/cm, the steady-state positive-charge-level was subtracted from the net-trapped-charge observed.

types of electron traps can drop to $<$ 60 % at the fields necessary for bandgap ionization (7 MV/cm).[8] The negative charge in these traps adds to the positive charge found in the hole traps by the superposition of the two separate distributions. At higher injected electron fluences, this electron trap creation and trapping gives a net negative-charge-state. This negative charging is usually reported for most experiments where high-field Fowler-Nordheim tunneling is used for oxide characterization.[39,49] In Fig. 10, the steady-state positive-charge-component has been subtracted from data similar to that in Fig. 7 for 24.5 nm thick oxide structures under Fowler-Nordheim tunnel injection at fields $\geq$ 7.0 MV/cm. The resulting amount of trapped-negative-charge from this procedure is compared to trapped-electron areal-densities at lower fields (4 and 6 MV/cm injection from the substrate over the Si-SiO_2 energy barrier) where net positive charging is not observed. These data show the onset of trap creation near 0.001 Coul/cm². The weak field-dependence observed is expected from the slowly increasing average-energies of the electron distribution above 4 MV/cm (see Fig. 2) and the decreasing trap occupancy at higher fields.

Besides their dependence on injected charge level, the two defect generation regimes caused by hot electrons can be separated from each other by varying temperature, oxide thickness, and field. Trapped electrons and interface-state generation due to trap creation both show a strong temperature dependence (decreasing with decreasing temperature), but a weak oxide thickness dependence (for films with oxide thickness between 10.0 and 100.0 nm).[2-4] However, trapped holes and interface-state creation due to electron-hole recombination show the opposite trend: a weak temperature dependence, but a strong dependence on oxide thickness over the same range as shown in Figs. 5 and 8. This is mainly due to the behavior of the generation rate of the holes from the bandgap impact-ionization-process. Trap creation starts at fields greater than 1.5 MV/cm for oxides thicker than 10.0 nm while holes produced by bandgap ionization do not appear until fields of over 7 MV/cm are attained (see Fig. 7).

On thinner oxide films (10 - 20 nm), significant numbers of trapped holes are only observed at the highest fields attainable (> 11 MV/cm for the 9.3 nm film shown in Fig. 8). This is due to the high energy tails (therefore, bandgap ionization) being suppressed. On very-thin-oxide films (5.0-10.0 nm), impact-ionization-induced defects are minimal, but trap creation by electrons with energy exceeding 2 eV still occurs.[2-4] Below 5.0 nm, neither mode of defect creation due to the hot electrons (either from the main portion of the distribution or from the high energy tail) is expected for similar fluences and fields. This has been recently demonstrated on $\approx$ 2.5 nm thick oxide films where electron fluences of > 1×10^4 Coul/cm^2 can be passed through the film without significant charging, defect creation, or destructive breakdown.[41,42]

DISCUSSION

From the studies reported here and the work of others, the unresolved question of destructive breakdown in SiO_2 can be addressed once again. Firstly, what is considered "extrinsic" as opposed to "intrinsic" must be considered. Here, extrinsic breakdown refers to destruction of the film triggered by grossly non-uniform electron injection due to surface asperities, metal-decorated stacking faults at Si-oxide interfaces, non-uniform doping along grain boundaries at poly-crystalline Si interfaces with SiO_2, gross inclusions incorporated during processing into the oxide film or at the oxide-contact interfaces, local non-uniformities in the oxide thickness, etc. Extrinsic breakdown events which are triggered at lower average fields can be reduced with careful processing. Purely intrinsic breakdown, as considered here, is due to an initially uniform process triggered by the electronic nature of the pure SiO_2 material. The electron heating and gap-ionization phenomena, which are intrinsic, should be key elements in this breakdown mode. However, the defect sites initially present in the as-fabricated material, such as the hole-trapping and decorated-hydrogenic sites which participate in the oxide degradation process (coupled with the hot carriers), are not "intrinsic" and can be reduced somewhat by processing. Although still not purely "intrinsic", it will be argued here that this latter mode currently determines the ultimate breakdown strength of the oxide. If the breakdown-strength of SiO_2 is limited by the effective "softening" of the interfacial regions through defect generation and charge trapping caused by hot electrons, then destructive breakdown studies should image the three oxide-thickness regimes discussed previously. Recent data of charge-to-breakdown as a function of electric field for capacitor structures above 9.0 nm in thickness show these trends.[43] From 9.0 to 13 nm, the charge-to-breakdown magnitude is relatively insensitive to oxide thickness and slowly decreasing with increasing field. This type of behavior would be expected if only trap creation was participating in the interface deterioration (see Figs. 3 and 10). The added defect generation due to impact ionization lowers the charge-to-breakdown with increasing field for oxides thicker than 13 nm with a transition region similar to that observed in Figs. 4 and 8 as the film thickness approaches 50 nm. Increasing the lattice temperature would be expected to increase defects caused by trap creation only, and lower the charge to breakdown and sensitivity to oxide thickness.

CONCLUSIONS

In this study, not only has the issue of positive charge generation at low fluences and high fields been resolved to be caused by impact ionization, but also a framework for analyzing SiO_2 deterioration for any fabrication condition has been provided. Since trap creation, hole trapping, and interface-state creation can each depend differently on processing, this framework can be crucial in process optimization. This is of particular value when standard high-field current-injection (Fowler-Nordheim) techniques are combined with C-V characterization to analyze oxide quality and reliability.

ACKNOWLEDGEMENTS

The authors wish to acknowledge the device preparation by the silicon processing facility at the IBM T. J. Watson Research Center and the critical reading of this manuscript by T.N. Theis, D.A. Buchanan, M.V. Fischetti, and J. Stathis.

REFERENCES

1. D.J. DiMaria, in *Insulating Films on Semiconductors 1991*, edited by W. Eccleston and M. Uren (Adam Hilger, New York, 1991), pp.65-72.
2. D.J. DiMaria, Appl. Phys. Lett. ***51***, 655 (1987).
3. D.J. DiMaria and J.W. Stasiak, J. Appl. Phys. ***65***, 2342 (1989).
4. D.J. DiMaria, J. Appl. Phys. ***68***, 5234 (1990).
5. D.J. DiMaria and J.H. Stathis, J. Appl. Phys. ***70***, 1500 (1991).
6. E. Harari, Appl. Phys. Lett. ***30***, 601 (1977).
7. M.M. Heyns, D. Krishna Rao, and R.F. DeKeersmaecker, Appl. Surf. Sci. ***39***, 327 (1989).
8. A.v. Schwerin and M.M. Heyns, in *Insulating Films on Semiconductors 1991*, edited by W. Eccleston and M. Uren (Adam Hilger, New York, 1991), pp. 263-266.
9. R. Gale, F.J. Feigl, C.W. Magee, and D.R. Young, J. Appl. Phys. ***54***, 6938 (1983).
10. F.J. Feigl, D.R. Young, D.J. DiMaria, S. Lai, and J. Calise, J. Appl. Phys. ***52***, 5665 (1981).
11. P. Solomon and N. Klein, Solid State Commun. ***17***, 1397 (1975).
12. T.H. DiStefano and M. Shatzkes, J. Vac. Sci. Technol. ***13***, 50 (1976).
13. M. Shatzkes and M. Av-Ron, J. Appl. Phys. ***47***, 3192 (1976).
14. P. Solomon, J. Vac. Sci. Technol. ***14***, 1122 (1977).
15. R.C. Hughes, Solid State Electron. ***21***, 251 (1978).
16. Z.A. Weinberg and T.N. Nguyen, J. Appl. Phys. ***61***, 1947 (1987).
17. D.J. DiMaria, T.N. Theis, J.R. Kirtley, F.L. Pesavento, D.W. Dong, and S.D. Brorson, J. Appl. Phys. ***57***, 1214 (1985).
18. S.D. Brorson, D.J. DiMaria, M.V. Fischetti, F.L. Pesavento, P.M. Solomon, and D.W. Dong, J. Appl. Phys. ***58***, 1302 (1985).

19. M.V. Fischetti, D.J. DiMaria, S.D. Brorson, T.N. Theis, and J.R. Kirtley, Phys. Rev. B *31*, 8124 (1985).

20. F.R. McFeely, E. Cartier, L.J. Terminello, A. Santoni, and M.V. Fischetti, Phys. Rev. Lett. *65*, 1937 (1990).

21. E. Cartier and F.R. McFeely, Phys. Rev. B *44*, 10689 (1991).

22. D. Arnold, E. Cartier, and D.J. DiMaria, Phys. Rev. B *45*, 1477 (1992).

23. J.N. Bradford and S. Woolf, Radiation Effects and Defects in Solids *117*, 227 (1991).

24. D.J. DiMaria, D.W. Dong, C. Falcony, T.N. Theis, J.R. Kirtley, J.C. Tsang, D.R. Young, and F.L. Pesavento, J. Appl. Phys. *54*, 5801 (1983).

25. C. Chang, C. Hu,and R.W. Brodersen, J. Appl. Phys. *57*, 302 (1985).

26. Z.A. Weinberg and M.V. Fischetti, J. Appl. Phys. *57*, 443 (1985).

27. I.C. Chen, S. Holland, K.K. Young, and C. Hu, Appl. Phys. Lett. *49*, 669 (1986).

28. Z.A. Weinberg, M.V. Fischetti, and Y. Nissan-Cohen, J. Appl. Phys. *59*, 824 (1986).

29. J.M. Aitken and D.R. Young, IEEE Trans. Nucl. Sci. *NS-24*, 2128 (1976).

30. D.A. Buchanan, M.V. Fischetti, and D.J. DiMaria, Phys. Rev. B *43*, 1471 (1991).

31. T.H. Ning, J. Appl. Phys. *47*, 3203 (1976).

32. M.M. Heyns and A.v. Schwerin, in *Insulating Films on Semiconductors 1991*, edited by W. Eccleston and M. Uren (Adam Hilger, New York, 1991), pp.73-82.

33. D.J. DiMaria, Z.A. Weinberg, and J.M. Aitken, J. Appl. Phys. *48*, 898 (1977).

34. D.A. Buchanan and D.J. DiMaria, J. Appl. Phys. *67*, 7439 (1990).

35. S.K. Lai, J. Appl. Phys. *54*, 2540 (1983).

36. A. Yankova, L. DoThanh, and P. Balk, Solid-State Electron. *30*, 939 (1987).

37. F.L. Terry, M.L. Naiman, R.J. Aucion, S.D. Senturia, IEEE Trans. Nucl. Sci. *NS-28*, 4389 (1981).

38. Y. Nissan-Cohen and T. Gorczyca, IEEE Electron Dev. Lett. ***EDL-9***, 287 (1988).

39. D.J. DiMaria, R. Ghez, and D.W. Dong, J. Appl. Phys. *51*, 4830 (1980).

40. P.L. Solomon, J. Appl. Phys. *48*, 3843 (1977).

41. I. Placencia, F. Martin, J. Sune, X. Aymerich, J. Phys. D: Appl. Phys. *23*, 1576 (1990).

42. K.R. Farmer, M.O. Andersson, and O. Engstrom, Appl. Phys. Lett. *58*, 2666 (1991).

43. Y. Ozawa and K. Yamabe, from "Ext. Abstr. 1991 Inter. Conf. Solid-State Dev. and Mat.", Yokohama, 1991, pp. 240-242.

HOT-ELECTRON DYNAMICS IN THIN SILICON DIOXIDE FILMS STUDIED BY PHOTON-INDUCED ELECTRON TRANSMISSION

E. Cartier, D. Arnold, E. Eklund, D.J. DiMaria and F.R. McFeely

IBM Research Division
T.J.Watson Research Center
Yorktown Heights, NY 10598

ABSTRACT

Photon-induced electron transmission experiments are used to measure absolute, energy dependent electron-phonon scattering rates and rates for electronic excitations in SiO_2 thin films. It is shown by Monte Carlo transport simulations that the measured scattering rates allow an accurate prediction of the hot electron dynamics. A new electron runaway phenomenon, acoustic phonon runaway, is found to be the key to a quantitative understanding of electron-hole pair generation by impact ionization and leads to hole related degradation phenomena in Si/SiO_2 thin film structures.

INTRODUCTION

Over the last decade, considerable progress has been made in the understanding of hot-electron phenomena in SiO_2 thin films on Si. This progress was to a large extent stimulated by increasingly stringent demands on the performance of silicon-based electronic devices accompanying new developments in device miniaturization. After the discovery of strong electron heating in SiO_2 at relatively moderate electric fields, a connection between the hot electron dynamics and the degradation of thin SiO_2 gate insulators could be established. Hot electrons with energies larger than 2 eV were shown to produce defects and charge trapping sites which directly affect the performance of field effect transistors and floating gate memory cells. A detailed understanding of the hot electron dynamics in SiO_2 is therefore a precondition for the understanding and reliable modeling of degradation phenomena.

In this contribution, we will discuss novel internal photoemission-based electron transmission techniques which yield information on the electron dynamics. The experiments allow measurements of *absolute, energy dependent* electron-phonon scattering rates and the total rate for electronic excitations in SiO_2 over a wide energy range.[1-3] The measured scattering rates are then used as input parameters for Monte Carlo transport simulations to predict the hot electron dynamics over a wide electric field range. We show that the measured energy dependence of the acoustic phonon scattering rate provides the bases for a quantitative understanding of band-to-band impact ionization and hole generation in SiO_2 thin films and for quantitative modeling of hole-related defect generation.

ELECTRON SCATTERING IN SiO_2

A thermal electron in the conduction band has to gain at least 9 eV from an applied electric field before it can undergo an impact ionization event. During heating, the electrons constantly lose energy and momentum to the lattice via electron-phonon scattering, the only intrinsic energy and momentum loss processes at kinetic energies smaller than the gap energy, $E_g = 9eV$. An understanding of the hot electron dynamics and its relation to band-to-band impact ionization therefore requires an accurate knowledge of the energy dependence of the electron-phonon scattering rates from thermal energies up to about $2E_g$. In principle, these rates can be calculated. However, thin SiO_2 films are amorphous and the band structure is not known, making such calculations difficult. Many different theoretical approaches to this problem have been used.[4-10] To overcome some of these difficulties, we have developed techniques to obtain more direct information on the scattering of hot electrons with phonons experimentally.

Calculation of scattering rates

Initially, only the interaction with the longitudinal optical (LO) phonons was considered. The scattering rate can be calculated from the optical constants within the Fröhlich approximation.[11, 12] The LO-phonon scattering rate has a maximum near the energy of the dominant mode, $\hbar\omega_{LO} = 153meV$, and it decreases with E^{-1} towards higher electron energies, reflecting the Coulombic nature of the interaction. Also, with increasing energy, electrons are primarily scattered forward, providing little momentum relaxation. The E^{-1} fall off leads to the phenomena of "LO-phonon runaway" at electric fields above 1.5 MV/cm.[13] For many years, this instability was believed to be the origin of dielectric breakdown.[14]

In the eighties, it was shown experimentally that LO-phonon scattering alone cannot account for hot electron transport at high fields. In contrast to LO-phonon scattering, acoustic phonon scattering becomes stronger with increasing electron energy and it is the the dominant scattering mechanism at electron energies of a few eV. Furthermore, large angle scattering is favored, especially when Umklapp processes begin to play a role. Theoretically, acoustic phonon scattering has been calculated within the deformation potential approximation including Umklapp scattering.[9, 12, 15]

Following the calculation of Sparks *et al* for alkali halides,[15] Fischetti *et al*[4] implemented the above electron-phonon scattering processes into a Monte Carlo transport simulation and demonstrated that acoustic scattering leads to a restabilization of the electron energy distribution at average energies of 2-4 eV at fields above the LO-phonon runaway threshold.[12] Impact ionization, however, was not considered. This was consistent with the electron-phonon scattering model used since the acoustic rate in the deformation potential approximation continuously increases with electron energy, preventing electron heating up to E_g. We will show experimentally that the deformation potential approximation breaks down above 6 eV. The measured deviation of the acoustic phonon scattering rate from the deformation potential calculation allows a quantitative prediction of impact ionization in thin films and accounts for previously unexplained high field transport phenomena in Si/SiO_2 device structures.

Estimates for the impact ionization rates can be obtained[7] within the parabolic band approximation using the Keldysh expression,[16] $1/\tau_{ii}(E) = P \times ((E - E_{th})/E_{th})^2$. A more

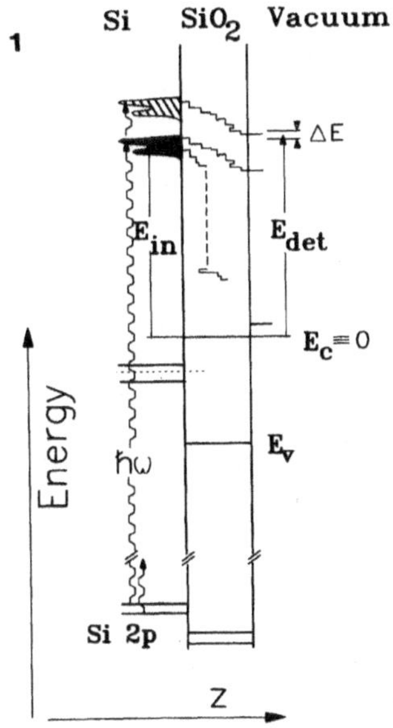

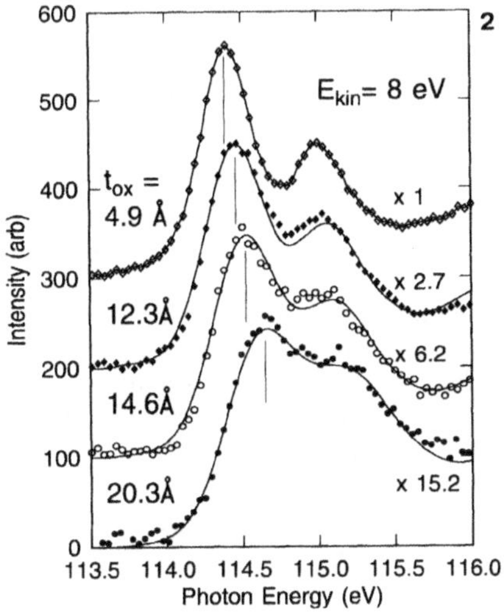

Figure 1. Schematic energy-space diagram illustrating the principle of soft x-ray induced low energy electron transmission for transport analyses.

Figure 2. Evolution of the substrate Si*2p* core level line shape with SiO_2 overlayer thickness measured at 8 eV energy. The multiplication factors quantify the signal attenuation. Symbols are measured spectra. Lines are reconstructed Si*2p* line shapes using Monte Carlo transport simulations.

realistic treatment requires the knowledge of the actual band structure.[17] Here, we use the Keldysh expression with experimental values for the prefactor, P, and the ionization threshold, E_{th}.

Experimental determination of scattering rates

Experimental information on the absolute magnitude of the electron scattering rates can be obtained from electron transmission through thin films. The large absolute values of the scattering rates in SiO_2 necessitate the use of very thin oxide films. Since free standing films in the 1-10 nm range are difficult to obtain, photon-induced low energy electron transmission techniques have proven to be useful.[1-3, 18-21] These techniques are based on internal photoemission in metal/SiO_2- or Si/SiO_2-substrate/overlayer structures. Photoelectrons from the substrate are injected into the SiO_2 overlayer using UV[1, 18-20], vacuum-UV[22], or soft X-ray[2, 3, 21] radiation. The electron energy distributions after transport through the overlayer are measured in vacuum. The energy dependence of the rates can be obtained by changing the injection energy via the photon energy. A schematic representation of such an experiment, using soft X-ray radiation for electron injection from the Si*2p* core level, is shown in Fig. 1. The emitted electron distributions are measured by scanning the photon energy at a constant detection energy, E_{det}.. This constant final state (CFS) spectroscopy allows a rigorous separation of the photo electrons within the *2p* substrate core line from secondary electrons being scattered down from higher kinetic energies.[2, 3, 21]

An absolute calibration of the scattering rates is obtained from the transport induced changes in the transmitted electron energy distributions as a function of the overlayer thickness. Typical experimental results are shown in Fig. 2. With increasing oxide thickness, the substrate Si*2p* line continuously shifts and broadens towards higher photon energies, which is the direction of energy loss in CFS-spectroscopy. Simultaneously, the line is strongly attenuated as quantified by the multiplication factors in Fig. 2.

The physical processes leading to these two effects are schematically shown in Fig. 1. Multiple phonon scattering leads to a continuously increasing energy loss with thickness. The line broadening is thus a measure for the phonon scattering rates. The line attenuation arises primarily from deep inelastic scattering processes which have energy losses much larger than the linewidth of the core level. Such processes completely remove the electrons from the *2p* spectral region in one single scattering event as illustrated in Fig. 1. Thus, the signal attenuation is a measure of the total deep inelastic rate which will contain the impact ionization rate above the ionization threshold.

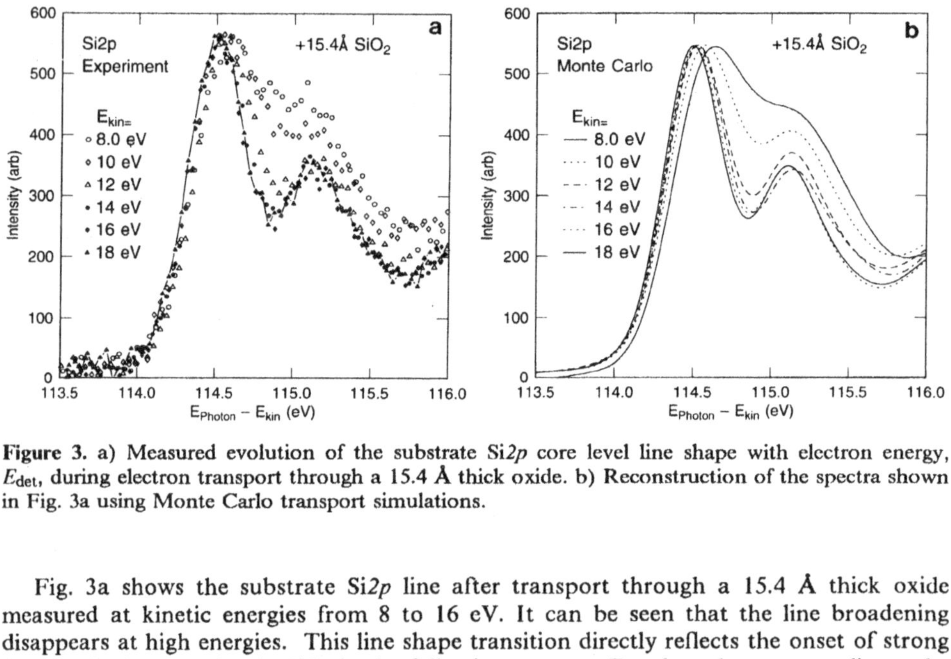

Figure 3. a) Measured evolution of the substrate Si*2p* core level line shape with electron energy, E_{det}, during electron transport through a 15.4 Å thick oxide. b) Reconstruction of the spectra shown in Fig. 3a using Monte Carlo transport simulations.

Fig. 3a shows the substrate Si*2p* line after transport through a 15.4 Å thick oxide measured at kinetic energies from 8 to 16 eV. It can be seen that the line broadening disappears at high energies. This line shape transition directly reflects the onset of strong deep inelastic scattering in SiO_2 in the following manner. For the substrate core line to be strongly broadened by phonon scattering, a significant fraction of the electrons must undergo many (large-angle) phonon scattering events. These electrons with long transit times have also a high probability of being removed from the spectral region in a single deep inelastic event. As the deep inelastic rate increases with electron energy, only electrons with direct trajectories have a chance of traversing the film within the core line. Such electrons emit few phonons and do not broaden the core line.

For a quantitative analyses, the core lines are reconstructed by Monte Carlo transport simulations. The unknown scattering rates are thereby adjusted until an accurate reproduction of the line evolution with thickness *and* energy is achieved. Typical results of such simulations are shown in Fig. 2 and Fig. 3b. A quantitative reproduction of the substrate core line evolution with thickness is possible with only two adjustable quantities at each energy (for details see Ref. 3). The resulting absolute acoustic phonon scattering rates and the deep inelastic rates are shown in Fig. 4 and Fig. 5, respectively.

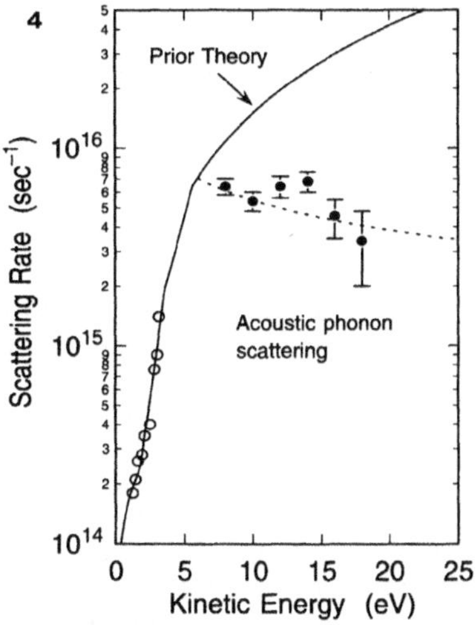

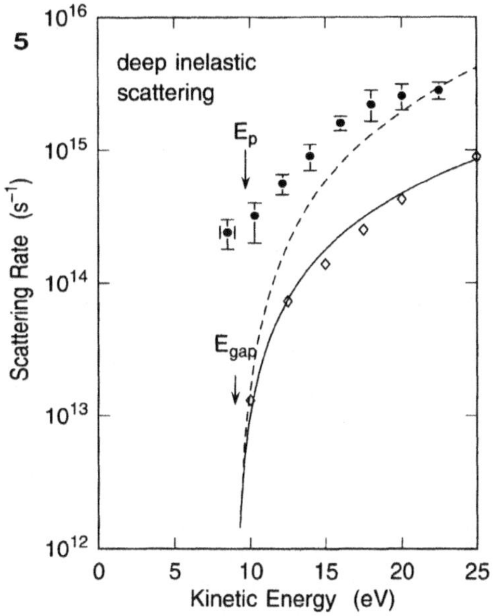

Figure 4. Acoustic phonon scattering rates in SiO_2 derived from photon-induced electron transmission experiments. Rates calculated within the deformation potential approximation are shown for comparison (solid line). The dotted line shows the rate used for high field transport simulations.

Figure 5. Deep inelastic scattering rate (full circles) derived from electron transmission through thin SiO_2 films on Si. For comparison the impact ionization rates derived by Bradford *et al*[9] from the dielectric loss function using a threshold of 9 eV (diamonds, solid line) are shown. The dashed line shows the impact ionization rates (Keldysh expression) used for high field transport simulations.

In Fig. 4, the acoustic phonon scattering rates derived from various transmission experiments are summarized.[1, 3] The rates at low energies agree well with calculated values within the deformation potential approximation using a deformation potential of 6 eV and an electron effective mass of $0.5m_0$. At energies above 2 eV, a non-parabolic band is required in order to account for the measured fast increase of the acoustic rate. At energies above 6 eV, the acoustic rate saturates and dramatically deviates from the deformation potential result. This deviation may arise from a maximum in the electron density of states around 6 eV, from screening of the electron-phonon interaction at high electron energy, or from quantum mechanical interference due to the short average time between collisions. For transport simulations, the experimental scattering rates as shown by the dashed line in Fig. 4 are used at energies $>6eV$. It is this decrease of the acoustic rate which leads to a quantitative understanding of impact ionization in SiO_2.

In Fig. 5, we compare the deep inelastic rates derived from our thin film experiments (solid circles) with the impact ionization rates derived from the measured dielectric loss function in α-quartz by Bradford *et al*[8, 9] (squares). Clearly, the thin film experiments yield much larger rates than expected for impact ionization. Most likely, the two methods measure different deep inelastic loss processes. The thin film experiments can be expected to be sensitive to interface processes. The energy of the Si/SiO_2 interface plasmon is $E_p = 9.4eV$ with an excitation line width of 4.1 eV. This numbers show that plasmon excitation is possible below $E_g \leq E_{th}$ as observed in our experiments. We believe that interface plasmon excitation is responsible for the high rates in our thin film experiments. An experimental separation of the two processes is difficult. At high fields, the number of electrons with energies above E_{th} rapidly increases with film thickness and impact ionization which is a bulk process can be expected to become the dominant inelastic scattering mechanism. In thin films, plasmon excitation may be important and lead to hole injection, for example. In the transport simulations below, we consider only impact ionization by using the Keldysh approximation with a value of 1.3×10^{15} sec^{-1} for the prefactor, P, and a value of 9 eV for the ionization threshold, E_{th}, as shown by the dashed line in Fig. 5. Most likely, this choice overestimates the ionization rates at high energies. We have shown however, that such an error has only a moderate influence on calculated impact ionization probabilities in thin films.[23]

TRANSPORT SIMULATIONS AND IMPACT IONIZATION

In this section, we summarize the basic aspects of high field electron transport as it results from Monte Carlo simulations[4, 5] in which the scattering rates derived in the previous section are used.[23, 24] In Fig. 6, the calculated average electron energies in a 50 nm thick oxide versus electric field are shown for three different simulations. The dotted line shows the result of a simulation which considers the interaction with LO-phonons only. The dashed line shows the results with acoustic scattering included in addition, and the solid line is the result with impact ionization included as well. The first two simulations reveal two distinct electron runaway phenomena which are directly related to the total energy and momentum loss rates, $dE/dt \times 1/E$ and $dp/dt \times 1/p$ shown in Fig. 7. These loss rates include LO-phonon scattering in the Fröhlich approximation, while acoustic phonon scattering and impact ionization are included as discussed above.[23, 24] Above 1.5 MV/cm, the electrons gain more energy from the field than they lose to the lattice by LO-phonon emission. The electrons run away from the maximum in the LO-phonon energy loss rate at 150 meV (see Fig. 7). Without acoustic phonon scattering, the average energy rapidly increases (dotted line in Fig. 6) since the LO-phonon rate decreases towards higher energy. However, acoustic phonon scattering prevents this instability. It leads to a drastic increase of the momentum relaxation rate with increasing energy (see Fig. 6) and to dispersive transport with long meandering electron trajectories, allowing for additional LO-phonon emission and larger energy losses per unit path traveled in the field direction. The electron energy distribution becomes restabilized at average energies of 2-4 eV. This combined energy dissipation process becomes too weak around 10 MV/cm. The electron distribution once more begins to run away (dashed line in Fig. 6), but this time from the maximum in the momentum relaxation rate at 6 eV (see Fig. 7). In analogy to "LO-phonon runaway", we call this phenomenon "acoustic phonon runaway". Inevitably, impact ionization sets in. This second instability gets removed by impact ionization which is a highly inelastic process leading to large energy relaxation rates (see Fig. 7). Electrons running away from the acoustic phonons rapidly gain energy in excess of E_{th} and become reset to energies of a few eV in an impact ionization event. For this reason, no clear indication for acoustic phonon runaway and for the onset of impact ionization can be obtained from measured average electron energies, as can be seen from the solid line in Fig. 6. An average electron cannot account for impact ionization.

COMPARISON WITH HIGH FIELD EXPERIMENTS

The hot electron dynamics presented above is consistent with a large number of high field experiments.[4, 5, 23-27] We only list a few important examples. The average electron energies, calculated on the basis of the measured rates, are in excellent agreement with values measured by DiMaria *et al*[25]. Good agreement is also found between calculated and directly measured electron energy distributions over a wide electric field range.[23, 24] This allows an accurate calculation of absolute impact ionization probabilities in SiO_2 thin films. Agreement with measured values was found for oxides in the thickness range from 25 to 95 nm and for electric field values up to 12 MV/cm.[23] In addition, positive charge formation due to hole trapping, and interface state built-up via electron-hole recombination have been shown to scale with the calculated ionization probabilities.[24]

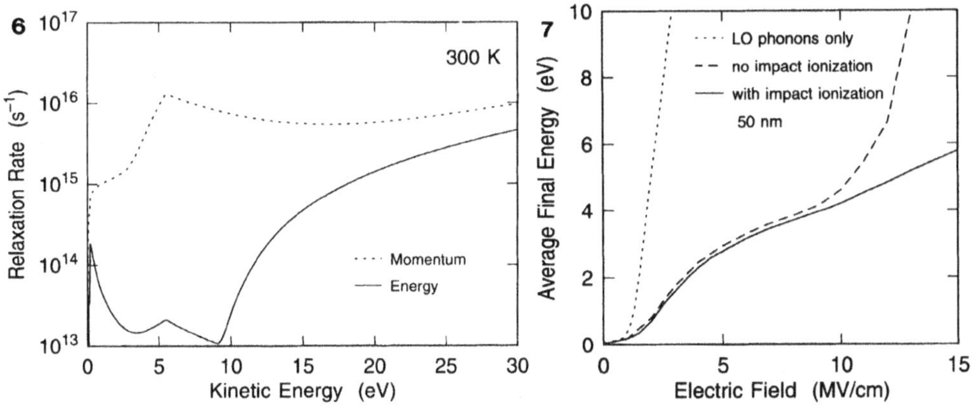

Figure 6. Calculated average electron energies in a 50 nm thick oxide versus electric field using Monte Carlo transport simulations.

Figure 7. Calculated total energy (solid line) and momentum (dashed line) relaxation rates versus electron energy in SiO_2 as calculated from the scattering rates shown in Fig. 4 and 5.

HOT ELECTRON DYNAMICS AND DEGRADATION

The two electric field values at which the LO-phonon runaway and the acoustic phonon runaway occur, have been shown to be threshold fields for hot electron-induced degradation of the Si/SiO_2 structure. Above the LO-phonon runaway threshold of 1.5 MV/cm, the majority of the electrons in the oxide film have energies exceeding 2 eV. It has been show that such energetic electrons create defects which act as charge traps.[28] Above the acoustic-phonon runaway threshold of approximately 7 MV/cm, some hot electrons in the high energy tail of their energy distribution are capable of producing electron-hole pairs by impact ionization leading to substrate hole currents in n-channel FETs and also to device degradation via trapped hole-electron recombination.[23, 24] An extensive discussion of the relationship between the hot electron dynamics and the degradation of Si/SiO_2 based device structures can be found in the contribution by DiMaria *et al* in this book. We therefore refrain here from a discussion of the field, thickness, temperature and processing dependence of the hot electron induced degradation processes.

CONCLUSIONS

The use of photo-induced electron energy transmission techniques provided us with a set of electron scattering rates which allow a quantitative prediction of the hot electron dynamics in SiO_2 thin films under high electric fields. The hot electron dynamics and its impact on device degradation can be characterized in terms of two distinct electron runaway phenomena. The first one is "LO-phonon runaway" which occurs at fields above 1.5 MV/cm and is responsible for the trap creation phenomena. The second one is "acoustic phonon runaway" which occurs at fields above 7 MV/cm and is responsible for hole production by band-to-band impact ionization and for hole related defect generation. The hot electron dynamics thus provides the basis for the understanding of degradation phenomena in Si/SiO_2 device structures. Quantitative modeling of device degradation becomes possible on the basis of the presented hot electron dynamics.[23, 24]

ACKNOWLEDGMENTS

The authors would like to thank M.V. Fischetti, D. Buchanan and J.H. Stathis for many stimulating discussions.

REFERENCES

1 E. Cartier and P. Pfluger, Physica Scripta **T23**, 235 (1988).
2 F. R. McFeely, E. Cartier, J. A. Yarmoff, and S. A. Joyce, Phys. Rev. B **42**, 5191 (1990).
3 E. Cartier and F. R. McFeely, Phys. Rev. B **44**, 10689 (1991).
4 M. V. Fischetti, D. J. DiMaria, S. D. Brorson, T. N. Theis, and J. R. Kirtley, Phys. Rev. B **31**, 8124 (1985).
5 M. V. Fischetti, Z. A. Weinberg, and J. A. Calise, J. Appl. Phys. **57**, 418 (1985).
6 W. Porod and D. K. Ferry, Phys. Rev. Lett. **54**, 1189 (1985).
7 D. K. Ferry, in *The Physics and Technology of Amorphous Silicon Dioxide*, edited by Roderick A. B. Devine (Plenum Press, New York, 1988), p. 365.
8 J. N. Bradford and S. Woolf, Radiation Effects and Defects in Solids **117**, 227 (1991).
9 J. N. Bradford and S. Woolf, J. Appl. Phys. **70**, 490 (1991).
10 R. L. Kamocsai and W. Porod, Solid-State Electronics **32**, 1825 (1989).
11 H. Fröhlich, Proc. R. Soc. London, Ser. A **160**, 230 (1937).
12 M. V. Fischetti, Pys. Rev. Lett. **53**, 1755 (1984).
13 H. J. Fitting and J. U. Friemann, Phys. Status Solidi A **69**, 349 (1982).
14 R. C. Hughes, Solid-State Electronics **21**, 251 (1978).
15 M. Sparks, D. L. Mills, T. Holstein, A. A. Mardudin, L. J. Sham, E. Loh Jr., and D. F. King, Phys. Rev. B **24**, 3519 (1981).
16 L. V. Keldysh, J. Sov. Phys. JETP **37**, 509 (1960).
17 E. O. Kane, Phys. Rev. **159**, 624 (1967).
18 P. Pfluger, H. R. Zeller, and J. Bernasconi, Phys. Rev. Lett. **53**, 94 (1984).
19 J. Bernasconi, E. Cartier, and P. Pfluger, Phys. Rev. B **38**, 12567 (1988).
20 E. Cartier, P. Pfluger, J. J. Pireaux, and M. R. Vilar, Appl. Phys. A **44**, 43 (1987).
21 F. R. McFeely, E. Cartier, L. J. Terminello, A. Santoni, and M. V. Fischetti, Phys. Rev. Lett. **65**, 1937 (1990).
22 E. Cartier and P. Pfluger, Phys. Rev. B **34**, 8822 (1986).
23 D. Arnold, E. Cartier, and D. J. DiMaria, Phys. Rev. B **45**, 1477 (1992).
24 D. J. DiMaria, D. Arnold, and E. Cartier, Appl. Phys. Lett. (April 27, 1992) (1992).
25 D. J. DiMaria, M. V. Fischetti, E. Tierney, and S. D. Brorson, Phys. Rev. Lett. **56**, 1284 (1986).
26 D. J. DiMaria, M. V. Fischetti, M. Arienzo, and E. Tierney, J. Appl. Phys. **60**, 1719 (1986).
27 D. J. DiMaria and M. V. Fischetti, Appl. Surf. Sci. **30**, 278 (1987).
28 D. J. DiMaria and J. W. Stasiak, J. Appl. Phys. **65**, 2342 (1989).

CONSTANT CURRENT STRESS BREAKDOWN IN ULTRATHIN SIO_2 FILMS

Pushkar P. Apte[1], Taishi Kubota[2], and Krishna C. Saraswat[1]

[1]Center for Integrated Systems, Stanford University
Stanford, CA 94305
[2]On leave from NEC Corp., 1120 Shimokuzawa, Sagamihara
Kanagawa, 229 Japan

INTRODUCTION

Ultrathin dielectric films play an important role in Integrated Circuits (ICs), both as gate dielectrics for MOS technologies, and as tunnel dielectrics for erasable memory (e.g. EEPROM) technologies. Therefore, significant effort has been directed towards understanding degradation and dielectric breakdown in these films [1, 2, 3]. One of the primary causes of degradation is electrical stress that the dielectric is subjected to during device operation, either by design (as in the case of tunnel dielectrics), or as an undesired effect (as in the case of hot-electron stressing of gate dielectrics in MOS transistors). Film thickness and temperature of operation also sensitively affect dielectric breakdown. Understanding the impact of the former becomes important as dielectrics are scaled to lower thicknesses, while the latter is important for ICs in automobiles, and other applications where the operating temperature is high.

Thus thickness, temperature, and stress-current density (J_{ox}) are important parameters in dielectric degradation and breakdown. We have characterized charge-to-breakdown (Q_{bd}) using constant-current stress for a range of each of these three parameters. To the best of our knowledge, this is the first such comprehensive characterization, and it reveals several trends not observed before. Primarily, we observe significant variations in both the absolute value and the sensitivity of Q_{bd} in different regions of this parameter space. We introduce some new concepts that may help explain some of these regimes which are not amenable to explanation using conventional models alone.

DEVICE FABRICATION AND CHARACTERIZATION

MOS capacitors were fabricated on low-resistivity (11–20 Ω-cm), < 100 > oriented, p-type, Si wafers using standard LOCOS isolation, with a field oxide thickness of 7000 Å. A 250 Å sacrificial oxide was grown and stripped before actual gate oxide growth. The gate oxides were thermally grown at temperatures of 800–950 °C in dry O_2, with thicknesses ranging from 30–250 Å. Finally, polysilicon gates were deposited by LPCVD and doped n^+ using phosphorus from a $POCl_3$ source. No forming gas anneal or metal wiring was used. The forming gas anneal was eliminated to minimize H_2 incorporation into the oxide. The oxide thicknesses

The Physics and Chemistry of SiO_2 and the Si-SiO_2 Interface 2
Edited by C.R. Helms and B.E. Deal, Plenum Press, New York, 1993

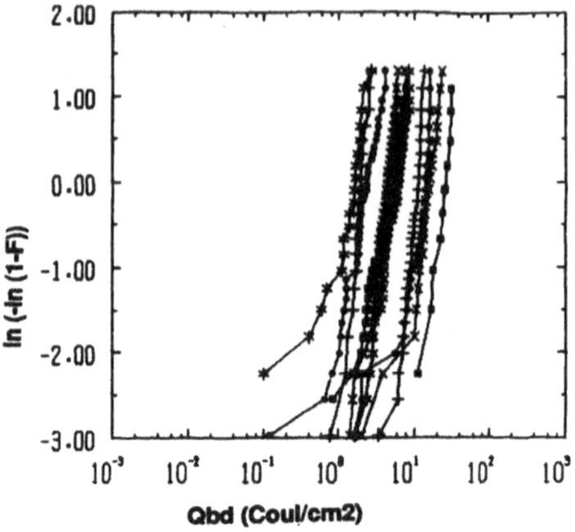

Figure 1. Cumulative failure percentage vs Q_{bd} for a 70 Å oxide at 25°C: the quantity ln $(\ln(1\text{-}F)^{-1})$, which represents cumulative failure percentage, arises from statistical considerations (Ref [2], Appendix B). Each line represents a specific J_{ox}.

were determined from C-V measurements. Q_{bd} was measured by the time dependent dielectric breakdown (TDDB) technique using constant-current stress, over the temperature range of 25–300°C. The gate was biased negative for all the measurements, so that the stressing was done with the p^- substrate in accumulation.

Several capacitors (5–20) were measured for each condition that we describe in the course of the paper. Interpreting this data to determine a unique Q_{bd} value can be non-trivial due to the presence of defect-related breakdowns. We have used a method of data representation described by Wolters *et al.* [2] that resolves the difference between defect-related and intrinsic breakdowns on a wafer. The details of this method are available in the reference—we will simply describe its use in our context. A sample plot for the 70 Å oxide at room temperature is shown in Figure 1. The physical significance of the quantity on the y-axis is the cumulative failure percentage of the capacitors. Each line in Figure 1 represents a specific J_{ox}, and the fact that they are closely bunched together indicates that the dependence on J_{ox} is small. This is also reflected in Figure 2, which is discussed in detail in the next section. The inclined parts of the curves in Figure 1 refer to the defect-related component, since the inclination implies a spread over Q_{bd}. On the other hand, the upright parts of the curves represent the intrinsic breakdown strength, since a straight vertical line implies constant Q_{bd}. Figure 1 shows that we have very few defect-related breakdowns for the 70 Å oxide at room temperature, since most of the lines have a near-vertical slope. This trend was quite typical for most of our measurements. Each data point in *all* our results has been picked to be the Q_{bd} value at the upright portion of the curves, and hence represents the intrinsic Q_{bd} value.

Although we have used only constant-current stressing, we also discuss the role of the oxide field. During constant-current stress, the applied electric field is essentially constant, with any variations resulting from charge-trapping within the oxide. The amount of trapping decreases with decreasing oxide thickness. Variation in the applied voltage was not more than 10% for *any* of our measurements. The value that we quote as "E_{ox}", refers to a time-averaged field over the course of the measurement, calculated as:

$$E_{ox} = \frac{V_g - (\phi_{poly} - \phi_{Si})}{t_{ox}} \simeq \frac{V_g - 0.85}{t_{ox}} \qquad (1)$$

where V_g is the time-averaged gate voltage, ϕ_{poly} is the work function of the n^+ polysilicon gate, ϕ_{Si} is the work function of the p^- silicon substrate, and t_{ox} is the oxide thickness. It should be noted that the field through the entire duration of the stressing, and at breakdown, is within ±5% of the value we use as E_{ox}.

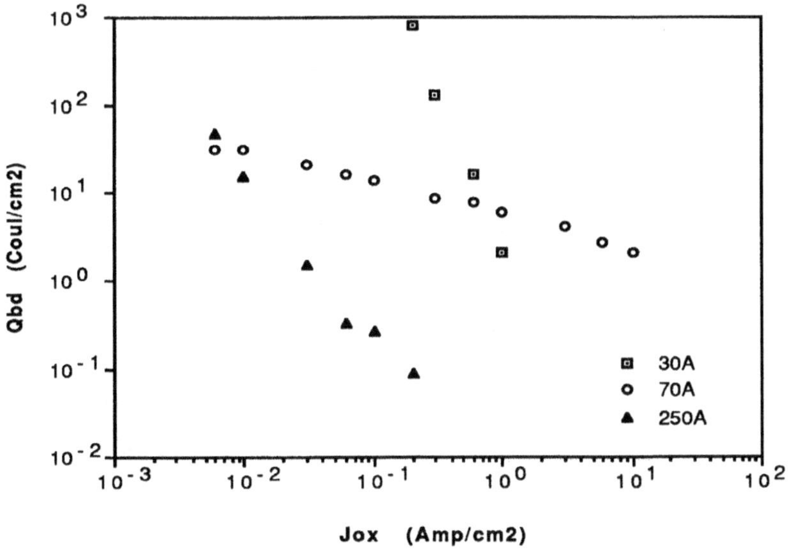

Figure 2. The variation of Q_{bd} with J_{ox}, at 25°C for thicknesses of 30, 70 and 250 Å. As discussed in the text, the data in this Figure are derived from plots such as Figure 1. The curve for the 70 Å oxide for instance is directly from Figure 1, and the gentle slope observed here is due to the lines for the different J_{ox}'s being closely bunched together in Figure 1.

RESULTS AND DISCUSSION

Q_{bd} as a Function of J_{ox} and Thickness

The variation (at room temperature, for three different thicknesses) of Q_{bd} vs J_{ox} and the corresponding E_{ox} is shown in Figures 2 and 3 respectively. Two trends are striking: one is the variation in sensitivity for different thicknesses as indicated by the different slopes, and the other is the different absolute values of Q_{bd} at a given J_{ox} for the different thicknesses. In general, Q_{bd} for the 30 and 250 Å oxides is much more sensitively affected by J_{ox} and E_{ox}, while the effect on the 70 Å oxide is relatively small. Thus Q_{bd} for the 30 Å oxide has a much sharper drop (steeper slope) compared to the 70 Å oxide. For $J_{ox} < 1$ A/cm^2, Q_{bd} is higher for the 30 Å oxides compared to the 70 Å ones, but due to the sharper slope, this is reversed for $J_{ox} = 1$ A/cm^2. Q_{bd} of the 70 Å and 250 Å are almost equal at low J_{ox}, but as we increase J_{ox}, Q_{bd} drops much more rapidly for the 250 Å oxide compared to the 70 Å oxide. Figure 3 shows a qualitatively similar dependence to Figure 2—i.e. E_{ox} scales with J_{ox}, and the two have a similar effect on Q_{bd} in this regime. DiMaria *et al.* [3] have shown that the *average* electronic energy increases with increasing E_{ox}, which together with Figures 2 and 3 implies that the *average* electronic energy also increases with increasing J_{ox}. The following discussion will show that this point is important in the context of breakdown mechanisms.

For the 250 Å oxide the injected electrons tunnel through a triangular barrier via the Fowler-Nordheim (F-N) tunneling mechanism, then travel through the conduction band of the oxide while undergoing scattering/trapping events, and finally "fall" into the anode (Figure 4). For the 30 and 70 Å oxides, two possibilities exist depending on E_{ox}. One is that the electronic transport occurs similar to the thicker 250 Å oxide, and the second is that ballistic transport occurs, either after F-N tunneling, or by direct tunneling through a trapezoidal barrier. The latter mechanisms are shown in Figure 5. The nature and number of energetic interactions

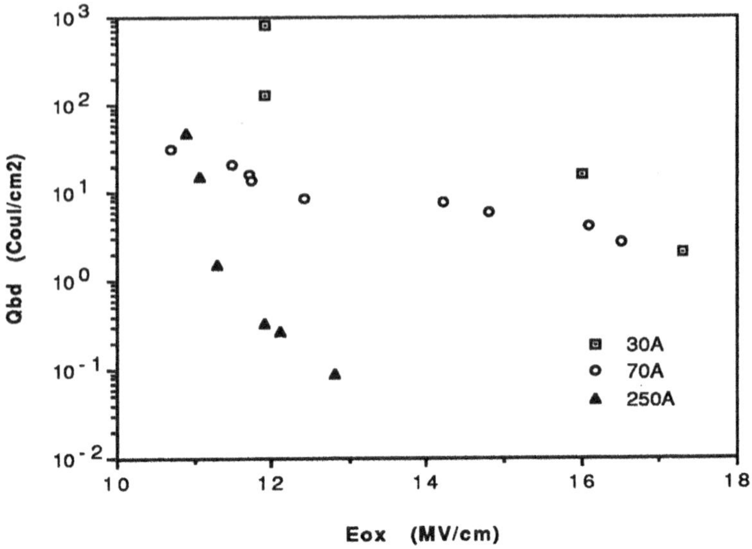

Figure 3. The variation of Q_{bd} with E_{ox}, at 25°C for thicknesses of 30, 70 and 250 Å.

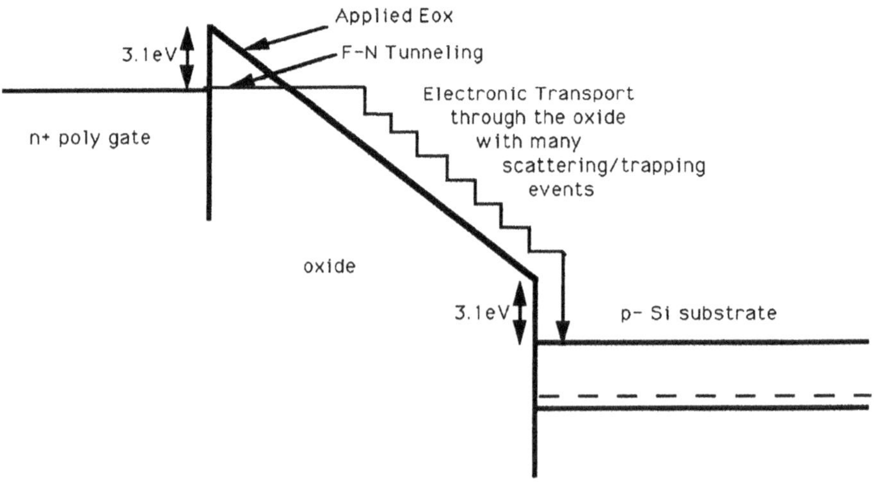

Figure 4. In a "thick" oxide (> 80 Å), electrons are injected through a triangular barrier via F-N tunneling and undergo multiple scattering/trapping events before falling into the anode.

between the lattice and the electrons thus depend on thickness, on whether the tunneling mechanism is F-N or direct, and in the case of F-N tunneling, on what distance remains to be traversed in the oxide conduction band. The energetic interactions in the bulk of the oxide involve the kinetic energy (K.E.) of the electrons, while those at the anode interface involve the potential energy (P.E), where the P.E. is the energy released by the electrons "falling" from the oxide conduction band to the bottom of the silicon conduction band. We believe that these energetic interactions hold the key to explaining the behaviour observed in Figure 2.

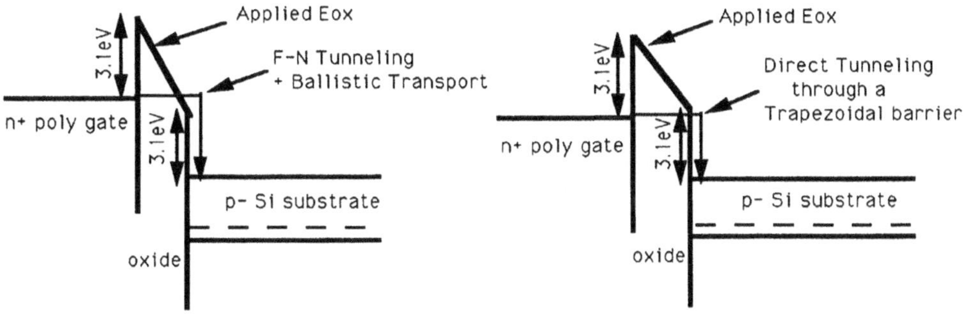

Figure 5. In a thin oxide (< 80 Å) at low fields, electrons can undergo ballistic transport after F-N tunneling, or can tunnel directly to the anode.

In the regular F-N tunneling regime, electrons undergo a number of energetic exchanges with the lattice throughout the bulk of the oxide (Figure 4). If the energy imparted by the electrons during such exchanges is larger than some threshold sufficient to cause "damage" (where damage may be trapping, band-gap ionization, or even bond-breaking), then such damage would be caused throughout the bulk of the oxide. The K.E. of the electrons scales with both J_{ox} and E_{ox} as mentioned earlier, and this means that with increasing J_{ox}, a greater fraction of the electrons being transported across the oxide have higher energies. Hence Q_{bd} will decrease with increasing J_{ox} in this regime, since "hotter" electrons will cause more damage. Also, energetic exchanges would be more for the 250 Å oxide compared to the 70 Å oxide, since the the electrons have to travel a larger distance and hence undergo more scattering/trapping events in the former. This would cause the 250 Å oxide to have lower Q_{bd} as well as a sharper dependence on J_{ox} and E_{ox}, as observed. At low J_{ox} ($= 0.003$ A/cm^2), the values of Q_{bd} for the 70 and 250 Å oxides are comparable (Figure 2). This may be because in this regime the K.E. of the electrons is low, and the P.E. loss—which, for a fixed J_{ox} and regular F-N tunneling, is the same for any thickness—may be the main energetic interaction. Such a mechanism has been used by Wolters *et al.* [2] to explain Q_{bd} staying constant with J_{ox} and E_{ox} at low values of these two parameters. At the fields examined here, the 30 Å oxides are on the threshold of switching from regular F-N tunneling, to F-N tunneling with ballistic transport and direct tunneling. Therefore it is likely that the transition between the different tunneling mechanisms and the consequent change in the energetic interactions may be the source of the high sensitivity of the 30 Å oxides to J_{ox} and E_{ox}. We hope to further

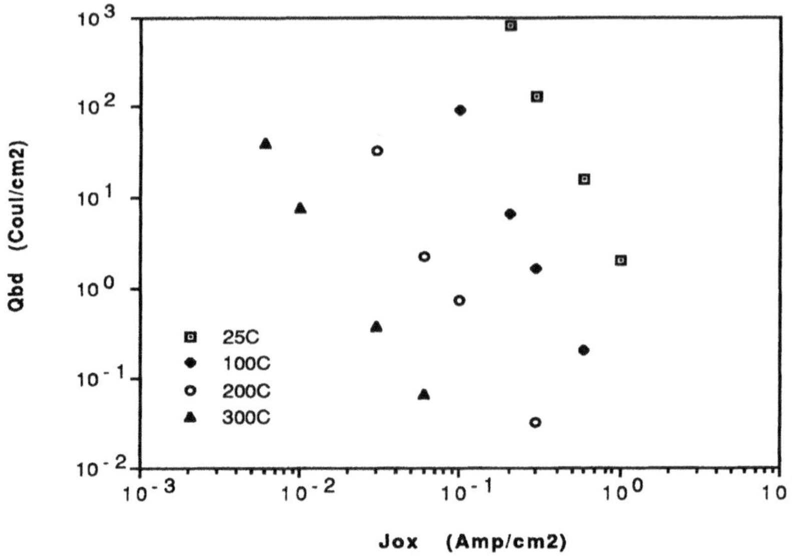

Figure 6. The variation of Q_{bd} with J_{ox} at 25, 100, 200 and 300°C for 30 Å oxide.

elucidate these issues in our ongoing work on a detailed model. Recently significant progress has been made in understanding both energy distributions, as well as transport phenomena of electrons passing through the dielectric [3, 4], and this should help in the model development.

Q_{bd} as a Function of J_{ox} and Temperature

Some observations that Q_{bd} decreases with temperature have been published [5], and our results are consistent with this general trend. We have characterized the specific variation of Q_{bd} for *each* thickness for four different temperatures. The trends for each thickness are quite unique. Figures 6 depicts the variation of Q_{bd} vs J_{ox} for the 30 Å oxide. The plot indicates that Q_{bd} decreases rapidly with temperature. Also, the lines for the four temperatures are displaced parallel to each other, suggesting that a similar mechanism may be active for each temperature. The 70 Å oxide (Figure 7), in addition to the trend of decreasing Q_{bd} with increasing temperature, shows that Q_{bd} becomes increasingly independent of J_{ox} as we go to higher temperatures. The 250 Å oxide (Figure 8) shows a similar behaviour to the 70 Å oxide at low J_{ox}, with Q_{bd} becoming independent of J_{ox} as we increase the temperature. At high J_{ox} though, the four temperature curves for the 250 Å oxide essentially come together, suggesting that the temperature effect is minimized.

An interesting point (not indicated in the Figures) is that for a given thickness, E_{ox} consistently decreases as we go to higher temperatures. The fact that Q_{bd} decreases, and E_{ox} also decreases with increasing temperature, indicates that a field-based mechanism *alone* might not be sufficient in explaining degradation and breakdown. Since the temperatures at which we have measured Q_{bd} are not large enough to significantly change the thermal energy of the electrons, we believe that it is the enhanced interaction with the lattice that causes the degradation as we increase temperature. Briefly, as the temperature increases, the lattice vibrations increase, and it is more likely that an electron will interact with the lattice, leading to a lowering of Q_{bd}. This phenomenon may be viewed in terms of an increase in a "collision cross-section" as a function of temperature. The term "collision" in this context is simply an interaction involving some energy transfer between the energetic electrons and the lattice, and the "cross-section" is the probability of such an interaction happening. We do

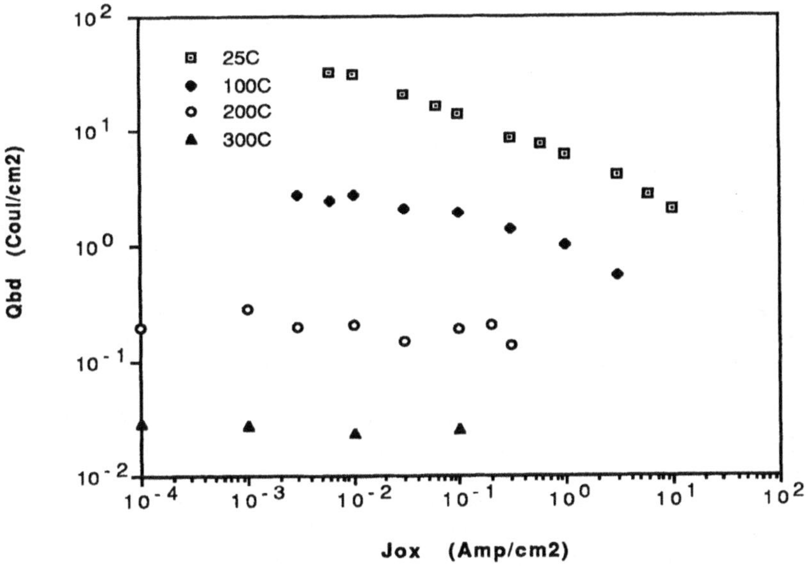

Figure 7. The variation of Q_{bd} with J_{ox} at 25, 100, 200 and 300°C for 70 Å oxide.

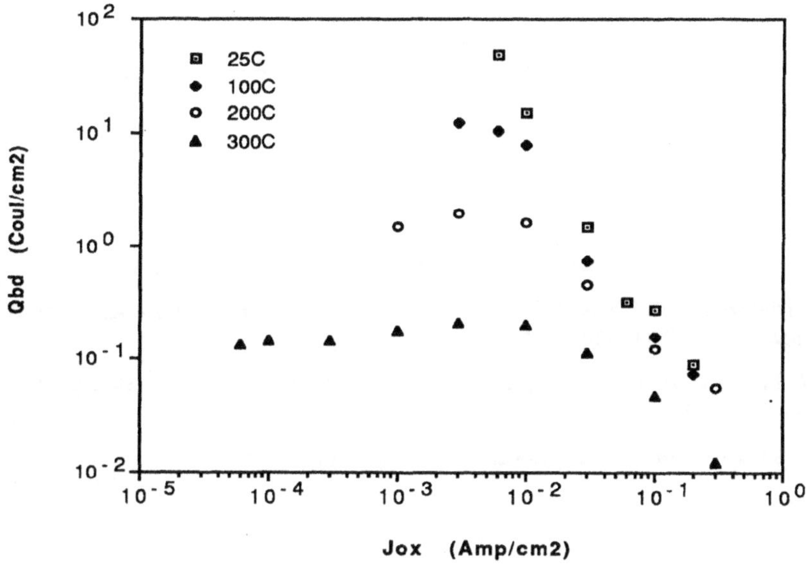

Figure 8. The variation of Q_{bd} with J_{ox} at 25, 100, 200 and 300°C for 250 Å oxide.

not currently have a detailed explanation of all the regimes that we observe in Figures 6–8. However, we believe that our ongoing work based on the concept of a collision cross-section may help provide the necessary explanations.

SUMMARY

We have characterized the differences in the charge that can be transported across a dielectric before breakdown occurs, as a function of stress-current density, oxide thickness and temperature, and we observe different regimes within this parameter space. We believe that such comprehensive characterization has not been performed before, and that some of the trends we observe have not been observed before. At room temperature, Q_{bd} for the 30 and 250 Å oxides depends much more sensitively on J_{ox} and E_{ox} compared to the 70 Å oxides. All Q_{bd} vs J_{ox} curves for different temperatures (25–300°C) plotted for a given thickness show degradation with increasing temperature, but in different ways. For the 30 Å oxide the degradation shows up as a parallel shift in the Q_{bd} curves for increasing temperatures, while for the 70 Å oxide the trend is towards a lower Q_{bd} that is independent of J_{ox} as temperature is increased. For the 250 Å oxide the trend is similar to the 70 Å oxide, except that at high J_{ox} the curves for all temperatures come together indicating a minimization of the temperature effect. A detailed model is currently under investigation, but we have introduced some of the concepts that may help explain the observed data. Specifically, we believe that the number and nature of the energetic interactions between the electrons and the lattice hold the key to understanding degradation. This is strongly influenced by the mode of electronic transport at a particular field and thickness. To explain the effect of temperature, we propose the concept of a collision cross-section that represents the probability of an energetic interaction between the electrons and the lattice. This collision cross-section increases with increasing temperature, since the atoms in the lattice vibrate more at higher temperatures.

ACKNOWLEDGEMENTS

This work was supported by the Semiconductor Research Corporation (SRC).

REFERENCES

[1] Ih-Chin Chen, Stephen E. Holland, and Chenming Hu. *IEEE Trans. Electron Dev.*, ED-32:413, 1985.

[2] D. R. Wolters and J. J. van der Schoot. *Philips J. Res.*, 40:115, 1985.

[3] D. J. DiMaria, T. N. Theis, J. R. Kirtley, F. L. Pesavento, D. W. Dong, and S. D. Brorson. *J. Appl. Phys.*, 57(4):1214, 1985.

[4] M. V. Fischetti, D. J. DiMaria, L. Dori, J. Batey, E. Tierney, and J. Stasiak. *Phys. Rev. B*, 35:4404, 1987.

[5] M. M. Moslehi and K. C. Saraswat. *IEDM*, page 157, 1984.

THE ROLE OF HYDROGEN IN INTERFACE TRAP CREATION BY RADIATION IN MOS DEVICES--A REVIEW

Nelson S. Saks and Dennis B. Brown

Naval Research Laboratory
Washington, DC. 20375

ABSTRACT

Many different models have been proposed to explain how interface traps N_{it} are created by radiation in MOS devices since this degradation problem was first observed in the mid-1960s. In this paper, we review recent research which provides strong support for one of these models, the H^+ transport model. In this model, the rate-limiting step of N_{it} creation is drift of radiation-induced hydrogen ions (H^+) through the oxide to the Si-SiO_2 interface. Extensive time-dependent experimental data have been obtained and are shown to agree well with qualitative predictions of the model and with detailed time-dependent calculations based on H^+ dispersive transport in SiO_2.

INTRODUCTION

When MOS (metal-oxide-semiconductor) devices are exposed to ionizing radiation, interface traps N_{it}, which are electrically active defects, are created at the Si-SiO_2 interface. These traps degrade the electrical properties of the MOS device. Degraded characteristics include changes in threshold voltage, reduced inversion layer mobility, increased minority carrier generation, and increased noise. Despite the technological importance of these effects, the mechanism(s) of the N_{it} creation process remain controversial. For recent review articles, see Oldham *et al.*[1] and Winokur.[2]

Electron spin resonance experiments have demonstrated that the interface trap is a trivalent silicon atom which is located at the Si-SiO_2 interface and which has one dangling (unsaturated) bond.[3,4] After device fabrication, most of these dangling bonds are passivated by hydrogen. During irradiation, the silicon-hydrogen bonds are broken, leading to the re-appearance of silicon dangling bonds and interface traps. Carriers in the silicon semiconductor interact with these defects, leading to degraded electrical characteristics.

How this silicon-hydrogen bond is broken is the point of some controversy. We know that the radiation does not interact directly with atoms at the interface, e.g. by atomic

The Physics and Chemistry of SiO_2 and the Si-SiO_2 Interface 2
Edited by C.R. Helms and B.E. Deal, Plenum Press, New York, 1993

displacement, as shown by non-penetrating photon experiments.[5,6] Thus, interface traps must be created by a more indirect interaction. It is generally agreed that N_{it} creation begins when ionizing radiation is absorbed in the oxide, creating electrons and holes. The radiation-induced holes have low mobility and a high probability of being trapped in the oxide. Many studies have suggested a correlation between radiation-induced holes trapped in the oxide and subsequent interface trap creation.[7-12] Yet many other experiments have suggested that hydrogen is also involved.[13-26]

A key difference between these various models involves the rate-limiting step in the trap creation process. In the trapped-hole models, the trap creation time is determined by the time required for a hole trapped near the interface to be converted to a broken silicon-hydrogen bond. Typically this process is not well-described in these models, but it does make intuitive sense that a trapped hole (i.e, a "missing" electron), could become a broken Si-H bond. In contrast, in the more complex hydrogen ion (H^+) model, most of the chemistry involved is well-described, as will be shown below. In this model, the trap creation time is determined by the time required for H^+ ions created by the radiation to drift through the oxide to the interface. In recognition of this key difference, a series of experiments on the time-dependence of N_{it} creation have recently been performed.[14-23] In this invited paper, we review the results of these experiments and selected H^+ transport calculations which clarify the nature of the interface trap creation kinetics and chemistry. We show that these experiments clearly support the hydrogen model, and are generally inconsistent with the trapped-hole models.

REACTION CHEMISTRY

The interface trap creation process in the hydrogen transport model involves the following radiochemical reactions (assuming positive applied gate bias):[13,14]

$$\equiv Si-H \; + \; H^+ \; + \; e^- \; \rightarrow \; \equiv Si-H \; + \; H^0 \; \rightarrow \; \equiv Si- \; + \; H_2. \qquad (1)$$

where $\equiv$Si-H is the hydrogen passivated trap precursor located at the $Si\text{-}SiO_2$ interface. We note the following points concerning these reactions:

(a) The H^+ ions in Eq. (1) are created throughout the oxide by the irradiation. The possible chemistry of this process has been discussed Griscom *et. al.*[14,27]

(b) In the presence of a positive applied oxide field, the H^+ ions drift through the oxide to the $Si\text{-}SiO_2$ interface. **This drift is the rate-limiting step of the N_{it} creation process.** The use of dispersive H^+ transport calculations to provide quantitative analysis and predictions of the time-dependence has been a key to our understanding of the drift process.

(c) Once they are close enough to the interface, the H^+ ions combine with electrons (e^-) which tunnel from the silicon substrate (for positive gate bias) to produce H^0. The H^0 is highly reactive and reacts with the trap precursor ($\equiv$Si-H). This reaction produces a silicon dangling bond ($\equiv$Si-), which is the interface trap, and H_2, which is free to diffuse away.[14]

(d) The energetics of Equation (1) is supported by recent studies by Brower[28] on interface trap passivation and de-passivation with hydrogen. Equation (1) is also supported by isochronal anneal experiments which have recently been reviewed by Griscom.[27]

TIME DEPENDENCE OF INTERFACE TRAP CREATION

In this section, we discuss the results of selected experiments which help elucidate the nature of interface trap creation. All of these experiments involve measurements of the **rate** of trap creation, which, as discussed above, provides critical information about the trap creation process.

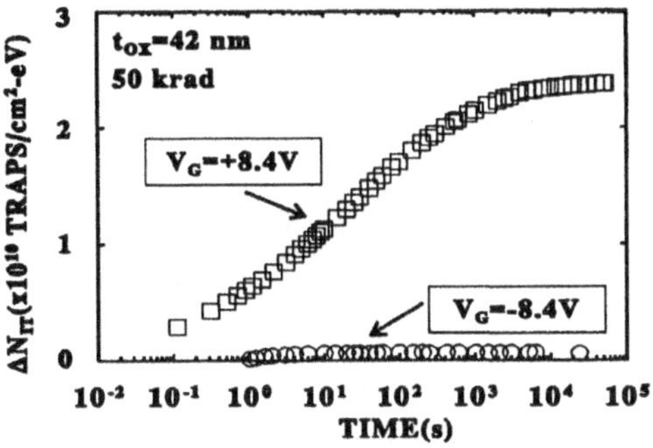

Fig. 1. Measurement of the slow increase in interface trap density N_{it} for two different applied dc gate biases.

A. Dependence on Applied Gate Bias

A typical measurement of the time dependence of interface trap N_{it} growth after irradiation is shown in Fig. 1.[15] The time dependence is measured using fast charge pumping. The radiation source is a short ($\sim$1 μs) pulse of high energy ($\sim$40 MeV) electrons from a LINAC. In Fig. 1, one device is biased at +2 MV/cm oxide field during and after irradiation, while a second device is biased at -2 MV/cm. Under negative bias, N_{it} creation after the radiation pulse is almost completely suppressed. In contrast, in the all positive bias case, a significant increase in N_{it} is observed. This growth in N_{it} takes place very slowly and is stretched out over more than four orders of magnitude in time. These results are consistent with the *dispersive transport* of a radiation-induced positively-charged species through the oxide. Under negative bias, this positively-charged species moves away from the Si-SiO_2 interface towards the gate, where it cannot participate in interface trap creation. Under positive gate bias, the measured ΔN_{it} is proportional to the integral of the species which reach the Si-Sio_2 interface. We emphasize that the time dependence in Fig. 1 cannot be due to transport of radiation-induced holes through the oxide, because the known hole transport time ($\sim$1 ms) is much too fast to explain the slow time dependence observed ($\sim$0.1-1000 s).[29] McLean[13] first suggested that this positive species is the hydrogen ion (H^+).

B. Deuterium Isotope Experiment

It appears to be clear from the above data that the transporting species must be positively charged, but what is the evidence that it is the *hydrogen* ion? This issue is addressed in recent work by Saks and Rendell[20] wherein the post-irradiation trap build-up was measured in oxides annealed in either hydrogen or deuterium at 900°C. Since

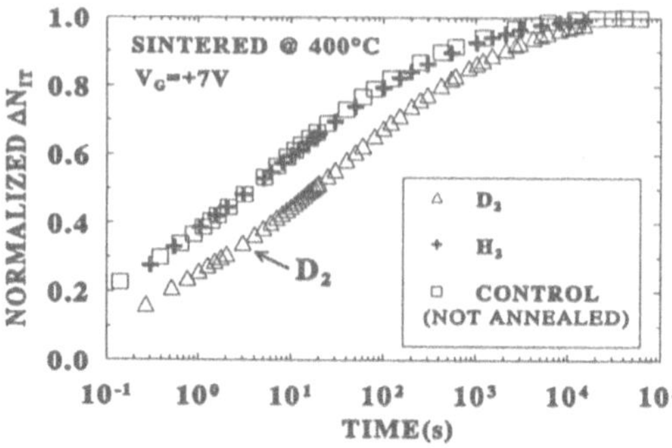

Fig. 2 The time-dependent N_{it} build-up in a deuterium-annealed oxide is slower than hydrogen-annealed oxides.

deuterium is an isotope of hydrogen with twice the mass, it would be expected to drift more slowly through the oxide under equal applied oxide fields. The normalized trap build-up rates in these annealed oxides and an unannealed control are shown in Fig. 2. The deuterium annealed oxide shows a significantly retarded rate of trap creation, as would be expected for transport of a heavier ion. This result strongly indicates that the transporting species is hydrogen. The N_{it} growth in the deuterated oxide was found to be retarded by a factor between 2.5 and 4.5, depending on experimental conditions. The magnitude of this retardation factor has been explained from the dispersive nature of H^+ hopping transport.[20] In contrast, the control and hydrogen-annealed oxides show nearly identical time dependencies. This result indicates that the N_{it} creation time in the control oxide, which was not purposely annealed in hydrogen at high temperatures, is still controlled by hydrogen drift. This conclusion is not unexpected, since the oxide was unavoidably exposed to hydrogen at high temperatures during fabrication of the complete MOS transistor test device.

C. Bias Switching Experiment

We next consider a set of experiments which help to differentiate between the trapped-hole and hydrogen models of N_{it} creation. In these experiments, the applied oxide field for one sample set is -2 MV/cm for a short time ($\sim$10 ms) after irradiation. This time is sufficient for transport of radiation-induced holes through the oxide to be complete (for example, the hole transport time is less than 1 ms in a 35 nm thick oxide biased at $\pm$2 MV/cm at 295K).[29] After this initial short time, the oxide field was switched to +2 MV/cm, which allows the N_{it} creation process to proceed.

These experiments for two different bias protocols are shown in Fig. 3.[18] One set of data ("+") was obtained with a constant positive gate bias throughout the experiment. For the second data set ("Δ"), the bias was negative during irradiation and the following 10 ms, after which it was switched to the same positive value. During the initial 10 ms, all radiation-induced holes move to either the gate or the substrate, and the hole transport phase is complete.[16,19] N_{it} creation during this initial 10 ms time is negligible. For both bias protocols in Fig. 3, a high (and almost equal) density of N_{it} is created at long times after irradiation. The fact that any substantial N_{it} creation is observed for the -2/+2 bias case is very difficult to explain with the trapped-hole model, since all radiation-induced

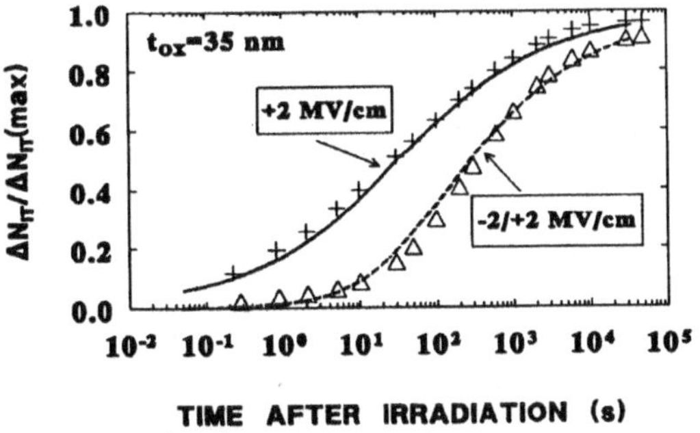

Fig. 3 Build-up in interface trap density N_{it} for two different gate bias protocols (see text).

holes are swept to the gate where they should be removed from the oxide. In this case, the amount of hole trapping *at the Si-SiO$_2$ interface* should be severely reduced, and thus the N_{it} build-up should have been almost eliminated. In contrast, these results are readily explained using the hydrogen model. In that model, the H^+ ions are created throughout the oxide for any bias during irradiation, and do not have a chance to move a significant distance during the initial 10 ms. Thus high N_{it} densities are ultimately created for both bias conditions. Qualitatively similar results first led McLean to suggest the H^+ model.[13]

In the -2/+2 bias case in Fig. 3, the time scale for N_{it} creation is substantially slower (by about an order of magnitude) than for the all positive bias case. This result may be explained using the hydrogen model and the concept that the hydrogen ions are created farther from the Si-SiO_2 interface when the gate bias is negative during hole transport. In this picture, the longer N_{it} creation time then results from the fact that the average H^+ ion must travel farther to reach the Si-SiO_2 interface.[16,30] Calculations based upon a dispersive H^+ transport model[14] are shown in Fig. 3 (solid and dashed lines). All parameters were derived by fitting the model only to the all-positive bias data. Then, for the -2/+2 bias case, the model prediction (dashed line) is obtained using the same parameters, and only assuming a different initial H^+ distribution.[16,30] Agreement between the model and experiment is excellent for both bias conditions in Fig. 3. This agreement shows that H^+ drift can be successfully modeled using dispersive transport.

D. Oxide Thickness Dependence

Measurement of the dependence of trap creation time on the oxide thickness, t_{ox}, has also been very useful in clarifying the mechanism of trap creation. In the H^+ model, the H^+ ions are created throughout the oxide. As t_{ox} increases, H^+ ions will be created farther from the Si-SiO_2 interface on average. Thus the trap creation time should increase with t_{ox}. On the other hand, the trapped-hole models[7-12] predict negligible dependence on t_{ox}. The trap creation rate in these models is related to the rate of converting trapped holes to N_{it}, which is determined by the characteristics of the interfacial oxide, which includes the nature and location of the hole traps, but not by the oxide thickness. The dependencies of N_{it} creation rates on oxide thickness have been measured and discussed by Brown and Saks.[18] In these experiments, all oxides were grown to the same thickness (108 nm), and

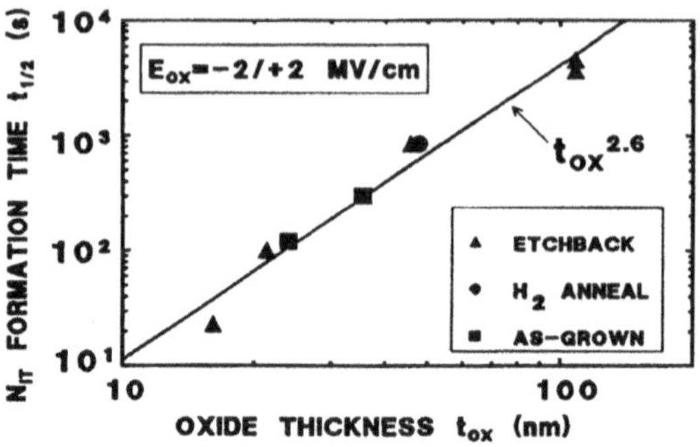

Fig. 4 The trap build-up time $t_{1/2}$ increases dramatically with increasing t_{ox}, in agreement with the H^+ model (solid line).

then thinner oxides were obtained by etching in buffered HF. This etch-back technique was chosen in an attempt to produce oxides with different oxide thicknesses but, as nearly as possible, identical interfacial regions and hole-trap characteristics. Experimental N_{it} creation times as a function of oxide thickness are shown in Fig. 4. The N_{it} creation time is represented by $t_{1/2}$, the time corresponding to creation of half the saturated ΔN_{it}. Identical electric fields, which for these data correspond to the -2/+2 MV/cm switched bias case in Fig. 3, were applied across the oxides to insure comparable results. As shown in Fig. 4, the normalized time dependence increases strongly with t_{ox}, in qualitative agreement with the H^+ model and in disagreement with the trapped-hole models. The oxide thickness dependence for a similar switched bias case has recently been measured independently by Shaneyfelt *et. al.*[23] Good agreement with the data in Fig. 4 was obtained.

Brown and Saks[18] have shown that in some cases the predicted oxide thickness dependence of the dispersive transport model is given by:

$$t_{1/2} \propto t_{ox}^{1/\alpha}. \tag{2}$$

where α is a parameter which quantifies the "dispersiveness" of the transport. A value for α may be obtained by fitting the transport model to the *shape* of the time dependent N_{it} build-up (see Fig. 3). Using this procedure, a value for α of 0.38 was obtained. This implies a strong thickness dependence given by:

$$t_{1/2} \propto t_{ox}^{1/\alpha = 1/0.38} = t_{ox}^{2.6}. \tag{3}$$

As shown in Fig. 4, this prediction (solid line) is in excellent agreement with the experimental oxide thickness dependence (data symbols). Furthermore, the value of α in Eq. (3) was obtained from the time dependence of a single sample, and is therefore unrelated to the thickness dependent data. Such a good correlation between an independent prediction of the model and experimental results again provides strong evidence in favor of the hydrogen model. For the all positive gate bias case, a weaker dependence on t_{ox} has been observed. This can be explained using the H^+ model by assuming significant creation of H^+ ions in the $Si\text{-}SiO_2$ interfacial region.[18,23]

E. Other Related Experiments

There are a number of other experiments whose results also provide useful insight into N_{it} creation and H^+ transport. Due to limited space, these can only be listed here:

(1) New theoretical calculations of the energy and bonding of H^+ in amorphous SiO_2 support the concept that the H^+ ions move by hopping between nearest-neighbor oxygen sites.[31] This is analogous to dispersive hole motion in SiO_2[29] and shows that the proposed concept of H^+ dispersive transport in SiO_2 is not unrealistic. It may be possible to exploit this analogy between hole and H^+ motion in future research to learn more about the nature of the dispersive transport process in SiO_2, which is inadequately understood at present.

(2) In the H^+ model, H^+ ions drift through the oxide on the same slow time scale as the interface traps are being created. Thus, H^+ drift should cause slow variations in the MOS threshold voltage. This effect has been observed and shown to correspond very well with calculations from the dispersive transport model.[17] About a 1:1 correlation is measured between the number of moving H^+ ions and the number of interface traps eventually created. This effect is generally small and difficult to observe because annealing of trapped holes usually dominates the time dependence.

(3) Several independent groups have demonstrated that when MOS devices are exposed to hydrogen gas *after* the radiation-induced increase in N_{it} has saturated, another large slow time-dependent increase in N_{it} is observed.[22,31-33] This result demonstrates yet another clear link between hydrogen and N_{it} creation.

(4) The measured time-dependence of N_{it} creation is found to depend on gate bias during irradiation and other parameters including oxide fabrication technology in complex ways. Efforts have been made to understand these results in terms of where the H^+ ions are created in the oxide by the radiation.[21,23,34,35] These results presumably reflect the initial profiles of hydrogen bonded in the oxide before irradiation and the mechanism(s) of H^+ creation. For example, the thickness dependence for the all positive bias case appears to be strongly process dependent.[18,21,23]

(5) In the H^+ model, H^+ ions are slowly swept to the gate under negative gate bias, where they are effectively removed from the N_{it} creation process. The number of ions reaching the gate increases with the time that the bias is held negative. Experiments of this type have shown that the magnitude of the subsequent N_{it} creation due to the remaining ions can be accurately predicted by the model.[19]

F. A Few Qualifications

There are several important cases where our understanding of N_{it} growth is less certain, including:

(1) The work reported in this paper exclusively describes results on devices with polycrystalline silicon gates. There is evidence that the time dependence of interface trap growth in aluminum gate devices may be somewhat different.[36]

(2) There exists a significant body of work exploring the defects produced by electron injection into gate oxides. (See for example Buchanan and DiMaria[37] and the references therein.) It appears likely that the mechanism for N_{it} creation is somewhat different for this process, although the mechanisms may be related.

(3) Likewise, it has been demonstrated in the technologically important case of channel hot carrier injection in MOSFETs that the mechanisms of N_{it} creation are similar at 77 and 295K.[38] Since H^+ drift is strongly thermally activated at room temperature (~ 0.8 eV)[13,15] and should be completely frozen out at 77K, it appears that the H^+ model cannot account for N_{it} creation by hot carriers at low temperatures.

SUMMARY

We have summarized recent experiments on the time-dependence of interface trap creation. Separate experiments show that the transporting species is positively-charged, and that hydrogen must be involved. All results are consistent with the concept that dispersive transport of radiation-induced H^+ ions is the rate-limiting step in the N_{it} creation process. We have also shown that H^+ transport can be accurately modeled by dispersive transport calculations. Indeed, the H^+ model is the only one for which detailed time-dependent calculations have been performed and successfully compared with experiment. In contrast, the time-dependent N_{it} build-up experiments appear not to agree very well with the only qualitative predictions of the alternative trapped hole models. There still remain other aspects of N_{it} creation in MOS devices, for example results on aluminum gate devices and hot carrier damage, which are not yet as well understood. We believe that the results summarized here represent a major step forward in our understanding of radiation-induced interface trap creation in MOS devices, an understanding which has eluded MOS researchers for about 25 years.

ACKNOWLEDGEMENTS

This work was supported by the Office of Naval Research and the Defense Nuclear Agency. We would also like to thank M. Ancona, E. Boesch, Jr., C. Dozier, A. Edwards, D. Griscom, R. Klein, S. Lyon, B. Mrstik, T. Oldham, R. Rendell, M. Shaneyfelt, and R. Stahlbush for their help and stimulating discussions during this work.

REFERENCES

1. T.R. Oldham, F.B. McLean, H.E. Boesch, Jr., and J.M. McGarrity, Semicon. Sci. Tech. 4, 986-999 (1989).
2. P. S. Winokur, in *Ionizing Radiation Effects in MOS Devices and Circuits*, T.P. Ma and P.V. Dressendorfer, Eds. (John Wiley and Sons, New York, 1989), pp. 193-255.
3. P.J. Caplan, E.H. Poindexter, B.E. Deal, and R.R. Razouk, J. Appl. Phys. 50, 5847 (1979).
4. P.M. Lenahan and P.V. Dressendorfer, J. Appl Phys. 55, 3495 (1984).
5. P.S. Winokur and M.M. Sokoloski, Appl. Phys. Lett. 28, 627 (1976).
6. P.S. Winokur, J.M. McGarrity, and H.E. Boesch, Jr., IEEE Trans. Nucl. Sci. NS-23, 1580 (1976).
7. P.S. Winokur, H.E. Boesch, Jr., J.M. McGarrity, and F.B. McLean, J. Appl. Phys. 50, 3492 (1979).
8. S.K. Lai, Appl. Phys. Lett. 39, 58 (1981).
9. F.J. Grunthaner and P.J. Grunthaner, Mat. Sci. Rep. 1, 65 (1986).
10. G. Hu and W.C. Johnson, Appl. Phys. Lett. 36, 590 (1980).
11. L. Do Than, M. Aslam, and P. Balk, Solid-St. Electron. 29, 829 (1986).

12. S.J. Wang, J.M. Sung, and S.A. Lyon, Appl. Phys. Lett. 52, 1431-1433 (1988).
13. F.B. McLean, IEEE Trans. Nucl. Sci. NS-27, 1651 (1980).
14. D.L. Griscom, D.B. Brown, and N.S. Saks, in *The Physics and Chemistry of SiO_2 and the Si-SiO_2 Interface*, C.R. Helms and B.E. Deal, Eds. (Plenum Press, New York, 1988), p. 287.
15. N.S. Saks, C.M. Dozier, and D.B. Brown, IEEE Trans. Nucl. Sci., **NS-35**, 1168 (1988).
16. N.S. Saks and D.B. Brown, IEEE Trans. Nucl. Sci. **36**, 1848 (1989).
17. N.S. Saks and D.B. Brown, IEEE Trans. Nucl. Sci. **37**, 1624 (1990).
18. D.B. Brown and N.S. Saks, J. Appl. Phys. **70**, 3734 (1991).
19. N.S. Saks and D.B. Brown, IEEE Trans. Nucl. Sci. **38**, 1130 (1991).
20. N.S. Saks and R.W. Rendell, to be presented at IEEE Nuclear and Space Radiation Effects Conf., New Orleans, LA, July 13-17, 1992.
21. D.B. Brown and N.S. Saks, to be presented at IEEE Nuclear and Space Radiation Effects Conf., New Orleans, LA, July 13-17, 1992.
22. J.R. Schwank, D.M. Fleetwood, P.S. Winokur, P.V. Dressendorfer, D.C. Turpin, and D.T. Sanders, IEEE Trans. Nucl. Sci. NS-34, 1152 (1987).
23. M.R. Shaneyfelt, J.R. Schwank, D.M. Fleetwood, P.S. Winokur, G.L. Hash, and K.L. Hughes, to be presented at the IEEE Nuclear and Space Radiation Effects Conference, New Orleans, LA., July 13-17, 1992.
24. R.A. Kohler, R.A. Kushner, and K.H. Lee, IEEE Trans. Nucl. Sci. NS-35, 1492 (1988).
25. Y. Nissan-Cohen, H.H. Woodbury, T.B. Gorczyca, and C-Y. Wei, 1988 VLSI Technical Symposium. Also, Y. Nissan-Cohen, private communication.
26. Y. Nissan-Cohen, Appl. Surf. Sci. 39, 511 (1989).
27. D.L. Griscom, "Hydrogen Model for Radiation-Induced Interface States in SiO_2-on-Si Structures: A Review of the Evidence", to be published in Journal of Electronic Materials, 1992.
28. K.L. Brower, Physical Rev. B 42, 3444 (1990).
29. F.B. McLean, H.E. Boesch, Jr., and T.R. Oldham, in *Ionizing Radiation Effects in MOS Devices and Circuits*, T.P. Ma and P.V. Dressendorfer, Eds., John Wiley and Sons, New York, 1989, pp. 87-192.
30. H.E. Boesch, Jr., IEEE Trans. Nucl Sci. **35**, 1160, Dec. (1988).
31. R.E. Stahlbush, A.H. Edwards, D.L. Griscom, and B.J. Mrstik, "Post-Irradiation Cracking of H_2 and Formation of Interface States in MOSFETs", submitted to Journal of Applied Physics.
32. R.E. Stahlbush, B.J. Mrstik and R.K. Lawrence, IEEE Trans. Nucl Sci. **NS-37**, 1641 (1990).
33. B.J. Mrstik and R.W. Rendell, IEEE Trans. Nucl. Sci. **38**, 1101, Dec. (1991).
34. P.U. Kenkare and S.A. Lyon, Appl. Phys. Lett. **55**, 2330 (1989).
35. D.M. Fleetwood, J. Appl. Phys. **67**, 580 (1990).
36. J.R. Schwank, P.S. Winokur, F.W. Sexton, D.M. Fleetwood, J.H. Perry, P.V. Dressendorfer, and D.C. Turpin, IEEE Trans. Nucl. Sci. **NS-33**, 1178, Dec. (1986).
37. D.A. Buchanan and D.J. DiMaria, J. Appl. Phys. **67**, 7439 (1990).
38. P. Heremans, G. Van den Bosch, R. Bellens, G. Groeseneken, and H.E. Maes, IEEE Trans. Elec. Dev. **37**, 980 (1990).

HYDROGEN-RELATED E' CENTERS AND POSITIVE CHARGE IN IRRADIATED OXIDE FILMS

M. E. Zvanut*, R. E. Stahlbush, W. E. Carlos, and H. L. Hughes

Naval Research Laboratory
Washington D.C. 20375

ABSTRACT

Electron paramagnetic resonance and capacitance-voltage results indicate that oxygen vacancy related defects (E' centers) in hydrogen-treated SiO_2 films occur in concentrations ten times that found in non-hydrogen treated films and, unlike their counterparts in non-hydrogen treated films, they are not associated with an increase in bulk or interfacial radiation-induced positive charge. We discuss evidence which supports the model in which incident radiation breaks SiH bonds, rather than a neutral oxygen vacancy, and produces hydrogen related E' centers. Further, through a post-irradiation annealing study, we show that the E' centers generated in H_2 treated films are more stable than those generated in as-grown oxide films.

INTRODUCTION

Metal-Oxide-Semiconductor (MOS) device processing frequently involves exposure of an SiO_2 film to a hydrogen ambient and irradiation, conditions which are often associated with device degradation. In an attempt to understand the basic mechanisms involved in irradiation of hydrogen-incorporated oxide films, several investigators have employed electron paramagnetic resonance (EPR) spectroscopy and electrical measurements[1-6]. Here we use both techniques to examine the effect of an H_2 treatment prior to irradiation on the density of radiation-induced oxide traps, interfacial and bulk, and radiation-induced point defects, E' centers, in thermally grown oxides.

The first part of the paper addresses the defect density and net oxide charge produced in wet and dry thermal oxides subjected to a thermal treatment in H_2 and irradiation with no applied bias. In general, several types of oxide films and different methods of charge generation have been used to study the relationship between E' centers and oxide charge[2,6-10]. Much of the work has concentrated on two point defects, the E' center, a trivalently bonded Si atom in the bulk of the SiO_2 film and the P_b center, a trivalently bonded Si atom at the Si/SiO_2 interface. These have been correlated with bulk oxide

*NRC/NRL Research Associate

The Physics and Chemistry of SiO_2 and the Si-SiO_2 Interface 2
*Edited by C.R. Helms and B.E. Deal, Plenum Press, New York, 1993

trapped charge and charge located at the Si/SiO_2 interface, respectively[2,3,8,10,11]. Recently, there have been suggestions that H_2 interacts with bulk defects and subsequently diffuses to the interface defects, thereby linking the bulk and interfacial processes[5]. Our work here shows that the E' centers produced in thermal oxides by an 800° C treatment in H_2 and irradiation with no external bias are not associated with a change in the radiation-induced bulk or interface trap density. Further, by gathering results from several types of amorphous SiO_2, bulk and thin film, we conclude that the production of E' centers without a change in positive charge density is a process typical to the material rather than a fabrication related phenomena.

In the second part of the paper we concentrate on the generation and annealing of the E' centers in irradiated H_2 treated films. Several groups studying optical fibers and bulk silica have reported an increased sensitivity to E' production in SiO_2 exposed to H_2 prior to irradiation[12,13]. Comparing EPR results with optical absorption measurements, Imai and coworkers have shown that the E' precursor in hydrogen treated silica is most likely an SiH bond. The data discussed below support this earlier work. Also, we present data from post-irradiation Ar anneals which demonstrate that the E' centers in H_2 treated SiO_2 are more stable than those generated in non-hydrogen treated oxides and that the annealing process for the hydrogen-related center is most likely reaction limited.

To avoid confusion and simplify nomenclature, throughout the paper 'treatment' refers to a pre-irradiation exposure to H_2 or Ar, and 'Ar anneal' refers to a post-irradiation exposure to Ar. Both are described in more detail below.

EXPERIMENTAL DETAILS

Oxides were grown by dry oxidation at 1000° C for 7.5 hours yielding a thickness of 159 nm. Wet oxides were grown by pyrogenic oxidation at 900° C for 2 hours giving a thickness of 180 nm. The SIMOX wafers were prepared by 1.8×10^{18} cm^{-2} O implantation and annealed for 5 hr at 1325° C in an Ar + 1.5%O_2 ambient. This produced a SIMOX layered structure consisting of a Si layer approximately 160 nm thick and an SiO_2 layer (buried oxide) approximately 360 nm thick on top of the silicon substrate. For the thermal oxides, the H_2 or Ar treatment consisted of a one hour anneal at temperatures of 300° C or 1000° C; the SIMOX samples underwent a 1000° C H_2 treatment. After the one hour H_2 treatment, the samples were cooled to approximately 200° C in an H_2 ambient. The hydrogen gas was then turned off and the furnace was purged with N_2 for approximately 10 minutes. As previously reported, the 1000° C H_2 treatment produces a concentration of at least 1×10^{18} cm^{-3} hydrogen in the oxide film[14].

For the EPR measurements, the material was cut into 0.24 X 1.5 cm^2 rectangular samples. EPR spectroscopy was conducted at 4 K or room temperature using a Bruker-300 X-Band spectrometer. To obtain an absolute concentration for the number of centers, comparison measurements were made at 20 K between the amplitudes of the phosphorus donor lines in 0.3 ohm-cm Si and the E' signal in the 1000° C hydrogen treated irradiated SIMOX substrates. The defect density for the remaining samples was obtained by a comparison with this SIMOX sample. The absolute accuracy is 50% and the relative accuracy about 10%.

For the electrical measurements, 0.0045 cm^2 Al contacts were evaporated onto the wafer surface and the back of the wafer was coated with approximately 100 nm of Al. High frequency (HF) and quasistatic (QS) CV measurements were made using a Keithly System 82 simultaneous CV system. The density of occupied oxide traps was estimated from the shift of the CV curve at the flatband capacitance, and the charge density was extracted via the standard equation, $N_{ot} = C_{ox}\Delta V/(qA)$[15]. Here N_{ot} is the centroid-weighted areal trap density, C_{ox} is the maximum capacitance of the device, ΔV represents the shift of the CV

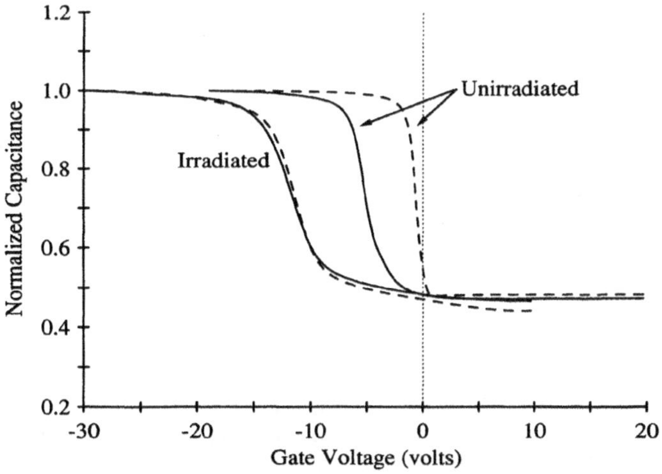

Figure 1. Normalized HF Capacitance (C/C_{ox}) vs. Gate Voltage for as-grown wet thermal oxides (solid) and H_2 treated wet oxides (dashed). C_{ox} for these films was 86 pF. Radiation dose was 10 Mrad.

curve induced by irradiation, q is the electronic charge, and A is the area of the Al contact. The energy distribution of the interface states, D_{it}, is not calculated here; rather, D_{it} is calculated at the voltage at which the fermi level in the Si is at midgap (V_m) using the relation $D_{it}(V_m) = (1/qA)[(1/C_l - 1/C_{ox})^{-1} - (1/C_h - 1/C_{ox})^{-1}]$. C_l and C_h are the QS and HF capacitances, respectively, evaluated at approximately the mid-gap voltage[15]. Since for the data analyzed here a majority of interface states form a uniform distribution in the gap, the areal density of states, N_{it}, may be approximated from $D_{it}(V_m) \times E_g$, where E_g is the Si bandgap, 1.12 eV.

After fabrication, samples were γ-irradiated to a total dose of 10 Mrad(SiO_2) or 100 Mrad(SiO_2) using a ^{60}Co source. No electric bias was applied during the irradiation. EPR and CV measurements were made before and after the radiation treatment.

A series of Ar anneals were performed on irradiated EPR samples. The step-anneal sequence began with a 10 min 80° C anneal followed by an EPR measurement. The samples were then returned to the furnace, annealed at a higher temperature, and the EPR measurement was repeated. EPR for hydrogen treated films was conducted at room temperature; due to a low signal-to-noise ratio, the non-hydrogen treated films were measured at 4K.

RESULTS AND DISCUSSION

CV Measurements

Figure 1 shows the effect of a pre-irradiation H_2 treatment on the high frequency CV characteristic of a wet thermal oxide. Similar results were obtained for the dry oxide. The dashed line represents normalized CV data obtained from a H_2 treated film; the solid line from an untreated film. When compared with the non-H_2 treated samples, the radiation-induced shift of the hydrogen treated sample is less than a factor of two larger. Table I lists the radiation-induced N_{ot} extracted from the CV curves of samples treated in H_2 or Ar at 300° C or 1000° C. The results show that N_{ot} appears to increase slightly with ambient and temperature.

Table 1. Radiation-Induced trap density ($1x10^{12}$ cm^{-2}) in thermal oxides subjected to pre-irradiation treatments.

	Untreated	Ar Treatment		H_2 Treatment	
		300° C	1000° C	300° C	1000° C
wet	1.2	1.5	1.6	1.7	1.9
dry	1.6	1.8	1.9	1.8	2.3

Figure 2 illustrates the effect of a pre-irradiation H_2 treatment on the quasistatic curve of a wet thermal oxide before (2a) and after (2b) irradiation. Note the different scales for the axes of Figures 2a and 2b. Based on calculations at approximately mid-gap, prior to irradiation the density of interface states in non-H_2 treated films is estimated to be $1x10^{12}$ cm^{-2}; for the H_2 treated films, the density is lowered to less than $1x10^{10}$ cm^{-2}. This decrease in N_{it} with H_2 treatment agrees with earlier work[16]. Also, as expected, the 10 Mrad γ-irradiation greatly increases the density of states[16]. Densities on the order of $1x10^{13}$ cm^{-2} are estimated from the quasistatic curve of both the H_2 treated and untreated films shown in Figure 2b. An analysis of the CV data shows that despite the factor of 100 decrease in interface states between unirradiated H_2 treated and non-H_2 treated films, the change in N_{it} induced by 10 Mrad γ-irradiation is the same to within 10%.

From the different shapes of the quasistatic curves of Figure 2b, it appears that in wet thermal oxides the energy distribution of the radiation-induced interface states is different for 1000° C H_2 treated films and untreated films. Analysis of the QS curves would yield a uniform distribution of interface states for the untreated film and a distribution with two peaks in the Si bandgap for the H_2 treated film. Thus, it appears that irradiation of H_2 treated wet oxides generates states at discrete energies. Note that this behavior was also seen in films treated with H_2 at temperatures as low as 300° C.

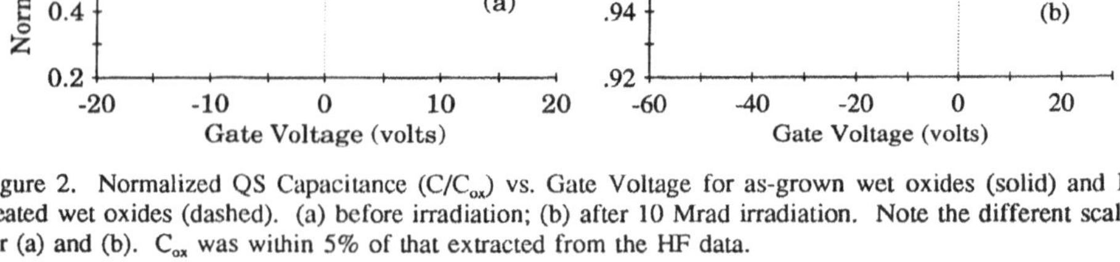

Figure 2. Normalized QS Capacitance (C/C_{ox}) vs. Gate Voltage for as-grown wet oxides (solid) and H_2 treated wet oxides (dashed). (a) before irradiation; (b) after 10 Mrad irradiation. Note the different scales for (a) and (b). C_{ox} was within 5% of that extracted from the HF data.

Overall, the quasistatic CV data show that the effect of radiation on the total density of interface states does not differ between those films treated with H_2 and those left untreated, and we tentatively conclude that in the H_2 treatment generates a radiation-induced interface state distribution with two peaks in the Si bandgap.

Previous EPR measurements of the E' concentration have shown that H_2 treated samples exhibit a dramatic sensitivity to radiation exposure[6,14]. Specifically, after a pre-irradiation H_2 treatment at temperatures above 600° C the density of radiation-induced E' centers approaches $1x10^{13}$ cm^{-2}, a factor of ten larger than that measured in samples exposed to H_2 at lower temperatures or to Ar at 1000° C[14]. On the other hand, the CV data reported here reveal that the density of radiation-induced bulk traps (table 1) and interface states does not change with exposure to either Ar or H_2 at temperatures of 300° C or 1000° C. This behavior indicates that in oxide films exposed to H_2 and irradiated without an external bias there exists no correlation between the E' centers and radiation-induced trap density or interface state density. Thus, since the trap density reflects the net charge state of the oxide film, we have shown that the generation of such centers is not accompanied by a net change in the oxide charge. (This statement assumes that there is not a large charge density near the metal/oxide interface to which the electrical measurement are insensitive.) Furthermore, assuming the extreme case that all the interface states are in the same charge state, we have shown that the hydrogen-related E' center concentration greatly exceeds the total radiation-induced charge density at the interface.

Although the lack of correlation between E' centers and trapped charge differs from the one-to-one correlation observed in thermal oxides irradiated under bias[2], it is similar to that seen in buried oxides treated in H_2 and deposited oxides[6,9,10]. Also, as noted in the introduction, the general relationship between H_2 treatment and E' production applies to bulk silica as well as to thin films[12,13]. In bulk SiO_2, the H_2-related E' centers must be produced without a build-up of charge because bulk material cannot support a net charge. Thus, the production of hydrogen-related E' centers without a corresponding increase in oxide charge appears to be a general characteristic of SiO_2 and is not dependent on the fabrication process.

It is important to realize here that in order to compare thin films with bulk silica we have irradiated the samples with no applied bias. Numerous reports show that biasing a sample during irradiation changes the density of radiation-induced trapped charge and E' centers[7,16,17]; thus, it is possible that irradiating the sample while applying an electrical bias may produce a different behavior between the E' centers and trapped charge.

To summarize, the CV measurements of H_2 treated thermal oxides irradiated with no applied bias indicate that hydrogen-related E' centers observed in wet and dry oxides are not related, in a simple way, to either the bulk oxide charge or interface charge. Since the behavior of E' centers and bulk oxide charge also applies to SIMOX buried oxides formed by a ion implantation as well as to bulk silica, we suggest that the formation of E' centers in irradiated H_2 treated oxides is a property of irradiated SiO_2 independent of the fabrication technique.

EPR Annealing Studies

Figure 3 illustrates the effect of an Ar anneal on irradiated hydrogen treated (open symbols) and untreated (filled symbols) thermal oxides (a) and SIMOX buried oxides (b). Plotted are the E' densities relative to that measured in the non-Ar annealed film. The data represent averages of 8-12 samples from each oxide type (thermal and SIMOX) and at least two separate anneal sequences. In Figure 3a, the circles and squares represent data obtained from dry and wet thermal oxides, respectively. To obtain a reasonable signal-to-noise ratio, the thermal oxides not treated with H_2 were irradiated to 100 Mrads(SiO_2).

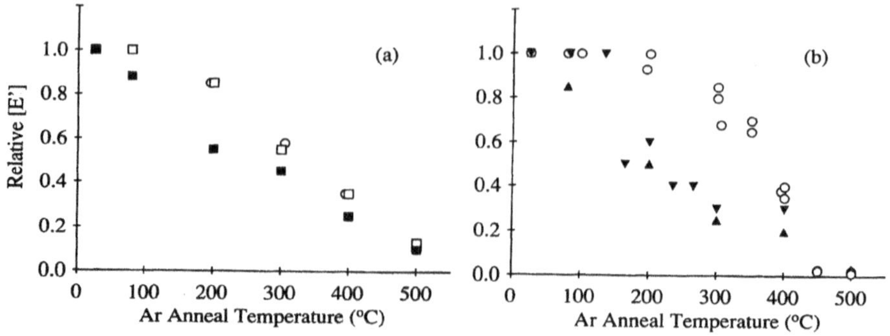

Figure 3. Relative E' concentration for 10 min Ar anneal. (a) 100 Mrad irradiated thermal oxides: filled squares, wet as grown oxides; open squares, wet H_2 treated oxides; open circles, dry H_2 treated oxides. (b) SIMOX: solid triangles, 100 Mrad untreated oxide; solid daggers, 10 Mrad untreated oxide; open circles, 10 Mrad H_2 treated oxide.

The data shown are for the wet thermal oxides. As expected the E' concentration in dry oxides saturates at doses above 10 Mrad[2] so that the signal-to-noise ratio remains insufficient to perform annealing studies in a reasonable amount of time. The hydrogen treated thermal oxides were studied after doses of 10 Mrad and 100 Mrad; to simplify the comparison, the 100 Mrad irradiated films are shown. The buried oxide data shown in Figure 3b were obtained after irradiation to 10 Mrad (open circles and solid daggers) and 100 Mrad (solid triangles).

The outstanding feature of the annealing data is that the hydrogen-related E' centers are more stable than non-H_2 related centers. This is particularly evident from a comparison of the points obtained after a 200° C anneal. For both the thermal and buried oxides, the non-hydrogen treated samples exhibit a 50% decrease in concentration while the H_2 treated films exhibit at most a 10% decrease. Further, the H_2-related E' center density does not decrease to half until temperatures greater than 300° C. The increased stability of E' centers in H_2 incorporated SiO_2 is consistent with that observed by Griscom in an extensive study of bulk silica[18]. However, the agreement is most likely fortuitous since in the work of Griscom, the silica contained OH and the annealing mechanism was dominated by diffusion. As discussed below, we attribute the behavior observed in Figure 3 to oxides containing SiH and reaction limited kinetics.

There is a variety of evidence suggesting that H_2 treatment of SiO_2 produces predominantly SiH bonds rather than OH bonds and that the SiH bonds are the pre-cursors for the radiation-induced E' centers[12,14]. The comparison of wet and dry oxides measured here strengthens the claim. As expected in as-grown films, infrared attenuated total reflectance (IR-ATR) measurements revealed a broad OH absorption at 3650 cm^{-1} equivalent to $6x10^{18}$ OH cm^{-3} in the wet thermal oxides and no absorption (less than $1x10^{18}$ OH cm^{-3}) in this region in the dry oxides. After irradiation, the E' concentration in the wet oxide is at most a factor of two greater than in dry oxides. A possible conclusion from this is that an OH concentration difference greater than a factor of six yields at most a factor of two difference in E' density. Now consider the wet and dry oxides after a 1000° C H_2 treatment. In both types of oxides, the OH density is less than $1x10^{18}$ cm^{-3}, yet irradiation produces a defect density on the order of ten times that seen in the untreated films. In this case, it seems that samples with less than $1x10^{18}$ cm^{-3} OH exhibit a greatly increased sensitivity to E' production. Further, in the wet thermal oxide the H_2 treatment decreased the OH concentration while it enhanced the radiation-induced E' centers. Considering the evidence sited above, it seems unlikely that OH is responsible for the ten-fold increase observed in the density of E' centers in H_2 treated films. Further, an IR-ATR measurement on SIMOX wafers implanted with $1x10^{18}$ O cm^{-2} revealed an SiH absorption but no OH absorption. We thus continue to support the model in which SiH is thought to be linked to the production of hydrogen-related E' centers.

Although the specific mechanism by which the annealing of E' centers occurs in the hydrogen treated and non-H_2 treated samples demands further investigation, the data of Figure 3 do indicate that the process for annealing the H_2-related E' centers is reaction limited rather than diffusion limited. The impurities which could diffuse through SiO_2 at these temperatures and which pertain to H_2 treated films are H_2, H_2O and O_2[1]. Applying the diffusion constants used for bulk silica and the anneal times of 10 minutes, we calculate that even the slowest diffusant, H_2O, can move 30 nm at 200° C. Assuming a uniform defect concentration of $5x10^{17}$ cm^{-3}, the furthest such species should have to move to reach a defect is 13 nm. Thus even at temperatures at which less than 10% of the annealing occurs in H_2 treated films, a diffusant could easily reach a defect. We conclude, therefore, that reaction at the site of the defect must be the limiting step in the anneal process. Contrary to intuition, note that the non-hydrogen-related E' centers anneal in a manner consistent with diffusion of a hydrogen-related species such as H_20; however, many other mechanisms are also possible.

In summary, in this second section we have addressed data concerning the generation and annealing behavior of the hydrogen-related E' center. Discussing IR-ATR results, we have supported the model linking SiH bonding and the generation of hydrogen-related E' centers. From the Ar annealing studies, we have demonstrated that E' centers produced in irradiated H_2 treated films are more stable than those produced in non-treated films. Also, we have shown that the annealing process of hydrogen-related E' centers is most likely reaction limited rather than diffusion limited.

Acknowledgements: The H_2 treatment and thermal oxidation were performed in the microelectronic processing facility at NRL by K. Foster and T. Swyden. The work was supported by Office of Naval Research and by the Defense Nuclear Agency.

REFERENCES

1. D.L. Griscom, J. Appl. Phys., Diffusion of radiolytic molecular hydrogen as a mechanism for the post-irradiation buildup of interface states in SiO_2-on-Si structures, 58:2524(1985).
2. P.M. Lenahan and P.V. Dressendorfer, Hole traps and trivalent silicon centers in metal/oxide/silicon devices", J. Appl. Phys. 55:3495(1984).
3. Y.Y. Kim and P.M. Lenahan, Electron-spin-resonance study of radiation-induced paramagnetic defects in oxides grown on (100) silicon substrates, J. Appl. Phys. 64:3551(1988).
4. J.L. Leray, J. Margail, and R.A.B. Devine, Electric field dependent paramagnetic defect creation in single step implanted SIMOX films, Proc. Mater. Sci. and Engr. B12:153(1992).
5. N.S. Saks and D.B. Brown, Observation of H^+ motion during interface trap formation, IEEE Trans. Nucl. Sci. 37:1624(1990). see also, R.E. Stahlbush, A.H. Edwards, D.L. Griscom, and B.J. Mrstik, J. Appl. Phys., Post-irradiation cracking of H_2 and formation of interface states in MOSFETs, J. Appl. Phys. submitted for publication, 1992.
6. M.E. Zvanut, R.E. Stahlbush, W.E. Carlos, H.L. Hughes, R.K. Lawrence, and R. Hevey; G.A. Brown, SIMOX with epitaxial silicon: point defects and positive charge, IEEE Trans. Nucl. Sci., 38:1253(1991).
7. K. Yokogawa, Y. Yajima, T. Mizutani, S. Nishimatsu, and K. Suzuki, Positive charges and E' centers formed by vacuum ultraviolet radiation in SiO_2 grown on Si", Jap. J. Appl. Phys. 29:2265(1990).

8. A. Kalnitsky, J.P. Ellul, E.H. Poindexter, P.J. Caplan, R.A. Lux, and A.R. Boothroyd, Rechargeable E' centers in silicon-implanted SiO_2 films", J. Appl. Phys. 67:7359(1990).
9. W.L. Warren, P.M. Lenahan, B. Robinson, and J.H. Stathis, Neutral E' centers in microwave downstream plasma-enhanced chemical-vapor-deposited silicon dioxide, Appl. Phys. Lett. 53:482(1988). see also, J.F. Conley, P.M. Lenahan, and P. Roitman, Electron spin resonance study of E' trapping centers in SIMOX buried oxides, IEEE Trans. Nucl. Sci. 38:1247(1991).
10. M.E. Zvanut, F.J. Feigl, W.B. Fowler, and J.K. Rudra; P.J. Caplan and E.H. Poindexter; J.D. Zook, Rechargeable E' centers in sputter-deposited silicon dioxide films, Appl. Phys. Lett. 54:2118(1989).
11. P. J. Caplan, E.H. Poindexter, B.E. Deal, and R.R. Razouk, ESR centers, interface states, and oxide fixed charge in thermally oxidized silicon wafers, J. Appl. Phys. 50:5847(1979).
12. H. Imai, K Arai, H.Hosono, Y. Abe, T. Arai, and H. Imagawa, Dependence of defects induced by excimer laser on intrinsic structural defects in synthetic silica glasses, Phys. Rev. B44:4812(1991).
13. E.J. Friebele, R.J. Ginther, and G.H. Sigel Jr., Radiation protection of fiber optic materials: effects of oxidation and reduction, Appl. Phys. Lett. 24:412(1974).
14. M.E. Zvanut, R.E. Stahlbush, and W.E. Carlos, Radiation-induced E' centers in H_2 annealed oxide films, Appl. Phys. Lett. June, 1992 (to be published).
15. E.H. Nicollian and J.R. Brews, "MOS (Metal Oxide Semiconductor) Physics and Technology," John Wiley & Sons, New York (1982).
16. P.S. Winokur, Radiation-induced interface traps, in: "Ionizing Radiation Effects in MOS Devices and Circuits," T.P. Ma and P.V. Dressendorfer, ed., John Wiley & Sons, New York (1989).
17. C.L. Marquardt and G.H. Sigel, Jr., Radiation-induced defect centers in thermally grown oxide films, IEEE Trans. Nucl. Sci. NS-22:2234(1975).
18. D.L. Griscom, Characterization of three E'-center variants in x- and γ-irradiated high purity a-SiO_2, Nucl. Instrum. Methods B1:481(1984).

REMOVAL OF TRAPS IN PROCESS-DAMAGED MOS STRUCTURES BY ROOM-TEMPERATURE HYDROGENATION

S. Kar

Department of Electrical Engineering
Indian Institute of Technology
Kanpur-208016, India

ABSTRACT

Silicon ions of 16 keV energy were implanted through the oxidized silicon samples to generate the electronic defects in the MOS structures. The oxide thickness was either 115 or 350 Å. Immediately after Si self-implantation, a Kaufman source was used to hydrogenate the radiation-damaged surface at room temperature. The experimental data from the admittance-voltage-frequency measurements suggest that, even when carried out at a low temperature, ion-beam hydrogenation is effective in neutralizing the effects of beam-induced damage.

INTRODUCTION

Fabrication of circuits with very or ultra large scale of integration is characterized by the presence of an ever-increasing amount of high energy particles, such as ions, electrons, and x-rays, in the processing environment. The larger particles, such as ions, generate both atomic displacement damage and ionization, in the layers these penetrate, while the smaller particles, only the latter. The basic electronic manifestation of beam/radiation induced damage is in the form of bonding defects, such as dangling or weak bonds. The electron traps, in turn, manifest themselves in electronic devices in a number of ways.

In MOS [metal-oxide-silicon] devices, considered to be very sensitive to beam/radiation induced damage, the electron/hole traps can be located in the silicon sub-surface [depending upon ion penetration], at the interface, and in the oxide. All of these traps manifest themselves in trap capacitance and conductance, and their characteristic signatures may also appear in such unusual features as low frequency dispersion of the accumulation capacitance, increase of the oxide dielectric constant, etc.[1].

Thermal annealing [if pre-metallization, then in N_2, and if post-metallization, then in H_2/N_2 ambient] has been a standard approach for removal of process-induced damage in MOS structures. However, with increasing circuit miniaturization, all processing temperatures need to be reduced to avoid wafer warpage, cross-diffusion, change of dopant profile, etc. The aim of this investigation has been to assess the

The Physics and Chemistry of SiO_2 and the Si-SiO_2 Interface 2
Edited by C.R. Helms and B.E. Deal, Plenum Press, New York, 1993

efficacy of ion beam hydrogenation for passivation of electronic traps, introduced by exposure to an energetic ion beam. In this first phase of the work, the subtrates were kept at 300 K during hydrogenation, so that mainly the interaction of atomic hydrogen with defects could be monitored, in isolation from the thermal effects.

EXPERIMENTAL DETAILS

The MOS samples were fabricated in the following manner. After wet chemical cleaning, p-type Si wafers oxidized in dry O_2 [containing 3 vol. % HCl] at 950 ^oC at atmospheric pressure to reach an oxide thickness of 115 or 350 Å. Immediately after oxidation, Si ions of 16 keV energy were implanted at 300 K into the oxidized Si surface in a Varian 350D ion implanter. Si ions were chosen to deemphasize the chemical effects of ion implantation. The range of Si ion dosage was $10^{12}/cm^2$ to $10^{14}/cm^2$.

Following Si ion implantation, hydrogenation was carried out at 300 K, by 400 eV hydrogen ions from a Kaufmann source. During ion beam hydrogenation, the ion current density was $1.2x10^{-3}$ A/cm^2, and the hydrogenation time was 5 min. This means that the incident hydrogen ion dosage was about $2x10^{18}/cm^2$. Following hydrogenation, Al front (1.0 mm dia dots) and Au back contacts were evaporated in an oil-free, ion-pumped Varian VT-112 ultra-high-vacuum system using filament sources. Care was taken to see that the samples were not subjected to any thermal treatment, after low energy hydrogenation, so that its effects alone could be observed.

The electrical characterization was carried out using the Hewlett Packard 4061S semiconductor/component test system, equipped with the 4192A impedance analyzer, the 4140B pA/DC voltage source, the 4083A switching controller, the 16057A switching module, the 7475 graphics plotter, and the 310M computer. Direct current-voltage [I-V], quasi-static capacitance-voltage [C-V], and sinusoidal small signal C-V, conductance-voltage [G-V], conductance-frequency [G-f] measurements were made.

The sinusoidal frequency was varied between 80 Hz and 3 MHz. The minimum ramp rate in the case of the quasi-static technique was 0.001 V/s. Since slower defect states are likely to be present in process-damaged devices, static C-V measurements[2] were made using a Keithley 595 meter, to access these states. In the static technique, a small step voltage is applied to the sample, and the resultant change in charge across the sample is measured after a time delay. For the static technique, the minimum step voltage was 0.01 V, and the maximum step delay was 200 s.

Ellipsometric data were obtained using a Rudolph Research A7905 automatic ellipsometer. All the admittance data acquisition and the subsequent analysis were carried out on the HP 310M computer, using a comprehensive software package developed by us. To determine the trap density, its capture cross-section, and its energy location, the standard approach was taken, i.e. it was assumed that all the traps are located at the $Si-SiO_2$ interface.

Values of the doping density and the effective series resistance were extracted from the measured high frequency C-V characteristic. From the measured static C-V characteristic the following parameters were extracted: the oxide capacitance, the experimental surface potential as a function of the applied bias, and the trap density as a function of the bandgap energy. From the measured G-f characteristics for various values of the applied bias, the following parameters were extracted: the trap density as a function of the bandgap energy, and the hole capture cross-section as a function of the bandgap energy.

RESULTS AND DISCUSSION

Table 1 displays data related to Si ion implantation and ion beam hydrogenation, for some of the samples investigated. Profiles of ion distribution [i.e. ion number density per unit dose versus target depth] and displacement distribution [i.e. total atomic displacements in number per Å per ion versus target depth] were calculated using the TRIM [Transport of Ion in Matter] program [3]. From these calculations, the mean ion range and the position of the peak target displacement were obtained. The data in Table 1 indicate that for the set A samples [with 115 Å thick oxides], the maximum atomic displacement damage by Si ions and their recoils was at the silicon-oxide interface, while for the set B samples [with 350 Å thick oxides], it was in the middle of the bulk oxide.

Table 1 also shows that for the set A samples, the hydrogen ion distribution had a significant overlap over the profile of atomic displacements by the Si ions/recoils, while for the set B samples, the overlap was much less. Further, the atomic displacements by H ions/recoils were far away from the oxide-silicon interface in the case of the set B samples, while these had a reach just up to the interface in the case of the set A samples.

Table 1. Data related to Si ion implantation and ion beam hydrogenation. Data on ion range and displacement damage were obtained by calculations.

Sample #	A15	A15/H	BC	BC/H	B12/H	B13	B13/H	B14/H
Oxide thickness [Å]	115	115	350	350	350	350	350	350
Si ion dosage [cm^{-3}]	10^{15}	10^{15}	0	0	10^{12}	10^{13}	10^{13}	10^{14}
Mean Si ion range [Å]	256	256	-	-	262	262	262	262
Peak of target displacements by Si ions [Å]	110	110	-	-	151	151	151	151
Mean H ion range [Å]	107	107	-	109	109	-	109	109
Peak of target displacements by H ions [Å]	52	52	-	63	63	-	63	63

The effects of ion beam hydrogenation, carried out at room temperature, were very pronounced on the admittance characteristics of the ion-beam-damaged MOS structures. The C-V, G-V, and G-f characteristics of the unpassivated samples deviated strongly from the standard MOS form.

Figure 1 presents the C-V characteristics of the unpassivated sample A15, measured at various frequencies, and Fig. 2 does the same for the hydrogenated sample A15/H, cf. Table 1. The Si ion dosage for samples A15 and A15/H was $10^{15}/cm^2$, which is much higher than what is needed for amorphization[4]. This very extensive damage is responsible for the unusual characteristics of Fig. 1. The G-V characteristics of sample A15 were also found to be flat, and exhibited strong frequency dispersion in all regions of bias, including accumulation. The unusual form of the admittance characteristics of sample A15 is the result of a combination of factors.

These include the generation of a very high density of traps in the oxide, at the interface, and in the silicon sub-surface. These traps, in turn, produce many secondary effects, such as pinning of the Fermi level; neutralization of the shallow dopants in the silicon sub-surface, leading to an intrinsic layer and the formation of a pi junction; and an increase in the oxide dielectric constant at low frequencies by space-charge effects in the oxide traps [1,5].

The C-V characteristics of the hydrogenated sample in Fig. 2 exhibit the standard form. Although a high density of traps is still present, Fig. 2 indicates a significant passivation of ion-beam-induced defects by hydrogenation. The efficacy of the latter can be better appreciated, by considering the fact that, post-metallization thermal annealing in pure

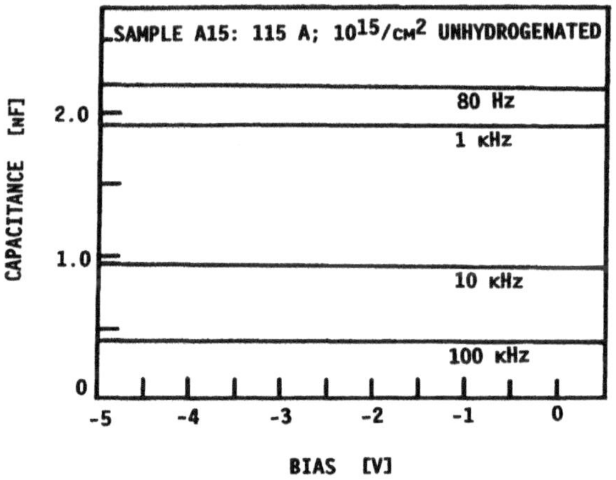

Figure 1. Capacitance-voltage characteristics of the unhydrogenated sample A15 [cf. Table 1], measured at different frequencies.

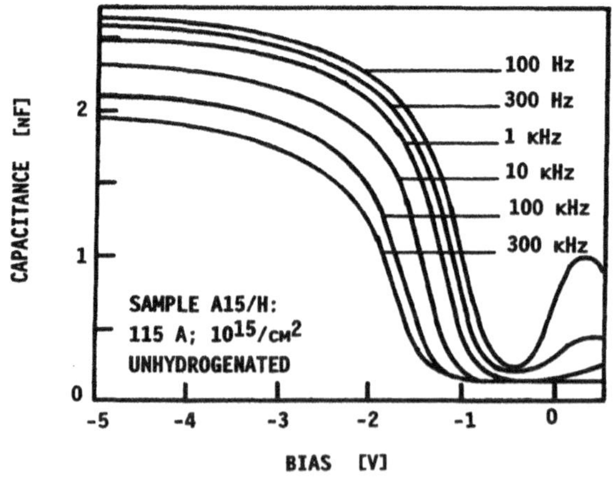

Figure 2. Capacitance-voltage characteristics of the hydrogenated sample A15/H [cf. Table 1], measured at different frequencies.

H_2, even at 450 °C up to 1 hour, failed to bring about the slightest change in the flat admittance characteristics of samples, identical to A15.

The experimental trap density, D_{it}, obtained as a function of the silicon bandgap energy, E, has been presented in Fig. 3, for the set B samples of Table 1. Figure 3 contains two values of the trap density, one obtained from the measured static C-V [solid lines], and the other from the G-f [broken lines] characteristics. The experimental data on trap density showed a significant reduction upon ion beam hydrogenation for the radiation-damaged samples, as can be seen from the results for samples B13 and B13/H. However, for the control [i.e. unexposed to Si ion beam] sample BC, there is an increase in the trap density after hydrogenation.

One significant feature of the $D_{it}(E)$ profiles of all the hydrogenated samples, including the control sample BC/H, is that these have nearly the same form, and display a peak in the upper half of the bandgap, i.e. in the energy range of 0.70-0.83 eV above the valence band edge. This result suggests the creation of some new traps during ion beam hydrogenation. In

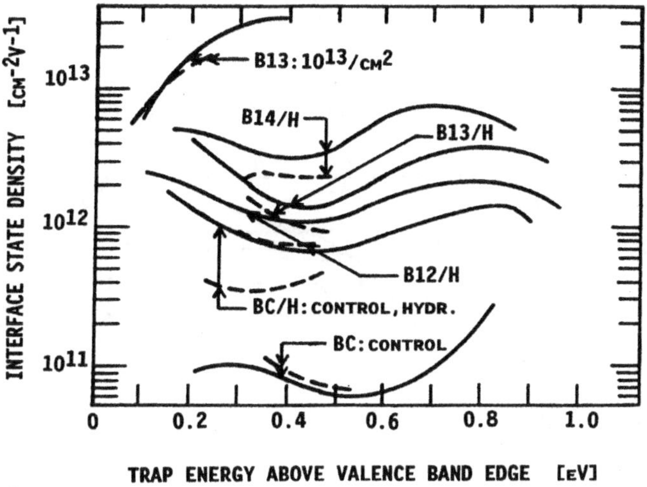

Figure 3. Interface trap density, D_{it}, obtained experimentally from the static C-V characteristics [solid lines], as well as from the G-f characteristics [broken lines], as a function of the bandgap energy, E, for the set B samples of Table 1.

principle, the new traps may be created by [i] atomic displacements by the H ion, [ii] formation of new defect complexes with H, or [iii] H-ion-attendant x-rays.

The first possibility seems remote in view of the fact that, according to the TRIM calculations, the atomic displacements by the H ions/recoils were concentrated in the outer half of the oxide for the set B swamples, cf. Table 1. The second possibility is dependent upon the extent of H diffusion to the interface and the silicon sub-surface, from the outer half of the oxide, at the low substrate temperature. In this connection, an interesting observation is that there was no evidence in the admittance characteristics, of acceptor neutralization by hydrogen, as has been earlier detected during hydrogenation of silicon [6]. This ia an indication of limited hydrogen diffusion into the bulk silicon and silicon sub-surface. The third possibility may be considered to be more likely, and will lead to the creation of traps at the interface [7].

The experimental hole capture cross-section, σ_h, obtained as a function of the bandgap energy, E, is presented in Fig. 4 for the set B samples of Table 1. The experimental data on unhydrogenated [i.e. unpassivated] set B samples indicated very small values for the capture cross-section, and a good correlation of the values of the trap density obtained from static capacitance with those obtained from ac conductance [1,8]. These two features [illustrated by the data for sample B13 in Figs. 3,4] are characteristics of slow oxide traps that are within a tunneling distance of the interface, and suggest that in the unpassivated samples, the oxide traps dominated [1]. This conclusion is supported by the results of TRIM calculations, which show that for the set B samples, the atomic displacements by the Si ions/recoils were confined mainly within the oxide.

After hydrogenation, it can be seen in Fig. 4 that, the traps exhibit much higher values of capture cross-section. Further, as Fig. 3 indicates, the value of the trap density obtained from ac conductance is significantly smaller than that obtained from the static capacitance. These features suggest that in the hydrogenated samples, the residual traps in the silicon sub-surface are dominating. This result appears to agree with the TRIM calculations, which as mentioned already, indicated implanted H ions to be confined mainly to the oxide. Hence, one could expect more effective passivation of the oxide traps than the ones in the silicon sub-surface.

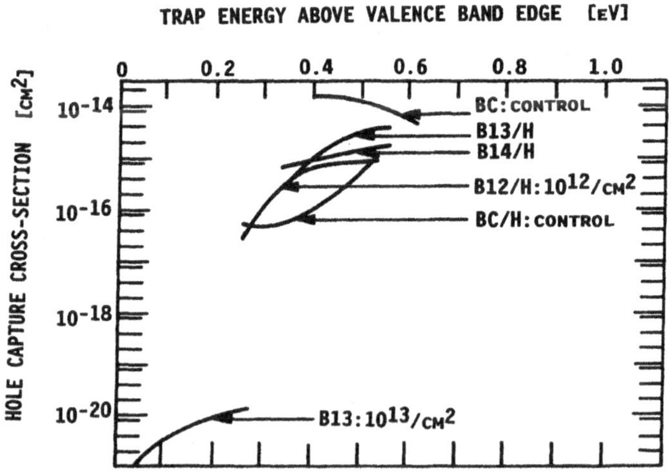

Figure 4. Hole capture cross-section, obtained experimentally from the G-f characteristics, as a function of the bandgap energy, E, for the set B samples of Table 1.

Typical examples of the experimental G_p/ω [G_p is the parallel conductance and ω the angular frequency], obtained as a function of the frequency, f, at some discrete values of the applied bias, for the unpassivated sample B13, is shown in Fig. 5, and the same for the hydrogenated sample is shown in Fig. 6. To facilitate analysis, the value of the surface potential corresponding to each bias, ψ_s, obtained experimentally from the static C-V characteristic, has been indicated in Figs. 5,6.

The G_p/ω versus f characteristic, especially the frequency at which the maximum, $[G_p/\omega]_{max}$, occurs, is useful in assessing the space location of traps[1]. It can be seen in Fig. 5, for the unpassivated sample B13, that even for bias corresponding to accumulation, the maxima are in the

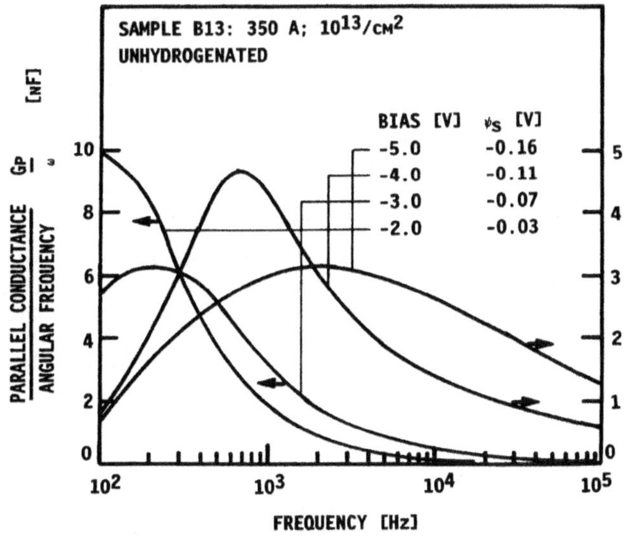

Figure 5. The experimental G_p/ω [G_p is the parallel conductance and ω the angular frequency], obtained as a function of the frequency, f, at some discrete values of the applied bias, for the unpassivated sample B13, cf. Table 1.

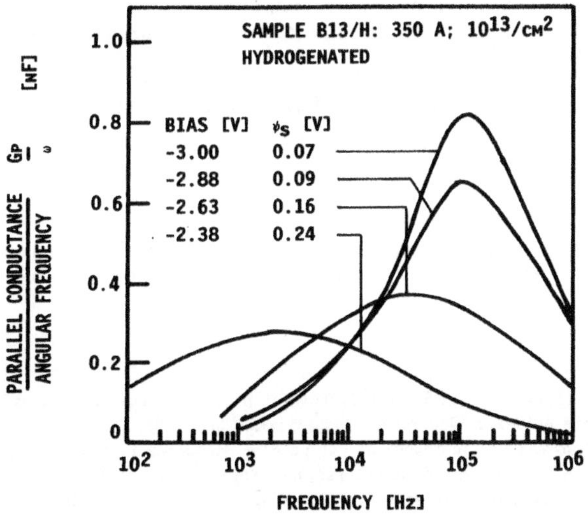

Figure 6. The experimental G_p/ω [G_p is the parallel conductance and ω the angular frequency], obtained as a function of the frequency, f, at some discrete values of the applied bias, for the unpassivated sample B13/H, cf. Table 1.

low frequency range, which suggests dominance of oxide traps, as has been mentioned earlier.

The G_p/ω versus f characteristics of the hydrogenated sample B13/H, in Fig. 6, are significantly different, from those of the unhydrogenated sample B13 in Fig. 5, on two counts. Firstly, the peaks, $[G_p/\omega]_{max}$, in Fig. 6, are located in the high frequency range, about two orders of magnitude higher in frequency. Secondly, the magnitudes of the peaks, which represent the trap density, are about an order of magnitude lower in Fig. 6, signifying the extent of reduction in trap density upon hydrogenation. The frequency location of the peaks in Fig. 6 suggests the dominant traps to be in the silicon subsurface.

It was observed that during ion beam hydrogenation, about 10 % of the silicon dioxide layer got etched. The net etching by H ions is determined by the the ion dosage, ion energy, and the ion current density. Future experiments will focus on optimization of these parameters to reduce the amount of etching.

Inspite of the oxide etching, it was observed that hydrogenation brought about a significant reduction in the oxide leakage current and oxide fragility. It has generally been observed that high dose implantation through the oxide renders it very fragile and prone to breakdown, and it is this damage which is very difficult to get rid of by thermal annealing.

CONCLUSIONS

In summary, the results of room-temperature passivation by atomic hydrogen appear to be very promising for the removal of ion-beam-induced defects in MOS structures, especially, when the oxide is thin. This technique should be applicable also for the removal of defects introduced by other energetic particles during device processing. Many interesting features have been observed that bear upon our understanding of the interaction of atomic hydrogen with silicon, its oxide, its interface, and defects.

Ion beam hydrogenation at room temperature was found to introduce new traps, and there was etching of the oxide [by about 10 %] by the H ions. Future experiments will focus on solution of these problems, and on reducing the trap levels to what is required for integrated circuits, if necessary, by low [< 300 ^{o}C] temperature substrate heating during hydrogenation.

ACKNOWLEDGEMENTS

This work has been supported by the National Science Foundation, Washington, D.C. and the Department of Science and Technology, New Delhi.

REFERENCES

1. S.Kar,A.Pandey,A.Raychaudhuri,and S.Ashok,Effects of process-induced damage on MOS structures with 115 Å thin gate oxides,J.Electrochem.Soc.139:xxxx(July 1992).
2. K.Ziegler and E.Klausmann,Static technique for precise measurement of surface potential and interface state density in MOS structures,Appl.Phys.Lett.26:400(1975).
3. J.F.Ziegler,J.P.Biersack,and U.Littmark."The Stopping Range of Ions in Solids," Pergamon Press,New York(1985).
4. J.F.Gibbons,Ion implantation in semiconductors-Part II: Damage production and annealing, Proc. IEEE 60:1062(1972).
5. S.Kar, A.Raychaudhuri,A.K Sinha,and S.Ashok,Ion beam modification of the dielectric properties of thin silicon dioxide films,Appl.Surf.Sci.48/49:264(1991).
6. J.I.Pankove."Semiconductors and Semimetals,"Vol. 34,Academic Press,New York(1991),p.45.
7. F.B.McLean,A framework for understanding radiation-induced interface states in SiO_2 MOS structures, IEEE Trans.Nucl.Sci.NS-27:1651(1980).
8. A.Raychaudhuri, S.Chatterjee,S.Ashok,and S.Kar,Ion-dosage-dependent room-temperature hysteresis in MOS structures with thin oxides,IEEE Trans.Electron.Devices 38:316(1991).

HOT-ELECTRON INDUCED HYDROGEN REDISTRIBUTION IN SiO_2

D.A Buchanan, A.D. Marwick, and D.J. DiMaria

IBM Research Division
T.J.Watson Research Center
P.O. Box 218,
Yorktown Heights, NY 10598

L.Dori

C.N.R. LAMEL - Institute
Via De' Castagnoli
1 Bologna, Italy

ABSTRACT

Hydrogen redistribution within hot-electron injected metal-oxide-silicon capacitors has been studied following electron injection by internal photoemission using hydrogen profiling by nuclear reaçtion analysis. Measurable hydrogen buildup (~ 10^{14} H/cm^2) at the Si/SiO_2 interface occurs for injection fluences of order 1 C/cm^2. The amount of buildup is larger for negative-gate injections than for positive gate, whereas the number of interface states, which appear for much lower fluences, is greater in the positive bias case. Since it is known that hydrogen is involved in the generation of interface states, these observations suggest that different mechanisms operate in the low and high-fluence regimes. Large amounts of hydrogen (~ 10^{15} H/cm^2) are found to be released from the Al/SiO_2 interface during the charge injection. Most of this hydrogen leaves the samples, but for positive gate bias there is evidence that some enters the SiO_2 as a species with a relatively low diffusion coefficient. In addition, some hydrogen enters the Si substrate and is detected by its passivation of boron atoms. The amount of passivation shows a better correlation with the interface state density than does the amount of hydrogen at the Si/SiO_2 interface.

INTRODUCTION

Hydrogen redistribution is believed to be the key process in the generation of interface states in metal-oxide-semiconductor (MOS) devices by hot electrons.[1-7] In the generally accepted model, hydrogen is released by hot electrons, diffuses to the Si/SiO_2 interface and reacts with precursor defects to produce electrically active interface states. Previous work[1-3] using secondary ion mass spectroscopy (SIMS) confirmed that large amounts of hydrogen segregate to the Si/SiO_2 interface in samples exposed to large electron fluences, though the absolute calibration was subject to uncertainties intrinsic to SIMS. The source of the hydrogen was not established, though it was suggested that it was the Al/SiO_2 interface. The data indicated that the initial ratio of the rates of hydrogen arrival at the interface to generation of interface states was about 2:1. However, the hydrogen arrival rate had to be extrapolated from much higher injection fluences due to the limited sensitivity of the measurements. The electrons were injected from the substrate using avalanche injection (positive gate bias).

In the present study we have extended the work of Gale et al.[1-3] by using nuclear reaction analysis[8] (NRA) for hydrogen profiling and by using internal photo-emission[7, 9] (IPE) for the electron injection. The NRA hydrogen profiling is not subject to the same calibration uncertainties as SIMS. By using IPE, we are able to compare the interface state generation with the hydrogen segregation for both positive and negative gate bias. We find that for positive gate bias, V_g^+, the interface state generation is higher than for negative gate bias, V_g^-. However, the opposite polarity dependence is found for the hydrogen buildup at the Si/SiO_2 interface. This apparent anti-correlation calls into question the validity of extrapolating from high electron fluence to the much lower fluences where interface state generation occurs. On the other hand, we find that the correlation of hydrogen passivation of boron in the silicon substrate with interface state generation previously noted by Gale et al.[1-3], still broadly holds for the wider range of conditions explored in the present work.

SAMPLE PREPARATION

The substrates used in this work were p-type <100> silicon wafers with a resistivity of 1-2 Ω-cm. After a standard clean, SiO_2 films were thermally grown with thicknesses varying from ~600 to 950 Å. Following the oxidation, some SiO_2 films were exposed at 200°C for 30 minutes to 2 liters/minute of nitrogen flowing through water at 55°C. Throughout this paper we shall denote the water-exposed oxides as "wet" SiO_2 films whereas those without the water exposure shall be denoted as "dry". The water exposures were performed to produce films with a higher interface state generation rate[7, 9] so that a comparison with the effects of hydrogen redistribution and interface state generation could be made. MOS capacitors were formed by thermal evaporation of ~150 Å of aluminum through a shadow mask to produce dots of diameter ~1.5 mm. Some samples were made with poly-silicon gates consisting of ~500 Å of in-situ arsenic doped poly-silicon deposited, which was annealed in Ar at 1000°C for 15 minutes prior to the evaporation of Al contacts.

ELECTRICAL AND NUCLEAR MEASUREMENTS

An ultra-violet mercury lamp was used to inject electrons into the SiO_2 films via internal photo-emission, IPE, from the thin Al-gate or from the silicon substrate. The details of this technique have been reported elsewhere[7, 9, 10]. With IPE, electrons are emitted from the cathode over the interfacial barrier. Samples were subjected to electron fluences ranging from $0.01 - 5\ C\ cm^{-2}$ at fields of 1-5 MV/cm. For the polysilicon-gate samples, Fowler-Nordheim tunneling was used, and the field was maintained at a much higher level (7-9 MV/cm) to obtain an appreciable current density. Interface state densities were measured at periodic intervals throughout the injection. The density was determined using the high-low frequency method where both the high frequency (100 kHz) and the quasi-static capacitance were measured as a function of the applied bias.

Nuclear reaction analysis was used to determine the hydrogen profiles in the samples. Electron-injected MOS dots were compared with uninjected dots and with hydrogen-free silicon blanks, which were used to establish the background level and hence the sensitivity of the measurements. We used the sharp resonance at 6.4 MeV in the ^{15}N reaction with hydrogen, which emits a 4.3 MeV gamma ray, and varied the beam energy to form a depth profile.[8] The beam was collimated to 1 mm diameter and aligned with the MOS dots using a coaxial telescope system. The typical beam dose per depth point was 20 μC. The gamma rays were detected with a 3×3 inch BGO scintillator. To minimize beam-induced redistribution of hydrogen during the measurement, the samples were cooled to 104 K and the beam current limited to 90 nA of $^{15}N^{2+}$. Under these conditions the change in the hydrogen profile during the measurement was found to be negligible. Raw data (gamma ray counts per μC of $^{15}N^{2+}$ vs beam energy) were converted to a depth scale using a simulation program.[11] Hydrogen concentrations were deduced by reference to a calibration sample of hydrogen implanted silicon. Previous measurements have confirmed the applicability of these techniques to the measurement of hydrogen profiles in MOS structures.[11-15]

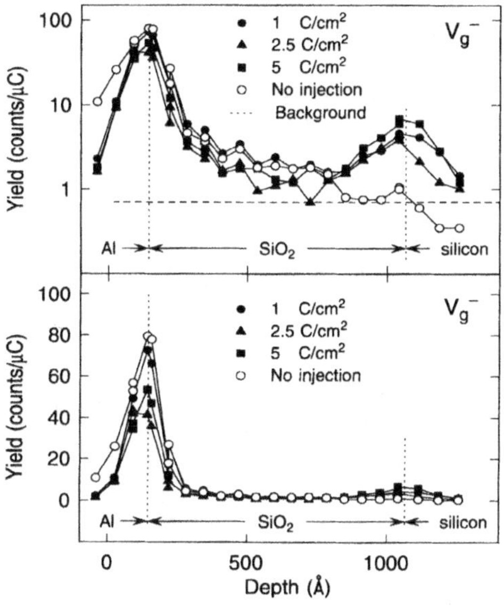

Figure 1. . Hydrogen profiles of the "dry", MOS structures. for V_g^- and electron fluences of 1, 2.5 and 5.0 C/cm². Note ordinate scale change.

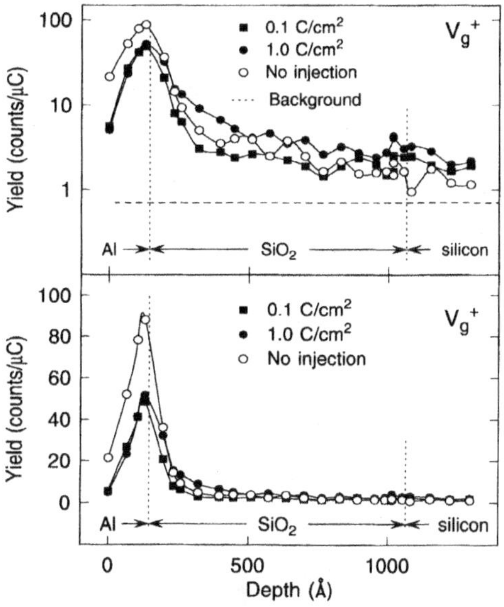

Figure 2. . Hydrogen profiles of the "dry", MOS structures for V_g^+ and electron fluences of 0.1, and 1.0 C/cm². Note ordinate scale change.

HYDROGEN PROFILES

In Figs. 1 and 2 we show the hydrogen profiles in "dry" MOS structures for injection from the Al gate, V_g^-, and the silicon substrate, V_g^+, respectively. Hydrogen concentrations deduced from this data are given in Table 1. We note the following : 1) prior to electron injection, there is a large hydrogen concentration at the Al/SiO_2 interface; 2) this hydrogen decreases after electron injection, and this decrease appears independent of bias polarity; 3) a peak of hydrogen appears at the Si/SiO_2 interface after injection of electrons; 4) this hydrogen peak increases with increasing electron fluence and is greater for negative bias than for positive bias at the same injected fluence level; 5) For V_g^+, the hydrogen concentration in the SiO_2 bulk near the Si/SiO_2 interface increases with increasing electron fluence.

Table 1. Hydrogen concentrations deduced from Figs. 1 and 2. The indicated accuracies reflect statistical uncertainties.

Electron fluence	**Al / SiO_2 (cm^{-2}) (±10%)**	**Bulk SiO_2 (cm^{-3}) (±30%)**	**Si / SiO_2 (cm^{-2}) (±20%)**
Negative gate bias, V_g^-			
Not injected	$1.5x10^{15}$	$7x10^{18}$	--
1 C/cm^2	$1.0x10^{15}$	$5.7x10^{18}$	$1.1x10^{14}$
5 C/cm^2	$1.0x10^{15}$	$4.6x10^{18}$	$2.0x10^{14}$
Positive gate bias, V_g^+			
Not injected	$1.8x10^{15}$	$6x10^{18}$	--
1 C/cm^2	$1.0x10^{15}$	$6.8x10^{18}$	$5.0x10^{13}$

In Fig. 3, the buildup of hydrogen at the Si/SiO_2 interface as a function of increasing fluence is shown for "wet" oxides under negative gate bias. These samples also show an increase in hydrogen concentration at the Si/SiO_2 interface as a function of increasing electron fluence. A comparison between the "wet" and "dry" SiO_2 films injected using V_g^- for various fields and a fluence of 0.5 C/cm^2 is shown in Fig. 4. The "wet" SiO_2 films show a slightly higher level of hydrogen at the Si/SiO_2 interface than the "dry" SiO_2 films. For the "wet" SiO_2 samples the hydrogen concentration is largest for the largest electric field.

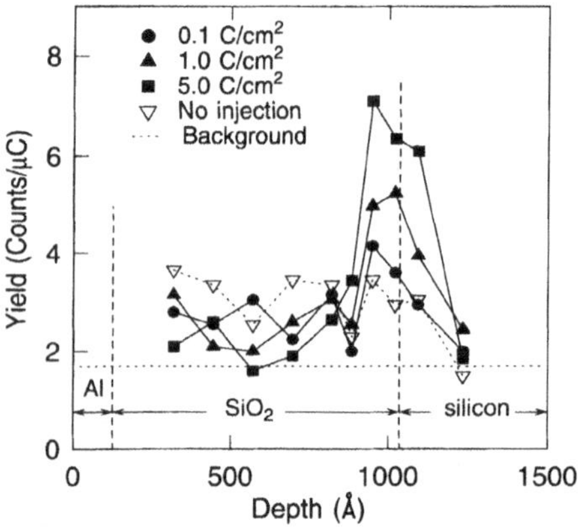

Figure 3. . Hydrogen profiles in "wet" films injected using V_g^- for electron fluence levels of 0.1, 1 and 5 C/cm^2.

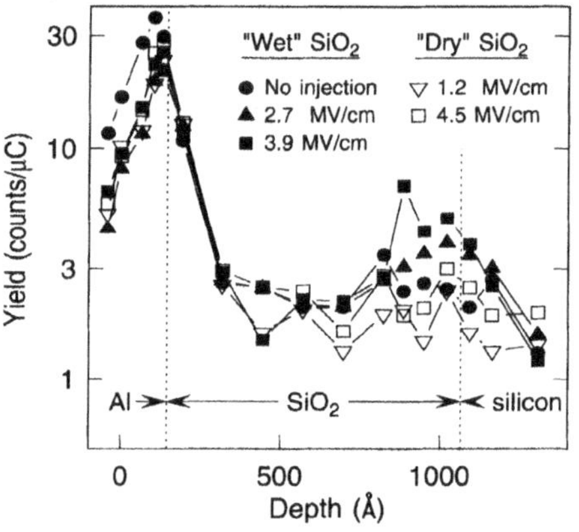

Figure 4. . Field dependence of the hydrogen profiles for V_g^- for injection fluences of 0.5 C/cm^2 comparing "dry" and "wet" MOS samples.

HYDROGEN REDISTRIBUTION

In Figs. 1 and 2 the hydrogen concentration at the Al/SiO_2 interface decreases substantially after electron injection, independent of the bias polarity. For V_g^+, the electrons are emitted from the silicon substrate over the interfacial barrier into the SiO_2 film where they gain energy from the applied field. After traversing the oxide, the electrons impinge upon the hydrogen-rich layer at the Al/SiO_2 interface where, given that they have gained sufficient energy, dislodge hydrogen.

Some of this hydrogen appears at the Si/SiO_2 interface and some contributes to the appearance of the *shoulder* in the profile near the Al/SiO_2 interface (see Fig. 2). Presumably, the shoulder results from the slow diffusion of a hydrogen-containing species. For V_g^-, electrons are emitted over the Al/SiO_2 interface. Their initial energy is determined by the spectrum of the light source and the density of states in the injecting contact. The electrons entering the SiO_2 do so with energies 1-2 eV above the conduction band edge. Some of them have enough energy to release hydrogen. For either injection polarity, therefore, these results, for the first time confirm that the Al/SiO_2 interface can act a a source of hydrogen within the MOS structure.

In Fig. 4, there is a change in the hydrogen concentration at the Si/SiO_2 interface as the field is increased, for both the "wet" and "dry" films. For low electric fields, ≤ 1.5 MV/cm, the electrons are unable to gain much energy from the field and therefore have energies close to the SiO_2 conduction band edge.[16, 17] For fields greater than ~1.5 MV/cm, referred to as the "heating threshold",[6, 16, 17] the average path length of the electrons increases, enabling them to gain energy from the applied electric field. Once the field within the SiO_2 exceeds the heating threshold, these "hot" electrons can dislodge hydrogen from an SiO_2 film. The data in Fig. 4, show the expected increase in the hydrogen concentration with increasing field.

INTERFACE STATE GENERATION

For all Al-gate samples, the generation of interface states was found to occur on a fluence scale much shorter than the appearance of the hydrogen peak at the Si/SiO_2 interface. At high fluences, where the hydrogen redistribution was observed, the interface state density saturated. Some typical interface state spectra are shown in Fig. 5.

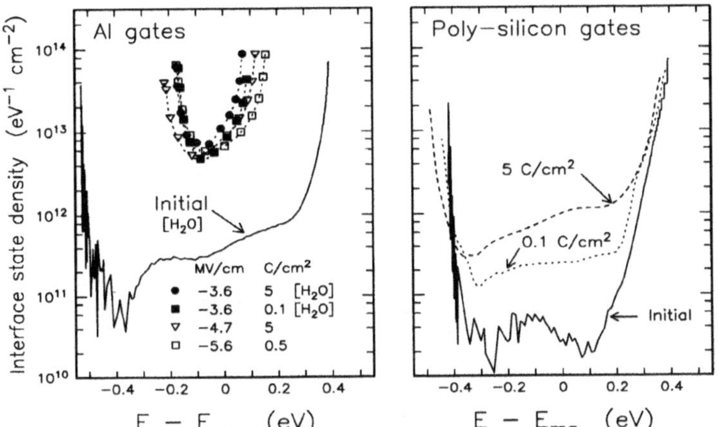

Figure 5. Interface state density, D_{it}, before and after electron injection for various films and bias conditions including "wet", "dry" and poly-gate samples.

Interface state densities, D_{it}, were measured for both "wet" and "dry" SiO_2 films, under various field and fluence conditions. The initial density of interface states at midgap, $D_{it}(mg)$, typically in the low $10^{11}\ eV^{-1}\ cm^{-2}$ prior to injection, was much higher in the Al-gate samples than in those with poly-silicon gates. After even the smallest fluences, and under virtually any bias condition we find that the density of interface states for the Al-gate structures was saturated, as shown in Fig. 5. The "wet" SiO_2 films showed slightly higher values. For comparison we have also shown the interface state generation for the poly-silicon gate samples. The generation of interface states in these samples is not nearly as rapid as for the the Al-gate samples. This is due to the higher hydrogen concentration in the latter.[2-7] The nuclear reaction profiles for the poly-gate samples, even for fluences of 5 C/cm^2, showed hydrogen concentrations indistinguishable from the uninjected films.

As reported in the literature[3, 18, 19], charge injection of MOS devices causes hydrogen to enter the Si substrate. The passivation of the boron atoms in the substrate, which occurs when B-H pairs form, was measured using capacitance-voltage profiling. Figure 6 shows the number of hydrogen atoms in B-H pairs (i.e. the number of passivated B atoms), as a function of injected fluence for samples prepared in different ways. Although there may be hydrogen in the silicon in other forms, in particular as neutral hydrogen complexes[20] which are not detected by the capacitance-voltage measurements, nevertheless these data give us some insight into the possible fate of some of the hydrogen released from within the MOS device. In particular it is of interest to compare the amount of hydrogen in B-H pairs with the amount of hydrogen at the Si/SiO_2 interface as measured both by direct hydrogen concentration profiling and by the buildup of interface states.

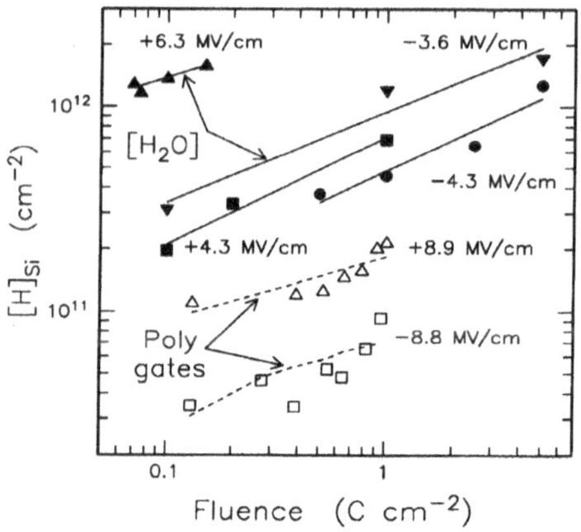

Figure 6. The hydrogen buildup in the silicon substrate is shown as a function of the injected electron fluence. The substrates are p-type silicon and the hydrogen is calculated by measuring the decrease in the electrically active boron concentration near the Si/SiO_2 interface from the high frequency capacitance-voltage data. Note the higher hydrogen concentration in the silicon for the "wet" SiO_2 and for V_g^+.

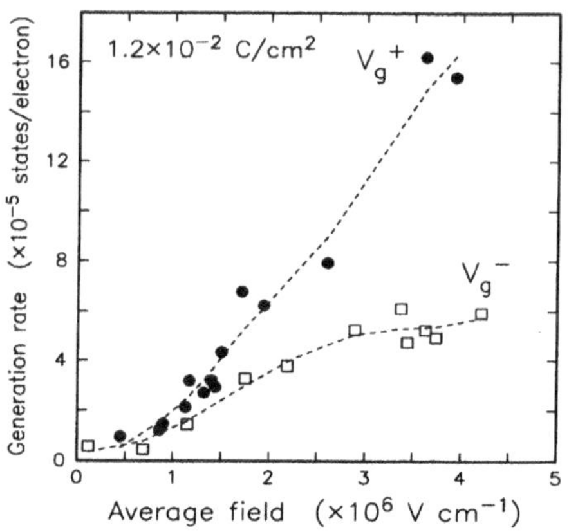

Figure 7. Field dependence of the interface state generation rate for V_g^+ and V_g^-. These data was obtained from a set of thin Al gate samples similar to those denoted here as "dry". This data demonstrates the more rapid generation of interface states for V_g^+ than for V_g^- once the average field exceeds the electron heating threshold.

This comparison shows that the amount of hydrogen in the substrate correlates better with the number of interface states than with the hydrogen buildup measured by NRA. Figure 6 shows that the amount of B passivation is highest for the "wet" MOS samples, which are also those in which the interface state density is highest, whereas the NRA-measured hydrogen concentration buildup is about the same in "wet" and "dry" SiO_2 films. Furthermore, the number of B-H pairs in the substrate has the opposite dependence on field polarity from that of the interface hydrogen concentration measured by NRA: the interface hydrogen is highest for V_g^-, whereas the number of passivated boron atoms is highest for V_g^+. For "dry" oxide Al-gate samples, the number of passivated B atoms is lower than in the "wet" SiO_2 case, and the polarity dependence is smaller (see Fig. 6). These variations again do not correlate well with the NRA measurement of the segregated hydrogen, but do show the same trend as interface state density. The overall correlation with interface state density suggests that hydrogen atoms in the SiO_2/substrate interface can either react to form interface states or can instead enter the silicon. The data shown in Fig. 7, from another study,[9] show that the interface state generation rate above the electron heating threshold, is substantially greater for V_g^+ than it is for V_g^+, thus supporting the correlation between the interface state buildup and the hydrogen within the silicon substrate. The field dependence of the substrate passivation may also be assisted by the field induced drift of H^+ in the substrate.[21]

DISCUSSION

We now turn to a comparison of the interface state generation with the number of hydrogen atoms accumulating at the Si/SiO_2 interface. The most striking feature of our results is that the long-range hydrogen diffusion as measured by NRA is much slower than the interface state buildup: the injection fluences needed to produce a measurable ($\geq 10^{14}$ H/cm^2) buildup of hydrogen are two orders of magnitude greater than those sufficient to saturate the number of interface states.

Although it is well established[6, 7, 9, 22] that the interface state generation correlates with the presence of hydrogen in the SiO_2, our present results show that the bias polarity dependence of interface state generation does NOT correlate well with that of the hydrogen redistribution measured at high fluence levels. This suggests that the hydrogen which participates in interface state generation reaches the interface by a process different from that which causes measurable hydrogen buildup at the interface. Therefore, a question arises regarding previous estimates of the hydrogen buildup[1-3] at low fluences extrapolated from high-fluence data. On the other hand, the present results support the correlation of interface state generation with the boron passivation in the substrate.

The decrease in hydrogen at the Al/SiO_2 interface that we see in injected samples occurs independent of bias polarity and gives for the first time experimental confirmation of the suggestion made by many previous workers that this interface is a source of hydrogen in charge-injected MOS structures. It is important to note that much of this hydrogen leaves the sample, at least in these samples which have very thin gates. The fate of the remaining hydrogen is different for samples stressed with different bias polarity, as already noted. The more rapid buildup of hydrogen at the Si/SiO_2 interface at high injection fluences for V_g^- perhaps suggests that some of the hydrogen is diffusing as a negative species. However, our observation of the shoulder or diffusion front near the A/SiO_2 interface for V_g^+ suggests the presence of a slower moving neutral or positive species.

CONCLUSIONS

This study of hot electron injection in Al-gate MOS capacitors has confirmed that large amounts of hydrogen ($\geq 10^{14}$ $atoms/cm^2$) segregate to the Si/SiO_2 interface at high injected fluences. The amount of segregated hydrogen showed an unexpected bias dependence; it was much larger in samples injected with negative gate bias. The source of the hydrogen appeared to be the Al/SiO_2 interface, which contained a high concentration of hydrogen, amounting to 1.5 - 2×10^{15} H/cm^2. Charge injection caused 0.5 - 1×10^{15} $atoms/cm^2$ of this hydrogen to be released, much of which left the sample. The remainder either appeared at the Si/SiO_2 interface (for negative bias), or for positive bias slowly entered the SiO_2 near the Al/SiO_2 interface.

Interface state generation was faster for positive gate bias, unlike the hydrogen segregation. Also, the number of interface states saturated for fluences < 0.1 C/cm^2, which was not enough to produce measurable hydrogen segregation. These differences imply that the measured segregation, which was detectable only at high fluences, was not caused by the same mechanism as interface state generation.

The amount of hydrogen entering the Si substrate during the charge injection was found to have a similar dependence on bias polarity and fluence as the interface state generation, suggesting that the two phenomena are related.

ACKNOWLEDGEMENTS

The authors would like to thank A. Parsons for his contributions to the sample preparation, M.V. Fischetti, E. Cartier and J. Stathis for valuable discussions concerning electron transport and defect generation in SiO_2 films and T.H. Zabel and P.A. Saunders for their assistance with the NRA measurements.

REFERENCES

[1] R. Gale, F. J. Feigl, C. W. Magee, and D. R. Young, J. Appl. Phys. **54**, 6938 (1983).
[2] F. J. Feigl, R. Gale, H. Chew, C. W. Magee, and D. R. Young, Nuclear Instruments and Methods in Physics Research **B1**, 348 (1984).
[3] R. Gale, H. Chew, F. J. Feigl, and C. W. Magee, The Physics and Chemistry of SiO_2 and the Si-SiO_2 Interface 177 (1988).
[4] Y. Nissen-Cohen, J. Shappir, and D. Frohman-Bentchkowsky, J. Appl. Phys. **60**, 2024 (1986).
[5] D. J. DiMaria, Appl. Phys. Lett **51**, 655 (1987).
[6] D. J. DiMaria and J. Stasiak, J. Appl. Phys. **65**, 2342 (1989).
[7] D. A. Buchanan and D. J. DiMaria, J. Appl. Phys. **67**, 7439 (1990).
[8] W. A. Lanford, H. P. Trautvetter, J. F. Ziegler, and J. Keller, Appl. Phys. Lett. **28**, 566 (1976).
[9] D. A. Buchanan, Appl. Phys. Lett. **60**, 216 (1992).
[10] D. A. Buchanan, M. V. Fischetti, and D. J. DiMaria, Appl. Surface Science **39**, 420 (1989).
[11] A. D. Marwick, J. C. Liu, T. H. Zabel, and J. P. Doyle, Nucl. Instrum. Methods **B64**, 408 (1992).
[12] A. D. Marwick and D. R. Young, J. Appl. Phys. **63**, 2291 (1988).
[13] M. A. Briere, F. F.Wulf, and D. Bräunig, Nucl. Instrum. Methods **B45**, 45 (1990).
[14] M. A. Briere and D. Bräunig, IEEE Trans. Nucl. Sci. **37**, 1658 (1990).
[15] M. A. Briere, R. Radke, and D. Bräunig, *Total dose effects in MOS systems: Influence of initial hydrogen system*, ESA Electronic Components Conference, Noordwijk, The Netherlands, 1990.
[16] M. V. Fischetti, D. J. DiMaria, S. D. Brorson, T. N. Theis, and J. R. Kirtley, Phys. Rev. B **31**, 8124 (1985).
[17] M. V. Fischetti and D. J. DiMaria, Phys. Rev. Lett. **55**, 2475 (1985).
[18] C. T. Sah, J. Y. Sun, and J. J. Tzou, Appl. Phys. Lett. **43**, 204 (1983).
[19] C. T. Sah, S. C. Pan, and C. C. Hsu, J. Appl. Phys. **57**, 5148 (1985).
[20] A. D. Marwick, M. Wittmer, and G. S. Oehrlein, Mat. Sci. Forum **83-87**, 39 (1992).
[21] C. Herring and N. M. Johnson, in *Hydrogen in Semiconductors*, edited by J. I. Pankove and N. M. Johnson (Academic Press, New York, 1991), p. 225.
[22] Y. Nissan-Cohen and T. Gorczyca, Electron. Dev. Lett. **EDL-9**, 287 (1988).

EFFECTS OF INTRODUCING H_2 INTO IRRADIATED MOSFET'S FROM ROOM TEMPERATURE TO 250°C

R. E. Stahlbush[1] and A. H. Edwards[2]

[1]Naval Research Laboratory, Code 6816
Washington, DC 20375
[2]Dept. of Elec. Eng., Univ. of North Carolina
Charlotte, NC 28223

ABSTRACT

The effects of introducing molecular hydrogen into the gate oxide of irradiated MOSFETs are studied from room temperature to 250°C using charge pumping. At room temperature the dominant reactions are initiated by H_2 cracking. This reaction sequence is characterized by the simultaneous interface state formation and trapped positive charge decay. Molecular orbital calculations indicate that the H_2 is cracked at oxygen hole traps. Above 100°C the introduction of H_2 forms anomalous positive charge (slow states). The annealing of fast interface states is more rapid under negative bias and is in qualitative agreement with the theoretical predictions of Edwards. At 250°C, there is annealing of all defects. The formation and annealing kinetics of fast states indicate that there are two distinct interface defects responsible for the fast states. The defect that forms more slowly and anneals more rapidly does not appear to be either a P_{b0} or a P_{b1} center.

INTRODUCTION

Hydrogen has dramatic and complex effects on MOS devices. For over twenty years, low temperature (~400° C) post-metallization anneals in forming gas have been used to reduce interface state densities by 1-2 orders of magnitude.[1] On the other hand, experiments also show that hydrogen can have deleterious effects. It can amplify hot electron effects[1,2,3] as well as radiation effects.[4,5] There is evidence of hydrogen's role in interface state formation during irradiation, provided by pulsed irradiation experiments.[6,7] All of these experimental data were obtained for hydrogen introduced during the fabrication of the MOS device. Recent experiments have examined the effects of introducing hydrogen after fabrication. Studies of hydrogen annealing are strongly dependent on temperature.[8] Also, hydrogen introduced into previously irradiated devices increases the interface state density several-fold.[9,10,11]

We have been performing hydrogen annealing studies on previously irradiated MOS devices. In this paper we report the results of a set of experiments in which molecular hydrogen is introduced into the gate at temperatures from room temperature to 250°C. At room temperature, a reaction sequence initiated by H_2 cracking which we have previously described dominates.[10,11] It is characterized by the simultaneous growth of interface states and decay of oxide trapped positive charge. Molecular orbital calculations of the cracking process indicate that the cracking occurs at oxygen hole traps. At 250°C, there is a reduction of all observed defects. At intermediate temperatures (~100 - 200°C) there is a

This work was supported by the Office of Naval Research.

complicated interplay among H_2 cracking effects, formation of anomalous positive charge (APC), and annealing of the various defects. The APC defect is near the Si/SiO_2 interface and can be charged and discharged by changing the surface potential, particularly at elevated temperature.[12,13,14] The behavior of MOS devices in the 100 - 200°C temperature range is complicated by the bias dependence of the fast state annealing. There is more rapid annealing of all interface states under negative bias compared to positive bias and this result is in qualitative agreement with the predictions of Edwards.[15] The formation and annealing of interface states differentiates two interface defects. The annealing properties are similar to those reported by Uren *et al.*[16] The defect that forms more slowly and anneals more rapidly does not appear to be either a P_{b0} or P_{b1} center.

The balance of the paper is organized as follows. In the next section the H_2 cracking reaction sequence is reviewed. It is followed by sections devoted to the experimental methods and results. We conclude with a discussion of the hydrogen related effects that contribute to the complicated temperature and gate polarity dependencies.

H_2 CRACKING REACTION SEQUENCE

In previous work we reported the results of a set of experiments in which molecular hydrogen is alternately introduced into and removed from a gate oxide of irradiated MOSFETs at room temperature by changing the ambient between hydrogen and nitrogen.[11] The presence of molecular hydrogen in the gate oxide causes the simultaneous growth of interface states and decrease of oxide trapped positive charge (created during irradiation). The results are explained by the following reaction sequence:

1) H_2 from the ambient diffuses into gate oxide.
2) H_2 is cracked at a radiation-induced oxygen hole trap. This reaction produces one bound hydrogen ion and a mobile H^+ species that carries the positive charge from the cracking site.
3) The H^+ drifts to the Si/SiO_2 interface under a positive gate bias.
4) At the interface, H^+ captures an electron and the resulting atomic hydrogen atom reacts with Si-H to produce a dangling bond defect and H_2.

The H_2 cracking is not bias dependent so that under negative gate bias the first two steps are the same. However, in step 3, the H^+ drifts to the gate and it does not produce defects at the Si/SiO_2 interface. Under negative bias, interface state formation is not completely suppressed suggesting that a fraction of the cracked H_2 forms mobile H^o.

The reaction sequence accounts for the experimental data. By changing the MOSFET ambient, H_2 is either introduced or removed from the gate oxide. (The MOSFETs studied had no silicon nitride capping layer, so that the gate oxide is in contact with the ambient near the channel edges.) In Fig. 1 the effect of the ambient is shown. When H_2 is present (0 - 4 hr and 48 - 76 hr) the number of interface states increases and the trapped positive charge decrease is accelerated. Before irradiation the cracking sites are not present and no changes to N_{it} and N_{ot} are observed when H_2 is introduced.

The reaction sequence also accounts for the gate polarity dependence of the trapped positive charge decrease and interface state formation. The trapped positive charge has a weak dependence whereas the interface state buildup depends more strongly on bias; it is several-fold greater under positive bias. For either polarity, H_2 reacts with the positively charged cracking site and carries away its charge. However, only under positive gate polarity does the H^+ drift to the Si/SiO_2 interface and create interface states. Under positive polarity, there is roughly a 2:1 ratio between the number of interface states formed and trapped positive charges neutralized during H_2 exposure suggesting that each H^+ produced reaches the interface and creates one dangling bond defect with two states in the silicon bandgap.

We have modeled cracking reaction using molecular orbital calculations. Previously, Griscom *et al.* have discussed H_2 cracking by a non-bridging oxygen atom.[17,18] The positively charged oxygen vacancy (E' center)[19,20] has also been considered. To explore the possibilities, we performed molecular orbital calculations on the E' center and on a variety of oxygen related hole traps (OHT's). Including the Si-O-Si core, all clusters used to represent the OHT's had three shells of oxygen atoms, all in two-fold coordination, and two shells of silicon atoms, all four-fold coordinated. The OHT's differ by the magnitude and sign of the strain placed on the cluster. One of these OHT's had sufficient tensile strain so that placing a positive charge on the cluster resulted in rupturing a Si-O

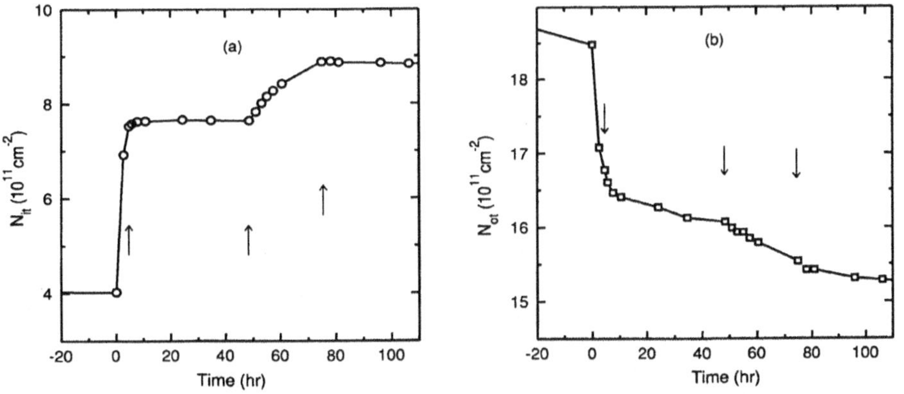

Fig. 1. The effect of a hydrogen ambient upon the number of interface states, N_{it}, and upon the number of trapped positive charges, N_{ot}, in a 5 μm long irradiated MOSFET. Hydrogen (10/90% H_2/N_2) is introduced at 0 and 48 hr. and nitrogen is introduced at 4 and 76 hr. Arrows point to the ambient changes at 4, 48 and 76 hr., respectively. The MOSFET is irradiated to 1 Mrad(SiO_2) 65 hr. before hydrogen is first introduced at t=0.

bond. For this reason we have called this defect a broken bond hole trap (BBHT). For the initial calculations, this was the only OHT exhibiting localized hole trapping and most of the quantitative results reported here were obtained from this OHT. Recently, we have come to understand self-trapping more completely and have obtained similar results on a variety of OHT's. All of the OHT's are much more likely candidates for the room temperature cracking of H_2 than the E$'$ center. (The calculated activation energy for H_2 dissociation at a BBHT was 1.1 eV and at an E$'$ center it was >1.7 eV.)

At this conference, and elsewhere, we have reported our calculations of a strongly localized, self-trapped hole.[11,21] Our calculations indicate that local strain, either tensile or compressive, stabilizes the self-trapped hole against migration. The fact that a self-trapped hole is localized on a compressively strained cluster is important, because most of the trapped positive charge in a MOS device is localized near the Si/SiO_2 interface,[22] in a region that is under compressive strain.[23] Furthermore, in recent ESR experiments on defects in compressed *a*-SiO_2, Griscom has observed strain-stabilized self-trapped holes at room temperature.[24] In *a*-SiO_2 that is not compressed the self-trapped holes are only stable below 200 K.[25]

The calculations also provide insight into the microscopic details of H^+ transport. A phenomenological model of dispersive H^+ transport has been used to successfully describe the buildup of interface states following pulsed irradiation.[6,7] However, the model does not address the microscopic nature the H^+. After H_2 is cracked, one of the hydrogen atoms is tightly bound and the second forms H^+. The tightly bound hydrogen atom forms Si-H during the E$'$ reaction and forms Si-O-H during the BBHT reaction. For either reaction, the second hydrogen is loosely bonded to a bridging oxygen atom and the complex is positively charged. With one of the oxygen valence electrons removed, the oxygen atom is similar to a nitrogen atom in that it prefers to be three-fold coordinated. In this case, the oxygen atom is bonded to two silicon atoms and a hydrogen atom. The similarity between this H^+ complex and an oxide hole suggest that their transport should be similar. In particular, both are expected to form small polarons and to undergo dispersive transport although details of the transport are different because of the mass difference between an electron and a proton. (For example, the time scale of the H^+ transport is much longer.)

There are strong similarities between interface states formed by irradiation and those formed by H_2 introduced after irradiation. Both processes yield the same energy distribution of interface states. Even more striking are comparisons of our results with pulsed irradiation experiments.[6,7] As in our experiments, the interface state formation has been attributed to H^+ transport through the oxide, and the buildup only occurs under positive bias. The similar effects observed in this study and in radiation studies adds strong support to the hydrogen models for radiation induced interface states.[6,7]

While the reactions reviewed in this section dominate at room temperature, as the temperature increases, additional effects due to anomalous positive charge formation and fast state annealing become important. The remainder of the paper addresses these effects.

EXPERIMENTAL DETAILS

N-channel MOSFETs with polysilicon gates were used. The channel width is 150 μm and measurements were made on devices with lengths of 5 and 10 μm. All devices exhibited the same behavior. A dry gate oxide was grown at 1000° C to a thickness of 770 Å. There was no capping layer deposited following metalization so that the gate oxide is exposed to the ambient at the channel edges. Molecular hydrogen in the ambient enters at the channel edges and diffuses through the gate oxide. Using bulk SiO_2 H_2 diffusion constants, the time for the hydrogen concentration to reach half of its equilibrium value in a 5 μm long MOSFET at room temperature is ~0.2 hr.[26,27] Thus, by changing the ambient, the hydrogen concentration in the oxide is controlled.

The transistors were x-ray irradiated in air with a tungsten tube (10 keV) to total doses of 0.5 or 1 Mrad(SiO_2) while biased at 7.7 V (1MV/cm). Following irradiation, the MOSFETs were mounted into a vacuum chamber and placed under bias. The chamber was evacuated or filled with an atmosphere of N_2 for several days before introducing forming gas (10/90% H_2/N_2 or H_2/Ar). In the temperature range studied, N_2 and Ar do not react in SiO_2 so we refer to these mixtures as hydrogen in the remainder of the paper.

The trapped positive charge, N_{ot}, the number of interface states, N_{it}, and the energy distribution of interface states, D_{it}, were monitored by charge pumping. The detailed method of determining these quantities has been described previously.[10,28,29] Some caution must be exercised in interpreting D_{it} results. The D_{it} determination of Ref. 28 assumes constant electron and hole capture cross sections. The values used here are $1.2\times10^{-15}cm^2$ and $4\times10^{-16}cm^2$ for the electron and hole capture cross sections.[30] A number of studies have suggested that either the capture cross section is energy dependent or that not all the interface states have the same capture cross section.[16,31] Different capture cross sections affect the positions of peaks in the D_{it} curve. There is evidence that peak positions in the D_{it} curves may be shifted and this is discussed later in the paper. However, the measurements of N_{it} are not significantly affected by the D_{it} shifts. All interface states whose capture times are longer than the rise or fall times of the gate pulse are measured. In our data, interface states between about 0.2 and 0.9 eV above valence band are measured. There is no indication that a significant number of interface states move between this measurement window and the regions of the bandgap close to the valence and conduction bands. Hence, the changes to N_{it} we report are primarily due to the formation or removal of interface states within the measurement window. We estimate that all but 10 to 20% of the states are counted. N_{ot} is calculated from the midgap voltage by assuming all the positive trapped charge is at the Si/SiO_2 interface. Previous work has shown that for our irradiation conditions most of the trapped charge is within roughly 100 Å of the interface.[22] N_{ot} includes anomalous positive charge and the trapped oxide charge. Because the centroid of charge is not exactly at the interface, the trapped charge density may be underestimated by up to 10%.

MOSFETs are mounted in ceramic DIP packages next to a small heater and a thermocouple for monitoring the temperature.

EXPERIMENTAL RESULTS

An overview of the effects due to H_2 in the gate oxide from room temperature to 250°C is shown in Fig. 2. Changes to both N_{it} and N_{ot} for two 10 μm long MOSFETs mounted in the same package are plotted. The devices were irradiated in air to 0.5 Mrad(SiO_2) with a gate bias of 7.7 V (1 MV/cm) and held under the same bias for 74 hr before introducing hydrogen. During this interval, N_{it} and N_{ot} change as expected (see Table I). The N_{ot} decrease is proportion to log(t), where t is the time after irradiation.[31] In the first few hours after irradiation, N_{it} increases[32] but changes in the 26 hr before introducing the hydrogen are negligible.

In Fig. 2, data before and after hydrogen exposure are shown. The temperature range of the hydrogen exposure is from room temperature to 250°C and devices were biased in inversion, 7.7 V (circles), or accumulation, -15 V (squares). The N_{it} and N_{ot} before the hydrogen ambient (74 hr after irradiation) are given by the solid diamond. The initial room temperature exposure to H_2 was for 49 hr and heating at all other temperatures was for 17 min. The initial room temperature H_2 exposure under positive bias was long enough to allow much of the N_{it} formation to occur; while N_{it} formation had not saturated after 49 hr, it had slowed significantly. During the first hour in the hydrogen ambient, N_{it} increased at a rate of $8\times10^{10}/cm^2hr$ compared with a rate of $0.1\times10^{10}/cm^2hr$ after 49 hr. Positive bias, 7.7 V, was maintained on both devices up to 125°C and the values of N_{it} and N_{ot} of the

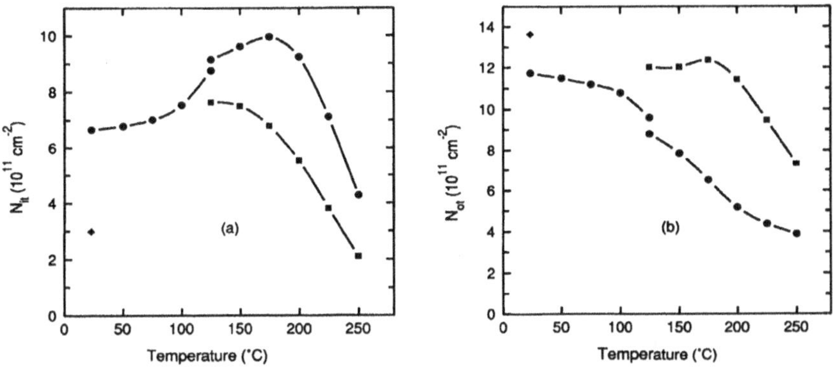

Fig. 2. Gate polarity dependence of changes to N_{it} and N_{ot} for 10 μm long irradiated MOSFETs in a hydrogen ambient from room temperature to 250°C. Values before introducing H_2 (74 hr after a 0.5 Mrad(SiO_2) irradiation) are shown by diamonds. Data with gate biases of 7.7 V (circles) and -15 V (squares) are displayed. The initial hydrogen exposure at room temperature is for 49 hr and all other exposures are for 17 min. Up to 125°C both MOSFETs are biased positively.

two MOSFETs agreed within 3%. The isochronal annealing was repeated at 125°C and went up to 250°C in 25°C increments. Above 125°C, the bias on one MOSFET was held at 7.7 V and on the other was changed to -15 V. These biases correspond to strong inversion and strong accumulation.

The results in Fig. 2 and Table I indicate that as the temperature is raised from room temperature to 250°C different reactions dominate. At room temperature, N_{it} increases whereas at the higher temperatures, N_{it} decreases. During the positive bias, room temperature, H_2 exposure, the $-\Delta N_{it}/\Delta N_{ot}$ ratio is approximately 2:1 which is characteristic of the H_2 cracking process described earlier. As already noted, this process had not been completed after 49 hr at room temperature. Under positive bias, N_{it} continues increasing up to 175°C. However, the $-\Delta N_{it}/\Delta N_{ot}$ ratio decreases as the temperature is increased indicating that other reactions are also becoming important. Fig. 2 shows that the temperature dependence of both N_{it} and N_{ot} are affected by the bias. At 125°C, a bias change from 7.7 to -15 V causes N_{it} to decrease and N_{ot} to increase. The N_{ot} increase in Fig. 2 is larger than the N_{ot} decrease during the 49 hr H_2 exposure at room temperature. Above 175°C both N_{it} and N_{ot} decrease for either bias.

The effects of bias are actually quite complicated as shown in Fig. 3. In this figure, results of alternating the bias between 4 V and -12 V at 105 and 175°C are shown. The odd frames are under positive bias and the even frames under negative bias (corresponding to strong inversion and strong accumulation). In each 100 min frame, the bias is held constant and the changes are plotted versus log(t). The data are from a 5 μm long MOSFET irradiated to 1 Mrad(SiO_2). Hydrogen is introduced at 105°C in frame 1. The time for H_2 to diffuse into and out of the gate oxide is about 1 min.[26,27] In frame 1, the H_2 cracking due to the H_2 cracking has not saturated. At this temperature it takes

Table I. The irradiation and hydrogen exposure history of the 10 μm MOSFET held under 7.7 V bias and shown in Fig. 2. The irradiation dose is 0.5 Mrad(SiO_2). N_{it} and N_{ot} are shown 1) after irradiation, 2) after H_2 exposure is started at room temperature and 3) after the isochronal annealing in Fig. 2 has extended to 125°C and 175°C. The ratio $-\Delta N_{it}/\Delta N_{ot}$ is calculated where ΔN_{it} and ΔN_{ot} are the changes from the previous line.

t(hr) after irradiation	t(hr) after H_2 ambient	T(°C)	N_{it} ($10^{11}/cm^2$)	N_{ot} ($10^{11}/cm^2$)	$-\Delta N_{it}/\Delta N_{ot}$
pre-rad		22	0.27	0	
0.12		22	2.45	18.1	
5.4		22	2.88	15.1	
47.6		22	2.98	14.0	
74.0	0	22	3.00	13.6	
	7.2	22	4.47	13.0	2.3
	49.3	22	6.66	11.7	1.7
		125	8.76	9.6	1.0
		175	9.96	6.5	0.4

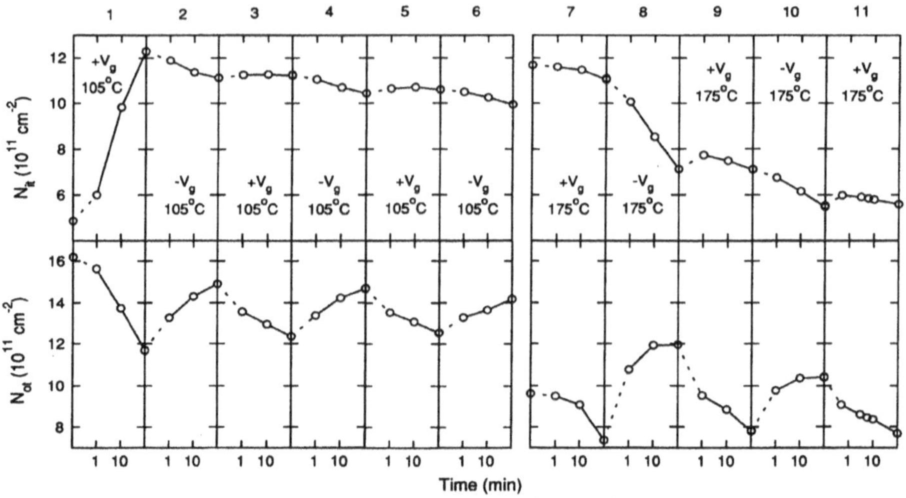

Fig. 3 A sequence of 100 min frames showing N_{it} and N_{ot} changes to a 5 μm long irradiated MOSFET (1 Mrad(SiO_2)). The temperature is 105°C in frames 1-6 and 175°C in frames 7-11. The gate bias is 4 V for odd numbered frames and -12 V for even frames. H_2 is first introduced in frame 1. The ambient is vacuum in frames 5 and 6 and is H_2 in all other frames. Between frames 6 and 7, the MOSFET is held at 105°C for 1000 min under positive bias.

roughly 1000 min to consume the cracking sites and saturate the N_{it} increase. Prior to introducing H_2, N_{it} and N_{ot} changed by less than $0.4\times10^{11}/cm^2$ and $1.2\times10^{11}/cm^2$, respectively, for the same bias reversal and temperature conditions. In frame 1, the effects of the H_2 cracking reactions are clearly visible. N_{it} increases by a factor of 2.5 and there is a drop of N_{ot}. However, $-\Delta N_{it}/\Delta N_{ot}$ is 1.6 indicating that reactions in addition to H_2 cracking are occurring. After introducing H_2, changes to N_{it} and N_{ot} with alternating bias are much larger.[10] While both N_{it} and N_{ot} respond to the bias, the detailed behavior of N_{ot} and N_{it} changes are different. At 105°C, N_{ot} changes within an envelop that slowly collapses in frames 2 - 6. This behavior is characteristic of anomalous positive charge, APC. In frames 5 and 6, the MOSFET is in vacuum and there is no noticeable change in the N_{ot} reversibility. Evidently, H_2 is involved in creating the APC, but H_2 does not affect the charging and discharging of the defect. The N_{it} behavior at 105°C is different. In frames 2 - 6, N_{it} decreases under negative bias and is essentially constant under positive bias. The N_{it} decreases under negative bias in frames 2, 4 and 6 saturate more rapidly than the envelop of N_{ot} changes.

Frames 7 to 11 show the annealing behavior at 175°C. (Between frames 6 and 7 the MOSFET was annealed for a 1000 min under positive bias.) The large swings of N_{ot} continue with an overall trend to decrease the average N_{ot}. N_{it} is roughly constant under positive bias and decreases under negative bias. The magnitudes of the N_{it} and N_{ot} changes do not appear to be correlated. At 210°C (not shown), reversible N_{ot} is still observed. However, after several bias reversals, the envelop of the N_{ot} collapses indicating that the APC defect has been annealed. The N_{it} changes at 210°C show the same pattern as at 175°C; under negative bias it anneals whereas under positive bias it is not changed.

The bias dependence of N_{it} is attributed to two factors. First, there is more formation of interface defects resulting from H_2 cracking under positive bias. While the cracking reactions are most important when H_2 is first introduced, the reactions continue at a reduced rate long after the H_2 is introduced. Once the cracking sites have been exhausted, there is no longer a bias dependence due to H_2 cracking. The second factor affecting the bias dependence of N_{it} is that the annealing of the interface defects is more rapid under negative bias. This difference is shown in frames 7 - 11 in Fig 3. The 1000 min heating prior to frame 7 is sufficiently long to complete the cracking process. In frame 7 the temperature is raised to 175°C under positive bias and there is a small decrease in N_{it}. When the bias is changed to negative in frame 8, the decrease is much larger. In subsequent positively biased frames (9 and 11) N_{it} is roughly constant and it decreases during the negatively biased frame (10). The same overall bias dependence is observed at 210°C. The bias dependence cannot be explained by only the bias dependence of the cracking reactions. It is possible that there is still some H_2 cracking occurring during frame 7. Note that any H^+ that is formed by cracking H_2

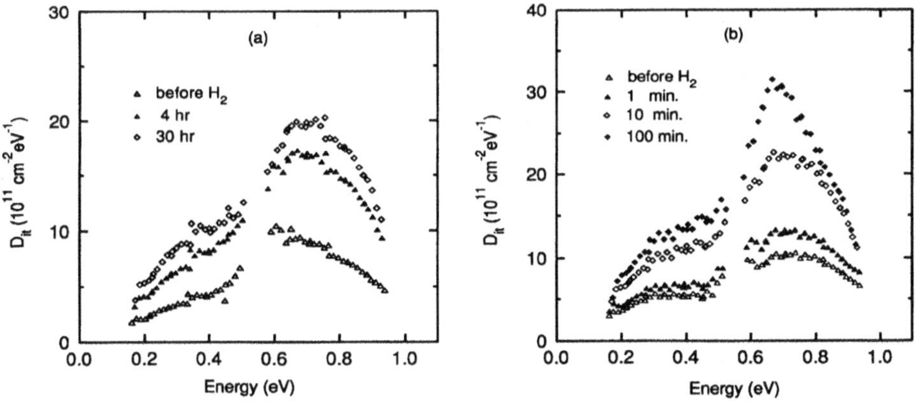

Fig. 4. Changes to the energy distribution of interface states, D_{it}, resulting from H_2 exposure under positive bias at room temperature (a) and 105°C (b). In each frame, D_{it} before H_2 exposure and during the course of interface defect formation is shown. In (a) the exposure times are 4 and 30 hr and in (b) the times are 1, 10 and 100 min. Data in both frames are from 5 µm long MOSFETs irradiated to 1 Mrad(SiO_2). Frame (b) corresponds to frame 1 of Fig. 3.

drifts to either the gate or the Si/SiO_2 interface very rapidly compared to the 100 min frame time. Based upon room temperature H^+ transport data and a 0.8 eV activation energy for H^+ transport,[6,33] over 90% of the H^+ can traverse the gate in a fraction of a second. Thus the annealing behavior cannot be affected by a time lag between H_2 cracking and H^+ reacting at the Si/SiO_2 interface. We conclude that the observed bias dependence of N_{it} annealing shows that defects responsible for the interface states anneal more rapidly under negative bias. These results do not contradict Fig. 2 in which there is annealing of N_{it} under positive bias at temperatures above 175°C. The results in Fig. 2 are for a MOSFET kept under positive bias as the temperature is raised. In contrast, in Fig. 3, the MOSFET is placed under both polarity biases at each temperature before raising the temperature. While annealing does occur more efficiently under negative bias, it also annealing occurs under positive bias for a sufficiently high temperature.

Examining the density of interface states, D_{it}, provides more information about the interface defects. In particular, it is evident that two different defects with different formation and annealing properties are present. Fig. 4 shows the post-irradiation growth of D_{it} due to H_2 at room temperature, (a) and at 105°C, (b). Fig. 4(b) corresponds to frame 1 of Fig. 3. Both MOSFETs are 5 µm long and were irradiated to 1 Mrad(SiO_2) with a 7.7 V bias. The general shape of the curves are typical of

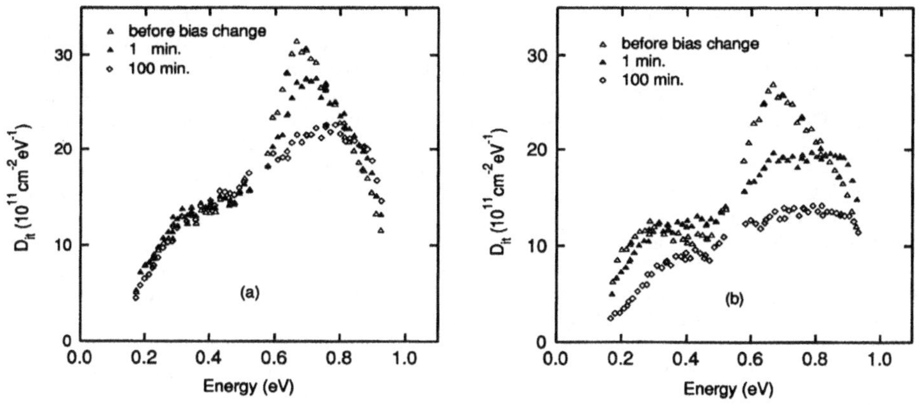

Fig. 5. Changes to D_{it} resulting from the annealing under a negative bias at 105°C (a) and 175°C (b) with H_2 present. These data correspond to frames 2 and 8 in Fig. 3. D_{it} before annealing and after 1 and 100 min at each temperature are shown. Note that the peak near 0.7 eV anneals more rapidly than the remainder of the interface state density.

post-irradiation data with a broad peak at 0.7 eV above the valence band and states extending throughout the bandgap. The D_{it} curves have the same shape before and after introducing H_2 at room temperature. The shape after 1 and 10 min at 105°C is also the same. However, between 10 and 100 min at 105°C a more narrow peak near 0.7 develops. The annealing of the extra peak near 0.7 eV is also different than the broader background. Annealing under negative bias at 105 and 175°C is shown in Figs. 5(a) and 5(b), (frames 2 and 8 of Fig. 3). At 105°C the reduction of the 0.7 eV peak is evident while the remainder of the bandgap is unchanged. The 175°C data show annealing of both D_{it} components with a more rapid annealing of the 0.7 eV peak. After 1 min the primary decrease is in this peak. With additional annealing, D_{it} drops uniformly across the bandgap. In the subsequent 100 min frame under positive bias, D_{it} across the bandgap does not change. The spectral features of D_{it} that are formed and anneal at different rates are interpreted as two different interface defects that are formed and anneal at different rates.

DISCUSSION OF HYDROGEN EFFECTS

The data show that there is a complicated interplay of effects when H_2 is introduced into the gate oxide of an irradiated MOS device. Over the temperature range from room temperature to 250°C, we identify four different effects.

1) Cracking of H_2 by an oxide hole trap to form H^+. Under positive bias H^+ drifts to the Si/SiO_2 interface and creates the interface defects responsible for fast states.
2) There are two different defects with different formation and annealing rates. The defect that forms more slowly and anneals more rapidly does not appear to be either a P_{b0} or a P_{b1} center.
3) Both of these defects are annealed more rapidly under negative bias (the surface potential at the Si/SiO_2 interface is near the valence band).
4) The defect responsible for slow states or anomalous positive charge, APC, is formed during irradiation and is increased several-fold when H_2 is introduced above room temperature.

The H_2 cracking reactions have been reviewed earlier in the paper. These reactions dominate at room temperature. At this temperature, changes in N_{ot} due to APC for a MOSFET in H_2 are small. Compared to the N_{ot} changes at 105°C in Fig. 3, changes at room temperature are an order of magnitude smaller leaving the MOSFET at each bias for a day. This is in agreement with earlier room temperature APC observations.[34] Also, there is no evidence for interface defect annealing at room temperature.

As the temperature is raised, H_2 cracking reactions continue, but effects due to APC formation and annealing of fast states are also observed. The increase of N_{it} visible in Fig. 2 up to 175°C under positive bias indicates that for the time and temperature profile used in this figure, N_{it} formation due to H_2 cracking is more important than N_{it} annealing under positive bias up to 175°C. That is, the cracking sites are not consumed during the 49 hr at room temperature and the 17 min anneals up to 175°C. In contrast, the cracking sites are consumed in 1000 min at 105°C. The combination of H_2 cracking and annealing reactions makes it difficult to determine a precise value for the activation energy of H_2 cracking. However, the observations from room temperature to 175°C are consistent with our earlier estimate determined from room temperature data.[11] There is a distribution of activation energies near 1 eV. We believe that the activation energy for the cracking reaction is broadened inhomogeneously and that the high energy tail of the distribution of activation energies is responsible for the additional H_2 cracking at 175°C even after the lower temperature annealing.

The 2:1 ratio of $-\Delta N_{it}/\Delta N_{ot}$ that is characteristic of the H_2 cracking reactions at room temperature drops at higher temperatures indicating other reactions are also occurring. The additional reactions lead to smaller N_{it} increases and/or larger N_{ot} decreases. It is likely that there is annealing of interface defects because above 175°C N_{it} decreases even for a MOSFET kept under positive bias as it is heated. The formation of APC may also play a role. These samples exhibit the properties associated with APC.[13,14] Above 100°C the APC are alternately charged and discharged by changing the bias polarity. After a number of charge reversals the effect is damped and as the temperature is raised, the damping increases. However, APC is probably not involved with decreasing the $-\Delta N_{it}/\Delta N_{ot}$ ratio. APC is believed to be donor-like.[14] Hence APC can add to the amount of positive charge, N_{ot}, but cannot contribute to a N_{ot} decrease that would account for a smaller $-\Delta N_{it}/\Delta N_{ot}$ ratio.

The structure of the APC defect is still in question. Lelis *et al.* have suggested that the dangling bond of a normally positively charged E′ center can be either singly or doubly occupied, producing a donor-like defect.[35] Molecular orbital calculations by Fowler support this possibility and suggest that the positive charge on an E′ center is stabilized by the formation of a three-fold coordinated oxygen atom.[36] However, Trombetta *et al.* have not seen any ESR signal associated with APC.[37] Any APC model must also account for the role that H_2 plays in its formation and for the temperature dependence of the charging and discharging. Based on molecular orbital calculations, Edwards has suggested the formation of a three-fold coordinated oxygen atom on the oxide side of the Si/SiO_2 interface during negative bias H_2 annealing [i.e. formation of $(Si_2OH)^+$].[15] A defect such as this may be responsible for APC although the details of the charging and discharging of this defect have not yet been thoroughly examined. The temperature dependence of the APC may be due to tunneling that is accompanied by an activated process in which hydrogen is transferred from one site to another.

Those same calculations compare the passivation of P_b centers by reacting with H_2 in positive, neutral and negative charge states. The calculated activation energy for the neutral case is 1.3 eV and is in good agreement with the 1.66 eV activation determined by Brower's ESR studies.[38] Compared to the neutral case, the calculations also predict increased passivation under negative bias and strongly suppressed annealing under positive bias. The experimental annealing results show that annealing is more rapid under negative bias than under positive bias, but annealing under positive bias is not as strongly suppressed as predicted.

There is a striking similarity between the evidence for two different interface defects in our measurements and the evidence presented by Uren *et al.* using AC conductance.[16] They identify two defects based on changes to the shape of the conductance curve with annealing and find that the defect with the smaller capture cross section anneals more rapidly over the temperature range from 125 to 275°C. A similar separation based on the shape of AC conductance curves has been made by Haneji *et al.*[31] The presence of more than one defect may help explaining complicated post irradiation D_{it} behavior.[10,39] Both groups extract a capture cross section that is roughly energy independent for the defect with the larger capture cross section. However, their results for the defect with the smaller capture cross section do not agree as well. Uren *et al.* determine that the capture cross section of this second defect also roughly energy independent, whereas Haneji *et al.* determine a strong energy dependence. The defect observed by Uren *el al.* which anneals more quickly has more than one order of magnitude smaller capture cross section. In the charge pumping analysis to determine D_{it} above midgap, a capture cross section used is $1.2\times10^{-15}cm^2$. The smaller of the two values, determined by Uren *et al.*, is about an order of magnitude smaller. Using a smaller capture cross section has the effect of moving the position of the more rapidly annealing peak slightly to the right (by less than 0.1 eV). This is within the measurement window of the charge pumping measurements. It appears that the defect that we observe forming more slowly with H_2 exposure at 105°C and annealing more rapidly is the same as the defect Uren *et al.* observe annealing more rapidly.

Both groups raise the possibility that the two defects are the P_{b0} and P_{b1} defects identified by ESR experiments[40] although Uren *et al.* conclude that this assignment is probably incorrect. Our results indicate that the defect which anneals more rapidly is neither P_{b0} nor P_{b1}. While the ac conductance can only measure defects in the top or bottom half of the bandgap, charge pumping measures both halves with the same device. Figs. 5 and 6 show that the bottom half of the bandgap is not changed by the decrease of the more rapidly annealing defect. This is not consistent with the commonly accepted +/0 and 0/- levels of P_{b0} or P_{b1}.[40] Both P_b defects introduce interface states below midgap.

SUMMARY

The complex effects of introducing H_2 into the gate oxide of irradiated MOSFETs have been studied from room temperature to 250°C. At room temperature, the dominant reactions involve cracking of H_2 and result in the simultaneous interface state formation (fast states) and decay of trapped positive charge. Molecular orbital calculations indicate that H_2 is cracked at oxygen hole traps and the mobil H^+ that is produced is a small polaron. At 250°C there is annealing of all observed defects. At intermediate temperatures there is a complicated interplay among H_2 cracking reactions, formation of anomalous positive charge (slow states) and annealing of fast states. The fast states anneal more rapidly and at a lower temperature under negative bias. The formation and annealing kinetics of the fast states indicates that there are two different interface defects. The defect that forms more slowly and anneals more rapidly does not have states within the silicon bandgap that are consistent with either the P_{b0} or P_{b1} centers.

REFERENCES

[1]E. H. Nicollian, C. N. Berglund, P. F. Schmidt, and J. M. Andrews, Electrochemical charging of thermal SiO_2 films by injected electron currents, J. Appl. Phys. **42**, 5654 (1971).

[2]E. H. Nicollian, A. Goetzberger, and A. D. Lopez, Expedient method of obtaining interface state properties from MIS conductance measurements, Solid-State Electronics **12**, 937 (1969).

[3]3. D. J. DiMaria, and J. W. Stasiak, Trap creation in silicon dioxide produced by hot electrons, J. Appl. Phys. **65**, 2342 (1989).

[4]Y. Nissan-Cohen, The effect of hydrogen on hot carrier and radiation immunity of MOS devices, Appl. Surf. Sci. **39**, 511 (1989).

[5]P. S. Winokur, Radiation-induced interface traps, in "Ionizing Radiation Effects in MOS Devices & Circuits," T. P. Ma and P. V. Dressendorfer, ed., John Wiley & Sons, New York, NY, (1989), p. 229.

[6]F. B. Mclean, A framework for understanding radiation-induced interface states in SiO_2 MOS structures, IEEE Trans. Nucl. Sci. **NS-27**, 1651 (1980).

[7]N. S. Saks and D. B. Brown, Interface trap formation vis the two stage H^+ process, IEEE Trans. Nucl. Sci. **NS-36**, 1848 (1989).

[8]L. Do Thanh and P. Balk, Temperature hydrogen annealing, J. Electrochem. Soc. **135**, 1797 (1988).

[9]R. A. Kohler, R.A. Kushner and K. H. Lee, Total dose radiation hardness of MOS devices in hermetic ceramic packages, IEEE Trans. Nucl. Sci. **NS-35**, 1492 (1988).

[10]R. E. Stahlbush, B. J. Mrstik and R. K. Lawrence, Post-irradiation behavior of the interface state density and the trapped positive charge, IEEE Trans. Nucl. Sci. **NS-37**, 1641 (1990).

[11]R. E. Stahlbush, A. H. Edwards, D. L. Griscom and B. J. Mrstik, Post-irradiation cracking of H_2 and formation of interface states in MOSFETs, J. Appl. Phys. (submitted) 1991.

[12]R. A. Gdula, J. Electrochem. Soc. **123**, 42 (1976).

[13]D. R. Young, E. A. Irene, D. J. DiMaria, R. F. DeKeersmaecker and H. Z. Massoud, Electron trapping in SiO_2 at 295 and 77K, J. Appl. Phys. **50**, 6366 (1979).

[14]S. K. Lai and D. R. Young, Effects of avalanche injection of electrons in silicon dioxide - generation of fast and slow interface states, J. Appl. Phys. **52**, 6231 (1981).

[15]A. H. Edwards, Interaction of H and H_2 with the silicon dangling orbital at the <111>Si/SiO_2 interface, Phys. Rev. B **44**, 1832 (1991).

[16]M. J. Uren, K. M. Brunson and A. M. Hodge, Separation of two distinct fast interface state contributions at the (100)Si/SiO_2 interface using the conductance technique, Appl. Phys. Lett. **60**, 624 (1992).

[17]D. L. Griscom, M. Stapelbroed and E. J. Friebele, ESR studies of damage processed on x-irradiated high purity *a*-SiO_2:OH and characterization of the formyl radical defect, J. Chem. Phys. **78**, 1638 (1983).

[18]D. L. Griscom, Thermal bleaching of x-ray-induced defect centers on high purity fused silica by diffusion of radiolytic molecular hydrogen, J. Non-Cryst. Solids **68**, 301 (1984).

[19]R. A. Weeks, Paramagnetic spectra of E_2' centers in crystalline quartz, Phys. Rev. **130**, 570 (1963).

[20]F. J. Feigl, W. B. Fowler and K. L. Yip, Oxygen vacancy model for the E_1' center in SiO_2, Solid State Comm. **14**, 225 (1974).

[21]A. H. Edwards, Molecular orbital studies of oxygen-related hole traps and of their interactions with hydrogen atoms and molecules, this conference.

[22]T. R. Oldham, A. J. Lelis, and F. B. McLean, Spacial dependence of trapped holes determined from tunneling analysis and measured annealing, IEEE Trans. Nucl. Sci. **NS-33**, 1203 (1986).

[23]F. J. Grunthaner, P. J. Grunthaner and J. Maserjian, Radiation-induced defects in SiO_2 as determined with XPS, IEEE Trans. Nucl. Sci. **NS-29**, 1462 (1982).

[24]D. L. Griscom (private communication).

[25]D. L. Griscom, Self-trapped holes in amorphous silicon dioxide, Phys. Rev. B **40**, 4224 (1989).

[26]J. E. Shelby, Molecular diffusion and solubility of hydrogen isotopes in vitreous silica, J. Appl. Phys. **48**, 3387 (1977).

[27]J. Crank, "The Mathematics of Diffusion," Claredon Press, Oxford (1975) p. 50.

[28]G. Groeseneken, H. E. Maes, N. Beltran, and R. F. dekeersmaecker, A reliable approach to charge pumping measurements in MOS transistors, IEEE Trans. Elec. Dev. **ED-31**, 42 (1984).

[29]N. S. Saks and M. G. Ancona, Generation of interface states by ionizing radiation at 80K measured by charge pumping and subthreshold slope teçhniques, IEEE Trans. Nucl. Sci. **NS-34**, 1348 (1987).

[30]E. H. Nicollian and J. R. Brews, "MOS (Metal Oxide Semiconductor) Physics and Technology," Wiley and Sons, New York (1982) p. 297.

[31]N. Haneji, L. Vishnubhotla and T. P. Ma, Possible observation of P_{b0} and P_{b1} centers at irradiated (100)Si/SiO_2 interface from electrical measurements, Appl. Phys. Lett. **59**, 3416 (1992).

[32]P. S. Winokar, H. E. Boesch, Jr., J. M. McGarrity, and F. B. McLean, Two-stage process for buildup of radiation-induced interface states, J. Appl. Phys. **50**, 3492 (1979).

[33]N. S. Saks, C. M. Dozier and D. B. Brown, Time dependence of interface trap formation in MOSFETs following pulsed irradiation, IEEE Trans. Nucl. Sci. **NS-35**, 1168 (1988).

[34]M. V. Fischetti, R. Gastaldi, F. Maggioni and A. Modelli, Slow and fast states induced by hot electrons at Si-SiO_2 interface, J. Appl. Phys. **53**, 3136 (1982).

[35]A. J. Lelis, T. R. Oldham, H. E. Boesch, Jr. and F. B. Mclean, The nature of the trapped hole annealing process, IEEE Trans. Nucl. Sci. **NS-36**, 1808 (1989).

[36]W. B. Fowler, (private communication).

[37]L. P. Trombetta, G. J. Gerardi, D. J. DiMaria and E. Tierney, An electron paramagnetic resonance study of electron injected oxides in metal-oxide-semiconductor capacitors, J. Appl. Phys. **64**, 2434 (1988).

[38]K. L. Brower, Passivation of paramagnetic Si-SiO_2 interface states with molecular hydrogen, Appl. Phys. Lett. **53**, 508 (1988).

[39]T. P. Ma, Interface trap transformation in radiation or hot-electron damaged MOS structures, Semicond. Sci. Technol. **4**, 1061 (1989).

[40]G. J. Gerardi, E. H. Poindexter and P. J. Caplan, Interface traps and P_b centers in oxidized (100) silicon wafers, Appl. Phys. Lett. **49**, 348 (1986).

AUTHOR INDEX

SUBJECT INDEX

GPSR Compliance

The European Union's (EU) General Product Safety Regulation (GPSR) is a set of rules that requires consumer products to be safe and our obligations to ensure this.

If you have any concerns about our products, you can contact us on ProductSafety@springernature.com

In case Publisher is established outside the EU, the EU authorized representative is:

Springer Nature Customer Service Center GmbH
Europaplatz 3
69115 Heidelberg, Germany

Batch number: 10761599

Printed by Printforce, the Netherlands